MW01157016

URBAN TRAVEL DEMAND MODELING

URBAN TRAVEL DEMAND MODELING

From Individual Choices to General Equilibrium

NORBERT OPPENHEIM

Department of Civil Engineering
The City College and Institute for Transportation Systems
and Graduate School
City University of New York

A Wiley-Interscience Publication

JOHN WILEY & SONS, INC.

New York / Chichester / Toronto / Singapore / Brisbane

This text is printed on acid-free paper.

Library of Congress Cataloging in Publication Data:

Oppenheim, Norbert.
Urban travel demand modeling: from individual choices to general equilibrium/Norbert Oppenheim.
p. cm.
A Wiley-Interscience Publication
Includes bibliographical references and index.
ISBN 0-471-55723-4 (acid-free paper)
1. Choice of transportation—Mathematical models. 2. Traffic estimation—Mathematical models. 3. Travel time (Traffic engineering)—Mathematical models. 4. Urban transportation—Mathematical models. I. Title.
HE336.C5066 1994
388.4—dc20 94-13852

Printed in the United States of America

10 9 8 7 6 5 4 3 2 1

To Leslee, in gratitude and love

FOREWORD

Urban transportation planning is approaching its fortieth year as a field of professional practice as well as academic research and teaching. From its beginnings in the mid 1950s, forecasting of urban travel demand as a basis for decisions about investment in transportation facilities has been a prominent dimension of this endeavor. For various institutional reasons, the models and computer methods for preparing such forecasts for large U.S. metropolitan areas were established and rather firmly prescribed within the first decade, well before their properties were fully understood and alternative approaches were explored and evaluated.

Although these developments enabled transportation planning professionals to make relatively early contributions to investment decisions, the long-term effects may be seen in a less positive light. The modeling concepts that emerged from this period are unduly mechanistic and generally lack a strong behavioral basis. Procedures central to the forecasting process lack a solid foundation or rationale; moreover, unnecessary jargon is pervasive.

Academic research during the second decade of the field, roughly 1966–1975, sought to address these shortcomings and made very important contributions and advances. The basis for reinventing the urban travel forecasting process was laid, and some valiant efforts at implementation were modestly successful. By and large, however, the professional community continued through the 1980s to use the traditional concepts and methods.

Now, it seems, change is upon us. One impetus is the Clean Air Act Amendments of 1990 and the Intermodal Surface Transportation Efficiency Act of 1991. Another is the efforts of environmental action groups to challenge the status quo of the urban transportation planning process. While the needed changes are often evident from a research perspective, they may appear revolutionary from the vantage of professional practice.

As our field proceeds with the task of re-inventing itself methodologically, it is truly fortuitous that Norbert Oppenheim has contributed this textbook on travel demand modeling. Why do I believe this is so? First, he has provided a new framework for the forecasting process, which is a synthesis of many advances of the past two decades, but at the same time is novel in its breadth of application. In this framework, travelers are considered "consumers" of trips. Accordingly, single, as well as combined, travel demands reflect their budget-constrained choices. Network user equilibrium for both the uncongested and congested cases is shown to correspond to individual, as well as collective, utility maximization. This result provides a solid basis for addressing such contemporary issues as urban road pricing. Second, he has succeeded in bridging the ever-widening gap between practice based on traditional thinking of the early 1960s and the fruits of academic research ever since. Third, he has delivered a product which I believe will be useful and understood, both in the classroom and in the planning agency. Moreover, he has done so with originality and insight into the difficulty of presenting the concepts involved.

Given the convoluted history and traditions of our field, no textbook will be regarded as just right for everyone, either instructor or professional. Traditions will tend to continue, as will seemingly arcane theory. In the current milieu, however, we have an opportunity to move our field substantially forward. If we succeed, I believe we will owe a substantial debt of gratitude to Norbert Oppenheim for the foundation he has given us in this book.

DAVID E. BOYCE

University of Illinois at Chicago
June 1994

PREFACE

The purpose of this book is to present a methodology for predicting urban travel demands which is based on the concept of the traveler as a consumer of urban trips. In contrast with "traditional" methodology, which is essentially statistical in nature and thus only descriptive, the urban travel demand models presented here explicitly describe the choices individual travelers make when faced with given travel alternatives, and when limited by budgetary or other constraints.

The basic common principle which rationalizes all travel demand models presented here is the same as in standard microeconomics, namely utility maximization. That is, travel activity, at the individual traveler level, as well as the community or social level, is a reflection of explicit preferences and limitations. This is the case not only for single travel demands, such as origin-destination, but for joint demands, such as for destinations by mode. Most importantly, it is also the case in the presence of congestion, that is, for equilibrium demands.

All standard aspects of urban travel demand analysis (e.g., travel generation, modal choice, destination, and route selection) are explicitly and consistently formulated within this framework. Rigorous numerical algorithms for solving the models are systematically presented. These models may then perhaps be useful in tackling major transportation issues facing urban areas, such as the mounting costs of congestion, both in environmental and economic terms. More generally, because urban travel is directly interacting with a variety of urban phenomena, including land use, these models are also potentially relevant to a variety of other analytical and planning activities, including city planning, traffic engineering, environmental protection, and economic development.

The coverage of this book is limited to static models, for average or steady conditions. Dynamic models, in which time plays an explicit role, while more general and useful for "real time" applications, pose significant analytical challenges, particularly in terms of application to realistic size networks, which have not been entirely resolved as of this writing. More generally, issues of traffic control and travel guidance are outside the scope of this book.

This book is conceived as a textbook for graduate students, a reference for researchers, and a primer for practitioners. This being a book on modeling, it is necessary to use, to the minimum extent feasible, various forms of mathematics. It is assumed in this regard that the reader has some basic knowledge of introductory calculus, including optimization theory, as well as probability theory and statistics, and basic consumer demand theory. In any event, this material is reviewed in Appendix A, in an effort to make the text as self-contained as possible.

The contents of the book are organized as follows. The first chapter introduces, in qualitative terms, basic concepts in urban travel demand modeling. Also, the basic dimensions of travel demand, and of the transportation system, are discussed. The traditional ("four-step") approach to urban travel demand modeling is briefly discussed, as background to the development of the adopted "travel consumer" framework. This chapter may be skipped by readers with some familiarity with urban travel demand modeling.

Chapter 2 then reviews the concepts and results of "discrete choice, random utility theory." This is a major analytical tool which is utilized in the subsequent developments. Its concepts, the resultant "logit" model and its analytical properties, are reviewed in some detail, as a prerequisite to their subsequent application to travel demand model development. The concept of the "representative traveler" (R.T.), which plays a central role in the text, is also introduced.

Chapter 3 then addresses the formulation of the first level of travel demand, route choice, in the simplest case, when network and destination congestion are not present. The deterministic car case is first considered, and the stochastic case is then presented as a first type of probabilistic generalization. The case of transit is then presented as a second case of probabilistic generalization. In all cases, aggregate travel demands are shown to be produced by the R.T.'s choices.

Chapter 4 combines the other dimensions of travel demand, destination, mode, and travel choice, with route choice, still in the absence of congestion. This expands the "travel consumer" framework. Chapter 5 then revisits route choice modeling, but this time under the more general, and often more realistic, assumption of network and destination congestion. Similarly, Chapter 6 parallels Chapter 4, under the assumption of travel demand externalities.

Chapter 7 then examines how the various models developed in the first six chapters may be "estimated," how their parameters may be given numerical values reflecting specific, observed conditions. Methods for assessing the performance of the resultant models with respect to actual conditions are also discussed.

Chapter 8 presents an integrated model of personal travel and goods movements. This is then generalized to urban activity/trading systems, in which personal travel is motivated by the conduct of various activities, which in turn create the need for the transportation of various supporting commodities.

Finally, Chapter 9 briefly addresses the supply side, the demand-responsive determination of optimal transportation. Equilibrium between travel supply and demand is first discussed. Various examples of transportation supply problems are also surveyed. The basic network design problem, for the car and transit modes, under both uncongested and congested conditions, is formulated next.

Illustrative examples of the models' application are provided throughout the text. Also, problems requiring the application of the methodology, or examining some aspect of the theory, are provided at the end of each chapter. The solutions to selected exercises are presented in Appendix B.

Two other appendixes supplement the text. Appendix A presents a background review on mathematics, including calculus, optimization theory, probability theory, and consumer demand theory, which may be useful to some readers. Appendix C presents a list of symbols used throughout the text.

This book is an outgrowth of my research work at the City University of New York (CUNY) Institute for Transportation Systems (ITS), as well as of lecture notes for various courses at CUNY's Graduate Center. Some of this work was supported by CUNY's Professional Staff Congress, whose support is gratefully acknowledged. Also, a sabbatical leave provided me with much-needed writing time. This support from CUNY is gratefully acknowledged. Support in various forms from ITS and the Federal University Transportation Research Center (Region II) is also gratefully acknowledged.

Among the many colleagues who have been helpful in this endeavor, I particularly wish to thank David E. Boyce for his active support, from beginning to end, as well as for his many specific suggestions for improving the manuscript. I also thank Tony Smith for pointing me in the right analytical direction and also providing a proof (page 245). Thanks also to several classes of graduate students who endured and tested the various evolutionary versions of my notes. I wish to thank in particular Hesham Ali, Qi-Feng Zheng, Ju Ding, Mansoor Qureshi, How Sheen Pau, Daniel Baah, Fleur Hartmann, and Hershel Yeres, for providing solutions to several exercises. The standard disclaimer applies, both to numerical and analytical derivations.

My colleagues at ITS, Fred Brodzinski, Claire McKnight, Robert (Buz) Paaswell, Neville Parker, and Mitsu Saito provided a supportive and helpful environment.

Also, lack of time did not allow a more concerted effort in identifying the accurate origins of some of the results presented in the text, other than bibliographical references to them. Some other unattributed results are new, at least to my knowledge.

Ultimately, the work of two seminal researchers made the developments in this book possible. They are Martin Beckmann, who first applied optimization theory to modeling route demands under congestion, and Moshe Ben Akiva, who first applied random utility theory to modeling combined travel choices.

I also wish to thank the staff at John Wiley, including Allison Morvay and Charles Schmieg, for their supportive collaboration throughout the production of this book.

In closing this preface, my most heartfelt gratitude goes to my family for their patience and understanding in spite of the innumerable evenings and weekends sacrificed to this project. More than anybody else, my wife Leslee made this book possible, not only during its writing, but also much, much earlier. To her it is dedicated.

NORBERT OPPENHEIM

New York City
November 1994

CONTENTS

CHAPTER 1

INTRODUCTION: MODELING URBAN TRAVEL DEMAND

1.1 THE PURPOSE OF THIS BOOK AND THE ISSUES ADDRESSED

As its title indicates, this book is about urban travel demand modeling. Urban travel in this context means travel by people, using transportation modes typically available in urban areas. While for purposes of simplicity we will focus on the major urban transportation modes, such as private car and public transportation (including bus and rail), other commonly used urban modes, including "paratransit" (taxis, jitneys, etc.) and walking and biking, may also be included.

Modeling means the development of mathematical formulations which represent observed travel patterns by travel mode, as well as the volume, travel speeds, and congestion level on the links of the transportation network.

This will require the examination of several issues. First, the behavior of individual travelers must be described as being the basic, elemental determinants of travel demands. Behavior means the travel-related choices individuals make under given circumstances, for example, choosing whether to travel in a given time period, and choosing a destination, mode, and route of travel.

Second, individual choices must be aggregated to the level of travel flows, or volumes. This requires the examination and formulation of how individual behaviors interact. Specifically, congestion, a major fixture of urban travel, causes each individual traveler's options to be affected by the travel choices of all others. As a result, aggregate travel patterns reflect an equilibrium between all these individual choices, which must be identified.

Thus, the two main conceptual themes of the book, which are reflected in its title, are individual traveler choices and the resulting general equilibrium across the various dimensions of travel demand listed above.

In practical terms, the ultimate purpose of urban travel demand modeling is to provide a tool with which one may predict, or forecast, urban travel patterns under various conditions. These conditions may, for instance, represent the expected, or planned, state of the transportation network, or more generally that of the urban area itself, at a future time.

These predicted travel patterns may then provide useful information in managing and planning the transportation system, and by extension, the urban area at large. For instance, the impacts on travel congestion (and its attendant negative effects, including environmental pollution, economic impacts, and other societal effects) of actions such as increasing the capacity of certain links, changing transit fares, or locating an activity center at a certain place, may be estimated through application of travel demand models.

Moreover, knowledge of the effects on travel conditions, including those on travel times, of various traffic management actions, either in the medium term, such as redirecting traffic flows, or in real-time, such as guiding individual vehicles during their actual trip through the network, may also be useful to individual travelers.

Thus, urban travel demand modeling is potentially useful in analyzing some of the present critical urban transportation problems. However, measuring, modeling, and predicting human behavior will always remain more difficult than modeling physical processes. The challenges posed by the sheer size of the systems these models replicate (e.g., transportation networks with tens of thousands of links, and the obvious difficulties of obtaining complete and accurate data about them) are also significant.

Nevertheless, urban travel demand modeling has made substantial progress recently. It is now possible to present under a unified, coherent behavioral framework integrated models which predict all the major dimensions of travel demand, namely origin demand, destination demand, cross (origin-destination) demand, and route demand, by all travel modes. Moreover, these models deal rigorously with the central issue of congestion, so that the aggregation of individual travel demands is performed correctly. They also deal formally with the issue of congestion. Finally, these models are exercised with formal numerical procedures, whose eventual convergence to the correct solution is guaranteed, and whose performance may be assessed explicitly, so that their results may be utilized reliably.

The main purpose of this book is to present such models. While many of the earlier models are still in widespread use today, they suffer from several pitfalls, as will be discussed in more detail in section 1.5. Briefly, they are not based on any explicit (e.g., behavioral) rationale, but are instead mostly statistical in nature. Moreover, their application is overly dependent on ad-hoc numerical techniques.

It will hopefully become apparent that the behavioral, combined equilibrium methodology presented in this book alleviates these pitfalls, and that its advantages in this regard are well worth the modest increase in complexity over the traditional approach. In particular, two specific analytical tools are now used that were not required previously, namely nonlinear programming and random utility theory. It is hoped that this will not be perceived as a drawback, or worse, as a deterrent.

In the next section, we begin to formalize how we look at the phenomenon of urban travel, starting with the spatial structure on which it takes place.

1.2 REPRESENTING THE SPATIAL STRUCTURE

Travel is an activity that takes place from one given geographic location to another, over a transportation network. Thus, most of the concepts we will be using in the course of the analytical developments have a spatial dimension. We must specify how the spatial structure is to be represented.

Individual travelers are in general distributed throughout the urban area of interest, each in principle at a different location. For our purposes, traveler locations are limited to residential locations, and/or employment locations. Thus, the exact spatial representation of the demand side of urban travel would require as many demand locations as there are travelers. This, although conceivable in theory, would be rather cumbersome in practice. For this reason, travelers' locations will be aggregated into "origin zones," for example, city blocks, census tracts, or "traffic zones" defined by the analyst. Each zone will then be represented by its centroid. This may be defined in various alternative ways, depending on the specific situation, for instance, as the point with coordinates equal to the average of individual travelers' locations' coordinates.

The abstracted spatial structure corresponding to a given geographic structure may be visualized in a hypothetical area, as in Figure 1.1, in which each white dot represents the individual traveler's locations, and black dots the zones' centroids.

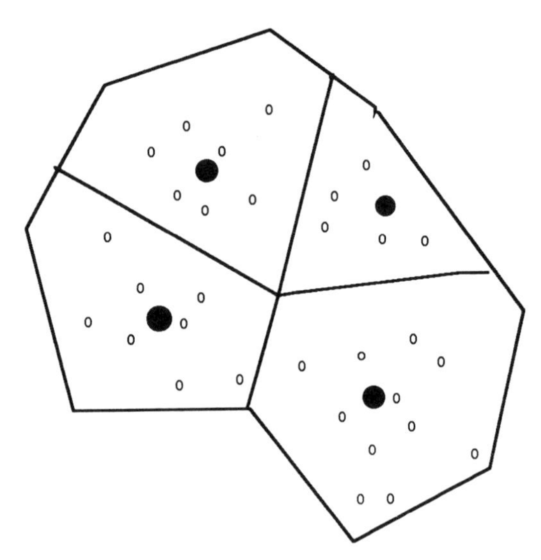

Figure 1.1 Geographic aggregation of individual travelers into trip origins.

What the optimal number of origin zones should be, their typical size, shape, and other such characteristics, must be decided on an empirical basis (e.g., by experimenting) in specific situations. The final choice will in general represent a compromise between available resources, in terms of data, computational resources, level of detail required, and so on. A convenient choice of spatial zones for practical purposes is Zip Code areas, or some aggregation of them, since many traveler data are available on that basis.

In general, the choice of a zonal system has critical effects on the performance of the models subsequently developed on its basis (see, for instance, Fotheringham and Wong, 1991, for a discussion of some of the issues in aggregating spatial data). Unfortunately, there are still no formal methods for designing spatial systems. A commonly used, general principle is to use as many zones as possible, in light of the above constraints, while at the same time maximizing the internal homogeneity of the resultant zones, as well as the differences between them.

In any case, there is a total number I of origin zones/locations, numbered $1, 2, \ldots, i, \ldots, I$. All demand-side variables will be subscripted with index i. Index i will thus refer to the origin of the trip travelers may make to perform the activity. For instance, the number of residents in origin zone i will be noted as N_i while the number of travelers will be T_i. Symmetrically, there will be a total of J destination zones/locations, numbered $1, 2, \ldots, j, \ldots, J$. All destination-related variables will symmetrically be referred to by index j. For instance, the cost (e.g., parking) at destination j will be s_j. A given location may be either an origin or a destination, or both. Note that in general $I \neq J$. This is graphically represented in Figure 1.2, which corresponds to Figure 1.1. In this case, $I = J = 4$.

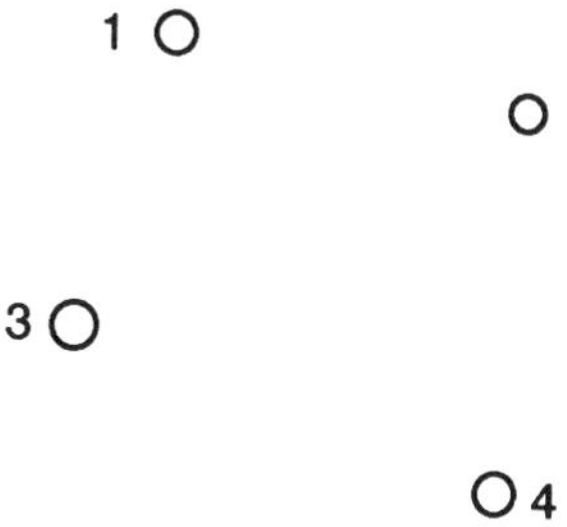

1 O

O 2

3 O

O 4

Figure 1.2 Spatial structure represented as nodes.

Each zone/location will be characterized by physical, and/or economic characteristics, such as number of residents in origin zone, and the size of parking facilities in destination zones. These characteristics may be numerically represented by the value of variables X_{ki}, as well as the socioeconomic characteristics Y_{kj} of zonal residents, for example, income and age.

The urban transportation system over which travel takes place is generally composed of various modal networks, such as the street network for private cars (the most important one in many cases), and the transit network. Each transportation network is in general composed of a set of nodes and a set of links connecting them. The nodes may include the centroids of origin and destination zones defined above, as well as other nodes that merely constitute the beginning or end of links. The links are numbered from 1 to A, with general index a. A link may represent the length of a street between two or several consecutive blocks, or of a stretch of parkway between two exits or between transit stops, depending on the geographical scale of analysis and the data available. Links constitute the basic element of a transportation system, since this is the level at which generalized travel costs, which include travel time, and possibly tolls and other such charges, are known by the analyst. However, from the viewpoint of individual travelers, the relevant travel costs are origin-destination costs, or more precisely, route costs.

If there are several distinct modal networks connecting the same links, individual links may be referred to by the index a_m. As in the case of the definition of spatial zones, the optimal definition of the network, in terms of number of links, typical length and so on, if left to the modeler, must be made empirically in a given practical situation. In any case, there is a total of A links, numbered $1, 2, \ldots, a_m, \ldots, A_m$. For instance, a network corresponding to Figure 1.2 is represented in Figure 1.3, in which there are ten links, $a = 1, 2, \ldots, 10$. In general, two-way links will be represented as two separate one-way links. Node numbers a are represented in italics alongside the links.

The connection between links and nodes may be described in several alternative ways. For instance, each link may be referred to by a beginning and an ending node, as in Table 1.1

Another possible representation, which may be more convenient in some cases, is to describe the respective links leaving and entering each node i. The set of links entering node i will be referred to as i^+ while the set of links leaving

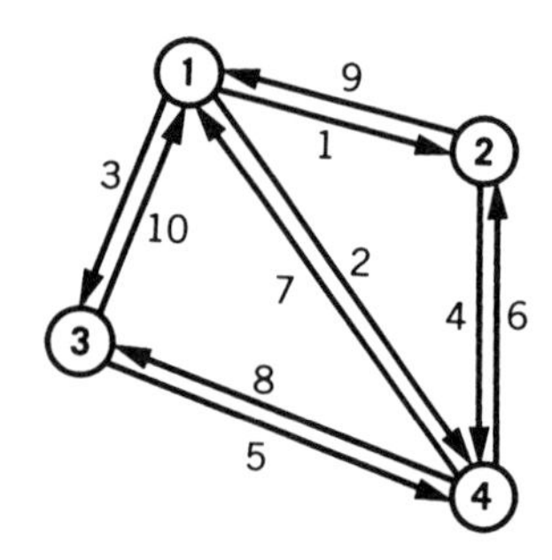

Figure 1.3 Links on a transportation system.

node i will be referred as to as i^-. The contents of these sets for each node i in the network above is described in Table 1.2.

Other representations of a network's topology may be developed, in particular for purposes of saving computer memory space, including the "forward star" (see Sheffi, 1985).

Each of the network links may be described by a number of physical characteristics, specifically those that influence the relationship between link volume and the cost of travel on the link. They include the link's length, the time it takes to travel the link under specified conditions, the frequency with which the link is served by scheduled transit or paratransit service, and so on. Link-related variables will be subscripted with the index a. For instance, for the reference network above, the time t_a^0 required to travel over link a under the best conditions (e.g., when there is no traffic) is described in Table 1.3.

In addition to the link-based description of a network, we will need to refer to routes, or paths between given locations/nodes, each composed of a different series of links. In general, there are several possible routes between given locations i and j. Routes will be indexed as r. For instance, in the case of the reference network above, there are 38 individual routes. Their origin, destination, and component links are described in Table 1.4.

TABLE 1.1 **"*i-j*" Link Reference**

Link #	Beginning Node i	Ending Node j
1	1	2
2	1	4
3	1	3
4	2	4
5	3	4
6	4	2
7	4	1
8	4	3
9	2	1
10	3	1

TABLE 1.2 Links in Sets i^+ and i^-

Node i	$\{i^+\}$	$\{i^-\}$
1	7, 9, 10	1, 2, 3
2	1, 6	4, 9
3	3, 8	5, 10
4	2, 4, 5	6, 7, 8

This information may also be presented equivalently in the form of the set R_{ij} of routes between a given origin and destination. This is represented in Table 1.5.

Routes may then be alternatively numbered internally to the origin destination pair they connect. For instance, in the present case, the correspondence between the absolute route number and its relative route number is as represented in Table 1.6.

As mentioned above, a network's characteristics are determined by those of its component links, both in terms of topological structure, and quantitatively, for example, in terms of capacity and travel times. However, from the viewpoint of the traveler, or network user, the relevant entities are at the route level, for example, travel time on a given route between a given origin and destination, as opposed to travel times on individual links. Route costs must then be estimated from the costs of their component links. To that effect, we need to specify the topological connection between links and routes. This is done through the "link-route incidence" variable δ^a_{ijr}, whose value is defined as

$$\delta^a_{ijr} = \begin{cases} 1, & \text{if link } a \text{ is part of route } r \text{ between } i \text{ and } j; \\ 0 & \text{otherwise} \end{cases} \qquad \forall\ i, j, r, a \qquad (1.1)$$

For example, in the reference network above, the values of combinations (i, j, r, a) for which $\delta^a_{ijr} = 1$ are as described in Table 1.7. All other δ^a_{ijr}'s are equal to zero.

TABLE 1.3 Link "Free Flow" Travel Time t^0_a

1	5
2	15
3	6
4	8
5	7
6	8
7	15
8	7
9	5
10	6

TABLE 1.4 Network Routes

Route #	Origin/Destination	Component Links
1	1, 2	1
2	1, 2	3, 5, 6
3	1, 2	2, 6
4	1, 3	3
5	1, 3	2, 8
6	1, 3	1, 4, 8
7	1, 4	2
8	1, 4	1, 4
9	1, 4	3, 5
10	2, 1	9
11	2, 1	4, 7
12	2, 1	4, 8, 10
13	2, 3	9, 3
14	2, 3	4, 8
15	2, 3	9, 2, 8
16	2, 3	4, 7, 3
17	2, 4	4
18	2, 4	9, 2
19	2, 4	9, 3, 5
20	3, 1	10
21	3, 1	5, 7
22	3, 1	5, 6, 9
23	3, 2	10, 1
24	3, 2	5, 6
25	3, 2	10, 2, 6
26	3, 2	5, 7, 1
27	3, 4	5
28	3, 4	10, 2
29	3, 4	10, 1, 4
30	4, 1	7
31	4, 1	6, 9
32	4, 1	8, 10
33	4, 2	6
34	4, 2	7, 1
35	4, 2	8, 10, 1
36	4, 3	8
37	4, 3	7, 3
38	4, 3	6, 9, 3

The same information may alternatively be presented in the form of an "incidence matrix" (see Exercises 1.1 and 1.2). Route-related variables will be referred to with the three indices i, j, and r. For instance, the volume on a given route will be noted T_{ijr}. One important such variable is the route travel time/cost. This may be estimated from the specification of link travel times/

TABLE 1.5 Routes Serving Origin-Destination Pairs

(i, j)	$R(i, j)$
1, 2	1, 2, 3
1, 3	4, 5, 6
1, 4	7, 8, 9
2, 1	10, 11, 12
2, 3	13, 14, 15, 16
2, 4	17, 18, 19
3, 1	20, 21, 22
3, 2	23, 24, 25, 26
3, 4	27, 28, 29
4, 1	30, 31, 32
4, 2	33, 34, 35
4, 3	36, 37, 38

costs t_a. Using the "link-route incidence" variable δ^a_{ijr} defined in Formula (1.1), the travel time on route r between i and j is equal to the sum of the travel times for all links that constitute the route:

$$t_{ijr} = \sum_a t_a \delta^a_{ijr}; \qquad \forall\ i, j, r \tag{1.2a}$$

Similarly, the volume v_a on a given link may be estimated as the sum of all volumes for routes that use the link:

$$v_a = \sum_i \sum_j \sum_r T_{ijr} \delta^a_{ijr}; \qquad \forall\ a \tag{1.2b}$$

TABLE 1.6 Alternative Route Numbering

(i, j)	r	$r_{(i,j)}$
1, 2	1, 2, 3	1, 2, 3
1, 3	4, 5, 6	1, 2, 3
1, 4	7, 8, 9	1, 2, 3
2, 1	10, 11, 12	1, 2, 3
2, 3	13, 14, 15, 16	1, 2, 3, 4
2, 4	17, 18, 19	1, 2, 3
3, 1	20, 21, 22	1, 2, 3
3, 2	23, 24, 25, 26	1, 2, 3, 4
3, 4	27, 28, 29	1, 2, 3
4, 1	30, 31, 32	1, 2, 3
4, 2	33, 34, 35	1, 2, 3
4, 3	36, 37, 38	1, 2, 3

TABLE 1.7 Nonzero δ^a_{ijr} for Network (i.j.r.a)

1.2.1.1; 1.2.2.3; 1.2.2.5; 1.2.2.6; 1.2.3.2; 1.2.3.6; 1.3.1.3; 1.3.2.2; 1.3.2.8; 1.3.3.1; 1.3.3.4; 1.3.3.8; 1.4.1.2; 1.4.2.1; 1.4.2.4; 1.4.3.3; 1.4.3.5; 2.1.1.10; 2.1.2.4; 2.1.2.7; 2.1.3.4; 2.1.3.8; 2.1.3.9; 2.3.1.3; 2.3.1.10; 2.3.2.4; 2.3.2.8; 2.3.3.10; 2.3.3.2; 2.3.3.8; 2.3.4.4; 2.3.4.7; 2.3.4.9; 2.4.1.4; 2.4.2.10; 2.4.2.2; 2.4.3.10; 2.4.3.3; 2.4.3.5; 3.1.1.9; 3.1.2.5; 3.1.2.7; 3.1.3.5; 3.1.3.6; 3.1.3.10; 3.2.1.1; 3.2.1.9; 3.1.2.5; 3.1.2.6; 3.2.3.9; 3.2.3.2; 3.2.3.6; 3.2.4.5; 3.2.4.7; 3.2.4.1; 3.4.1.5; 3.4.2.9; 3.4.2.2; 3.4.3.9; 3.4.3.1; 3.4.3.4; 4.1.1.7; 4.1.2.6; 4.1.2.10; 4.1.3.8; 4.1.3.9; 4.2.1.6; 4.2.2.7; 4.2.2.1; 4.2.3.8; 4.2.3.9; 4.2.3.1; 4.3.1.8; 4.3.2.7; 4.3.2.3; 4.3.3.3; 4.3.3.6; 4.3.3.10;

For instance, for the reference network, the routes travel times corresponding to the link travel times in Table 1.3 are given in Table 1.8.

Finally, we will also be using another type of variable, which characterizes demand between an origin-destination pair. Such variables will be called "O/D" variables. They will be subscripted with the double index (ij). For instance, the *minimum* travel time between locations i and j will be noted t_{ij}, and the number of travelers from origin i to destination j will be noted T_{ij}.

1.3 THE RESPECTIVE DIMENSIONS OF TRAVEL DEMAND

Having specified the spatial structure over which travel is going to take place, as well as some of the variables that describe it, we now define the various dimensions of travel demand. These quantities, which are the central focus of the subsequent analysis, are respectively:

- The number T_i of person trips originating in each zone. Obtaining these numbers is sometimes referred to as performing "trip generation."
- The number of trips T_{ij} originating in a given zone i *and* terminating in a given zone j. Obtaining these numbers is sometimes referred to as performing "trip distribution."
- The number of trips T_{ijm} originating in a given zone i and terminating in a given zone j, *and* using a given mode m. Obtaining these numbers is sometimes referred to as performing "modal split."
- The number of trips T_{ijmr} originating in a given zone i, terminating in a given zone j, using a given mode m, and following a given route m.

TABLE 1.8 Route Travel Times (r, t_r)

(1,5) (2,20) (3,23) (4,6) (5,22) (6,20) (7,15) (8,13) (9,13) (10,5) (11,23) (12,21) (13,11) (14,15) (15,27) (16,29) (17,8) (18,20) (19,18) (20,6) (21,22) (22,20) (23,11) (24,15) (25,29) (26,27) (27,7) (28,21) (29,19) (30,15) (31,13) (32,13) (33,8) (34,20) (35,17) (36,7) (37,21) (38,19)

Obtaining these numbers is sometimes referred to as performing "trip assignment."

It is clear that, to be consistent with one another, these respective quantities must observe some balance or conservation of flow requirements:

$$\sum_j T_{ij} = T_i; \qquad \forall\, i \tag{1.3a}$$

$$\sum_i T_{ij} = T_j; \qquad \forall\, j \tag{1.3b}$$

$$\sum_m T_{ijm} = T_{ij}; \qquad \forall\, i, j \tag{1.3c}$$

$$\sum_r T_{ijmr} = T_{ijm}; \qquad \forall\, i, j, m \tag{1.3d}$$

The object of urban travel demand modeling is to develop methodology with which one may predict the values of these quantities in a given situation, that is, given the characteristics of the urban area, in terms of its spatial structure, and the personal characteristics of its inhabitants.

1.4 THE TRADITIONAL, FOUR-STEP APPROACH TO URBAN TRAVEL DEMAND MODELING

Urban travel demand modeling has progressively evolved over the last thirty to forty years into an established methodology, commonly referred to as the traditional, or classical approach (Ortuzar and Willumsen, 1990). In this section, we briefly describe the basic features of the traditional approach. This is sometimes called the "four-step" process, as it successively addresses each of the respective dimensions of travel demand described above sequentially.

1.4.1 Trip Generation

The first phase in the traditional approach, "trip generation," is designed to estimate the T_i's and/or T_j's, that is, the numbers of person-trips originating in, and/or ending in given zones. There are two main approaches to this phase of the travel demand estimation process. It may be performed either at the aggregate level of the zones i (i.e., when the number of trips is assumed to be a function of zonal characteristics), or alternatively, at the disaggregate level of the household (i.e., when it is assumed to be a function of households' characteristics).

In either case, trip generation is effected mainly using various statistical methods, at differing levels of complexity (see, for instance, Ortuzar and Willumsen, 1990). The simplest and crudest such method is to use observed zonal, or household, "trip rates." There are various sources from which such figures

TABLE 1.9 Trip Rates per Household per Day

	Number of Cars Owned			
	0	1	2	3^+
Household Size				
1	1.1	2.5	4.2	—
2	1.7	4.8	6.6	—
3	2.5	6.2	8.4	11.1
4	2.7	7.4	12.0	14.2
5^+	5.2	9.3	14.4	17.6

may be obtained (see, for instance, ITE, 1990). If need be, simple extrapolations based on the estimated future value of "growth factors" may be performed. These factors may in turn be a function, either numerically, or qualitatively, of prevailing socioeconomic conditions.

For instance, a (hypothetical) trip rate table might be as in Table 1.9 (see Exercise 1.7).

Another, somewhat more formal statistical approach to trip generation is econometric modeling, including linear or quasi-linear regression, in which the dependent variable is either T_i or T_j, the independent variables are zonal and/or traveler characteristics (ITE, 1990). The form of such relationships may be specified as:

$$T_i = a_i^0 + \sum_k a_i^k X_i^k; \qquad \forall\, i \tag{1.4}$$

in which the X_i^k's are given trip "production" factors (or "attraction" in the case of the T_j's), and the a_i^k's are parameters whose numerical value is produced by linear regression, using the "minimum sum of squares" method (Wonnacott and Wonnacott, 1977).[1]

For instance, a (hypothetical) model might be, in a specific locale

$$T = 4.3 + 3.9X_1 + 0.005X_2 + 0.13X_3 + 0.012X_4$$

where X_1 is the average level of car ownership per household, X_2 the number of residential dwelling units per acre, X_3 the distance from the Central Business District in miles, and X_4 the household's income in thousand of dollars. The performance of such regression models may be evaluated from the coefficient of correlation R^2, as well as tests on the estimated values of the coefficients a_i^k.

Several techniques, including analysis of variance, factor and cluster analysis, contingency tables, and discriminant analysis, may also be used to clas-

[1]The topic of model estimation is addressed in Chapter 7.

sify travelers into a few homogeneous and distinct groups, each having a characteristic trip rate. These methods are referred to as cross-classification (Stopher and McDonald, 1983), or category analysis (Wooton and Pick, 1967). Standard statistical techniques used in this connection, including some of the remedies that may be used to deal with various technical difficulties, (e.g., dependency between independent variables, etc.), are presented in all introductory texts on statistics, including Wonnacott and Wonnacott (1977).

Whatever the particular technique used, the factors of trip generation in the traditional approach are typically limited to zones and travelers attributes, and do not include attributes such as those of travel modes, and/or travel routes, including travel times and costs. More precisely, origin/destination attributes are used only during trip distribution, modal attributes only during modal split, and route attributes only during route assignment. This issue is discussed further in section 4.6 on direct demand models.

1.4.2 Trip Distribution

The next stage of travel demand estimation in the traditional approach is trip distribution. This is typically performed after trip generation is completed, and consists of distributing across various destinations each of the trip origins T_i obtained in the first phase above. Typically, there is no feedback between these two phases, that is, trip generation is not itself affected by the attributes of travel destinations, or travel modes, or travel routes. Moreover, this procedure is performed separately for each origin zone i.

The typical traditional approach to trip distribution is to use "synthetic" models (i.e., models based on analogies with models describing phenomena other than urban travel). One of the most commonly used models is the "gravity" model, which is adapted from Newton's "gravitational" law of physics. In its "production constrained" form, the trip origins T_i are given, and consequently the T_{ij}'s for a fixed value of i must add up to the T_i's, that is, must meet the constraints

$$T_i = \sum_j T_{ij}; \qquad \forall\, i \tag{1.5a}$$

In this case, the gravity model takes the form

$$T_{ij} = T_i \frac{X_j^{\alpha} f(t_{ij})}{\sum\limits_{j=1}^{J} X_j^{\alpha} f(t_{ij})}; \qquad \forall\, i, j \tag{1.6a}$$

where R_j is some measure of attractiveness of destination zone j, and the function $f(t_{ij})$ of the distance, or cost c_{ij} between origin i and destination j, may take several alternative forms.

In its "attraction constrained" form, when the trip ends T_j are given, and consequently the T_{ij}'s must meet the constraints

$$T_j = \sum_i T_{ij}; \qquad \forall\, j \tag{1.5b}$$

the form of the model is

$$T_{ij} = T_j \frac{X_i^{\alpha} f(t_{ij})}{\sum_i X_i^{\alpha} f(t_{ij})}; \qquad \forall\, i, j \tag{1.6b}$$

Finally, a third version, in which *both* the T_i's and the T_j's are given, so that both constraints (1.5a) and (1.5b) must be met *simultaneously*, may be formulated as

$$T_{ij} = a_i\, b_j T_i T_j X_i^{\alpha} Y_j^{\beta} f(t_{ij}); \qquad \forall\, i, j \tag{1.7}$$

in which the unknown values a_i and b_j must consequently solve the system of equations (see Exercise 1.4)

$$a_i = \sum_j b_j T_j Y_j^{\beta} f(t_{ij}); \qquad \forall\, i \tag{1.8a}$$

$$b_j = \sum_i a_i T_i X_i^{\alpha} f(t_{ij}); \qquad \forall\, j \tag{1.8b}$$

The numerical value of β and α, as well as those of the parameters intervening in function $f(t_{ij})$, are determined using various techniques with varying degrees of formalism, from "trial and error," possibly based on a specific criterion such as "minimum sum of squared errors," to "maximum likelihood" (Wonnacott and Wonnacott, 1977); (see Exercise 1.10).

1.4.3 Modal Split

The third stage in the traditional approach, modal split, is typically performed after trip distribution. Specifically, each of the origin-destination volumes T_{ij} obtained in the trip distribution phase are now "split," or distributed, into the various alternative modes. In some cases, if more appropriate, the order of these two respective phases may be inverted. There is again typically no feedback between them, in the sense that modal distributions are only a function of modal attributes, and not origin-destination attributes. Furthermore, modal split is effected separately for each origin-destination combination *i-j*.

The models used in this connection, like those used for trip distribution, are also synthetic models. In this case, the underlying analogy is with an empirical "diversion curve," which represents the percentage P_t of travelers diverted to the use of transit, as a function of transit's attributes relative to those of the private car, for example, the difference or ratio between their respective "attractiveness,'' A_t and A_c. These quantities may in turn be specified as (e.g., linear) functions of the mode's attributes, including cost, and ''level of service,'' and the average characteristics of travelers, including income and car ownership.

These empirically determined functions have an S-shaped curve of the form represented in Figure 1.4, which is similar to that of the ''logistic,'' or ''Gompertz'' function in calculus.

The corresponding model is accordingly

$$P_c = \frac{1}{1 + e^{\beta(A_t - A_c)}} \tag{1.9a}$$

where parameter β specifies the curvature of the plot (see Exercises 1.8, 1.9).

When not given, the values of the parameters may in this binary case be determined from observations of the modal split, by noting that Formula (1.9a) may be rewritten as

$$P_t/P_c = \exp\{\beta(A_t - A_c)\} \tag{1.9b}$$

so that, again equivalently,

$$Ln(P_t/P_c) = \beta(A_t - A_c) \tag{1.9c}$$

If functions A_m are specified as

$$A_m = a_m + \sum_k a^k X^k_m; \qquad m = c, t$$

a linear regression of the values of the ratio (P_t/P_c) as observed in various surveys against those of the variables X^k_m will provide the numerical values of parameters a (see Exercise 1.11).

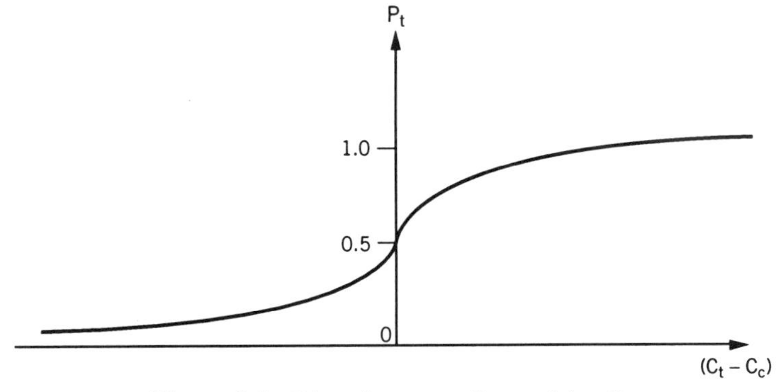

Figure 1.4 Diversion curve for modal split.

In the case of several available modes, the above synthetic model may be generalized to

$$P_m = \frac{e^{\beta A_m}}{e^{\beta A_1} + e^{\beta A_2} + \ldots + e^{\beta A_M}}; \quad \forall\, m \qquad (1.10)$$

where P_m is the share of the ridership of mode m, and A_m its attractiveness.

1.4.4 Trip Assignment

Finally, in the last phase, "trip assignment," the modal trips T_{ijm} from a given origin to a given destination on a given mode obtained in the preceding phase are now assigned to the network's links, or more precisely, routes between a given origin and destination. A route, as will be formalized soon, is composed of a series of links. The rationale for trip assignment is to assign all trips to the route with minimum cost, on the basis that these are the routes travelers would (rationally) want to use. That is

$$\left.\begin{array}{lll} T_{ijmr^*} = T_{ijm}; & \text{for the minimum cost route } r^*; \\ T_{ijmr} = 0; & \text{for all other routes} \end{array}\right\}; \quad \forall\, i, j, m; \qquad (1.11)$$

where T_{ijm} is the given modal demand between i and j obtained in the preceding stage. This procedure is referred to as "all-or-nothing" assignment, for obvious reasons.

Finding the minimum cost route is a straightforward task, which may be solved with the application of readily available "shortest path" algorithms (see, for instance, Pape, 1974). As for the three preceding phases, trip assignment is performed independently of the other phases.

1.4.5 Treatment of Congestion

When there is no congestion on the respective modal networks, that is, when all factors of travel demand are fixed and independent of travel volumes, the above procedure is straightforward, and its four steps are applied sequentially, as described above.

When there is congestion, however, the situation becomes significantly more complex. Specifically, in that case, demand factors, including link and destination travel costs, are functions of the demands. For instance, the higher the volume is on a given route, the lower the travel speed on it, and thus, the higher its travel time, decreasing the attractiveness of the route, and thereby the demand for its use.

This, in turn, may influence the demand for the origin-destinations served by the route, which may in turn influence the demand for the modes serving these origin-destinations, which may ultimately influence the origin and destination demands themselves. This creates complex feedback relationships between travel demands at all four levels, which must be resolved simultaneously to obtain the levels of demands that are in "balance" or equilibrium, that is, consistent with the variable factors underlying them.

Typically, the treatment of congestion in the traditional approach to network assignment consists of enacting these feedbacks through repeated application of the above procedures, within each stage of the four-stage process, as well as between them, in a variety of ways, as described below.

In the case of route assignment, the last stage on the procedure, assigning any number of trips to any given link changes the link's costs, and therefore, the identity of the shortest paths. Two main methods are used to resolve this situation. The first is to start from some initial, guessed values for the link costs, then assume, temporarily, that they will remain constant, find the minimum cost routes, and assign trips to them. Then, based on the corresponding new link volumes, revise link costs, and repeat this cycle until there is no need (i.e., the new costs are not appreciably different from the previous values), or no time to continue. This procedure is sometimes called "capacity restraint" assignment.

The other approach is to load the network incrementally, and to keep updating link travel costs in the process. Specifically, a given fraction (i.e., 10%, 20%, etc., depending on the number of iterations desired), of the origin-destination volumes will be loaded with the Shortest Path Algorithm on the basis of travel times estimated at zero volumes. Given the resulting link volumes, travel costs are then reevaluated on the basis of the current volumes. Then, the next 10% (or 20%, etc.) of the origin-destination volumes are assigned to minimum cost routes, and on the basis of the current travel costs. The current link volumes are the cumulation of the previous link volumes and the current link volumes. Travel times are then updated on the basis of the current link volumes, and this procedure is iterated until 100% of the origin-destination volumes are assigned.

In turn, modal split may itself be revised iteratively on the basis of updated modal costs or attractiveness, either in parallel with each iteration of the network assignment procedure, or sequentially, after the latter has been performed in its entirety. The other two procedural stages may similarly be embedded as two additional outer "loops" in a generalized iterative procedure.

1.4.6 Direct Demand Models

As mentioned earlier, the main characteristic of the four-stage approach is the compartmentalized treatment of the various aspects of travel demand. An alternative, in which all phases are addressed simultaneously, has been used in the past (Kraft, 1968, Quandt and Baumol, 1966; Domencich et al., 1968). The form of these "direct demand" models, as they are sometimes called, is essentially that of linear, or quasi-linear statistical regressions:

$$T_{ijmr} = \alpha_{ijmr} \prod_k X^k_{ijmr} \beta^k_{ijmr}; \qquad \forall\ i, j, m, r; \tag{1.12}$$

where α^k_{ijmr} and β^k_{ijmr} are parameters to be calibrated. The X^k_{ijmr} measure various attributes of demand zones, destination, modes and routes. One advantage of such a formulation is that the calibrated value of the β^k parameters measures the elasticity of the demand T_{ijmr} with respect to the corresponding attribute X^k (see Exercise 1.12).

However, the main theoretical disadvantage of direct demand models, which they share with the synthetic models above, is the fact that they are purely descriptive. A practical disadvantage is their large number of variables, due to their multiplicative form, and the resultant data needs. The direct demand model, on the other hand, may be calibrated in a straightforward fashion (se Exercise 1.3). Nevertheless, in some situations, and as a simplified tool for "quick response" application, the direct demand model may be useful.

1.5 CRITIQUE OF THE FOUR-STEP APPROACH

The four-step process as described in the previous section is widely used in current practice. It is in particular implemented in the Urban Transportation Planning System (UTPS), which was distributed for many years by the U.S. Department of Transportation (FTA, 1977), as well as in most commercially available transportation planning software. Nevertheless, it presents some significant drawbacks. In this section, we briefly discuss them as background to the development of the models in this text.

First, the conventional approach is not based on any single unifying rationale that would explain or legitimize all aspects of demand jointly, and in the presence of *congestion*. Individually, trip distribution, modal split, and network assignment may be given a "behavioral" interpretation. Specifically, trip distribution models of the gravity Formulas (1.6 and 1.7) have been shown to be strictly equivalent (i.e., functionally, as well as numerically), to "behavioral" (i.e., "logit") models of destination choice (Anas, 1983).[2]

Similarly, modal split models of the Formula (1.10) are also functionally equivalent to logit models, and can thus be interpreted as the result of "utility maximizing" choices by individual travelers,[3] that is, they reflect the same principle that will be systematically assumed to underlie all travel demands in this text (see section 1.6). However, trip distribution and modal split models are typically not integrated, reflecting the separate conduct of these respective phases, as discussed above. Finally, minimum cost route trip assignment in the four-stage approach does, by definition, reflect a behavioral principle of cost minimization. On the other hand, trip generation, as discussed above, is performed using linear regression, the "descriptive" technique par excellence. However, taken as a whole, the traditional four-step approach to travel demand modeling lacks such behavioral interpretation, let alone rationale.

[2]The logit model is introduced in Chapter 2.

[3]Interestingly, modal split was formulated early on as a case of binary choice by individual travelers (Warner, 1962).

The second major pitfall in the traditional approach to travel demand modeling is the adhoc treatment of demand externalities, that is, the effects of network and destination congestion. The traditional approach proceeds from the top of the structure, the decision to travel, on through destination and modal choice, ending with route choice. Each level is thus treated serially and independently of the others, and its output is passed through to the next lower level. The T_i's are first obtained at the first level, and are then "distributed" among destinations, providing the T_{ij}'s. These are then "split" between modes, providing the T_{ijm}'s. Finally, these are "assigned" to routes, providing the T_{ijmr}'s. When congestion is not present, as when travel times are independent of travel volumes and thus fixed, a single pass through the respective levels is therefore sufficient.

When congestion is present, however, travel costs are dependent on travel volumes, and vice versa. To take this phenomenon into account, travel demands are periodically used to revise current estimates of travel costs at the link level, leading to revised estimates of modal route costs, which in turn imply new demand estimates at all levels, and so on, until some form of convergence is attained, in which both demands and travel times have attained stable, "compatible" values.

This "feedback" process may be structured in various alternative ways. For instance, car route volumes may be fed back to convergence, and then transit route volumes, given the stable car volumes. Alternatively, both car and transit volumes may be fed back simultaneously. Combining destination demand estimation with these can lead to further schemes.

However, the various feedback loops introduced to represent congestion effects may or may not converge to a stable distribution of the respective demands. Consequently, the nature of the travel demands obtained after performing various "feedbacks" may not be entirely clear. For instance, in the case of network assignment (route demands), the solution obtained with the incremental approach clearly constitutes an approximation to the exact solution, that is, the estimated solution that would be obtained with increments of one traveler, were that feasible. The precision of such an approximation is usually not known. The same difficulties are compounded further when several dimensions of travel demand are combined, for example, multimodal network assignment.

Another significant drawback is that such iterative techniques are typically inefficient computationally; they require a large number of iterations to converge. This is not surprising, given that these iterations do not follow a guided search for a clearly specified state of the system. This is an important consideration, given the size of real-world networks.

There is thus a need to improve the traditional methodology of urban travel demand modeling. Over the last several years, various improvements have been made by many researchers in the field, which have contributed to meet this need. As a result, it is now possible to present methodology that does not suffer from the above problems. Specifically, it is consistently and systematically based on the same explicit, "behavioral" principles of utility maximization that underlie established microeconomic theory. In addition, it deals with the presence of congestion in a rigorous manner, which is compatible with this principle. Finally, it may be implemented with efficient, exact numerical methods, without resort to artificial, or computationally prohibitive, numerical procedures.

As mentioned in the Preface, the main object of this book is to present this methodology. It might also be mentioned that the new methodology does not represent a radical departure from the traditional approach, but rather an evolutionary improvement, in which the "behavioral" potential of the traditional models discussed above has been formalized, and their operation cast in the form of powerful techniques of mathematical programming. In the next section, we introduce the main postulates underlying this particular approach, which may be referred to as the behavioral, combined equilibrium approach.

1.6 THE TRIP CONSUMER APPROACH

The basic, systematic viewpoint we will adopt in modeling urban travel demand is that travel in urban areas is best understood as the result of various decisions by *individual* travelers, such as whether to travel, where to go, which mode to use, and which route to follow.

While hardly revolutionary or arguable, this viewpoint implies that the best approach to modeling *aggregate* travel demands is not descriptive (e.g., based on statistical modeling techniques), but rather, "behavioral" (i.e., reflecting *disaggregate, or individual*, demands). These in turn may be estimated on the basis of assumptions about individual objectives, opportunities, constraints, and so on, similar to those used in microeconomic analysis. In this framework, the individual traveler may be considered a consumer of urban trips, just as he/she is a consumer of other "goods."

Such an approach has several advantages. First, it grounds travel demand analysis in an explicit, rigorous framework, in which each aspect of individual travel behavior must be explained as the rational outcome of an explicit decision-making process on the part of individual travelers, under specific conditions. The resulting formulations are thus endowed with an "explanatory" ability that is absent from descriptive ones. With the latter, in fact, travel demand estimation is predicated not on the permanence of human behavior (i.e., "first principles"), but on that of the "patterns" used to represent its manifestation.

Second, the behavioral approach to travel demand modeling allows one to be guided by and/or use well-established microeconomics theory and methods. This significantly enlarges and enriches the range and efficiency of available methodology to tackle urban travel demand modeling. For instance, travel pricing, equity, welfare, and other issues relevant to urban transportation policy may be addressed rigorously.

Third, a behavioral approach to travel demand modeling allows one to address in a totally consistent manner the supply side, in which the behavior of suppliers of activity as well as travel, as opposed to performers of urban travel, is considered. This is in fact the topic of Chapters 8 and 9. In this case, it is even more clear that a behavioral approach is required, one in which suppliers attempt to attain specific objectives, be they profit, community welfare, or given stated constraints (e.g., on resources, or from regulation, etc.). The integration of the supply and demand sides of travel, which is necessary to identify the resulting equilibrium, is best effected through the use of a common unifying rationale, that of "benefit" maximization.

In taking this behavioral approach, it should be remembered that travel usually supports the conduct of some type of activity, be it short term (e.g., shopping or obtaining some service, recreating, socializing, etc.), or long term (e.g., working, residing, etc.). This implies that individual travelers optimize their choices not only with respect to travel, but also *jointly*, with respect to such an activity. In other words, if one chooses to think of urban trips as a "commodity," it is in most if not all cases consumed in conjunction with the conduct of such activities.

Consequently, individual travel decisions in general concern not only travel, but also the associated activity that travel supports. Being made jointly, these respective decisions may potentially influence one another. For instance, the choice of a residence may determine the mode of travel to work, and vice versa. Thus, travel demand analysis should ultimately, at least in theory, be related to the analysis of the demand for all such activities. In this perspective, land use–transportation or "activity systems" models may be considered long-term travel demand models. In most of the models developed in this text, however, travel demand is typically analyzed *given* the demand for these activities, that is, assuming a known value for such demands. Nevertheless, a possible approach to integrated activity and travel demand modeling is described in Chapter 8.

The other fundamental feature of the models developed in this text is that aggregate travel demands in the presence of congestion will reflect demand externalities accurately. That is, they will correspond to a *combined* equilibrium in which the values of the T_i, T_{ij}, T_{ijm}, and T_{ijmr}'s are all *simultaneously* compatible with the values of the factors which depend on them, including cost of traveling, destination costs, mode costs, and modal link/route costs. In addition, numerical techniques for model operation, including parameter estimation and demand prediction, will be based on formal, efficient numerical methods.

1.7 SUMMARY

In this introductory chapter, the book's main purpose was stated as the development of behavioral, combined equilibrium models of urban travel demand, and the stage set to that effect. First, the elements of the generic urban spatial system, including the transportation system, were defined. The major dimensions of travel demand were next identified. This, in particular, introduced notation that will be used systematically throughout the text. Next, the traditional four-step approach to travel demand modeling was surveyed briefly, and its weak points discussed. In contrast, the advantages of the behavioral modeling and combined equilibrium approach were outlined.

In the next chapter, the "random utility" framework for modeling traveler choices, which offers these advantages, will be developed as a prerequisite to the development of the travel demand models themselves.

1.8 EXERCISES

1.1 For the network in Figure 1.5, construct the link-route incidence matrix, i.e., identify the values of the δ_{ijr}^{a}'s in Formula (1.1).

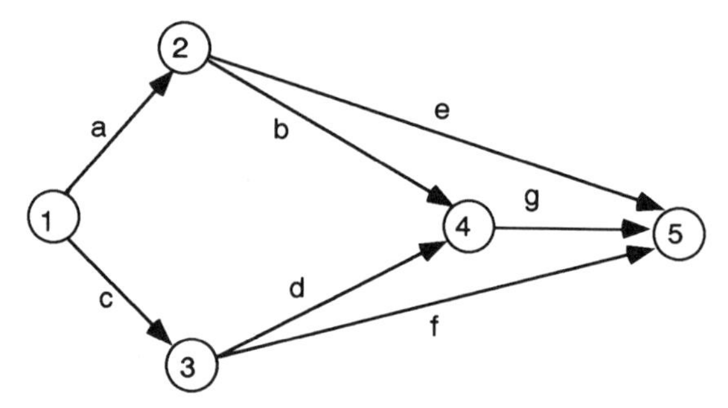

Figure 1.5 Network for Exercise 1.1.

1.2 For the reference network represented in Figure 1.3, compute the value of the link-route incidence variables δ_{ijr}^{a}, and present this information in the form of an "incidence matrix," i.e., as

Link	1	2	3	4	5	6	7	8	9	10
Route										
1.2.1	1	0	0	0	0	0	0	0	0	0
1.2.2	0	0	1	0	1	1	0	0	0	0
etc.					etc.					

1.3 Show how the direct demand model, Formula (1.12) may be calibrated using the technique of linear regression. (Hint: Use the logarithms of the variables instead of the variables themselves.)

1.4 Devise an iterative technique for solving the equation system, Formula (1.8) for the identification of the "balancing factors" a_i and b_j in the doubly constrained gravity model, Formula (1.7).

1.5 To illustrate the application of the "production constraint" gravity model, Formula (1.6a), assume that the values of the T_i's are (in thousands)

$$T_1 = 2, T_2 = 3, T_3 = 4$$

and that the values of the X_j's are

$$X_1 = 3, X_2 = 2, X_3 = 2$$

The value of parameter $\alpha = 1$ and the functions $f(t_{ij}) = t_{ij}^{-1}$ with t_{ij} are presented in Table 1.10.

TABLE 1.10 Travel Times

t_{ij}	$j =$	1	2	3
$i =$	1	2	8	6
	2	8	3	4
	3	6	4	3

a. Given these data, determine the trip distribution, i.e., the values of the T_{ij}'s, as well as the trip ends, i.e., the values of the T_j's.

b. What would the estimated values be if $\alpha = 2$ and the function

$$f(t_{ij}) \equiv t_{ij}^{-2} \tag{1.13}$$

were used instead? Compare the respective answers under either assumption.

1.6 To illustrate the application of the "double constraint" gravity model, Formula (1.7), assume that the values of the T_i's are (in thousands)

$$T_1 = 5, T_2 = 5, T_3 = 45$$

and that the values of the T_j's are (in thousands)

$$T_1 = 20, T_2 = 30, T_3 = 5$$

The function $f(t_{ij})$ is defined as

$$f(t_{ij}) \equiv t_{ij}^{-2} \exp(-0.5t_{ij}) \tag{1.14}$$

Finally, all X_i and Y_j values are equal to 1, i.e., these factors are not present. Given these data, determine the trip distribution, i.e., the values of the T_{ij}'s.

TABLE 1.11 Data for Exercise 1.7

Size =	1–2 persons	2–3	4^+
Number of cars 0–1	0.25	0.20	0.15
1–2	0.15	0.15	0
3+	0.1	0	0

1.7 In a given hypothetical area, there are 2,000 households, which are distributed in terms of percentages in various classes of size and car ownership, as in Table 1.11. Using the data in Table 1.11, estimate the number of trips this area will generate on a daily basis.

1.8 To illustrate the application of modal split Formula (1.9a), assume that

$$\beta(A_t - A_c) = -0.7 + 0.3(c_c - c_t) + 0.2(t_c - t_t)$$

where c_m is the modal cost of travel in dollars, and t_m is the modal travel time, in hours.

a. If the difference in travel time is $(t_t - t_c) = 0.4$, and if the cost of travel by car is $c_c = 1.50$, what should be the transit fare so that transit could attract one-third of the ridership?

b. By how much should transit travel time be lower than that with the car, in order for transit to attract the same percentage of the ridership, at a fare of one dollar? Compare the result with the answer in part *a* above, and derive an estimate of the ''value of time'' to travelers.

1.9 The choice between two travel modes is assumed to be represented by the model

$$P_m = \frac{e^{\beta A_m}}{\sum_m e^{\beta A_m}}; \qquad \forall\ m \tag{1.15}$$

where

$$A_1 = -2 - t_1$$

$$A_2 = -4 - t_2$$

where t is the travel time on the mode. Plot, as in Figure 1.4, the probability of choosing mode 1 as a function of the difference in travel times for values of $\beta = 0.025, 0.5, 0.75, 1$ and 1.25. Interpret the findings in connection with the role of parameter β.

TABLE 1.12 Observed Values $\hat{T}_{ij}$ for the Calibration of Doubly Constrained Gravity Model (in Thousands)

$\hat{T}_{ij}$	$j =$	1	2	3
$i =$	1	1.8	3.1	0.1
	2	3.1	1.35	0.4
	3	15.1	25.4	4.5

1.10 The observed trip pattern to be used in the calibration of the doubly constrained model in Exercise 1.6 above is shown in Table 1.12. The function $f(t_{ij})$ is defined as

$$f(t_{ij}) \equiv t_{ij}^{-\alpha} \exp(-\beta t_{ij}) \qquad (1.16)$$

where α and β are to be given the values most compatible with the observed values above. To that effect, conduct a "grid search," combining values for α and β in the range -1 to $+1$ by increments of 0.2. For each combination, estimate the T_{ij} according to the resulting model, and compare with the observed values in Table 1.12, on the basis of the sum of *squared* differences between predicted and observed value. (The square is to prevent negative and positive errors from obliterating one another.) Retain the combination of parameter values which leads to the minimum sum of squared errors. (Hint: You may want to program a computer to perform the corresponding calculations.)

1.11 The respective travel times and ridership for both the car and transit in five selected corridors are given in Table 1.13. Calibrate the binary modal split model

$$P_m = \frac{e^{-\beta t_m}}{\sum_m e^{-\beta t_m}}; \qquad \forall\, m \qquad (1.17)$$

TABLE 1.13 Data for Binary Modal Split Model Calibration

Corridor #	1	2	3	4	5
car travel time (min.)	15	20	25	40	55
transit travel time	25	35	40	70	75
car ridership (×1,000)	3	8	4	1	10
transit ridership	1	2	1.5	0.1	3

by first putting it in the form of Formula (1.9a), and then plotting the values of $y = Ln(P_t/P_c)$ against those of $x = (t_c - t_t)$ to estimate the value of the slope β of the straight line.

1.12 Show that parameters β_k in the direct demand model represented by Formula (1.12) in fact represent the elasticities of demand with respect to the corresponding variable X_k. (For a definition of elasticity see section 2.7 in Chapter 2.)

CHAPTER 2

MODELING TRAVELERS' DECISIONS AS DISCRETE CHOICES

2.1 THE INDIVIDUAL TRAVELER'S DECISION PROCESS

The approach taken in this book to predicting the various components of travel demand described in the previous chapter is based on modeling individual travelers' decisions regarding the travel choices, or alternatives, facing them. This is sometimes referred to as the "behavioral approach," because, in contrast with the "traditional" approach, it is not based on descriptive modeling, but rather on an explicit principle of human behavior. Specifically, individual travelers are assumed to make travel choices which are "the best" for them, in a sense we'll define more precisely.

Accordingly, the individual traveler's decision process, together with various related probabilities which will play a central role in subsequent developments, is structured as follows.

a. Given his or her geographic location i, a given time period (hour, day, etc.), and an activity (e.g., shopping, work, recreation, etc.), a given individual first decides whether to travel or not. The *unconditional* probability that he or she makes *one* trip in the time period is then P_i, and, of course, that he or she does not is $(1 - P_i)$.

b. Given the choice made at the first level of decision above, and given his or her present location i, the traveler then chooses a location i for the conduct of the given activity. The *conditional* probability of this decision is $P_{j/i}$.

c. Given the outcomes of the first two decisions above, the traveler then decides which transportation mode m to use, among the various alternative modes available between his or her location i and the location j chosen for the conduct of the activity. The conditional probability of this decision is $P_{m/ij}$.

d. Given the outcomes of all the preceding decisions, the traveler finally chooses a route r among those available for the trip as so far decided. The conditional probability of this decision is $P_{r/ijm}$.

This hierarchical process may be represented as in Figure 2.1.

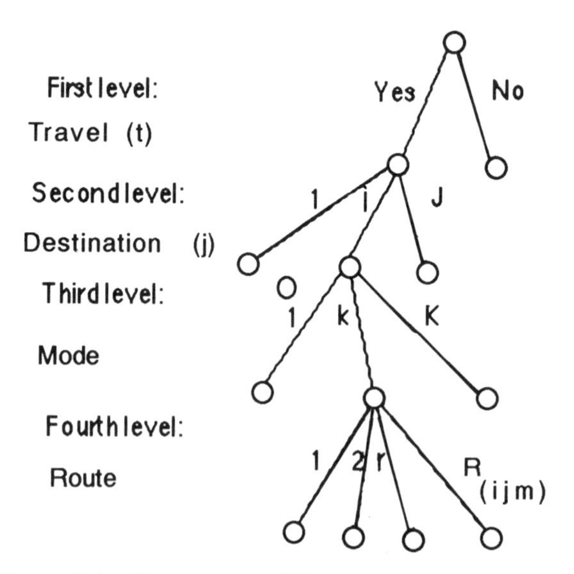

Figure 2.1 The structure of the traveler choice process.

Several points should be noted. First, individual travelers in trip origin i are not distinguishable in terms of the probabilities of their travel choices. This, however, as will be explained in Section 3, does not imply that all individual travelers are similar, only that the probabilities above are average probabilities for travelers in a given zone i.

Also, even though for analytical purposes the respective choices above are structured sequentially, each being conditioned on those preceding it, in reality they are made simultaneously. That is, none of these choices is separable from the other in the mind of the traveler. In fact, the order above is somewhat arbitrary. For instance, for residents who do not own a car, or cannot drive one, the choice of mode may be paramount and might take precedence over the other choices. In such cases, it may be more logical to order the choice of mode before that of destination.

We now specify the relationships between the individual traveler's probabilities of choice at each of the respective levels above, and the respective travel demands T_i, T_{ij}, T_{ijm}, T_{ijmr} defined in the preceding chapter. In each origin zone i, there is a number of *potential* travelers N_i (e.g., residents). As the demand prediction process must start somewhere, this number is assumed given. If not, it would have to be estimated for instance from models of residential location, or employment, and so forth, which, for our purposes, are external to models of travel demand.[1]

Since we have assumed that all individual travelers in a given zone i are similar in terms of their probabilities of choice, the proportion of these individual travelers who will decide to travel during the given period of time of analysis is equal to the probability P_i. Thus, the demand for travel originating in demand zone i, in number of trips, or travelers,[2] is equal to

$$T_i = N_i P_{t/i}; \qquad \forall\ i \tag{2.1}$$

Within these travelers, the proportion of those who choose zone/location j as their destination, is equal to $P_{j/i}$. Consequently, the demand for travel from origin i to destination j, again measured in number of trips, or travelers, is equal to

$$T_{ij} = T_i P_{j/i} = N_i P_{t/i} P_{j/i}; \qquad \forall\ i, j \tag{2.2}$$

Next, the proportion of these T_{ij} travelers who choose mode m as their transport mode is equal to $P_{m/ij}$. Consequently, the demand for travel from zone i

[1]However, models in which these variables are determined endogenously, i.e., internally, together with the travel variables, may be developed. In such cases, as noted earlier, N_i reflects the outcome of a prior decision by individual travelers to choose i as an origin location. (See, for instance, Anas, 1984, or Boyce, 1988.)

[2]As mentioned above, the decision concerns one trip per time period. In particular, we do not consider tours.

to zone j on mode m is

$$T_{ijm} = T_{ij}P_{m/ij} = N_i P_{t/i} P_{j/i} P_{m/ij}; \quad \forall\ i, j, m \tag{2.3}$$

Finally, the proportion of these T_{ijm} travelers who choose a given route r between i and j on mode m as their itinerary is equal to $P_{r/ijm}$. Consequently, the demand for travel from zone i to zone j on mode m and route r is:

$$T_{ijmr} = T_{ijm}P_{r/ijm} = N_i P_{t/i} P_{m/ij} P_{r/ijm}; \quad \forall\ i, j, m, r \tag{2.4}$$

Laying out the general framework for modeling travel demand in the above fashion means that the respective probabilities $P_{t/i}$, $P_{j/i}$, $P_{m/ij}$, and $P_{r/ijm}$, play a central role. This is particularly convenient as there exists a well developed, widely used theory in microeconomics to which we may resort to evaluate these probabilities. This theory has been developed to deal precisely with analytical situations such as the present one, in which the choices facing the traveler at each of the four levels of decision above are *discrete*, or qualitative, and can be referred to with integer indices. This theory is called "discrete choice" theory, or "random utility" theory. Thus, in the next section, we briefly review its central concepts as prerequisite to the development of models of travel demand.[3]

Before pursuing this exposition, it should be pointed out that the definition of the choices facing a traveler may not be obvious. For instance different travelers may have different opportunities for travel. These aspects will not, however, be addressed here. (For a further discussion see, for instance, Fotheringham, 1988; Swait and Ben Akiva, 1987; Horowitz, 1991).

We first begin with the concept of utility.

2.2 THE CONCEPT OF UTILITY

The "utility" to a given individual traveler offered by a given travel choice, or alternative (i.e., of making a trip, or of a given destination, mode, or route, or combinations thereof), may be thought of as measuring the preference the traveler attaches to that particular choice, or combination of choices. For instance, the utility of a given mode of transportation for a given trip might be measured by the total "bundle" of the mode's attributes, such as speed, comfort, safety, and cost, translated into its monetary value, or worth, to the traveler. Similarly, the utility of a given travel destination might depend on that location's number and/or size of parking facilities, average cost of the activity to be performed (e.g., retail prices in the case of shopping travel), accessibility, and so on.

[3]For a more detailed discussion of the theoretical foundations of random utility theory, and of its role in microeconomics, see, for instance, Anderson et al. (1992).

The specific manner in which the various attributes of a given alternative combine to define the overall, or total utility is specified by the "utility function." For instance, we might stipulate that *if* alternative j (e.g., a given destination) is chosen, then the utility *received* by a given individual traveler in demand zone i from choosing j, for one trip, is a linear function[4]

$$\tilde{U}_{ij} = b_i - c_{ij} + \sum_k a_j^k x_j^k + \sum_l b_i^l Y_i^l + \sum_m d_{ij}^m Z_{ij}^m; \qquad \forall\, i, j \tag{2.5}$$

where b_i is the traveler's budget or income, c_{ij} is the alternative's cost or price from i, and the X_j's measure other attributes, in their own units (e.g., the destination's number of stores), the Y_i's measure other travelers' attributes (e.g., income, etc.), and the Z_{ij}'s characterize the combination traveler/alternative, including in particular the distance, or travel cost, from origin zone i to destination j. The utility of a given alternative thus depends not only on the attributes of the alternative, but also on the characteristics of the traveler, because two individual travelers may value the same set of attributes differently. For instance, lower-income travelers might be more sensitive to travel cost, and higher-income travelers more sensitive to travel time.

The presence of the traveler's budget b_i, as well as the alternative's cost, c_j, should be particularly noted, as these quantities will play important roles later. The tilde symbol (~) over U reminds us that the utility is a conditional (or indirect) utility, that is, the utility received *given* that the corresponding choice has been made. It is thus a "posterior" utility. We shall also use unconditional ("prior" or direct) utilities.

It is important to note that this utility specification, which is common to all individual travelers in origin zone i, implies that all individual travelers in a given demand zone i are represented by a "typical," or average traveler. This assumption is very convenient when developing models of aggregate demands, since it allows for the aggregation of individual travelers into geographically defined demand zones i.

The a_j^k's, b_i^l's and d_{ij}^m's are parameters whose value must be identified from observations of the actual choices of a sample of travelers. (See Chapter 7 for a detailed discussion of the process of "model estimation".)

The reason for being concerned with the concept of utility is that it is the key to the prediction of individual traveler choices, and consequently, the determination of travel demands. Specifically, we will assume that travelers always choose the travel alternative (or combination of alternatives when making several, joint choices) which offers them maximum utility. Applying this principle, as obvious as it may seem, means that the travel alternative(s) which will be chosen by a given traveler may be identified, if the individual condi-

[4]While linear specifications are most efficient in connection with estimation of the numerical values of the parameters of the utility function a_j^k, quasi-linear, (i.e., linear in some predefined transformation of the variables, e.g., their logarithms), or even nonlinear forms are, of course, possible. Without loss of generality, we will therefore use linear formulations for simplicity.

tional utilities of all the various alternatives may be measured, or at least ranked, for example, compared numerically. Then, if all utilities have different numerical values, only the alternative with the highest utility should be chosen.

Translated in terms of probabilities of choice of the various alternatives, this means

$$P_{j/i} = 1.0; \qquad \text{for } j \text{ such that } \tilde{U}_{ij} = \operatorname*{Max}_{k} \{\tilde{U}_{ik}\}$$
$$P_{k/i} = 0; \qquad \text{for } k \neq j \tag{2.6}$$

where $P_{j/i}$ is the conditional probability of choosing j given that the traveler is in zone i. If m alternatives have the same, highest, utility, the probabilities of choice may be specified as $P_{i/j} = 1/m$ for these alternatives, and 0 for the other alternatives with smaller utility. This result means that if the modeler has perfectly accurate information about travelers' utilities $\tilde{U}_{ij}$, then he or she can deterministically, and with certainty, predict their choices.

This sounds rather good until one realizes that it also implies that *every individual* traveler in a given demand zone i would be predicted to make the same choice. This, however, is clearly unrealistic. For instance, it is apparent that not every traveler in a given geographic unit, however small it is, uses the same transportation mode to go to work. Also, from a theoretical standpoint, the resulting aggregate demands for the various alternatives are *discontinuous*. That is, a given alternative does not attract any individuals until it offers the highest utility, at which point it attracts them all. Clearly, then, the assumption that the utility specification (2.5) is common to all individual travelers in origin zone i is flawed, and needs to be revised.

There are two alternative approaches to this. The first one, obviously, is not to aggregate travelers into groups, that is, perform travel demand on an individual basis, traveler by traveler. In other words, the demand locations i defined above would, in this case, be equated to the location (e.g., residential address) of each individual traveler. This is clearly unfeasible in practical terms.

The other is to revise the above formulation of utility. This is done in the next section.

2.3 RANDOM UTILITY AND THE MULTINOMIAL LOGIT MODEL

The specification of the received (conditional) utility in Formula (2.5) may then be refined, and generalized, by introducing a random term, which represents the difference, unknown to the modeler, between an individual traveler's utility specification and the average traveler's.[5] We specify the utility which the average traveler in zone i receives when choosing alternative j as

$$\tilde{V}_{ij} = \tilde{U}_{ij} + \epsilon_{ij}; \qquad \forall\, i, j \tag{2.7}$$

[5]This term represents the heterogeneity in individual utilities.

where $\tilde{U}_{ij}$ is the average traveler's utility, as specified in Formula (2.5), and the term ϵ_{ij} represents the uncertainty, on *the part of the modeler*, about the specification, or value of the utility function for an individual traveler. This term may, for instance, represent the unobservable, or unmeasurable factors of utility (e.g., "force of habit"), or errors in the measurement of the factors which have been included.

If the mean of the random term ϵ_{ij} is set equal to zero, for convenience, the utility component $\tilde{U}_{ij}$, which is sometimes called the "systematic" or "fixed" utility, equals the expected value or mean of the random utility $\tilde{V}_{ij}$. It represents the observable part of the individual utility. The value of the random utility term ϵ_{ij} may not be observed, so that the value of the total utility $\tilde{V}_{ij}$ of alternative j is not known, but is also random, as the sum of a constant, $\tilde{U}_{ij}$, and a random term ϵ_{ij}.

It is clear that the modeler may no longer be certain about which alternative a given individual traveler will choose, since he or she has only probabilistic information about the traveler's various utilities. It is also clear that the prediction of choice by the modeler will now take the form of the probabilities that a given individual traveler is observed by the modeler as choosing a given alternative.[6] Furthermore, these probabilities will now depend not only on the values of the $\tilde{U}_{ij}$'s, but also on the random nature of the terms ϵ_{ij}, that is, the characteristics of the joint probability distribution function of $\epsilon_{i1}, \epsilon_{i2}, \ldots,$ and ϵ_{ij}. Finally, these probabilities must be consistent with the same principle of utility maximization as stated above.

2.3.1 Illustrative Example: Empirical Estimation of Choice Probabilities

Because this "random utility" concept is central to our approach to modeling travel demand, we now illustrate it in a specific case. The key to the estimation of the probabilities of choice is to make specific assumptions about the probability distribution function of the random terms ϵ_{ij}. Let us assume, without loss of generality, that there are two alternatives $j = 1, 2$ facing each individual traveler in a given zone i.[7,8] The corresponding "average traveler" random utilities, common to all individual travelers in the zone are specified as

$$\begin{aligned} \tilde{V}_1 &= \tilde{U}_1 + \epsilon_1 = 3 + \epsilon_1 \\ \tilde{V}_2 &= \tilde{U}_2 + \epsilon_2 = 2.75 + \epsilon_2 \end{aligned} \tag{2.8}$$

Which alternative an individual traveler will be assigned to by the modeler depends on the respective realizations of the ϵ_j's. Let us assume that the values

[6]In this interpretation, the choices are certain from the point of view of the individual traveler, i.e., the individual knows exactly what his or her utilities are, but the modeler does not. (Manski, 1977)

[7]The same concepts and derivations may be extended to the case of any number of alternatives.

[8]For the simplicity of the exposition, the index i will be dropped, as it is implicitly understood that the utilities referred to are for an individual traveler in a given zone i.

of ϵ_1 and ϵ_2 are each distributed according to a Bernouilli probability distribution function (p.d.f.), and that they are independent:[9]

$$
\begin{aligned}
P(\epsilon_j = 0.5) &= 0.6 \qquad j = 1, 2 \\
P(\epsilon_j = -0.5) &= 0.4 \qquad j = 1, 2
\end{aligned}
\tag{2.9}
$$

A given individual traveler will then choose the first alternative whenever its utility to him or her is greater than that of the second alternative. The probability P_1 of this event, in terms of the random variables ϵ_1 and ϵ_2, is

$$P(\tilde{V}_1 > \tilde{V}_2) = P(3 + \epsilon_1 > 2.75 + \epsilon_2) \tag{2.10}$$

which is equal to

$$P(\epsilon_2 - \epsilon_1 < 0.25)$$

This probability may be estimated analytically in this particularly simple case. The four possible combinations of values for ϵ_2 and ϵ_1 are graphically represented as points A, B, C, and D (the "sample space") in Figure 2.2.

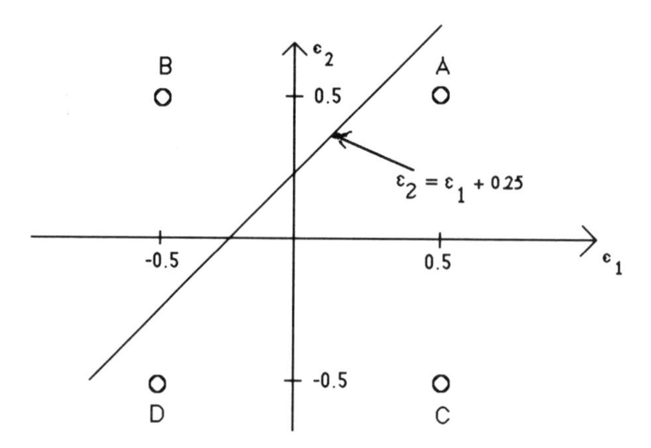

Figure 2.2 Probabilities of outcome in repeated Bernouilli experiment.

The probability that event A takes place is the probability that $\epsilon_2 = 0.5$ *and* that $\epsilon_1 = 0.5$. Since ϵ_1 and ϵ_2 are independent random variables, the probability of event A is equal to the product of the corresponding probabilities (see section 3.2 of Appendix A).

$$P(A) = P(\epsilon_2 = 0.5)P(\epsilon_1 = 0.5) = (0.6)(0.6) = 0.36$$

[9]This is the "Heads/Tails," or "0/1" distribution, which is reviewed in section 3.5.1 of Appendix A.

Similarly:

$$P(B) = P(\epsilon_2 = 0.5)P(\epsilon_1 = -0.5) = (0.6)(0.4) = 0.24$$

$$P(C) = P(\epsilon_2 = -0.5)P(\epsilon_1 = 0.5) = (0.4)(0.6) = 0.24$$

$$P(D) = P(\epsilon_2 = -0.5)P(\epsilon_1 = -0.5) = (0.4)(0.4) = 0.16$$

It is clear from Figure 2.2 that the condition

$$\epsilon_2 - \epsilon_1 < 0.25$$

is only met at points A, D, and C, so that

$$P(\epsilon_2 - \epsilon_1 < 0.25) = P(A) + P(D) + P(C) = 0.76$$

Therefore, the probability P_1 that an individual traveler chosen at random by the modeler is observed by him or her to choose alternative $j = 1$ is equal to 76%, and thus the probability P_2 that choice $j = 2$ is observed is equal to 24%. It is worth noting that even though the utility of alternative 2, averaged over all individual travelers, is smaller than that of alternative 1, the former does not attract all of the demand. In other words, the demands are no longer of the "all-or-nothing" type. This is precisely the feature we were trying to avoid.

However, as $(\tilde{U}_2 - \tilde{U}_1)$ becomes infinitely large, alternative 2 should attract an increasingly large fraction of the total demand. Also, since the ϵ_j's are assumed to be identically and independently distributed (i.i.d), when $\tilde{U}_2 = \tilde{U}_1$ the respective probabilities of the two alternatives being chosen should be equal, and thus equal to 0.50 each.

The variations of the probability of choice of, say, alternative 2, as a function of the difference in systematic utilities $(\tilde{U}_2 - \tilde{U}_1)$ are represented graphically in Figure 2.3.

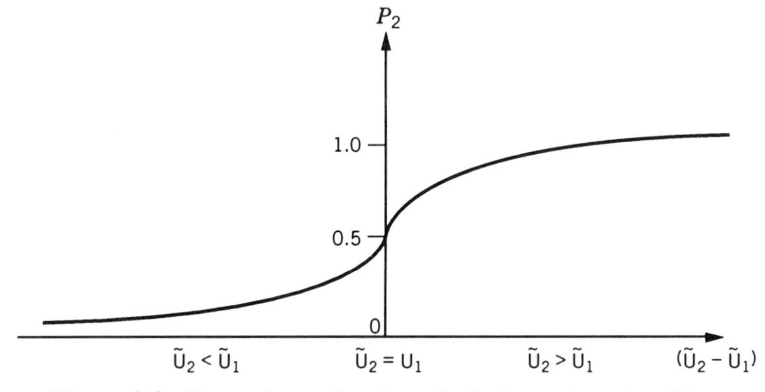

Figure 2.3 Demand as a function of relative systematic utility.

It is clear that the probabilities estimated above depend, in addition to the values of the systematic utilities, on the probabilistic nature of the random utility terms ϵ_j. For instance, assuming that the ϵ_j's are uniformly distributed over the interval $[-0.5; 0.5]$ would lead to different values for the P_j's, given the same systematic utilities $\tilde{U}_j$. The curve in Figure 2.3, while keeping the same general S shape, would have a different curvature. In the limit case when the random utility terms are no longer random variables, when their variances are equal to zero, as expected, alternative 2 would be selected 100% of the time whenever $\tilde{U}_2 > \tilde{U}_1$. The curve in Figure 2.2 would then have the form of the "step" function in Figure 2.4.

2.3.2 Derivation of the Model

This leads to the issue of specifying the probability distribution function for the ϵ_j's. Since these random terms may be considered the result of the interaction of a large number of factors related to an individual traveler's socio-geographic characteristics, the Normal probability distribution function might be used, as suggested by the "Central Limit Theorem." (See, for instance, Wonnacott and Wonnacott, 1970.) This is the same justification for the application of the Normal probability distribution function to measurement errors. The probabilities of choice in this case will be of the "probit" type. (See, for instance, Finney, 1964; Daganzo, 1979.) However, these probabilities may not be expressed analytically, except in the "binary" case of two alternatives. (See Exercise 2.2.)

In the general case of multiple alternatives, the probabilities of choice may be numerically *approximated* using a sequential procedure (see Appendix I), which quickly becomes cumbersome when there are more than a few alternatives. (See Exercise 2.3.) Alternatively, they may be estimated *numerically* using "Monte Carlo" statistical simulation procedures. This is very cumbersome, and often prohibitive, in practical work. Consequently, it is standard practice to use another distribution as a substitute for the Normal density function. This is the "Gumbel," "double exponential," "Extreme Value, type I"

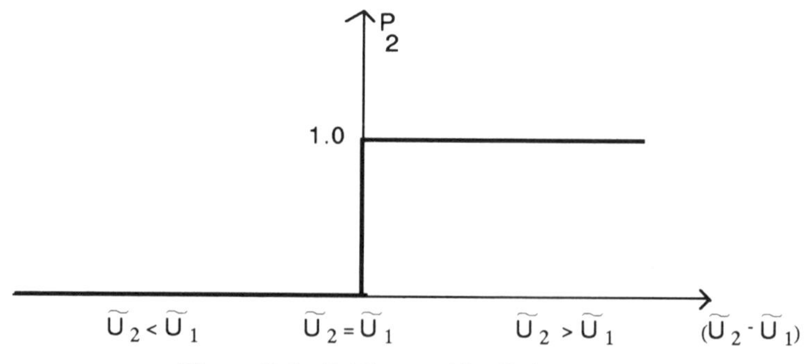

Figure 2.4 "All-or-nothing" demand.

probability distribution function (p.d.f.), which is described in subsection 3.5.5 of Appendix A (Gumbel, 1958). The advantage of using this particular probability distribution function is that the resulting probabilities of choice may be expressed in closed form.

THEOREM 2.1

When the random utility terms ϵ_j in the utility function (2.7) are i.i.d with a Gumbel p.d.f., the probabilities an individual chosen at random will select alternative j are given by

$$P_j = \frac{e^{\beta \tilde{U}_j}}{\sum_k e^{\beta \tilde{U}_k}}; \qquad \forall j \tag{2.11}$$

This result is demonstrated in Appendix 2 at the end of this chapter.[10] This particular expression for the probability P_j that an individual traveler, chosen at random, will be observed by the analyst to choose alternative j, is usually referred to as the "multivariate logit" model since, in general, $\tilde{U}_j$ is a multivariate function of several variables, as seen in section 2.2. For instance, in terms of the general specification of the utility $\tilde{U}_j$ as specified by Formula (2.5), the logit demands are equal to

$$P_j = \frac{e^{\beta \tilde{U}_j}}{\sum_j e^{\beta \tilde{U}_j}} = \frac{\exp\left[\beta\left(b - c_j + \sum_k a_j^k X_j^k\right)\right]}{\sum_j \exp\left[\beta\left(b - c_j + \sum_k a_j^k X_j^k\right)\right]}$$

$$= \frac{e^{\beta(-c_j + A_j)}}{\sum_j e^{\beta(-c_j + A_j)}}; \qquad \forall j \tag{2.12a}$$

after simplification by the term $e^{\beta b}$, and where, for short

$$A_j = \sum_k a_j^k X_j^k; \qquad \forall j \tag{2.13}$$

For N travelers, the corresponding average aggregate demand is equal to

$$T_j = N \frac{e^{\beta(A_j - c_j)}}{\sum_j e^{\beta(A_j - c_j)}}; \qquad \forall j \tag{2.12b}$$

[10]This theorem is apparently due to Holman and Marley, and first appeared in an unpublished document (Anderson et al. pg. 39).

As an illustration of the application of the logit model, let us assume in the example of discrete choice above that the error terms are Gumbel distributed, with a value for parameter β equal to 0.64. The probability of an individual choosing alternative 1 is then

$$P_1 = \frac{e^{0.64(3)}}{e^{0.64(3)} + e^{0.64(2.75)}} = \frac{6.82}{6.82 + 5.81} = 0.54$$

The probability of the other alternative being chosen is of course $1 - 0.54 = 0.46$, since there are only two alternatives. In the case of more than two alternatives, the computations are similar.

Thus, once the random utility terms are assumed to be i.i.d. Gumbel, the estimation of the probabilities of choosing the respective alternatives is quite straightforward, in marked difference with using models based on other assumptions. As we shall discuss later in more detail, however, such an assumption may not always be entirely justified, in particular regarding the independence of the random utility terms. In this case, as in fact in all modeling situations, the price of analytical convenience must be weighted against that of approximating reality to various degrees.

2.4 RANDOM AND EXPECTED CHOICES

The result obtained in the preceding section may be translated in terms of the observed demand d_{nj} of an individual traveler n for travel alternative j. This individual demand function is random, and can only take the two values 0 or 1; either traveler n chooses the alternative, or does not. Thus,

$$d_{nj} = \delta_j; \qquad \forall\, n, j \tag{2.14a}$$

where the binary random variable δ_j has the values

$$\delta_j = \begin{cases} 1 & \text{if choice } j \text{ is made} \\ 0 & \text{if not} \end{cases}; \qquad \forall\, j \tag{2.14b}$$

and the Bernouilli p.d.f.[11]

$$\begin{matrix} P(\delta_j = 1) = P_j \\ P(\delta_j = 0) = 1 - P_j \end{matrix}; \qquad \forall\, j \tag{2.14c}$$

For instance, in the case of the above example, the random demand for alternative 1 from each and any individual k may be obtained as the outcome

[11] See section 3.5i. of Appendix A.

of the toss of a (biased) coin with probabilities of "Heads" equal to 0.54 and probability of "Tails" equal to 0.46. A "Heads" outcome means alternative 1 is chosen, and so on. If there were six alternatives, the process of their choice by an individual traveler could be simulated by the toss of a die with probabilities of each outcome equal to the respective P_j's and so on.

On the average, in the long run, individual travelers with the same *random* utility specification would be observed to choose alternative j, P_j percent of the time, in the present case with frequency 54%. This implies that in a sample of N such individual travelers chosen at random, the expected value for the random, total number N_j of travelers choosing alternative j is N times the mean of the Bernouilli distribution above:

$$E(N_j) = NP_j; \qquad \forall\, j \tag{2.15}$$

In the above case, for instance, in a sample of 100 individual travelers, 54 would be expected to choose alternative 1 *on the average*, and 46 alternative 2.

Similarly, the variance of the random variable N_j is equal to N times the variance of the Bernouilli distribution above:

$$\text{Var}\,(N_j) = NP_j(1 - P_j); \qquad \forall\, j \tag{2.16}$$

This value provides an indication of the fluctuations in the observed demand from sample to sample. For instance, continuing with the same numerical example above, the variance of N_1 would be equal to

$$\text{Var}\,(N_1) = 100P_1(1 - P_1) = 100 \cdot 0.54 \cdot 0.46 = 25$$

so that the standard deviation of N_1 is equal to 5.[12]

Similarly, the covariance of the random variables N_j and N_k is equal to

$$\text{Cov}\,(N_j, N_k) = -NP_jP_k; \qquad \forall\, j, k \tag{2.17}$$

The negative sign is due to the fact that, since the sample size is fixed and equal to N as specified above, when the demand for a given travel alternative goes up, the demand for other alternatives must necessarily go down. The magnitude of the covariance provides an indication of the degree of (linear) correlation between the numbers of travelers choosing two given alternatives,

[12]Because of the fact that a multinomial random variable with a "large" N (here 100) may be approximated by a Normal random variable (see section 3.5.4 of Appendix A), the p.d.f. of N_1 may be approximated by a normal distribution with mean μ equal to the expected value of N_1, i.e.,

$$\mu = 100P_1 = 100(0.54) = 54$$

and a standard deviation $\sigma = 5$.

as they vary jointly, from sample to sample. For instance, in the same situation as above, the covariance between N_1 and N_2 is equal to

$$\text{Cov}\,(N_1, N_2) = -100P_1P_2 = -100 \cdot 0.54 \cdot 0.46 = -25$$

Consequently, the coefficient of linear correlation between N_1 and N_2 is equal to

$$\rho = \text{Cov}\,(N_1, N_2)/\sigma_{N1}\sigma_{N2} = -25/5^2 = -1$$

As expected, there is a perfect negative correlation between N_1 and N_2 since the sum of their random values must always add up to 100, so that knowing the value of one implies perfect knowledge of the value of the other. With more than two alternatives, this, of course, would no longer be true.

We may also want to estimate the probabilities that the random numbers N_j be equal to given values X_j. Assuming that the modeler's uncertainty about one individual traveler n is independent of that about any other (i.e., the vectors ϵ_k are statistically independent), the numbers of travelers $N_1, N_2, \ldots, N_j, \ldots$, and N_J observed choosing the given J travel alternatives are distributed with a multidimensional p.d.f. (See section 3.5.3 of Appendix A.) Consequently, the joint probability that these numbers are equal to given values $X_1, X_2, \ldots, X_j, \ldots$, and X_J, is

$$\text{Prob.}\,(N_1 = X_1, \ldots, N_j = X_j, \ldots, N_J = X_J)$$

$$= N!\left(\frac{P_1^{X_1}}{X_1!}\right), \ldots, \left(\frac{P_j^{X_j}}{X_j!}\right), \ldots, \left(\frac{P_J^{X_J}}{X_J!}\right) = N! \prod_{j=1}^{J} \frac{P_j^{X_J}}{X_j!} \tag{2.18a}$$

with, of course:

$$N = \sum_j X_j \tag{2.18b}$$

For instance, in the above situation, the joint probability that out of 100 individual travelers selected at random 50 choose the first alternative *and* 50 choose the second is equal to:[13]

$$\text{Prob.}\,(N_1 = 50, N_2 = 50) = 100!\left(\frac{0.5^{50}}{50!}\right)\left(\frac{0.5^{50}}{50!}\right)$$

From this formula, the *marginal* probabilities that the numbers of travelers N_j observed choosing a given alternative j are equal to X_j, without specifying

[13]Value of large factorials such as 100! may be approximated with the use of "Stirling's formula," i.e., $\ln x! \approx x(\ln x - 1)$.

the numbers choosing the other alternatives, may be computed as

$$P(N_j = X_j) = \frac{N!}{X_j!(N - X_j)!} X_j^{P_j}(N - X_j)^{1 - P_j}; \qquad \forall j \qquad (2.19)$$

2.5 EXPECTED RECEIVED UTILITY

2.5.1 Illustrative Example: Empirical Estimation of Expected Received Utility

In this section, we discuss the concept of expected received utility. As we have seen above, the utility *received* by individual travelers depends on which choice they make, and therefore depends on the value of the random utility terms ϵ_j. Form the modeler's viewpoint, this quantity is random. Let us illustrate this concept with a practical example.

Going back to the Example in section 2.3.1 and to the assumption that the ϵ_j's have a Bernouilli distribution, we can simulate the random utilities for two alternatives specific to ten individual travelers by tossing a (biased) coin, as described in section 2.3. Let us imagine that the results are as represented in Table 2.1.

The received utility is, by definition, the largest of the two utilities $\tilde{V}_1$ and $\tilde{V}_2$, or the random variable $M = \text{Max}(\tilde{V}_1, \tilde{V}_2)$. For this sample of ten individual travelers, the total utility received by the ten travelers is equal to 34.25, corresponding to an average received utility of 3.425. The estimation of the exact, or theoretical expected value of the received utility requires the determination of the p.d.f. of M. The only possible values for M are, respectively, 2.5, 3.25, and 3.5. Indeed, in no case will 2.25 be the maximum, since $\tilde{V}_1$ is always at least equal to 2.5. The value $M = 2.5$ corresponds to the only event $\tilde{V}_1 = 2.5$ and $\tilde{V}_2 = 2.25$, whose probability is equal to $P(\epsilon_1 = -0.5)P(\epsilon_2 = -0.5) = (0.4)(0.4) = 0.16$ since ϵ_1 and ϵ_2 are assumed independent.

TABLE 2.1 Random Utility Assessments

Individual	Value of $\tilde{V}_1$	Value of $\tilde{V}_2$	Alternative Chosen	Received Utility = Max $(\tilde{V}_1, \tilde{V}_2)$
1	3.50	2.25	1	3.50
2	2.50	3.25	2	3.25
3	3.50	2.25	1	3.50
4	3.50	3.25	1	3.50
5	2.50	3.25	2	3.25
6	2.50	3.25	2	3.25
7	3.50	2.25	1	3.50
8	3.50	2.25	1	3.50
9	3.50	2.25	1	3.50
10	3.50	3.25	1	3.50

Similarly, the value $M = 3.25$ corresponds to the only event $\tilde{V}_1 = 2.5$ and $\tilde{V}_2 = 3.25$, whose probability is equal to $P(\epsilon_1 = -0.5)P(\epsilon_2 = 0.5) = (0.4)(0.6) = 0.24$. Finally, the value $M = 3.5$ can result from either the event $\tilde{V}_1 = 3.5$ and $\tilde{V}_2 = 2.25$, or the event $\tilde{V}_1 = 3.5$ and $\tilde{V}_2 = 3.25$, so that its probability is the sum of these two mutually exclusive events, $(0.6)(0.4) + (0.6)(0.6) = 0.6$. This might be expected, since this is the last possible value, and the probabilities for M already add up to 0.40.

Consequently, the expected value of M is

$$\tilde{W} = \mu_M = 0.16(2.5) + 0.24(3.25) + 0.6(3.5) = 3.28 \qquad (2.20)$$

2.5.2 Derivation and Properties of Expected Received Utility Function

In general, the average utility received by an individual traveler is equal to

$$\tilde{W} = E\{\operatorname*{Max}_j \tilde{V}_j\} = E\{\operatorname*{Max}_j (\tilde{U}_j + \epsilon_j)\} \qquad (2.21)$$

where the expected value $E\{\ \}$ is taken with respect to the random variables ϵ_j.

THEOREM 2.2

When the random utility terms ϵ_j in the utility function (2.7) are i.i.d. with a Gumbel p.d.f., the expected utility an individual chosen at random receives from utility maximizing choices is given by

$$\tilde{W} = \frac{1}{\beta} \ln \sum_j e^{\beta \tilde{U}_j} \qquad (2.22)$$

This result is demonstrated in Appendix III. It is worth noting that, as expected, $\tilde{W}$ has the same dimension that the utilities $\tilde{U}_j$. In the numerical example above, the average received or indirect utility is equal to

$$\tilde{W} = \frac{1}{0.64} \ln [e^{0.64(3)} + e^{0.64(2.75)}] = 3.96$$

in the same units (e.g., monetary) as the original utilities of 3 and 2.75 respectively. The fact that it is higher than any of these respective values reflects the maximization of random utility. It should be noted that the average received utility $\tilde{W}$ is *not* equal to

$$\sum_j P_j \tilde{U}_j = 0.54(3) + 0.46(2.75) = 2.88$$

nor to

$$\frac{1}{J}\sum_j \tilde{U}_j = (3 + 2.75)/2 = 2.87$$

nor to

$$\operatorname*{Max}_j \{\tilde{U}_j\} = \text{Max }(3, 2.75) = 3.00$$

By definition, $\tilde{W}$ is the expected *indirect* utility a given individual traveler selected at random receives on the average from his/her repeated choices. This interpretation (Anderson et al., 1992. pg. 73) will be useful in the formulation of "nested" Logit models, which we take up next.

There is an important, simple relationship between the probabilities of choice and $\tilde{W}$ (see Exercise 2.6).

$$\frac{\partial \tilde{W}}{\partial \tilde{U}_j} = P_j; \qquad \forall\, j \tag{2.23}$$

Thus, the impacts on average received utility of a marginal change in the utility of given alternatives are proportional to their probability of choice. Another important property of the average received utility is that it is increasing in the number of alternatives:

$$\tilde{W}(\tilde{U};\, \tilde{U}_{n+1}) \geq \tilde{W}(\tilde{U}); \qquad \forall\, \tilde{U},\, \tilde{U}_{n+1} \tag{2.24}$$

This implies that $\tilde{W}$ cannot decrease when the choice set is enlarged. Other properties are also relevant (see Exercise 2.13.)

2.6 THE HIERARCHICAL (NESTED) LOGIT MODEL

The probabilities of choice above were derived in the case when the choice of an alternative was made as a *singe*-level choice. However, in general, the choice of alternative may be preceded by, followed by, or more generally made simultaneously with other choices. In fact, in the most general case we shall consider, a traveler might have to decide on whether to travel at all, which destination to go to, which mode to use (e.g., car, transit, telephone, etc.), and finally, which route to follow, as represented in Figure 2.1.

In general, an alternative's utility at a given level is dependent on the utilities at subsequent levels. For instance, the decision to travel may in the first place be influenced by the utility of the various possible travel destinations. In turn, the utility of a given destination may be influenced by the utility of the various modes which serve it, and so on down to the level of route choice. In order to be able to apply the principle of utility maximization to *joint* choices of alternatives, the random utilities must be specified in a systematically structured manner.

This will be illustrated in the case of two generic levels of choice, i and j, in that order.[14] The *total* utility of a first level i and second level j joint choice is specified as

$$\tilde{V}_{ij} = \tilde{U}_i + \tilde{U}_j + \tilde{U}_{ij} + \epsilon_i + \epsilon_{ij}; \qquad \forall\, i, j \tag{2.25a}$$

The *conditional* utility of a second-level choice j, *given* that first-level choice i has already been made, is equal to

$$\tilde{V}_{j/i} = \tilde{U}_j + \tilde{U}_{ij} + \epsilon_{ij}; \qquad \forall\, i, j \tag{2.25b}$$

The components $\tilde{U}_i$ and $\tilde{U}_j$ represent the systematic utilities specific to the first-level choice i and second-level choice j, respectively. The component $\tilde{U}_{ij}$ represents the expected utility specific to the combination of first-level choice i and second-level choice j. The random utility terms ϵ_i and ϵ_{ij} have a similar interpretation. Note, however, that there is no random components ϵ_j at the last level; it is assumed that $\sigma_{\epsilon_j} = 0$. This utility structure may be illustrated as in Figure 2.5.

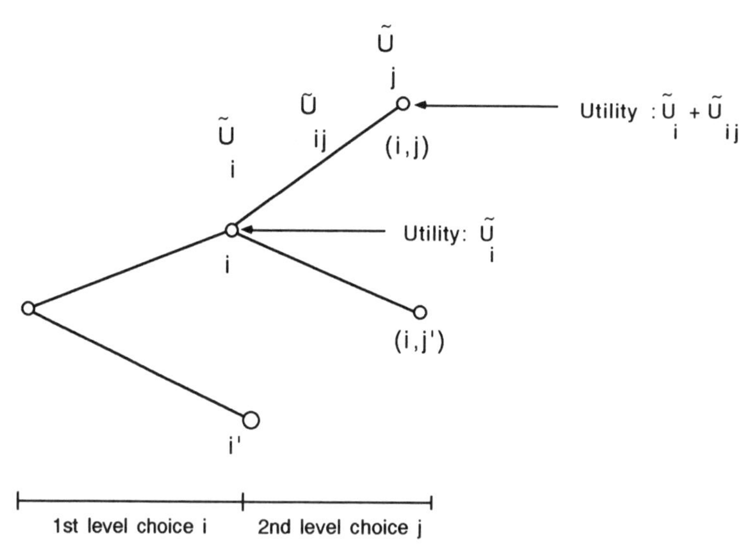

Figure 2.5 Structure of systematic utilities for a two-level hierarchical choice process.

[14]The same formulation applies in the same manner to any number of hierarchical levels.

Thus, $\tilde{U}_i$ is the systematic utility which is common to all choices which include first level choice i. Similarly, $\tilde{U}_{ij}$ is the systematic utility which is common to all choices which include first level choice i and second level choice j.

It is assumed that the random terms ϵ_i are Gumbel-distributed, each with a mean zero, and common variance σ_1^2, and that the random terms ϵ_{ij} are also Gumbel-distributed, each with a mean zero, and common variance σ_2^2. Also, in the same manner as above, it is assumed that the ϵ_i's are independent of one another, and that the ϵ_{ij}'s are independent of one another.

With this utility structure it is clear, in similarity to the single-level case above, that the *conditional* probability of selecting the second-level alternative j, *given* that the first-level alternative i has already been chosen, is equal to

$$P_{j/i} = \frac{e^{\beta_2(\tilde{U}_{ij} + \tilde{U}_j)}}{\sum_j e^{\beta_2(\tilde{U}_{ij} + \tilde{U}_j)}} = \frac{e^{\beta_2 \tilde{U}_{j/i}}}{\sum_j e^{\beta_2 \tilde{U}_{j/i}}}; \qquad \forall\ i, j \tag{2.26}$$

Furthermore, if the condition

$$\sigma_1 \geq \sigma_2 \tag{2.27a}$$

which implies

$$\beta_1 \leq \beta_2 \tag{2.27b}$$

holds, then the following result obtains (McFadden, 1981):

THEOREM 2.3

The marginal, or *unconditional* probability that first-level alternative i offers the maximum random utility, and is therefore chosen, is equal to

$$P_i = \frac{e^{\beta_1(\tilde{U}_i + \tilde{W}_i)}}{\sum_i e^{\beta_1(\tilde{U}_i + \tilde{W}_i)}}; \qquad \forall\ i \tag{2.28a}$$

where the term $\tilde{W}_i$ represents the expected *indirect* utility of the choices (i, j) which may be made at the second level j, after choosing alternative i at the first level, i.e.,

$$\tilde{W}_i = E_{\epsilon_{ij}}\{\operatorname*{Max}_j\ (\tilde{U}_{ij} + \tilde{U}_j + \epsilon_{ij})\} \tag{2.28b}$$

According to Formula (2.22), $\tilde{W}_i$ is equal to

$$\tilde{W}_i = \frac{1}{\beta_2} \ln \sum_j e^{\beta_2(\tilde{U}_{ij} + \tilde{U}_j)} = \frac{1}{\beta_2} \ln \sum_j e^{\beta_2 \tilde{V}_{j/i}} \tag{2.29}$$

The value of parameters β_2 and β_1 in Formulas (2.26) and (2.28), respectively, are related to the variances of the random components ϵ_i and ϵ_{ij} respectively. Specifically,

$$\beta_1 = \pi/\sqrt{6}\sigma\epsilon_i \tag{2.30}$$

$$\beta_2 = \pi/\sqrt{6}\sigma\epsilon_{ij} \tag{2.31}$$

The random utility terms for two alternatives (i, j) and (i, j') sharing the same first-level choice may be (positively) correlated through their common inclusion of the term ϵ_i. It may be shown (Daganzo and Kusnic, 1993) that

$$\text{Cor}\,(\epsilon_{ij}, \epsilon_{ij'}) = \text{Var}\,(\epsilon_i) = 1 - (\beta_2/\beta_1)^2; \qquad \forall\, i, j, j' \tag{2.32}$$

If Condition (2.27b) is not borne out by the results of the model's calibration (see Chapter 7), Formula (2.32) could not hold, since a correlation is less than 1. This would be an indication that the order of levels of choice should be inverted.

In any case, the *joint*, unconditional probability of a combined choice of alternatives (i, j) is, from the definition of conditional probability

$$P_{ij} = P_i P_{j/i}; \qquad \forall\, i, j \tag{2.33}$$

Replacing the respective probabilities by their expressions above we have

$$P_{ij} = P_{j/i} P_i = \frac{e^{\beta_1(\tilde{U}_i + \tilde{W}_i)}}{\sum_i e^{\beta_1(\tilde{U}_i + \tilde{W}_i)}} \frac{e^{\beta_2(\tilde{U}_{ij} + \tilde{U}_j)}}{\sum_j e^{\beta_2(\tilde{U}_{ij} + \tilde{U}_j)}}; \qquad \forall\, i, j \tag{2.34}$$

When the random utility of a second-level alternative j is independent of random utility of a first level alternative i, that is, when ϵ_i and ϵ_{ij} are uncorrelated, then $\beta_1 = \beta_2 = \beta$. In this case, the expression for P_{ij} in Formula (2.34) (see Exercise 2.20) becomes

$$P_{ij} = \frac{e^{\beta(\tilde{U}_i + \tilde{U}_j + \tilde{U}_{ij})}}{\sum_i \sum_j e^{\beta(\tilde{U}_i + \tilde{U}_j + \tilde{U}_{ij})}}; \qquad \forall\, i, j \tag{2.35}$$

This is the expression of a single-level, multinomial logit model, as described in section 2.3. Thus, in this case, the two levels of choice have collapsed into a single level. Correspondingly, the systematic part of the utility is the sum of the systematic parts of the first and of the second level's utilities.

2.7 PROPERTIES OF THE LOGIT MODEL

The logit model possesses several noteworthy analytical properties. These properties, which reflect the assumptions underlying the model, offer significant practical advantages, or conceptual disadvantages, depending on one's viewpoint. In this section, we examine them in some detail, in the simplest case of the multinomial model, that is, a single-level model, Formula (2.11). Other, more complex cases, including several levels of traveler choice, would be treated in a similar manner. Thus, the basic logit demand model we examine is given by

$$Y_j = \frac{e^{\beta \tilde{U}_j}}{\sum_{k=1}^{J} e^{\beta \tilde{U}_k}}; \qquad \forall j \tag{2.36}$$

where Y_j is the expected demand for alternative j for a single traveler, which is equal to his or her probability P_j of choosing the alternative.

First, it was mentioned above that the value of parameter β is inversely related to the standard deviation of the random errors on the utility's specification, as represented by Formula (A.59b). The role of parameter β may be illustrated by giving it extreme values. When $\beta = 0$, the variance of the random terms is infinite, implying complete lack of information by the modeler about the alternative's utilities. Consequently, one might expect that the respective probabilities of observing the individual traveler choose any given alternative should be equal, or independent of any traveler or alternative characteristics. This is, indeed, the case, as may easily be checked. (See Exercise 2.18).

$$Y_j = 1/J; \qquad \forall j \tag{2.37}$$

Conversely, an infinite value for β implies that the variance of the random terms is zero, that the modeler's information about alternatives' utilities is deterministic, or perfect. In this case, one might expect that the probability of observing the individual traveler select the alternative with the highest utility would be equal to 1.0, and the probabilities of observing him or her select any other alternative equal to zero.[15] The expected demands would then be nonzero for only one alternative. (See Exercise 2.19.) In the case where, say m alter-

[15]In terms of the solution to the utility maximization problem, this may be characterized as a "corner solution."

natives have the same, maximum utility, the respective probabilities of observing the individual traveler select them should be equal to $1/m$. Thus, the role of parameter β is to set the "dispersion" of the distribution of expected demands at some intermediate level between these two extremes.

In practical terms, according to Formula (2.36), the respective expected demands for two alternatives m and r are in the ratio

$$Y_m/Y_1 = e^{\beta \tilde{U}_m}/e^{\beta \tilde{U}_l} = e^{\beta(\tilde{U}_m - \tilde{U}_l)} \tag{2.38}$$

For a given range between the lowest and highest utility, the value of β thus sets the resulting ratio between the lowest and highest expected demands. This expression is only a function of the respective utilities of alternatives m and l, and is not affected by the introduction/removal of other alternatives. This analytical feature, the so-called Independence of Irrelevant Alternatives (IIA), implies that two given alternatives' relative shares of the total demand are independent of the composition of the set of alternatives. This may be unrealistic in some cases. It should be noted, however, that with a hierarchical formulation, the IIA property does not hold for two alternatives which do not belong to the same second-level choice set. (See Exercise 2.4.)

In any case, adding (or removing) a given alternative i from the alternatives' set results in the following changes in the demand for alternative j from Y_j^b, the demand "before" to Y_j^a, the demand "after":

$$Y_j^a = \frac{e^{\beta \tilde{U}_j}}{\left(\sum_{k=1}^{J} e^{\beta \tilde{U}_k}\right) - e^{\beta \tilde{U}_i}} = \frac{e^{\beta \tilde{U}_j}}{\sum_{k=1}^{J} e^{\beta \tilde{U}_k}} \frac{\sum_{k=1}^{J} e^{\beta \tilde{U}_k}}{\left(\sum_{k=1}^{J} e^{\beta \tilde{U}_k}\right) - e^{\beta \tilde{U}_i}}$$

or finally

$$Y_j^a = \frac{Y_j^b}{1 - P_i}; \qquad \forall\, i, j \tag{2.39}$$

Thus, Y_j^a is proportional to Y_j^b, with a proportionality coefficient being a simple function of the probability of the removed alternative.

Similarity, if the respective alternative utilities change from $\tilde{U}_j$ to $(\tilde{U}_j + \Delta \tilde{U}_j)$, the corresponding expected demands change from a prior value Y_j^b given by Formula (2.36) to a posterior value

$$Y_j^a = \frac{Y_j^a e^{\beta \Delta \tilde{U}_j}}{\sum_{k=1}^{J} Y_k^a e^{\beta \Delta \tilde{U}_k}}; \qquad \forall\, j \tag{2.40}$$

(See Exercise 2.16.) These relationships are sometimes used to perform "pivot" analyses.

We now examine the elasticities of logit demands. In general, the elasticity of a quantity Y with respect to a quantity X is defined as

$$e_{Y/X} = \frac{\partial Y}{Y} \bigg/ \frac{\partial X}{X} \tag{2.41}$$

The elasticity measures the percentage change in the value of variable Y resulting from a one percent change in the value of variable X. It is a dimensionless measure, as opposed to the rate of change of Y with respect to X, which is

$$r_{Y/X} = \frac{\partial Y}{\partial X} \tag{2.42}$$

It is worth noting that the elasticity may also be estimated as

$$e_{Y/X} = \frac{\partial \ln Y}{\partial \ln X} = \frac{\partial Y/Y}{\partial X/X} \tag{2.43}$$

If the expected demands for various alternatives j are given by the logit model, Formula (2.36), simple calculus shows that

$$\frac{\partial Y_j}{\partial \tilde{U}_j} = \beta Y_j(1 - P_j) \tag{2.44}$$

and that

$$\frac{\partial Y_j}{\partial \tilde{U}_k} = -\beta Y_j P_k; \qquad \forall\, j, k \tag{2.45}$$

Consequently, the "direct" elasticity of demand for alternative j, the elasticity with respect to its own utility, is equal to

$$e_{Y_j/\tilde{U}_j} = \beta \tilde{U}_j(1 - P_j); \qquad \forall\, j \tag{2.46}$$

Thus, the change in demand resulting from a change in "own" utility is proportional to the utility, and proportional to the demand for all the other alternatives. This property is related to Formula (2.23). The "cross" elasticity of demand for alternative j, that is, the elasticity with respect to alternative k's utility, is equal to:

$$e_{Y_j/\tilde{U}_k} = -\beta \tilde{U}_k P_k; \qquad \forall\, j, k \tag{2.47}$$

The cross elasticity is thus independent of j, that is, is the same for all alternatives. This feature translates the IIA property mentioned above.

It may be useful in some cases to estimate the elasticities, not with respect to the overall utility, but rather, with respect to a specific utility component. If the utilities are themselves linear functions of several factor X_k

$$\tilde{U}_j = a_{j0} + \sum_k a_j^k X_j^k; \qquad \forall j \tag{2.48}$$

the rate of change in the demand for alternative j, as a function of the change in the value of the kth factor in its own utility, is then, using the "chain rule" of partial derivatives (see section 1.3 of Appendix A)

$$\frac{\partial Y_j}{\partial X_j^k} = \frac{\partial Y_j}{\partial \tilde{U}_j}\frac{\partial \tilde{U}_j}{\partial X_j^k} = \beta a_j^k \tilde{U}_j (1 - P_j); \qquad \forall j, k \tag{2.49}$$

Similarly, the rate of change in the demand for alternative j, as a function of the change in the value of the kth attribute in the utility of the lth alternative is

$$\frac{\partial Y_j}{\partial X_l^k} = \frac{\partial Y_j}{\partial \tilde{U}_l}\frac{\partial \tilde{U}_l}{\partial X_l^k} = -\beta a_l^k \tilde{U}_k P_k; \qquad \forall j, k, l \tag{2.50}$$

These various formulas provide the basis for an approximation for the variation in demand for a given alternative corresponding to a change in the value of one of its factors. For instance, using Formula (2.49), the change in the value of Y_j which would result from a change in the value of the kth attribute of the jth alternative may be estimated as

$$dY_j = \beta a_j^k \tilde{U}_j (1 - P_j)\, dX_j^k; \qquad \forall j, k \tag{2.51}$$

while using Formula (2.50), the change in demand resulting from a change in the value of the kth attribute for the lth alternative is approximately equal to

$$dY_j = -\beta a_l^k \tilde{U}_k P_k\, dX_l^k; \qquad \forall j, k, l \tag{2.52}$$

2.8 DERIVATION OF THE LOGIT MODEL AT THE AGGREGATE LEVEL: THE "REPRESENTATIVE TRAVELER"

The logit model, Formula (2.11), was derived at the disaggregate, individual level, from principles of random utility maximization. It may alternatively, and *equivalently*, be derived at the aggregate level, as follows. Consider a group of N travelers, each with the same indirect (conditional, or received) utility function, specified, for instance, as

$$\tilde{U}_j = b - c_j + A_j + \epsilon_j; \qquad \forall j \tag{2.53}$$

where c_j is the alternative's cost.

Given the results of section 3, it is clear that when considering N travelers the expected aggregate demands for each of the alternative choices j are equal to N times the expected individual demands:

$$T_j = N \frac{e^{\beta(A_j - c_j)}}{\sum_j e^{\beta(A_j - c_j)}}; \qquad \forall j \tag{2.54}$$

The same aggregate demands in the present case may be obtained alternatively as the result of the utility maximizing choices of a representative traveler (R.T.), subject to an aggregate budgetary constraint.

THEOREM 2.4

The direct utility function of the representative traveler corresponding to aggregate demands (2.12b) is

$$U = \sum_j T_j A_j - \frac{1}{\beta} \sum_j T_j \ln T_j + T_0 \tag{2.55}$$

in which T_0 is the amount spent on nontravel items.

The proof is provided in Appendix 2.4. Consequently, the aggregate demands (2.54) may be retrieved as the maximization of this utility with respect to the T_j's and T_0, subject to the constraints

$$\sum_j T_j c_j + T_0 = Nb \tag{2.56a}$$

$$\sum_j T_j = N \tag{2.56b}$$

$$T_j \geq 0; \qquad \forall j \tag{2.56c}$$

T_0 may be thought of as a "slack" variable for Budgetary Constraint (2.56) written as

$$\sum_j T_j c_j \leq Nb$$

This fact is demonstrated in Appendix 2.5. It is easy to show (see Exercise 2.17) that the R.T.'s utility maximization (U.M.) problem in this case is a convex problem (see section 1.7 of Appendix A), and therefore has a unique solution.

Thus, the maximization of function U subject to budgetary and consistency constraints retrieves the aggregate demands T_j. Consequently, function U represents the *unconditional* (or "direct") utility which is a hypothetical "representative'' traveler (''R.T.'' for short) faces, and which he or she maximizes through an optimal choice of aggregate demands T_j's. To reinforce this distinction, direct utilities do not have tilde symbols over them.

It is important to note that utility function U in Formula (2.55), in contrast to the conditional (indirect, or received) individual utility (2.7), is *deterministic*. The optimal value U^* of function U at the solution to Problem (2.55)-(2.56) is equal to $N\tilde{W}$, the total received utility for all the travelers the R.T. represents. (See Exercise 2.15.)

Utility function U as specified by Formula (2.55) contains two terms. The first represents the weighed combination of the alternatives' indirect utilities, net of budget, with weights equal to the demands for them. The second term is equal to the ''entropy'' of the distribution of demands T_j, times a weight equal to $1/\beta$. (The concept of entropy is discussed in section 3.6 of Appendix A.) This latter term may be taken as measuring the ''utility'' to the R.T. of a diversity of choices, as represented by the entropy of their distribution. Indeed, the larger the value of this entropy term, the more dispersed the distribution of T_j's is. (See section 3.6 of Appendix A.) Constraints (2.56a) translate the requirements that the amount spent by the R.T. is equal to the total budget, and Constraint (2.56b) that the total number of trips is equal to the given number of travelers.

The results of this section are important, as they show that, when properly formulated as maximization of utility functions of the type in Formula (2.55) the mathematical programming approach has rigorous behavioral meaning, and is not just a mathematical construct. The mathematical programming approach to travel demand modeling has been used extensively in the past, but without rigorous grounding in microeconomics.

We shall systematically resort to the mathematical programming approach throughout the course of the development of travel demand models. In particular, describing the behavior of the ''representative'' traveler is necessary in the presence of congestion, when the individual behaviors of travelers are affected by demand externalities, and when simple (arithmetic) addition of demands is not possible, as above. In this case, individual behaviors must be combined, in a less than intuitive manner, into that of the representative traveler. Also, this approach may be used in connection with model estimation, that is, the determination of the model's parameter values, using aggregate observations (i.e., of the R.T.'s choices), as an alternative to using observations of individual traveler choices. Both methods are described in Chapter 7. This approach may also be used in the case of multiple levels of choice.

This approach of maximizing direct utility subject to a monetary budget may be contrasted, for instance, with the ''maximum entropy'' approach (Wilson, 1970), which is based on maximizing entropy subject to observed travel cost, or, alternatively, minimizing travel cost subject to observed entropy (Erlander, 1977).

Finally, another significant advantage of estimating the aggregate demands from the R.T.'s utility maximizing choices is that the effect of the monetary cost of travel on demands is modeled in a ''natural'' manner, through the presence of the budget constraint, and not, as is typically done, by treating travel cost as another factor in the utility function. As a consequence of this apparently minor point, the utility the R.T. receives from his or her choices (the aggregate travel demands) is an accurate measure of benefits to the traveling public as a whole, that is, the ''traveler surplus,'' or community travel welfare. This particular indicator of transportation system performance provides a useful assessment of the impacts of transportation pricing policies, for instance, including schemes based on congestion pricing.

2.9 SUMMARY

In this chapter, the hierarchical process of individual travel choices was structured, and related to the various dimensions of travel demand introduced in the previous chapter. The concept of utility was introduced and formalized. The fundamental behavioral principle of utility maximization was stated as the basic driving mechanism for traveler choices. Basic assumptions underlying random utility theory were then briefly reviewed.

The derivation and properties of the logit model of discrete choice as the outcome of random (indirect) utility maximization at the individual level were presented and illustrated. The alternative derivation of the logit model as the outcome of deterministic (direct) utility maximization at the aggregate level of the ''representative traveler'' (R.T.) was also presented. This introduces, and justifies from the behavioral standpoint, the mathematical programming approach to travel demand modeling, which will be prevalent in this book.

For all of its appeal, the random utility framework also presents some pitfalls. A major one is the property of ''independence of irrelevant alternatives,'' which results from the assumption of independent random utilities, which may be unrealistic in some situations. For instance, individual travel routes on a given transportation network in general share some common links. Thus, their random utility will automatically be correlated through inclusion of these links' individual utilities. The larger the ''overlap'' between routes, the higher the degree of correlation. As already noted above, this pitfall is alleviated by the use of the probit model. However, this entails substantially more complex computations. Also, the use of hierarchical, or nested, logit models mitigates this drawback to some extent, as we shall see in Chapter 4.

Thus, the use of random utility theory represents a compromise between analytical tractability and realism. In their defense, logit models have performed well in a wide variety of practical situations. In any event, the material in this chapter provides us with a basic conceptual framework and our analytical ''building blocks'' for the developments in the following chapters.

2.10 EXERCISES

2.1 Assume that there are two models serving the downtown area. On the average, a trip costs \$3, by either mode. However, the modal trip costs cannot be estimated exactly, and thus have a random component, so that the utility received from traveling by each mode is:

$$\tilde{U}_1 = -3 + \epsilon_1 \qquad \text{for the 1st mode}$$

$$\tilde{U}_2 = -2.5 + \epsilon_2 \qquad \text{for the 2nd mode}$$

Assume that ϵ_1 is uniformly distributed between -0.5 and $+0.5$ (i.e., it can be anything between plus and minus 50 cents with equal probability), and that ϵ_2 is uniformly distributed between $-\$0.25$ and $+0.25$. The probability density functions of ϵ_1 and ϵ_2 are then:

$$f_1(x) = 1 \qquad \text{for } -0.5 \leq x \leq 0.5\text{; and 0 otherwise}$$

$$f_2(x) = 2 \qquad \text{for } -0.25 \leq x \leq 0.25\text{; and 0 otherwise}$$

Assume also that these random utilities are independent of one another. What is the probability that an individual traveler will select the first mode?

2.2* Derive the probabilities of choice in the case of two alternatives, when the random utility terms both have a Normal distribution function with mean zero and standard deviation σ, and the coefficient of correlation between them is equal to ρ.

2.3 The utilities of three modes are given by the following utility function

$$\tilde{V}_m = -2t_m - 1.5c_m + \epsilon_m; \qquad \forall\, m$$

where t and c are the time and cost of travel, respectively. Assume that the random terms ϵ_m are Normally distributed, with means zero, and a variance-covariance matrix

$$[\sigma_{ij}] \equiv \begin{bmatrix} 4 & 0 & 3 \\ 0 & 4 & 2 \\ 3 & 2 & 4 \end{bmatrix}$$

The values of the travel times and costs are as represented in Table 2.2. Using the procedure described in Appendix 2.1 at the end of this chapter, determine the respective probabilities of choice of the three alternatives.

TABLE 2.2 Modal Costs for Exercise 2.3

$m =$	1	2	3
$c =$	2	1	1
$t =$	2	3	4

2.4 Show that in the case of a hierarchical choice structure, the IIA property no longer holds for two alternatives (i, j) and $(i'j')$ which do not belong to the same second-level choice set.

2.5 Demonstrate Formulas (2.44) and (2.45).

2.6 Demonstrate Formulas (2.23) and (2.24).

2.7 Going back to Exercise 2.1 above, assume now that the random terms have a Gumbel distribution, with mean 0 and standard deviation 0.2. Derive analytically, i.e., without resorting to Formula (2.11), the numerical value of the probabilities of choice. (Hint: You may want to follow the steps described in Appendix 2.1 at the end of this chapter for the derivation of Formula (2.11) in the case of $J = 2$.)

2.8* Going back to Exercise 2.1 above, estimate the value of the expected utility received as given by Formula (2.21). (Note: This is *not* given by Formula (2.22), which is only valid when the ϵ's have a Gumbel distribution.)

2.9 Show that the logit probabilities are not affected by adding or subtracting a constant to all utilities. Discuss the implications of this fact.

2.10 Obtain the probabilities of choice P_1 and P_2 in the illustrative example in section 2.3.1 for random utility with Bernouilli distributed ϵ_j's when $P(\epsilon_j = 0.5) = 0.5$.

2.11 Obtain the same answer experimentally, i.e., from observing the choices made according to the outcome of a large number of tosses of a fair coin.

2.12 Given the probabilities of choice below.

$j =$	1	2	3
$P_j =$	0.3	0.5	0.2

and a total number of travelers equal to 100, estimate:

a. The probability that the number of choosers of the first, second, and third alternative are respectively equal to $N_1 = 50$, $N_2 = 40$ and $N_3 = 10$.

b. The probability that the number N_2 of choosers of the second alternative are equal to 40.

c. The variance of the number N_1 of travelers choosing the first alternative.

d. The covariance of the numbers N_1 and N_3 of travelers choosing the first and third alternatives respectively.

2.13 Show that the indirect utility $\tilde{W}$ as specified in Formula (2.22) is only defined up to a constant. What does this imply about the sign of $\tilde{W}$ and of the "traveler surplus"? (Hint: Remember that utilities $\tilde{U}_j$ are themselves defined up to a common constant.)

2.14 Show that in the case of the illustrative example used in section 2.3.1, the numerical values of the probabilities of choice (0.54 and 0.46) may be obtained as the solution of the R.T.'s U.M. Problem (2.55)–(2.56).

2.15 Show that the average received utility $\tilde{W}$ as specified in Formula (2.22) may be obtained as the optimal value U^* of objective function U in Problems (2.55)–(2.56). (Hint: Simply replace the expression of the optimal demands in the objective function and simplify.)

2.16 Demonstrate Formula (2.40).

2.17 Show that the R.T.'s U.M. Problem (2.55)–(2.56) always has a unique solution (Hint: Evaluation the Hessian of the objective function and use the fact that all the constraints are linear).

2.18 Show that when the parameter β in logit model (2.11) is equal to 0, the probabilities are all equal.

2.19 Show that when parameter β in logit model (2.11) is equal to ∞, the probabilities are all equal to zero, except for one.

2.20 Demonstrate Formula (2.35).

APPENDIX 2.1 ESTIMATING PROBIT PROBABILITIES

This procedure is based on results due to Clark (1956), and will be illustrated in the case of three alternatives. In this case, one may write Max (V_1, V_2, V_3) = Max $\{\text{Max}\ (V_1, V_2), V_3\}$. If V_1 and V_2 are both normal variables, which in general may be correlated, then Max (V_1, V_2) is *approximately* normal, with a mean μ_{12} and variance σ_{12}^2. Furthermore, we have

$$\begin{aligned}\mu_{12} &= E\{\text{Max}\ (V_1, V_2)\} \\ &\approx \mu_1\Phi(\alpha_{12}) + \mu_2\Phi(-\alpha_{12}) + (\sigma_1^2 + \sigma_2^2 - 2\rho_{12}\sigma_1\sigma_2)^{1/2}\phi(\alpha_{12}) \qquad (2.57)\end{aligned}$$

where $\Phi(x)$ and $\phi(x)$ are respectively the values of the cumulative probability distribution and of the probability density function for the standard normal variable, estimated at x, and

$$\alpha_{12} = (\mu_1 - \mu_2)/(\sigma_1^2 + \sigma_2^2 - 2\rho_{12}\sigma_1\sigma_2)^{1/2} \qquad (2.58)$$

and

$$\begin{aligned}\omega_{12} = E\{\text{Max}^2\ (V_1, V_2)\} &\approx (Y_1^2 + \sigma_1^2)\Phi(\alpha_{12}) \\ &+ (Y_2^2 + \sigma_2^2)\Phi(-\alpha_{12}) + (V_1 + V_2)(\sigma_1^2 + \sigma_2^2 - 2\rho_{12}\sigma_1\sigma_2)^{1/2}\phi(\alpha_{12}) \\ &\qquad (2.59)\end{aligned}$$

These first two formulas enable one to estimate

$$\sigma_{12}^2 = \omega_{12} - \mu_{12}^2 \tag{2.60}$$

Finally, the covariance between Max (V_1, V_2) and the third variable is

$$\begin{aligned}\rho_3' &= \text{Cov}\ \{V_3, \text{Max}\ (V_1, V_2)\} \\ &\approx [(\sigma_1\,\rho_{13})\Phi(\alpha_{12}) + (\sigma_2\,\rho_{23})\Phi(-\alpha_{12})]/(\omega_{12} - \mu_{12}^2)^{1/2}\end{aligned} \tag{2.61}$$

Based on these approximations, the following probabilities may then be computed

$$P_3 = Pr\ (V_3 > \text{Max}\ (V_1, V_2)) = Pr(V_3 - \text{Max}\ (V_1, V_2) > 0) \tag{2.62}$$

in which $(V_3 - \text{Max}\ (V_1, V_2))$ is the difference between two normally distributed random variables with known parameters, using the above formulas.

This brings us back to the case of two alternatives (see Exercise 2.2). Specifically,

$$\begin{aligned}P_3 = Pr\{Z > (\mu_{12} - U_3)/(\sigma_3^2 + \omega_{12} \\ - \mu_{12}^2 - 2\rho_3'\sigma_3\sqrt{(\omega_{12} - \mu_{12}^2)})^{1/2}\}\end{aligned} \tag{2.63}$$

where Z is the standard normal variable.

For more than three alternatives, the same procedure is iterated, using the fact that

$$\text{Max}\ (V_1, V_2, V_3, V_4) = \text{Max}\ \{\text{Max}\ (V_1, V_2, V_3), V_4\}$$

so that

$$\begin{aligned}P_4 &= Pr(V_4 > \text{Max}\ (V_1, V_2, V_3)) \\ &= Pr(V_4 - \text{Max}\ (\text{Max}\ (V_1, V_2), V_3)) > 0)\end{aligned}$$

The application of this procedure is illustrated in the solution to Exercise 2.3. (See Appendix B.)

APPENDIX 2.2 DERIVING THE LOGIT MODEL

P_j, the probability of alternative j being chosen, is equal to the probability that its random utility is greater than any of the other alternatives' random utilities. It is thus equal to the probability of the joint events A_k:

$$A_k \equiv (\epsilon_k < \tilde{U}_j + \epsilon_j - \tilde{U}_k); \qquad \forall\ k \neq j$$

For a *given* value of ϵ_j, since the ϵ_k's are assumed to be independent, the (conditional) probability of this joint event is equal to

$$P_j = \prod_{k \neq j} P(\epsilon_k < \tilde{U}_j + \epsilon_j - \tilde{U}_k); \qquad \forall j$$

Given the definition of the cumulative Gumbel p.d.f. with mean zero and standard deviation σ, Formula (A.58a in Appendix A) this is equal to

$$P_j = \prod_{k \neq j} e^{-e^{-\beta(\tilde{U}_j + \epsilon_j - \tilde{U}_k) - E}}; \qquad \forall j$$

where the positive value of parameter β is related to the standard deviation of the ϵ_j's (Formula (A.59b)). From Formula (A.58a) again, the probability density function $f(x)$ for ϵ_j is the derivative of its cumulative probability function:

$$f(x) = \frac{d}{dx}\{e^{-e^{-\beta x - E}}\} = \beta e^{-\beta x - E} e^{-e^{-\beta x - E}} \tag{2.64}$$

The unconditional probability P_j of the above event may then be obtained as the expected value of the conditional probability above, with respect to the distribution of ϵ_j:

$$P_j = \int_{-\infty}^{+\infty} \prod_{k \neq j} e^{-e^{-\beta(\tilde{U}_j + x - \tilde{U}_k) - E}} \beta e^{-\beta x - E} e^{-e^{-\beta x - E}} \, dx; \qquad \forall j$$

With a change of variable $y = e^{-e^{-\beta x - E}}$, and after some simple manipulations, this is equal to

$$P_j = \int_0^{+\infty} e^{-ye^{-\beta \tilde{U}_j} \Sigma_{k \neq j} e^{\beta \tilde{U}_k}} \, dy = \int_0^{+\infty} e^{-\lambda y} \, dy$$

with

$$\lambda = \frac{\sum_{k=1}^{J} e^{\beta \tilde{U}_k}}{e^{\beta \tilde{U}_j}}$$

The value of this definite integral is equal to

$$\frac{1}{\lambda}[e^{-\lambda y}]_{\infty}^{0} = \frac{1}{\lambda}$$

Thus

$$P_j = \frac{e^{\beta \tilde{U}_j}}{\sum_k e^{\beta \tilde{U}_k}}; \qquad \forall j \tag{2.65}$$

APPENDIX 2.3 OBTAINING THE EXPECTED RECEIVED UTILITY IN THE LOGIT CASE

To evaluate the expected received utility, we must first obtain the cumulative p.d.f. of the maximum $\tilde{V}$ of the random variables $\tilde{V}_j$. Since the ϵ's are i.i.d. Gumbel with mean zero and a common variance σ, using the cumulative density function specified in Formula (A.58a) we may write

$$P(\tilde{V} \leq x) = P(\tilde{U}_j + \epsilon_j \leq x; \forall j) = \prod_j e^{-e^{-\beta(x - \tilde{U}_j) - E}}$$

$$= \exp\left[-\sum_j e^{-\beta x + \beta \tilde{U}_j - E} \right]$$

$$= \exp\left(-e^{-\beta x - E} \sum_i e^{\beta \tilde{U}_j} \right)$$

$$= \exp\left(-\exp\left(-\beta x - E + \ln \sum_j e^{\beta \tilde{U}_j} \right) \right) = \exp\left(-e^{-\beta(x-K)}\right)$$

with

$$K = \frac{1}{\beta}\left(-E + \ln \sum_j e^{\beta \tilde{U}_j} \right)$$

According to Formula (A.58b), this is the expression of the general cumulative Gumbel p.d.f. Consequently, using Formula (A.59c) the expected value is equal to

$$\mu = E/\beta + K$$

so that expected value $\tilde{W}$ is, after simplification:

$$\tilde{W} = \frac{1}{\beta} \ln \sum_j e^{\beta \tilde{U}_j} \tag{2.66}$$

APPENDIX 2.4 DERIVING THE R.T.'S DIRECT UTILITY FUNCTION

According to Formula (A.69) in Appendix A, given the R.T.'s indirect utility $N\tilde{W}$, where $\tilde{W}$ is the individual indirect utility as specified in Formula (2.22) above, the corresponding direct utility function may be derived as the solution of the minimization problem

$$\min_{(c_j)} \tilde{W}(c_j) \tag{2.67a}$$

such that

$$\sum_j c_j T_j + T_0 = Nb \tag{2.67b}$$

where T_0 is the amount spent on other things than travel, (numéraire), and Nb is the total (R.T.) budget. The problem is then

$$\min_{(c_j)} \left\{ Nb + N \frac{1}{\beta} \ln \sum_j e^{\beta(A_j - c_j)} \right\} \tag{2.68}$$

subject to budgetary Constraint (2.67b). The other constraint on the problem is the consistency Constraint[16]

$$\sum_j T_j = N \tag{2.69}$$

It is easy to show that Problem (2.68), (2.67b), (2.69) is convex, (see section 2 of Appendix A), and has a unique solution. With the change of variable

$$c_j' = e^{\beta(A_j - c_j)}; \qquad \forall j \tag{2.70a}$$

which implies

$$c_j = A_j - \frac{1}{\beta} \ln c_j'; \qquad \forall j \tag{2.70b}$$

Objective function (2.68) may then be re-written as

$$\min_{(c_j')} \left\{ Nb + \frac{N}{\beta} \ln \sum_j c_j' \right\}$$

[16]It may be noted that Constraints (2.69) are not formally part of the problem, since they do not involve variables c_j.

or Min $\Sigma_j c_j'$, since N, b and β are constants, and Constraint (2.67b) becomes

$$-\frac{1}{\beta}\sum_j T_j \ln c_j' + \sum_j T_j A_j + T_0 = Nb$$

The Lagrangean for this problem is (see section 2.1 of Appendix A)

$$L(c_j') = \sum_j c_j' - \lambda\left[-\frac{1}{\beta}\sum_j T_j \ln c_j' + \sum_j T_j A_j + T_0 - Nb\right]$$

Setting its partial derivatives with respect to the c_j's equal to zero, we get

$$1 + \frac{\lambda T_j}{\beta c_j'} = 0; \qquad \forall j$$

and

$$c_j'^* = -\frac{\lambda T_j}{\beta}; \qquad \forall j$$

or

$$c_j^* = A_j - \frac{1}{\beta}\ln\left(-\frac{\lambda T_j}{\beta}\right); \qquad \forall j \tag{2.71}$$

where T_j is as specified in Formula (2.54). Replacing these values in Constraint (2.67b), we have at optimality, after rearranging terms

$$Nb + \frac{N}{\beta}\ln\left(\frac{-\lambda}{\beta}\sum_j T_j\right) = -\frac{1}{\beta}\sum_j T_j \ln T_j + \sum_j T_j A_j + T_0 \tag{2.72}$$

The left-hand side of this equality is equal to the optimal value of objective function (2.68), i.e. the direct utility function. Indeed, replacing c_j^* by its optimal value above, Formula (2.71), we have

$$Nb + \frac{N}{\beta}\ln\sum_j c_j'^* = Nb + \frac{N}{\beta}\ln\left(\frac{-\lambda}{\beta}\sum_j T_j\right) \tag{2.73}$$

Thus, the R.T.'s direct utility function is equal to the right hand side of Equality (2.72).

APPENDIX 2.5 RETRIEVING THE AGGREGATE LOGIT DEMANDS FROM THE R.T.'S CHOICES

To solve the R.T.'s utility maximization problem, we form the Lagrangean L (see section 2.1 of Appendix A.)

$$L = \sum_j T_j A_j - \frac{1}{\beta} \sum_j T_j \ln T_j + T_0 - \lambda \left(\sum_j T_j - N \right) - \mu \left(\sum_j T_j c_j + T_0 - Nb \right) \tag{2.74}$$

Setting the partial derivatives with respect to the variables and the Lagrange coefficients, we have

$$\frac{\partial L}{\partial T_j} = A_j - \frac{1}{\beta} (\ln T_j + 1) - \lambda - \mu c_j = 0; \qquad \forall j$$

which implies

$$T_j = N e^{\beta(A_j - \lambda - \mu c_j) - 1}; \qquad \forall j$$

From Constraints (2.56b), we have

$$\sum_j e^{\beta(A_j - \lambda - \mu c_j) - 1} = e^{\beta\lambda} \sum_j e^{\beta(A_j - \mu c_j) - 1} = 1$$

and therefore

$$e^{\beta\lambda - 1} = \frac{1}{\sum_j e^{\beta(A_j - \mu c_j)}}$$

so that

$$T_j = N \frac{e^{\beta(A_j - \mu c_j)}}{\sum_j e^{\beta(A_j - \mu c_j)}}; \qquad \forall j \tag{2.75}$$

Finally, setting the partial derivative with respect to T_0 equal to zero would lead to $\mu = 1$, so that Formula (2.75) is precisely the expression of the expected average demands in Formulas (2.12b).

CHAPTER 3

ROUTE CHOICE ON UNCONGESTED NETWORKS

3.1 INTRODUCTION: THE CONDITIONAL APPROACH TO TRAVEL DEMAND FORECASTING

In the first chapter, the various choices facing the individual traveler were defined and the generic, logit model of probabilistic traveler choice was developed. We are thus ready to begin the development of the demand models for each of these respective dimensions of travel. As we have seen, since all decision levels are interdependent, all four levels of travel choices must be considered jointly, as a traveler, of course, never makes a *partial* decision, for example, which mode, but not which destination. This, however, may be complex.

A major advantage of the hierarchical traveler decision structure introduced in the first chapter is that it may be utilized in a "conditional" manner. That is, by assuming that preceding choices have been made, a given level of travel choice may be analyzed independently of those preceding it. For instance, since the route choice is the last one in the hierarchy, we may assume that the decision to travel, the mode, and the destination have been chosen. At the level of aggregate demands, this means that we assume that the analyses of trip generation, modal split, and trip distribution have already been conducted, resulting in *given* travel demands T_{ijm}. Given these demands, that is, *conditional* on them, we may focus exclusively on the problem of route choice, that is, the determination of the T_{ijmr}'s, whose values will be a function of the assumed values of the T_{ijm}.

Once this is done, we may "uncondition," or treat these assumed values as unknown, and in the same manner focus on the determination of the T_{ijm}'s, the mode choice, *given* the origin-destination demands T_{ij}. That is, in the same manner as above, we assume that the preceding choices of destination and of travel have been made, that the values of the T_{ij}'s are known. Consequently, the values of the T_{ijm}'s will be a function of the assumed values of the T_{ij}'s, and as a result, so will be the values of the previously determined route demands T_{ijmr}'s, which depend on those of the T_{ijm}'s.

In the next stage, we then treat the assumed values of the T_{ij}'s as unknowns, by unconditioning the destination choices, and so on, until, one level at a time, one backtracks to the first level, when all four dimensions of demand are determined *simultaneously*. Throughout this process, demands reflects the U.M. principle. This approach may be contrasted with the traditional "four-step" approach, which treats the individual levels of demand individually (unconditionally), and where the only connection between individual demand levels are the "control totals" represented by Formulas (1.3).

In this, and the next chapter, we consider the case when the utilities of the respective travel choices are fixed, and are not affected by travel demands, so that all utilities are taken as given by the modeler. This implies in particular that there is no congestion on any of the modal networks, or at the facilities they serve. This is the simplest case in analytical terms, since in this case the choices of a given individual traveler are not affected by the decisions of other travelers. In Chapter 5, we will consider a more general case, when utilities are variable, in particular because congestion is assumed present. The fixed utility case may, of course, be treated as a special case of variable utilities. However, examining the uncongested case serves as a useful introduction to the major issues in travel demand modeling. Throughout this text, analyses are conducted with respect to a given travel purpose, such as going to work, or shopping.

3.2 CAR ROUTE CHOICE: DETERMINISTIC CASE

3.2.1 The User Equilibrium and Utility Maximization Principles

In accordance with the above, in this section we model the individual choice of route by travelers who are known to have already chosen to travel from a given origin i to a given destination j, and on a given mode m, which in this section is the private car. In other words, the only travel choice we are considering is that of a route between i and j for car drivers. Consequently, the only travel demand we are estimating is T_{ijr}, the other demands, T_{ij}, and T_i, being already known and assumed given.[1]

In this section, we examine the case when route utilities are deterministic. This might seem to contradict the argument made earlier that utilizing random utilities alleviates some significant theoretical pitfalls in travel demand prediction. Thus, we might go directly to the next section, and consider the more general case of random utilities, corresponding to probabilistic route choice. The deterministic case would then be considered the special, limit case corresponding to zero variance in the random utilities.

However, examining the deterministic case in its own right is also important, for several reasons. First, traditionally, the deterministic and stochastic network assignment problems are considered separately, in particular when congestion is present. Also, although from a conceptual standpoint the two cases are the same, in practice, the deterministic case requires the application of different methodology than the stochastic case. Accordingly, we begin with the special case when the utility received from making a single trip on route r on mode m between zones i and j is not random, as in the general case discussed in Chapter 2, but is known by the modeler with certainty.

We assume that a given route's utility is primarily related to the route's travel time t_{ijr} and the travel cost c_{ijr}, respectively. Thus, according to Formula (2.7) in Chapter 2, the utility received from one trip made between i and j on the given mode, following route r, is

$$\tilde{U}_{ijr} = b_{ij} - c_{ijr} - \tau t_{ijr} = b_{ij} - g_{ijr}; \qquad \forall\ i, j, r \tag{3.1a}$$

with, for compactness,

$$g_{ijr} = c_{ijr} + \tau t_{ijr}; \qquad \forall\ i, j, r \tag{3.1b}$$

[1]The index of the mode will be dropped for this section, as it is understood that the mode is the car.

There is no random utility term, and b_{ij} is the budget or income of an individual traveler from i to j.[2] τ is a parameter whose value must be calibrated. Other potentially relevant factors, including delay at intersections along the route, may be included in the travel time.[3]

The graphical representation of the structure of systematic route utilities is shown in Figure 3.1.

The travel costs and times of a given route are, as seen in section 2 of Chapter 1, respectively the sum of the travel costs and times of the links which compose the route. Formally,

$$t_{ijr} = \sum_a t_a \delta^a_{ijr}; \qquad \forall\, i, j, r \tag{3.2}$$

$$c_{ijr} = \sum_a c_a \delta^a_{ijr}; \qquad \forall\, i, j, r \tag{3.3}$$

where t_a and c_a represent link a's travel time and user charge, respectively.

In the deterministic case, as discussed in section 2 of Chapter 2, the demand function specifies that all travelers choose the same route, that with the largest deterministic utility. That is, the probability that route r is chosen by an individual traveler between i and j selected at random is equal to

$$P_{r/ij} = \begin{cases} 1 & \text{if } \tilde{U}_{ijr} > \tilde{U}_{ijk}; \quad \forall\, k \\ 0; & \text{otherwise} \end{cases}; \qquad \forall\, i, j, r \tag{3.4a}$$

In terms of aggregate route demands, this implies the following:

FACT 3.1

When route choice is deterministic, aggregate route demands are specified as

$$T^*_{ijr} = \begin{cases} T_{ij} & \text{if } r = r^* \\ 0; & \text{otherwise} \end{cases}; \qquad \forall\, i, j, r \tag{3.4b}$$

where r^* is the index of the minimum cost route between i and j.

In the case when several routes all offer the same utility, they will all be chosen with the same probability, which is equal to 1 divided by the number

[2] b_i the average income of travelers in origin zones i, or b the average individual income, may be used alternatively.

[3] The incorporation of intersection delays is taken up in chapter 5, under congested conditions. (See Exercise 5.27.)

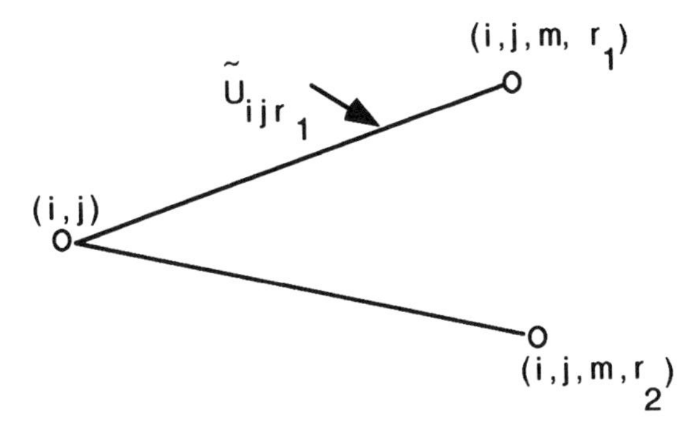

Figure 3.1 Systematic conditional utility for modal route.

of such routes. Formula (3.4a) states that the route(s) which offers the maximum utility will be used, and not the others, that is, those routes which offer a lower utility. This statement represents the "user equilibrium" (UE) principle of network assignment. It implies that individual travelers are assigned to network links in such a way that no individual traveler may increase further his or her received route utility, that is, decrease travel costs, by unilaterally changing routes, without cooperation from the other travelers. In other words, travelers maximize the utility of their choice of route, and consequently, the UE principle is equivalent in this case to the utility maximization principle.

It might seem at first glance that the expression of the T^*_{ijr} is not a valid demand function, as it does not possess some of the fundamental properties of any legitimate demand function. (See section 4.2 of Appendix A.) Specifically, it is not a continuous function of the utility, and is not continuously differentiable with respect to the utility for $T^*_{ijr} = T_{ij}$. However, as we shall see in section 3.3, when the deterministic case is considered the limit of the general probabilistic case, instead of independently, these difficulties no longer exist. This may be visualized as shown in Figure 3.2 for the case of two alternatives.

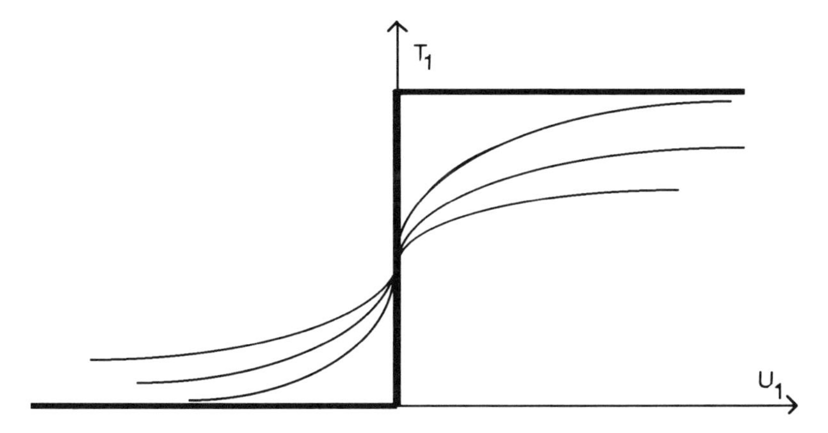

Figure 3.2 Deterministic demands as asymptotic limits of logit demands.

As we have seen in the previous chapter, the deterministic demand, represented by the thick line, is the limit to which the continuous, differentiable demand function tends when the variance of the random utilities tends to zero.

3.2.2 Illustrative Example: Deterministic Car Network Assignment

In order to estimate route demands from Formula (3.4b), we must then find the "minimum cost route" (MCR) between from any given node to any given node on a network, given individual links' generalized costs. Indeed, all individual travelers are supposed to choose the route with the maximum utility, or minimum generalized cost ($c_{ijmr} + \tau T_{ijmr}$). This problem is not very difficult to solve on small networks, where the routes may easily be enumerated. We illustrate its solution in the following example, the case of the prototype network introduced in the first chapter, whose structure is reproduced in Figure 3.3 for convenience.

The travel times t_a in minutes on the individual links of the network are also reproduced in Table 3.1.

We assume that there are no link charges, that is, $c_a = 0$; $\forall\ a$. The numbers of travelers T_{14} and T_{12} from node 1 to node 4 and from node 1 to node 2 respectively, are given, and are equal to

$$T_{14} = 150$$

$$T_{12} = 250$$

Given this information, we want to estimate the volumes on the respective alternative routes between from node 1 to node 4 and from node 1 to node 2, as well as on the individual network links.

In the case of the above network, it may be remembered from section 1.2 in Chapter 1 that 38 routes were enumerated. The three routes from node 1 to node 4 are routes number 7, 8, 9, while the routes from node 1 to node 2 are routes number 1, 2, 3. The respective costs for these routes were then determined from the link costs from Formula (1.2a), as represented in Table 1.8, which is partially reproduced in Table 3.2.

Since route $r = 1$ for $(i, j) = (1, 2)$ has the lowest cost, in the present deterministic case, all 250 travelers from node 1 to node 2 should use it. The

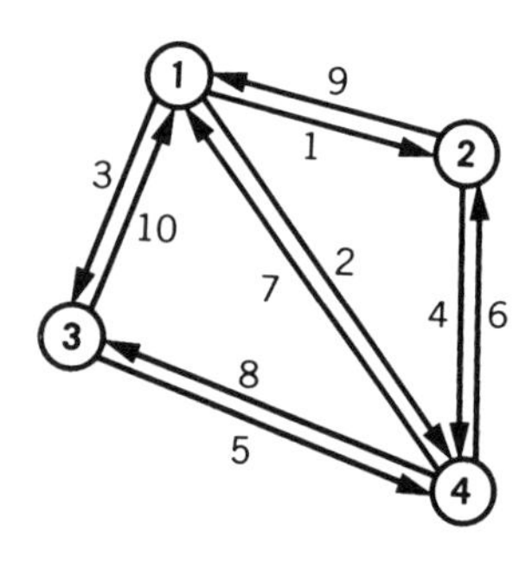

Figure 3.3 Example transportation network.

TABLE 3.1 Link Travel Times

Link #	t_a
1	5
2	15
3	6
4	8
5	7
6	8
7	15
8	7
9	5
10	6

corresponding volume on route 1, $(i, j, r) = (1, 2, 1)$ is then

$$T_{121} = 250$$

Similarly, since the costs on routes 8 and 9 from node 1 to node 4 are equal (i.e., 13 each), the 150 travelers from node 1 to node 4 should all use routes 8 and 9, with equal frequencies. The corresponding volumes should then be

$$T_{142} = T_{143} = 150/2 = 75$$

Having now the route demands T_{ijr}, the link volumes may easily be estimated with Formula (1.2b):

$$v_a = \sum_i \sum_j \sum_r T_{ijr} \delta_{ijr}^a; \qquad \forall\, a \tag{3.5}$$

In the present case, the values of the relevant δ_{ijr}^a's which are equal to 1 are given in Table 1.7, which is partially reproduced in Table 3.3.

Thus, the volume on link number 1 is equal to

$$v_1 = \sum_i \sum_j \sum_r T_{ijr} \delta_{ijr}^a = T_{121} \delta_{121}^1 + T_{142} \delta_{142}^1 = 250(1) + 75(1) = 325$$

TABLE 3.2 Route Costs

i, j	r	Absolute Route #	Route Cost
1, 2	1	1	5
	2	2	21
	3	3	23
1, 4	1	7	15
	2	8	13
	3	9	13

TABLE 3.3 Route-link Incidence δ^a_{ijr}

i, j, r	a
1, 2, 1	1
1, 2, 2	3, 5, 6
1, 2, 3	2, 6
1, 4, 1	2
1, 4, 2	1, 4
1, 4, 3	3, 5

since all other route volumes T_{ijr} are equal to zero, and since other δ^a_{ijr}'s are zero, since there are no other routes between 1 and 2 and between 1 and 4 of which link 1 is part of. The volumes on all the other links on the network may be estimated in a similar manner, and are given in Table 3.4, corresponding to Figure 3.4.

The total generalized traveler cost (in this case the total travel time) is equal to 3,200, and may be estimated as

$$t = \sum_a t_a v_a = \sum_i \sum_j \sum_r t_{ijr} T_{ijr} \tag{3.6}$$

As we shall soon see in a more formal way, this value may be taken as a measure of efficiency of the transportation system, or equivalently, of "welfare" of the travelers as a group.

Other origin destination volumes T_{ij} would similarly be cumulated with these volumes.

It may be noted that other optimal network assignments exist in this case, as represented in Figure 3.5.

Consequently, the aggregate route or link demands are not uniquely specified when travel times are constant. This unfortunate characteristic disappears, as we shall soon see, when again deterministic route choice is considered the limit of the probabilistic case.

TABLE 3.4 Link Volumes

Link #	Volume
1	325
2	0
3	75
4	75
5	75
6	0
7	0
8	0
9	0
10	0

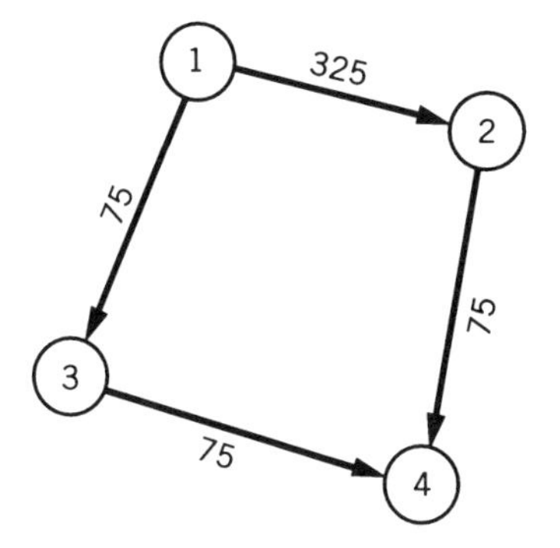

Figure 3.4 Network loadings.

3.2.3 Individual Demands: The Minimum Cost Route Algorithm

Although the network loading procedure above is straightforward, it requires exhaustive route enumeration, as well as the knowledge of the link-route incidence structure. However, because the number of links in a real-world urban transportation network is large (i.e., at least in the hundreds, more usually in the thousands), the number of alternative routes, which are combinations of links, will be too large to permit actual enumeration. These facts preclude the practical determination of the MCR from the systematic examination of all routes. More efficient procedures are thus clearly needed.

Several minimum cost route (MCR) algorithms have been devised to that

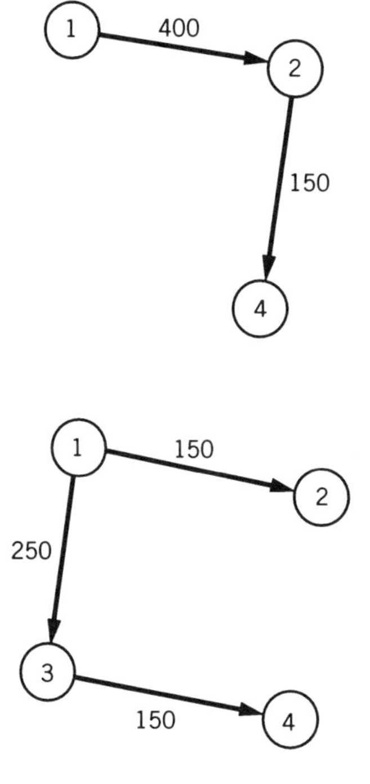

Figure 3.5 Alternative optimal solutions.

effect. Although there are numerous MCR algorithms, one of the most efficient in terms of computer implementations is the Pape and Moore algorithm (Pape, 1974). Given the costs on the individual links of a transportation network, the algorithm determines the sequence of links which constitutes the minimum cost route, from any given node to *all* other nodes, together with the corresponding (minimum) costs. The algorithm does not require route enumeration, nor the link-route incidence matrix. Only the knowledge of the sets k^+ and k^- for each node k in the network is required.

It is important to note that the algorithm only determines the shortest paths from a single node i to all the other nodes on the network. That is, the shortest paths from *other* origin nodes must be obtained from other, independent applications of the algorithm.

The procedure may be described as follows. At each step, each node k will labeled with two numbers. The first is the current minimum generalized cost g_{ik} from node i. The second is the node h which precedes it on the current minimum cost path from i. Also, at each iterative step, a list of links to be examined in the next iteration is constructed. This list consists of all the links which originate at the nodes whose label has changed during the current iteration. The procedure begins with the "initialization."

ALGORITHM 3.1 MINIMUM COST ROUTE (DETERMINISTIC)

Initialization

Set iteration counter at zero.

Set the label of node i equal to (0, 0).

Set the label of all the other nodes equal to (∞, blank).

Include in the list all the links which originate at node i.

Iterative step

Update iteration counter by one.

If the current list contains any links, go on; otherwise, stop.

For each of the links in the current list:

For a link a with origin node k and end node l, compare the current value of g_{il} for the end node l to g_{ik} for the origin node plus the link's time/cost t_a.

If $g_{ik} + g_a \geq g_{il}$, go to the next link in the list.

If $g_{ik} + g_a < g_{il}$, update the current g_{il} of the end node of the link under consideration to $(g_{ik} + g_a)$, and update the second label of node l to k, i.e., the index of link a's origin node.[3] Place all links originating from the links' end node l in the next list.

When all links the current list have been examined, iterate, i.e., go back to the top of the iterative step.

[3]This clearly means that node l may now be reached from node i in less time than before, using the link under consideration.

At the end of the procedure, when the current list is empty, the minimum cost path between the origin i and every other node may be traced by back-tracking through each node's predecessor node. Also, the generalized cost of the minimum cost path to a given node from the origin is the last value of the node's second label.

Once the MCR's from a given origin zone i to all destinations j are determined, the route demands T_{ijr} (in numbers of person trips) must be obtained. These, according to Formula (3.4b) are equal to

$$\begin{aligned} T_{ijr^*} &= T_{ij}; \quad \text{for the minimum cost route } r^* \\ T_{ijr} &= 0; \quad \text{for all other routes} \end{aligned} \quad ; \quad \forall\, i, j, r \qquad (3.7)$$

where T_{ij} is the given demand for car travel between i and j. In general, this may in particular come from an integrated, or combined, destination, mode, and route demand prediction, as described in the next chapter. Car trips may be obtained from person trips by dividing them by the average car occupancy, or number of passengers per car.

3.2.4 Illustrative Example: Application of the MCR Algorithm

We now illustrate the procedure in the determination of the MCR from node 1 to node 4 on the prototype network above. The outcome of the various iterations in terms of node labels, (c_j = minimum cost from origin to j; l = predecessor node) are presented in Table 3.5, which also contains the list of links to be examined.

After two iterations, no nodes are (re)labeled, and consequently, no links are placed in the list. The procedure has then terminated, and the MCR has been obtained. From node 4, the cost from node 1 is found as the first entry in node 4's last label, and is equal to 13. The predecessor of node 4 on the optimal route is the second label, nodes 2 or 3. From node 3, the minimum cost from node 1 is found as the first entry in node 3's last label, and is equal to 6, the predecessor being node 1, the origin. The optimal route would then be 1-3-4, at a cost of (6 + 7). Alternatively, from node 2, the cost is 5, the

TABLE 3.5 MCR Algorithm

Iteration	Nodes				List of	
#	1	2	3	4	Links a	$g_{ik} + g_a > g_{il}$?
0	0, 0	∞,	∞,	∞,	1	$0 + 5 > \infty$ N
					2	$0 + 15 > \infty$ N
					3	$0 + 6 > \infty$ N
1	0, 0	5, 1	6, 1	15, 1	4	$5 + 8 > 15$ N
					5	$6 + 7 > 15$ N
2	0, 0	5, 1	6, 1	13, 2 or 3	6	$13 + 8 > 5$ Y
					7	$13 + 15 > 0$ Y
					8	$13 + 7 > 6$ Y

predecessor again being node 1. The other optimal route would then be 1-2-4, at the same minimum cost of (5 + 8).

Travelers from node 1 to node 4 would then choose between these two routes with equal probability (50%), and choose the other available routes with probability zero. In terms of network loading, as already seen, the two routes would consequently be loaded equally with travelers from node 1 to node 4.

It is important to note that as part of the determination of the MCR from node 1 to node 4, the MCR's from all nodes other than 1 to node 4, specifically nodes 2 and 3, are also determined. Therefore, the assignment of T_{24} and/or T_{34} travelers would not necessitate another application of the MCR procedure. This feature of the method is a significant advantage in practical terms.

A minimal set of MCR's from a given origin to all destinations defines a "spanning tree" rooted at that origin. For instance, the two spanning trees from node 1 in the above example network are represented in Figure 3.6.

As mentioned earlier, alternative MCR algorithms have been developed. One should be mentioned in particular, since it corresponds to the version of the MCR algorithm used in connection with transit networks, which we take up in the next section. This algorithm is described in Exercise 3.10. In contrast with the above algorithm, it proceeds "backwards," that is, starts from the destination node, rather than the origin. The Pape and Moore algorithm may, however, be applied symmetrically from the destination. However, this alternative algorithm is not as efficient, in that it examines each and every link in the network, one link at a time.

3.2.5 Aggregate Demands: The R.T.'s U.M. Problem and the Traveler Surplus

As shown in section 2.8 of Chapter 2, an alternative way to derive the aggregate (or average) travel demands in the general case of deterministic choice is not from individual travelers' behavior, as above, but directly from the behavior of their "representative traveler" (R.T.). In this section, we show that the same approach may be utilized in the present case of deterministic route choice.

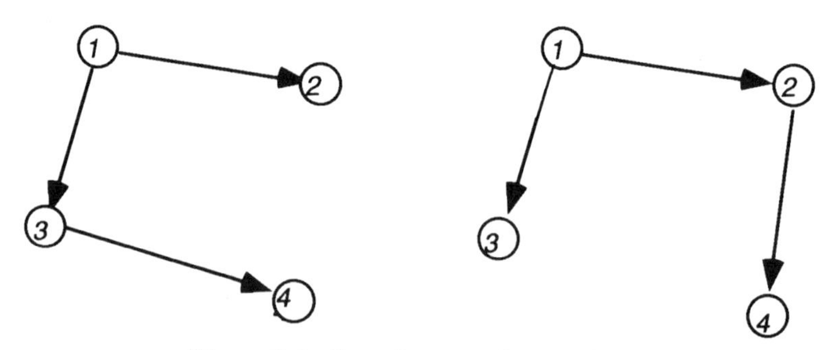

Figure 3.6 Spanning trees from node 1.

PROPOSITION 3.1

The R.T.'s direct utility in the deterministic route choice case is equal to

$$U_R = -\tau \sum_{ijr} t_{ijr} T_{ijr} + T_0 \quad (3.8a)$$

The symbol U is used, since this is a deterministic utility. Moreover, the absence of the tilde over the U is to signify that this utility is an unconditional, or direct utility. Subscript R refers to the aggregate, total utility for all travelers obtained from route choices. The aggregate route demands (car trip assignments) may then be retrieved from the maximization of this utility, subject to the budgetary and consistency constraints.

The R.T.'s U.M. problem is therefore

$$\underset{(T_{ijr}, T_0)}{\text{Max}} \left[-\tau \sum_i \sum_j \sum_r T_{ijr} t_{ijr} + T_0 \right] \quad (3.8b)$$

such that

$$\sum_r T_{ijr} c_{ijr} + T_0 = B \quad (3.9a)$$

$$\sum_r T_{ijr} = T_{ij}; \qquad \forall\, i, j \quad (3.9b)$$

$$T_{ijr} \geq 0; \qquad \forall\, i, j, r; \quad (3.9c)$$

where the unknown T_0 is the amount spent by the R.T. on other things than travel, and is in effect a "slack" variable for the budgetary Constraint (3.9a). B is the aggregate budget for all travelers, including for travel.

The fact that the aggregate utility is linear in the demands makes this optimization problem a convex, but not a *strictly* convex problem. (See section 1.7 of Appendix A.) Consequently, the program may in principle admit of several solutions, so that the aggregate route demands are not uniquely defined, as already noted. However, again by the same continuity argument as above, this problem no longer exists in the more general case of probabilistic route choice, as was seen in section 2.8 of Chapter 2.

Variable T_0 may readily be eliminated from the problem by replacing it in the objective function, using budgetary Constraint (3.9a), which may then be deleted. The new objective function for the R.T.'s U.M. problem is then

$$\underset{(T_{ijr})}{\text{Max}} \left[B - \sum_i \sum_j \sum_r T_{ijr}(\tau t_{ijr} + c_{ijr}) \right] = \underset{(T_{ijr})}{\text{Max}} \left[B - \sum_i \sum_j \sum_r T_{ijr} g_{ijr} \right] \quad (3.10)$$

subject to Constraints (3.9b)–(3.9c). Since B is a constant which may be deleted from the objective function, with Min F = Max $-F$, the objective function of the R.T.'s problem may also be written with a more familiar appearance as

$$\underset{(T_{ijr})}{\text{Min}} \sum_i \sum_j \sum_r T_{ijr} g_{ijr}$$

This utility maximization problem may alternatively be formulated in terms of link volumes v_a. (See Exercise 3.14.)

According to Formulas (3.1), the expected (average) value of the received utility from route choices by a randomly selected individual traveler between i and j is equal to

$$\tilde{W}_{m/ij} = b_{ij} + \underset{k}{\text{Min}} (-c_{ijk} - \tau t_{ijk}) = b_{ij} - c^*_{ijr} - \tau t^*_{ijr} = b_{ij} - g_{ij}; \qquad \forall\ i, j \tag{3.11a}$$

where, for compactness, g_{ij} is the minimum generalized travel cost between i and j:[4]

$$g_{ij} = g^*_{ijr} \qquad \forall\ i, j \tag{3.11b}$$

Correspondingly, the R.T.'s received utility on behalf of all travelers is equal to the sum of these individual received utilities:

PROPOSITION 3.2

The R.T.'s indirect utility in the deterministic route choice case is

$$\tilde{W}_R = B - \sum_i \sum_j \sum_r T^*_{ijr}(\tau t_{ijr} + c_{ijr}) \tag{3.11c}$$

where T^*_{ijr} are the aggregate route demands.[5] It is clear that the optimal value of the R.T.'s direct utility function (3.8a) is equal to the R.T.'s received utility:

$$\tilde{W}_R = U^*_R \tag{3.12a}$$

[4] m is the (so far implicit) index of the mode of interest, consistent with the notation in subsequent chapters.

[5] This expression, however, is not a continuous function of the travel costs, since the T^*_{ijr}'s are not. It does, however, possess all of the other required properties of indirect utility functions. (See section 4.3 of Appendix A.) Again, the same continuity argument may be invoked to resolve this technical difficulty, so that this expression does, in the limit, indeed constitute a legitimate indirect utility function at the aggregate level.

Given the expression of the T^*_{ijr}'s, Formula (3.4b), the R.T.'s received utility may also be specified as

$$\tilde{W}_R = B - \sum_i \sum_j T_{ij} g_{ij} \tag{3.12b}$$

The "traveler surplus" is the R.T.'s received utility, net of the budget.

PROPOSITION 3.3

The traveler surplus in the deterministic route choice case is equal to

$$TS_R = -\sum_i \sum_j T_{ij} g_{ij} = -\sum_i \sum_j \sum_r T_{ijr} g_{ijr} \tag{3.13}$$

This may be used as an indicator of social welfare, a measure of efficiency of transportation plans.[6] For instance, in the illustrative example of section 3.2.2, the total (minimum) travel time expended by all travelers was equal to 3,200 minutes. If now a user charge of $1 is imposed on link 1, and if parameter τ has a value of 0.5, the optimal network assignment under these conditions, the solution of the R.T.'s U.M. problem, is now as represented in Figure 3.7. It is the same as without user charges.

The corresponding traveler surplus is equal to

$$-(250(1) + 0.5(250)(5) + 0.5(150)(6 + 7)) = -1{,}850$$

Since the traveler surplus was previously equal to

$$TS = -(0.5)3{,}200 = -1{,}600$$

[6]It should be pointed out in this connection that, since utilities are defined up to an arbitrary constant, the traveler surplus must be used in a *comparative* fashion, for example, "before and after," to measure the impact of changes in the transportation system, or its operating policies.

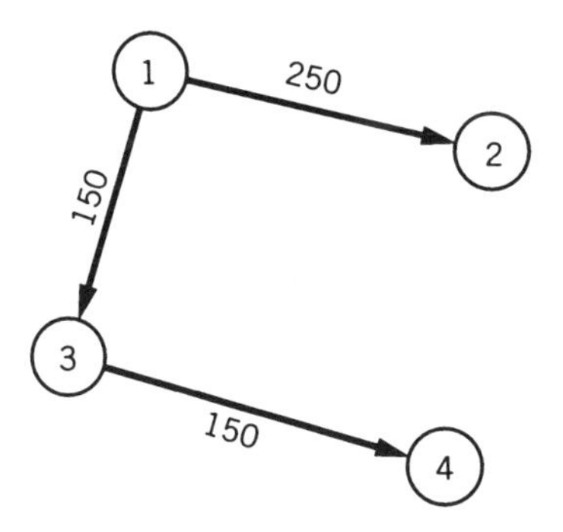

Figure 3.7 Network loadings with user charges.

the user charge has resulted in a decrease of 250 (dollars) in the traveler surplus. This is also the amount collected by the network operator.

In both formulations above, the problem of individual route choice from a given origin i to a given destination j may be formulated by setting

$$T_{ij} = 1; \qquad \forall\, i, j$$

$$T_{kl} = 0; \qquad \forall\, k \neq i,\, l \neq j$$

and, of course, the budget equal to the budget b_{ij} for an individual traveler. In particular, it may be shown that the MCR algorithm above in fact represents a systematic search for attaining the Karush-Kuhn-Tucker optimality conditions for the R.T.'s problem in link-based formulation.

Another related formulation of the R.T.'s U.M. problem may also be developed in terms of the R.T.'s decision of whether or not to include a given link in his or her route from a given origin to a given destination. (See Exercise 3.3.)

It should be pointed out that the R.T.'s problem in the present case would also solve the problem of loading the network, assigning individual travelers to travel routes, in an optimal manner, not only from the individual traveler's point of view, but also from a societal one. The fact that individual and social objectives coincide is due to the absence of demand externalities, and will no longer hold when congestion is present, as we shall see in Chapter 5. In this manner, the absence of congestion insures that the transportation system is being utilized optimally for the community through individual self-interest.

Absence of congestion also implies that travel volumes corresponding to various different travel purposes may be loaded on the network independently of one another. That is, individual R.T.'s representing different travel purposes do not compete for the use of the common network. Consequently, their individual behaviors, which may depend on different utility factors and parameter values, may be estimated separately, and the corresponding demands simply cumulated. When congestion is present, these individual behaviors and associated travel demands must be considered jointly, since they affect one another through affecting the common factors of travel choices, such as, travel times. The route choice problem is then significantly more complex in that case.

In practical terms, solving the deterministic, uncongested route choice/network assignment problem is much more efficiently done at the individual level with the MCR algorithm than at the aggregate level through solving the R.T.'s linear program. Indeed, generic algorithms for linear problems (e.g., the "Simplex" method) in general do not take specific advantage of the fact that the problem is a *network* problem, with a special structure, in contrast with the MCR algorithm. This is a significant consideration in the case of real-world networks with hundred of nodes and thousands of links.[7]

[7]Typical "real-world" networks have a ratio of links to nodes of about 4.

However, using the mathematical programming approach to represent the behavior of the R.T. provides a useful alternative, particularly in connection with model estimation at the aggregate level, as will be seen in section 7.5 of Chapter 7.

3.3 CAR ROUTE CHOICE: PROBABILISTIC CASE

3.3.1 Individual Route Demands

In this section, we now address the general case of car route choice, that is, when the random error terms in the utility specification are now present.

$$\tilde{V}_{ijr} = \tilde{U}_{ijr} + \epsilon_{ijr} = b_{ij} - c_{ijr} - \tau t_{ijr} + \epsilon_{ijr} = b_{ij} - g_{ijr} + \epsilon_{ijr}; \qquad \forall\ i, j, r \tag{3.14}$$

This formulation generalizes that in Formulas (3.1), through the addition of the random term ϵ_{ijr} and, as we shall soon see, will lead to a resolution of the technical problems discussed earlier about uniqueness, continuity, and differentiability of demands.

Consistent with the random utility framework developed in the last chapter, the random terms ϵ_{ijr}'s are assumed to be i.i.d. Gumbel variables. Consequently, the probability that an individual traveler traveling by car between i and j and selected at random is observed to choose route r is, according to Formula (2.12a), equal to

$$P_{r/ij} = \frac{e^{\beta_r \tilde{U}_{ijr}}}{\sum_r e^{\beta_r \tilde{U}_{ijr}}} = \frac{e^{-\beta_r(\tau t_{ijr} + c_{ijr})}}{\sum_r e^{-\beta_r(\tau t_{ijr} + c_{ijr})}}; \qquad \forall\ i, j, r \tag{3.15a}$$

The value of parameter β_r, which must be calibrated together with that of τ, is related to the value of σ_r, the standard deviation of the ϵ_{ijr}'s, as represented by Formula (A.59b).[8] The larger the value of β_r, the less random and imprecise the analyst's error on an individual traveler's route utility.

FACT 3.2

The aggregate route demands in the probabilistic route choice case are equal to

$$T_{ijr} = T_{ij} P_{r/ij} = T_{ij} \frac{e^{-\beta_r(\tau t_{ijr} + c_{ijr})}}{\sum_r e^{-\beta_r(\tau t_{ijr} + c_{ijr})}}; \qquad \forall\ i, j, r \tag{3.15b}$$

[8]Calibration procedures are described in Chapter 7.

The T_{ij}'s are given from observed or estimated data. In the general case, when the value of β_r is not infinite, as assumed in the previous section, all routes between a given origin and a given destination will be utilized, irrespective of their cost, with a probability given by the logit model (3.15b). It may be verified that, in contrast with the deterministic case, the route demand functions specified in Formulas (3.15b) now possess all of the required properties of demand functions, including uniqueness, continuity, and differentiability.[9]

In a prescriptive, rather than descriptive, alternative interpretation, Formula (3.15b) may be considered the optimal route assignment, in terms of individual as well as total generalized travel cost, when the modeler has uncertain information about these costs. As a result, to "hedge his bets," as it were, the best the analyst can do, under the present assumptions, is to assign travelers to all routes, in proportion to their cost measured in exponential scale, that is, the numerator of Formula (3.15b). Each individual traveler is thus assigned to a route which is not necessarily, and in fact unlikely to be, the actual MCR, and may thus incur more than the minimum cost during one given single trip.

In any case, given the definition of individual indirect utility in Formula (3.14), the expected received utility for each of the T_{ij} travelers from i to j may be obtained from Formula (2.22), and is equal to

$$\begin{aligned}\tilde{W}_{m/ij} &= E_{\epsilon_{ij}}\{\tilde{U}_{ijr} + \epsilon_{ij}\} = b_{ij} + \frac{1}{\beta_r} \ln \sum_r e^{\beta_r \tilde{U}_{ijr}} \\ &= b_{ij} + \frac{1}{\beta_r} \ln \sum_r e^{-\beta_r(\tau t_{ijr} + c_{ijr})}; \qquad \forall\ i, j \qquad (3.16a)\end{aligned}$$

This implies that the total utility received by the R.T. in this probabilistic route choice case is equal to

$$\tilde{W}_R = B + \frac{1}{\beta_r} \sum_i \sum_j T_{ij} \ln \sum_r e^{-\beta_r(\tau t_{ijr} + c_{ijr})} \qquad (3.16b)$$

where B is the total travel budget for all travelers.

$$B = \sum_{ij} T_{ij} b_{ij}$$

Again, it may be verified that, in difference with the deterministic case, this expression possesses all of the required properties of indirect utility functions. (See section 4.3 of Appendix A.)

When measured net of the budget, this quantity measures the "traveler surplus."

[9]Uniqueness will be apparent from the R.T.'s U.M. problem, examined in section 3.3.4.

PROPOSITION 3.4

The traveler surplus in this probabilistic route choice case is equal to

$$TS_R = \frac{1}{\beta_r} \sum_i \sum_j T_{ij} \ln \sum_r e^{-\beta_r g_{ijr}} \tag{3.17}$$

As mentioned earlier, the traveler surplus may be used as a rigorous indicator of the impacts of changes in transportation systems or their operation on community welfare. Its definition, which in the present case is limited to route demands, given origin-destination demands, will be extended to include all travel demands in Chapter 4. These respective expressions may be compared to their deterministic counterparts, Formulas (3.11)–(3.13).

In either case, the impacts of various urban transportation policies and "travel demand management" actions may be evaluated from the corresponding change in traveler surplus. Actions which raise its value are desirable, and that which raises it the most may be considered the "best." For instance, the design of transportation networks, in terms of link capacities and/or user charges, might be based on raising the traveler surplus.

Finally, two related results may be shown. (See Exercise 3.15.) First:

$$\frac{\partial \tilde{W}_{m/ij}}{\partial c_{ijr}} = -P_{r/ij}; \qquad \forall\, i, j, r \tag{3.18a}$$

and second (See Exercise 3.18):

THEOREM 3.1

The aggregate route demands in the stochastic route choice case may be retrieved from the R.T.'s received utility as

$$T^*_{ijr} = -\frac{\partial \tilde{W}_R}{\partial c_{ijr}} \Big/ \frac{\partial \tilde{W}_R}{\partial B}; \qquad \forall\, i, j, r \tag{3.18b}$$

Both results follow directly from Roy's identity, Formula (A.68) in Appendix A.

3.3.2 Illustrative Example: Stochastic Car Network Assignment

To illustrate the previous exposition, in the case of the example network above, the routes between nodes 1 and 4 have been enumerated, and are routes number 5, 6, and 7. Their travel times, as represented in Table 1.8 in Chapter 1, are

15, 13, and 13, respectively. Since there are no link charges, according to Formula (3.15a), the probabilities of choice of these routes by an individual traveler going from 1 to 4 are then

$$P_{r/ij} = \frac{e^{-\beta_r(\tau t_{ijr})}}{\sum_r e^{-\beta_r(\tau t_{ijr})}}; \qquad \forall\ i, j, r$$

Assuming for the purposes of illustration that the calibrated value of $\tau\beta_4 = 0.1$, these probabilities are numerically equal to

$$P_{1/14} = \frac{e^{0.1(-15)}}{e^{0.1(-15)} + e^{0.1(-13)} + e^{0.1(-13)}} = \frac{0.22}{0.76} = 0.28$$

$$P_{2/14} = P_{3/14} = 0.36$$

The 150 travelers from 1 to 4 considered above would then be distributed to the three routes between these points as in Figure 3.8.

Other origin-destination volumes would be assigned in a similar manner by applying Formulas (3.15b), and the corresponding link volumes cumulated with the volumes above. For instance, with the same origin-destination volumes as in the deterministic case, $T_{12} = 250$, $T_{14} = 150$, the volumes on routes 1, 2, and 3 between nodes 1 and 2 are 182, 38 and 30, respectively. The link loadings would then be as represented in Figure 3.9.

It is instructive to compare the stochastic network assignment above with the volumes obtained in the deterministic case of the previous example. First, it is apparent that more links are loaded in the former than in the latter case, since given origin-destination volumes are always distributed across all available routes between these points. As mentioned in Chapter 1, this is a desirable feature, in terms of realism. Another difference, in this illustrative example, is that costs on all used routes are *not* equal to the minimum route cost, as in the deterministic assignment above. Instead, probabilistic assignment of individual travelers implies that no one may reduce further his or her *random* route

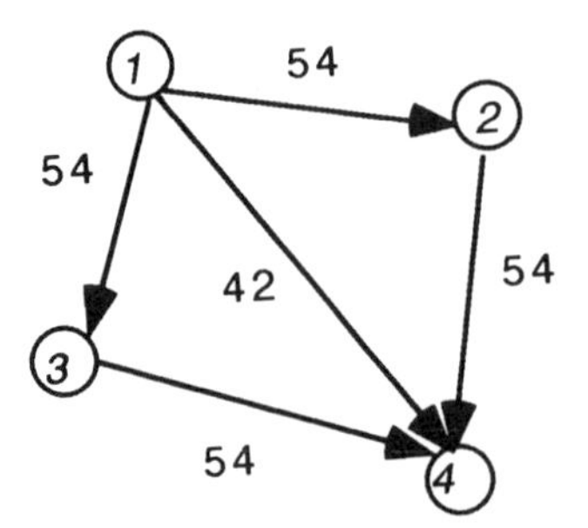

Figure 3.8 Stochastic route choice.

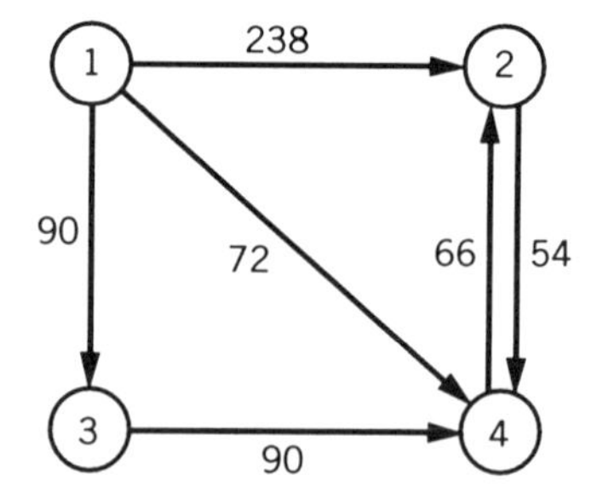

Figure 3.9 Stochastic route choice.

utility, as either perceived by, or attached to, him or her by the analyst, by unilaterally changing routes.

In the limit case when β_r is equal to 0, the variance of the random route utilities ϵ_{ijr} is infinite, which implies that the modeler has no information whatsoever about travelers' route utilities, and therefore, in effect, cannot distinguish between routes. On the other hand, as discussed in Chapter 2, the individual traveler knows precisely each route's utility. As a result, the analyst/model predicts that all routes are equally likely to be selected. This prediction of route choice may be considered the "worst" possible, again from the modeler's viewpoint, in the sense that it is "blind," as it cannot utilize the information represented by the knowledge of the systematic utilities $\tilde{U}_{ijmr}$.

In that case, route choice probabilities are given by

$$P_{r/ij} = 1/R_{ij}; \qquad \forall\ i, j, r$$

where R_{ij} is the number of routes from i to j. In the present example, if the value of β_r is equal to 0, each of the three available routes from node 1 to node 4 will be chosen with equal, 1/3, probability. The corresponding link volumes will then be as illustrated in Figure 3.10.

This is the most reasonable, least committal, prediction which could be made in the total absence of information. It also represents the "most stochastic" solution, at the other extreme from the deterministic solution in Figure 3.4.

3.3.3 The STOCH Algorithm

However, the hypothetical example above was predicated on the knowledge of all the available routes between nodes 1 and 4. In the case of "real-world" networks, it may be difficult, if not impossible, to enumerate all available routes between a given origin and destination. In any case, even if it were possible, their number might be very large, so that many of these routes would typically have a very low, negligible probability of being selected.

A possible pragmatic approach to alleviating this problem is to restrict the routes in the travelers' "choice set" to a few major routes, the few routes between any two points which a traveler would be most likely to know and/or

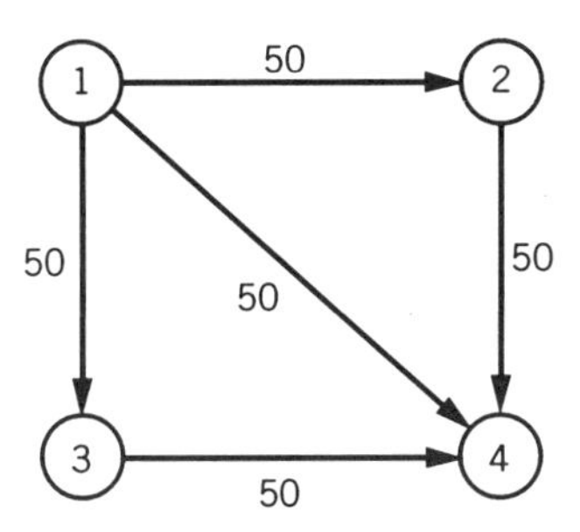

Figure 3.10 Equiprobable route assignment.

consider, since it is improbable that a traveler would be aware of *all* the possible routes between two given locations. Selecting these routes would obviously require some familiarity with the area.

Another possible approach is to formulate traveler choices at the *link* level, rather than at the *route* level, as above. To that effect, the utility of a single trip on given route (i, j, r) may alternatively be formulated in terms of link utilities, using Formula (1.2a):

$$\tilde{V}_{r/ij} = b_{ij} + \sum_a \tilde{V}_a \delta^a_{ijr}; \qquad \forall\, i, j, r \tag{3.19a}$$

with

$$\tilde{V}_a = -(c_a + \tau t_a) + \epsilon_a = \tilde{U}_a + \epsilon_a; \qquad \forall\, a \tag{3.19b}$$

If in addition we assume that the random variables

$$e_{ijr} = \sum_a \epsilon_a \delta_{ijr}; \qquad \forall\, i, j, r \tag{3.19c}$$

are i.i.d. Gumbel, the two formulations of route utility are equivalent.

Consequently, it should be possible to obtain the logit model of route choice from the probabilities of link choices. This is indeed the case, as is now shown. Given the assumptions above, the probability that an individual traveler at a given node chooses one of the links originating from the given node is given by a logit function

$$P_a = \frac{e^{-\beta(\tau t_a + c_a)}}{\sum_{a \in m} e^{-\beta(\tau t_a + c_a)}} \equiv K_m e^{-\beta g_a}; \qquad \forall\, a \tag{3.20a}$$

with

$$g_a = c_a + \tau t_a; \qquad \forall\, a \tag{3.20b}$$

Since we assumed that the random terms ϵ_a are independent, the probability that the particular sequence of links constituting route (i, j, r) is chosen is equal to the products of the respective probabilities that each of the individual links composing route r are taken, and may be written as

$$\begin{aligned} P_{r/ij} &= K_{ij} \prod_a e^{\beta(-\tau t_a - c_a)\delta^a_{ijr}} \\ &= K_{ij} \exp\left(\beta \sum_a (-\tau t_a - c_a)\delta^a_{ijr}\right) = K_{ij} e^{\beta(-\tau t_{ijr} - c_{ijr})} \end{aligned}$$

where K_{ij} is a proportionality constant which does not depend on r, since it is the product of K_m's which are node-related coefficients. Its value may be determined from the condition that the sum of the route probabilities for a given (i, j) combination sum up to 1.0:

$$\sum_r P_{r/ij} = 1; \qquad \forall\, i, j$$

This leads to

$$K_{ij} = \frac{1}{\sum_r e^{\beta(-\tau t_{ijr} - c_{ijr})}}; \qquad \forall\, i, j$$

and consequently

$$P_{r/ij} = \frac{e^{\beta(-\tau t_{ijr} - c_{ijr})}}{\sum_r e^{\beta(-\tau t_{ijr} - c_{ijr})}}; \qquad \forall\, i, j \tag{3.21}$$

which is the same as Formula (3.15b) as we expected. More generally, the following result holds (see Exercise 3.12).

FACT 3.3

Individual link utilities formulated as

$$\tilde{V}_a = -(c_a + \tau t_a) + f_{kl} + \epsilon_a; \qquad \forall\, a \tag{3.22a}$$

where f_{kl} is any specific function of the link's origin and destination nodes k and l, and the random variables

$$e_{ijr} = \sum_a \epsilon_a \delta^a_{ijr}; \qquad \forall\, i, j, r \tag{3.22b}$$

are i.i.d. Gumbel, produce logit route choice probabilities.

This property was first shown and utilized by Dial (1971) to reduce the number of potential routes in the choice set without changing the logit model for their probabilities, as follows.

Specifically, a link's utility in connection with a trip from a given node i to a given node j may, for instance, be specified as

$$\tilde{V}_a = \begin{cases} g_{il} - g_{ik} - g_a + \epsilon_a; & \text{if } g_{il} > g_{ik} \\ -\infty & \text{otherwise} \end{cases}; \qquad \forall\, a \in k^- \text{ and } l^+ \tag{3.23}$$

where g_{il} is the minimum generalized cost from i to link a's beginning node k, and similarly for l. The difference between the first two terms may be interpreted as the traveler's gain in using the link. The third term is the link's generalized cost.

Thus, this particular utility specification in effect insures that only "efficient" links are used between a given origin i and a given destination j, links (kl) such that their end node l is further away, in terms of generalized cost (travel time and cost) g_{il} from the origin node i, than their beginning node l. That is, links with a negative infinite utility have a zero probability of being used, thus reducing the number of potential routes which may be composed from such links. Such a specification then only requires, in addition to the knowledge of the link utilities, the computation of the minimum costs from the origin i to all nodes.

More generally, other restrictions are similarly possible, as long as they involve only node specification. For instance, the above condition may be made somewhat stricter by requiring that, in addition, the end node l should be closer from the destination than the origin node k. This, however, requires the computation of the minimum costs from all nodes to the destination j.

In any case, network loading reflecting the logit route choice model, Formula (3.15a) may be effected at the link level, using the following algorithm, STOCH, due to Dial (1971). The iterative step for the algorithm consists of a "forward" pass, during which the probability of individual routes from i to all destinations j are estimated, and a "backward" pass, when the corresponding origin-destination volumes are accordingly allocated to these routes. The algorithm is presented in the following, again for a given origin-destination pair. It must be applied separately to all trip origins i on the network. Also, link utilities are based on Formula (3.19b).

ALGORITHM 3.2 PROBABILISTIC MINIMUM COST ROUTE (STOCH)

Initialization

1. Estimate the minimum costs g_{ik} from the given origin node i to all other nodes k, using the MCR algorithm described above, with node i as the "root," and the generalized link costs

$$g_a = c_a + \tau t_a; \qquad \forall\, a \tag{3.20b}$$

2. For each link a with beginning node k and ending node l, estimate h_a as

$$h_a = \begin{cases} e^{\beta(g_{il} - g_{ik} - g_a)}; & \text{if } g_{il} > g_{ik} \\ 0 & \text{otherwise} \end{cases}; \qquad \forall\, a \in k^- \text{ and } l^+ \tag{3.24a}$$

Forward Iterative Step

Starting with the origin i, and in increasing order of cost from the origin g_{ik}, for each link a with beginning node k and ending node l estimate p_a as

$$p_a = \begin{cases} h_a; & \text{if } a \in i^- \\ h_a \sum_{b \in k^+} p_b; & \text{otherwise} \end{cases} \qquad \forall\, a \in k^- \text{ and } l^+ \tag{3.24b}$$

Stop when destination j is reached.

Backward Iterative Step

Starting with the destination j, and in decreasing order of cost g_{il} from the origin i, determine the volume of each link a between nodes k and l as

$$v_a = \left(T_{il} + \sum_{b \in l^-} v_b\right) \frac{p_a}{\sum_{b \in l^+} p_b}; \quad a \in k^- \text{ and } l^+; \qquad \forall\, a \tag{3.25}$$

where T_{il} is the given flow from origin i to node l.
Stop when origin i is reached.

3.3.4 Illustrative Example: Application of the STOCH Algorithm

The application of this algorithm is now illustrated, to assign the 150 travelers from node 1 to node 4, on the same network as used above. The first stage of the algorithm, initialization, consists of estimating the quantities h_a in Formula (3.24a), using the information in Tables 1.3 and 1.8. We have:

$$\begin{aligned} h(1) &= e^{0.1(5-0-5)} = 1 \\ h(2) &= e^{0.1(13-0-15)} = 0.82 \\ h(3) &= e^{0.1(6-0-6)} = 1 \\ h(4) &= e^{0.1(13-5-8)} = 1 \\ h(5) &= e^{0.1(13-6-7)} = 1 \end{aligned}$$

All other h_a are equal to zero, as links other than those above do not meet the condition that $g_{il} > g_{ik}$. It may be noted that in the present case, all links in the ''forward'' direction, from node 1 to node 4, are retained. Only those from 4 to 1 are dismissed.

The next step consists of performing the forward pass, and estimating the quantities p_a in Formula (3.24b). We have

$$p(1) = 1$$
$$p(2) = 0.82$$
$$p(3) = 1$$
$$p(4) = 1(1) = 1$$
$$p(5) = 1(1) = 1$$

Finally, we perform the backward pass, and estimate the link volumes, according to Formula (3.25). We have:

$$v_5 = 150 \frac{1}{1 + 1 + 0.82} = 53$$

$$v_4 = 150 \frac{1}{1 + 1 + 0.82} = 53$$

$$v_3 = 53 \frac{1}{1} = 53$$

$$v_2 = 150 \frac{0.82}{1 + 1 + 0.82} = 44$$

$$v_1 = 53 \frac{1}{1} = 53$$

As in the deterministic case, other origin-destination volumes, T_{12}, would be loaded separately onto the network using the same procedure. (See Exercise 3.13.)

This assignment may be compared to that obtained with the ''exact'' procedure, as represented in Figure 3.6. Because in the present case no links are eliminated on the basis of the condition in Formula (3.23), the two assignments are equal, within rounding errors. In general, if no condition is imposed on individual links to be used, the STOCH algorithm produces the stochastic route assignments represented by Formula (3.15b). Consequently, the STOCH algorithm can be used to load a network according to logit probabilities of route choice, without enumerating routes. It may be remembered in this connection that the counterpart MCR algorithm in the deterministic case did not require route enumeration either. These algorithms thus both resolve a critical issue, namely the fact that route choice from the viewpoint of the modeler's network information is link based, while from that of the individual traveler it is route based, since only route costs are relevant. Both algorithms manage to identify the best route(s) in deterministic and probabilistic terms, respectively, by considering one link at a time, in an efficient fashion.

3.3.5 Aggregate Demands: The R.T.'s Utility Maximization Problem

In this section, we pursue the R.T.'s approach to the determination of aggregate demands which we have already used in the previous deterministic case. In the stochastic case, the following result applies.

PROPOSITION 3.4

The R.T.'s direct utility in the probabilistic route choice case is equal to

$$U_R = -\tau \sum_i \sum_j \sum_r T_{ijr} t_{ijr} - \frac{1}{\beta_r} \sum_i \sum_j \sum_r T_{ijr} \ln T_{ijr} + T_0 \quad (3.26a)$$

Consequently, the maximization of this function with respect to the T_{ijr}'s and T_0, subject to the same Constraints (3.9a)–(3.9c) as in the deterministic case, produces the aggregate route demands (3.15b). This is demonstrated in Appendix 5.1 of Chapter 5 for the more general case of variable travel times and will therefore not be shown here. Again, variable T_0 and the budgetary constraint may be deleted from the problem's formulation, through substitution, as above.

The fact that the solution of the UM Problem with objective function (3.26) conforms to Formula (3.15b) may be proven by using the same approach as in section 2.8, using the Lagrangean of the problem. Another, equivalent approach is to use the Karush-Kuhn-Tucker conditions.[10]

In any case, this proof is presented in Appendix 1 of Chapter 5, in the more general case when the travel times are variable, due to the presence of congestion, and so will not be presented here. It is also easy to show that the R.T.'s utility function is concave (see Exercise 3.19), and that since the constraints are linear, there is a unique solution to the problem. (See section 1.7 of Appendix A.) Thus, the route demands in the stochastic case are unique.

In addition, it may be easily verified (see Exercise 3.16) that $\tilde{W}_R$, the R.T.'s received utility which was estimated in Formula (3.16b), is indeed equal to the optimal value of the R.T.'s direct utility:

$$\tilde{W}_R = U_R^* \quad (3.26b)$$

The traveler surplus is, as in the deterministic case, equal to the value of the welfare function $\tilde{W}_R$, net of the budget. For instance, in the illustrative example of section 3.3 above, the traveler surplus was equal to −22,155.

If, as in the deterministic case, a user charge of $1 is imposed on link 1,

[10]Both methods are reviewed in section 2.2 of Appendix A.

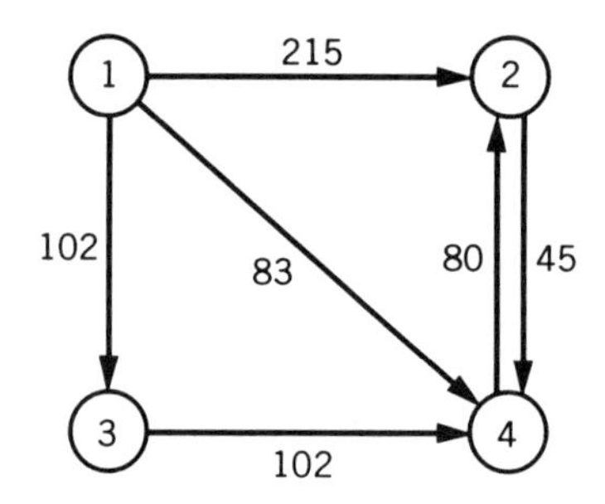

Figure 3.11 Stochastic route choice with user charges.

and $\tau = 1$, (corresponding to the same value of $\beta\tau = 0.1$ as in the example), the new equilibrium assignments are as represented on Figure 3.11.

The corresponding travel surplus is equal to -22720, translating a net decrease in traveler welfare of \$565. Note that this is *not* equal to traveler's monetary expense, which is equal to $2.5(236 - 215) = 52.5$. This fact, as well as the small impact of the user charge, are both attributable to the uncertainty in route utilities, as reflected in the value of parameter β, relative to that of τ.

The fact that Formula (3.18b),[11] which characterizes any demands derived from utility-maximizing behavior (see Formula A.6 in Appendix A), also holds in the present case justifies the concept of the R.T. That is, aggregate route demands are produced by the R.T.'s utility maximizing choices, in the same manner that the individual route demands are produced by the individual travelers' choices. The R.T.'s direct utility function (3.26a) does therefore represent a legitimate utility function.[12] This fact will be further exploited in section 3.4.

In conclusion of this section, the caveats about the use of the logit model which were mentioned at the conclusion of Chapter 2 should perhaps be reiterated. In the case of route choice, it is unlikely that all individual routes between a given origin and destination will have completely independent random utilities, as is assumed. This is because individual links are in general common to several routes, thus establishing a relationship between the routes' utilities. As pointed out earlier, other random utility models, including the probit model (Daganzo, 1979), are available which do not require the assumption of independence between individual choice utilities. However, the resulting travel demands may not be expressed in closed, analytical form, a theoretical disadvantage, and consequently must be obtained through numerical (computer) simulation, a practical one. In particular, this would preclude the identification of the R.T.'s direct and indirect utility functions, and consequently the application of the R.T. concept, which is so powerful, both conceptually and practically.

[11]This equality is known in microeconomics as Roy's identity. See, for instance, Varian (1992).
[12]This has sometimes been referred to as a "social preference function."

3.4 TRANSIT ROUTE CHOICE

In the case of transit networks, the route choice problem takes on a rather different form. The fundamental difference is that a transit traveler must in general wait for a vehicle, or train, to arrive, before boarding it, whereas the private car is always instantly available. In terms of network links, car links are served instantaneously, whereas transit links are served only periodically. The corresponding waiting time for transit (which is zero for cars), will then in general be random, if we assume that both the arrival time of individual passengers, and that of the vehicle, are random.[13] Transfers between different transit lines will also result in a random waiting time. In addition, travelers must also go to and from transit stations, before and after transit travel, resulting in access/egress time.

These various features of a transit trip in turn imply a random travel time, and consequently, a random route utility, *as assessed by the individual traveler*. This is a fundamental difference from the deterministic car case, in which the traveler, and consequently the modeler, knew the utility with certainty, but also from the probabilistic car case, where the traveler, but not the modeler, knew the utility with certainty.

This fundamental feature of transit networks has been addressed in conceptually different ways in various models of transit route choice. Many of these models are based on transforming the transit network, in one fashion or another, into a car network, so that the deterministic network assignment can be applied. This section is based on the work of (Spiess, 1984; Spiess and Florian, 1989), who first developed an approach which explicitly deals with this inherent feature of transit networks.

The first concept specific to transit networks is that of the "augmented" network, corresponding to the addition of "walk" links used for ingress, egress, and transfers between transit lines. Each of the transit lines available between two given consecutive stops i and j is represented by an *individual* link a. That is, two different lines between the same nodes will be represented by two different links. In this connection, it is convenient to think of a vehicle arrival at a given node as the corresponding link "being served" or becoming available. The given link travel times are t_a, and for generality, the given link costs are c_a.

The rate of arrival (or, if they are random, the expected or average headways) of transit vehicles for each of the lines serving a given node are assumed given, and constant over time for each line i, and equal to $1/f_i$, where f_i is the frequency of service of line i, in number of vehicle arrivals per unit time, usually a minute. Walk links are only served by the "walk line," which has

[13]We assume that travelers do not, or cannot, exactly time their arrival at any of the transit stops, including en route, to that of a vehicle, and/or similarly, that vehicles cannot exactly follow the fixed, planned schedule.

an infinite frequency of service. The rate of arrival of passengers at each of the nodes i is assumed known, and constant over time.

Origin-destination travel demands T_{ij} per unit time between given origins and destinations are given, for $i \neq j$. In addition, it will be convenient analytically to use a fictitious origin-destination demand from each node to itself:

$$T_{jj} = -\sum_{i \neq j} T_{ij}; \qquad \forall j \tag{3.27}$$

This will allow us to specify the same conservation of flow at each node, regardless of whether it is an origin, destination, or transfer node.[14]

Having described the transit network, we now consider the behavior of the individual traveler. Specifically, each individual traveler does not know the actual waiting time for vehicles at any of the transit stops, including the origin of their trip, since we have assumed that his or her arrival at a given transit stop is random, as is that of vehicles. That is, depending on the actual arrival times of both the passenger and the desired vehicle(s), the optimal choice of route between a given origin and a given destination may change.

Therefore, on the *average* (i.e., in the long run), there may be several "optimal" routes leading from a given node to the destination, each being optimal a certain proportion of the time, with a certain probability at any given arrival time of an individual traveler. This is in contrast with the car case where, since all links are always available, under stable conditions there can only be one optimal route for a given passenger at any time. The probability of such a route being optimal at any time, in other words, is 100%.

Under these circumstances, we assume that travelers' transit route choices are made so as to maximize the utility of their choices, when estimated on the basis of expected line-haul travel time and waiting time, rather on the basis of their probability distribution functions, which are difficult to estimate, as they must reflect the arrival patterns of passengers and vehicles at the various network nodes. The utilities facing the individual traveler are then defined as expected utilities, i.e., on the basis of average conditions. For a single trip on a given route r between i and j this is specified as

$$\tilde{V}_{ijr} = b_{ij} - \tau t_{ijr} - \overline{\omega} w_{ijr} - c_{ijr}; \qquad \forall i, j, r \tag{3.28}$$

where t_{ijr} is the route's expected line-haul travel time, w_{ijr} is the expected total waiting time on the route and c_{ijr} the monetary cost.[15]

In the same fashion as for the car case above, this route utility may be equated to the sum of the utilities for the links which compose the route. The expected waiting time for a given link is the average time a passenger has to

[14]Note that this implies that the network is "connected"; any node is reachable from any other node.

[15]Index t for transit, which is implicit, has been omitted, in the same fashion as was done for the car, above.

wait for the link to become available or be served. It will be convenient to attach this waiting time to the link's beginning node.

3.4.1 Individual Demands: The Minimum Expected Cost Route Algorithm (MECR)

The route choice (network assignment) problem for an individual traveler going from i to j consists of deciding which links to use as to incur, on the average, the minimum total travel cost, including waiting. The "state-of-nature" faced by individual travelers is now probabilistic. This problem may be thought of as a generalization of the MCR problem above, but where a random waiting time is now attached to any node, which was equal to zero in the car case. On the average, as seen above, there may be several optimal links at a given node. Which links are actually used in a given trip depends on the respective arrival times of the traveler and the transit vehicle at the transit stops. In the car case, only one link could be chosen.

In this connection, we assume for simplification that travelers will use the *next* optimal vehicle arriving at a given stop, that is, use the first optimal link which becomes served. More complex, and/or realistic traveler behavior may be assumed, based on observations of vehicle arrival during the trip. However, the introduction of such "sampling" rules would significantly complicate the problem.

It should be emphasized that the randomness inherent in transit travel has nothing to do with the uncertainty of the modeler with respect to utilities as, for instance, in the previous probabilistic route choice case. In fact, we will assume, again for simplification, that there is no uncertainty on the part of the analyst and/or the traveler with respect to the average utility, which is thus deterministic, so that the random utility terms ϵ_{ijr} in Formula (3.28) are assumed to have a zero variance, and are therefore all equal to zero.[16]

In this section, we describe an algorithm which was developed by Spiess (1984) specifically for solving this problem, and which is based on the identification of expected nodal waiting times. Accordingly, the algorithm generalizes the MCR algorithm presented above.

The algorithm has two phases, like the STOCH algorithm, consisting respectively of finding the optimal routing through the transit network, and then assigning the origin-destination demands along them. It is based on the specific assumption that the arrival times of individual passengers and individual transit vehicles at the transit stops are uniformly distributed random variables over the time period of analysis. Consequently, the waiting time of an individual traveler at the transit stops is equal to one-half (1/2) the average headway between vehicles. Other assumptions about arrival patterns may also be made without changing the basic nature of the algorithm. (See Spiess and Florian, 1989.)

[16]Note that the problems discussed above in connection with the car case may again be expected.

ALGORITHM 3.3 DETERMINISTIC TRANSIT ROUTE ASSIGNMENT

For each destination j, do each of the phases below:

FORWARD PASS. Determination of the optimal routes.

1. Initialization.

Set g_{ij} equal to ∞ (in practice a very large number) for all nodes $k \neq j$.

Set $g_{jj} = 0$ for the destination. g_{kj} represents the current expected generalized cost (travel time plus waiting time plus link charges) to reach the destination j from node k.

Set "node frequencies" f_k equal to 0 for all nodes.

Define the set $\{S\}$ of unexamined links as the whole network A.

Define the set A_j^+ of selected (optimal) links with respect to the given destination j as empty.

2. Iterative step.

2a. If $\{S\}$ is empty stop. Otherwise, find the link (k, l) such that the expected travel time from its *end* node plus the expected generalized travel cost (*cost*, for short, in the remainder) on the link is the smallest of any link in $\{S\}$, i.e., such that

$$(g_a + g_{lj}) < (g_{a'} + g_{l'j}); \qquad \forall\, l'$$

Take that link out of $\{S\}$, i.e., set

$$\{S\} = \{S\} - a$$

Compare the current g_{kj} for the *origin* node of the selected link a, to $g_{lj} + g_a$.

If the latter quantity is smaller than the former, i.e., if

$$(g_a + g_{lj}) < g_{kj}$$

then using link a from node k improves the present cost from k to the destination; then include link a in A_j^+, and go to 2b. Otherwise, go back to Step 2a.

2b. Update the current cost from node k, to the weighted average of two quantities. The first is node k's current cost g_{kj} to destination node j. The second is the sum of the cost from l to k, plus the cost from k to destination j, using link a. The weights on these two quantities are equal to the node frequency f_i, and the link frequency f_a, respectively.

$$g_{kj} = [g_{kj} f_k + (g_a + g_{lj}) f_a]/(f_k + f_a) \qquad (3.29)$$

Update the current node's frequency as the sum of the previous frequency plus link a's frequency.[17]

$$f_k = f_k + f_a \tag{3.30}$$

Go back to 2a and iterate.

BACKWARD PASS. Assignment of the transit demand to the transit network.

1. Initialization.

Set all node volumes V_k at the given demands T_{kj}.

$$V_k = T_{kj}; \qquad \forall\, k \tag{3.31}$$

2. Iteration.

For each link, and in order of decreasing values of the link's cost plus the cost from the link's end node to the destination, i.e., $(g_a + g_{lj})$

If a is included in A_j^+ go to 2a. Otherwise go to 2b.

2a. Set the link's volume to a percentage of its origin node's volume equal to the ratio between the link's frequency and the origin node's frequency.

$$v_a = V_k f_a / f_k \tag{3.32a}$$

Update the link's end node's volume to the previous node volume plus the link's volume.

$$V_l = V_l + v_a \tag{3.32b}$$

2b. Set the link's volume equal to 0.

3. Cumulation.

Add all link volumes corresponding to each of the destinations j.

3.4.2 Illustrative Example: Transit Network Assignment

We will now illustrate the above procedure in the case of a prototype transit network represented in Figure 3.12. This represents a hypothetical transit network between the same nodes served by the prototype road network described above.

[17] At the start of the algorithm, when the node frequencies f_k are equal to zero, and the node labels u_{ij} are equal to zero, the product $(g_{kj} f_k)$ should be set to 1/2 under the above assumptions. More generally, a weight $\overline{\omega}$ may be given to the frequencies in the algorithm, to reflect other assumptions. See objective function Formula (3.34).

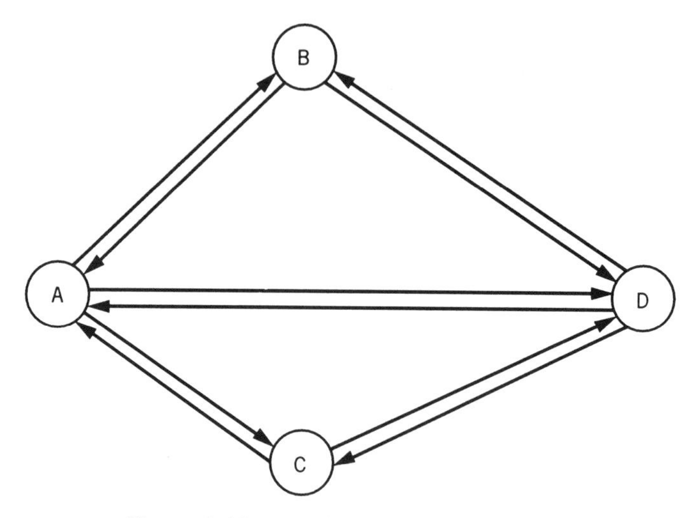

Figure 3.12 Transit Network representation.

The respective link travel times, in minutes, are presented in Tables 3.6a and 3.6b respectively.

The augmented network corresponding to this network is as represented in Figure 3.13.

The access links are represented as AA_x, BB_x, and so on. If we assume that the walking times from/to the stations (the squares) to/from the platforms (the circles) are all equal, their values will not affect the choice of lines. Consequently, there is no need to represent access/egress links. Accordingly, and to simplify computations without loss of generality, we may set all access/egress times to zero. Also, in this particular case, there are no transfer links.

With this new representation, the links costs and frequencies on the augmented network's links are as shown in Table 3.6c. The frequency on access links corresponds to the frequency on the line they serve. The infinite frequency on travel links (the original transit network links) represents the fact that there is no wait involved in using them once they have been accessed; the traveler is on board the vehicle.

TABLE 3.6a Link Travel Times

Link	*AB/BA*	*BD/DB*	*AC/CA*	*CD/DC*	*AD/DA*
Time	3	5	4	6	8

TABLE 3.6b Link Frequencies (/hr)

North line *ABDA*	South line *ACDA*	Middle line *ADA*
6	7.2	9

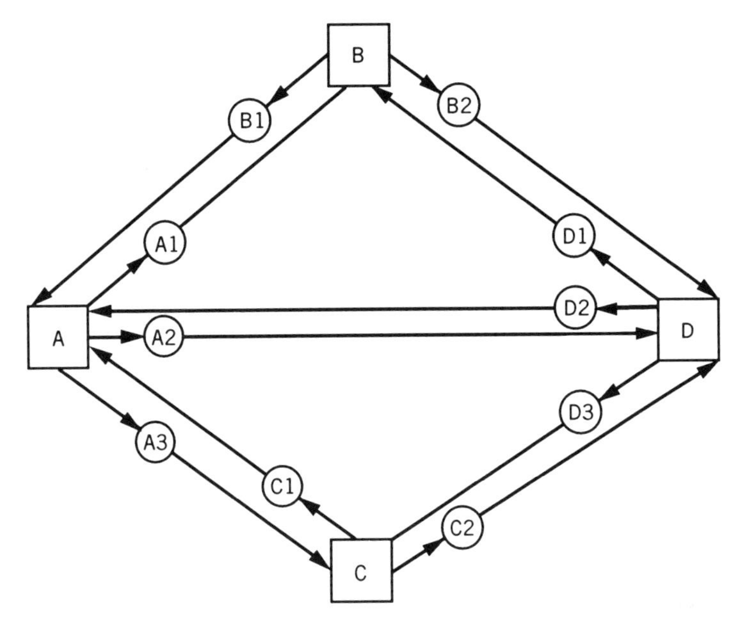

Figure 3.13 Augmented transit network.

TABLE 3.6c Link Data for Example Transit Network

Link	Travel Time	Frequency
AA1	0	0.1
AA2	0	0.15
AA3	0	0.12
A1B	3	∞
A2D	8	∞
A3C	4	∞
BB1	0	0.1
BB2	0	0.1
B1A	3	∞
B2D	5	∞
CC1	0	0.12
CC2	0	0.12
C1A	4	∞
C2D	6	∞
DD1	0	0.1
DD2	0	0.15
DD3	0	0.12
D1B	5	∞
D2A	8	∞
D3C	6	∞

We assume that there are only 200 travelers per hour going from A to D. Since in this case there are no travelers from B, C, or D, we may simplify the networks' representation by deleting unused links. The equivalent, simplified network, which is only valid for this assignment problem is shown in Figure 3.14, in which links' travel times and frequencies, respectively, are shown in that order along the links.

Generally, even though there is no systematic way to simplify a given augmented transit network, link elimination is beneficial, since the route choice algorithm requires the systematic examination of each and every link in the augmented network.[18]

In any event, the various iterations of the procedure are represented in Table 3.7. The first half of the table, on the left-hand side, is node-related, and presents node labels, in the form (g_i, f_i), as they are modified during the procedure. The second part of the table, on the right-hand side, is link-related, and identifies the link a with the current minimum value of $(g_a + g_{kj})$.

The detail of the computations for u_k and f_k is represented in Table 3.8.

At the fifth iteration, all links have been examined, and the procedure is therefore completed. The minimum expected total travel time from A to D is equal to the last value of g_A, 10 minutes. The average waiting time for the first optimal line to be boarded at node A is equal to $1/2(1/f_A) = 1/2(1/0.37) = 1.35$ minutes.

Since all links originating at A are in the optimal set A^+, the optimal plan for travel from A to D consists of taking the first available of either the North, Middle, or South lines, and consequently staying on it until D. The probabil-

[18]While this may not be as critical when the procedure is computerized, it will nevertheless save dynamic as well as static memory space. The same efficient network representations as were mentioned above in connection with road networks may also be used in the case of transit networks.

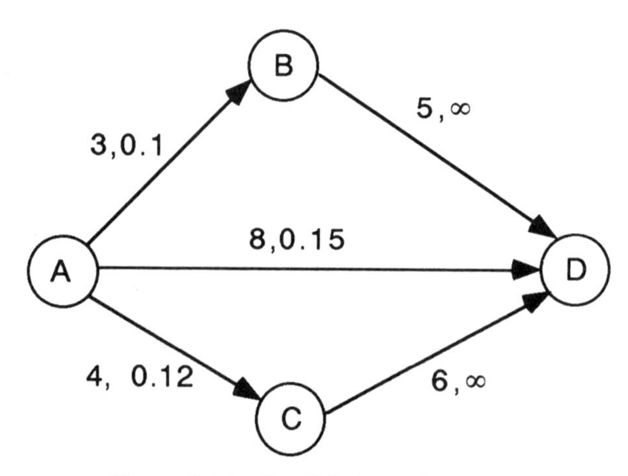

Figure 3.14 Simplified transit network.

TABLE 3.7 Transit Network Assignment Algorithm

Iteration	Nodes				Links			
#	A	B	C	D	(k, l)	f_a	$(g_l + g_a)$	a in A^+
0	∞, 0	∞, 0	∞, 0	0, 0	BD	∞	5	Yes
1	∞, 0	5, ∞	∞, 0	0, 0	DC	∞	6	Yes
2	∞, 0	5, ∞	6, ∞	0, 0	BA (or DA)	0.1	8	Yes
3	13, 0.1	5, ∞	6, ∞	0, 0	DA	0.15	8	Yes
4	10, 0.25	5, ∞	6, ∞	0, 0	CA	0.12	10	Yes
5	10, 0.37	5, ∞	6, ∞	0, 0				

ities of a vehicle from the respective lines arriving first at node A at any given individual traveler's arrival time at node A are proportional to the frequencies of service of the corresponding links

$$P_N/f_{AB} = P_M/f_{AD} = P_S/f_{AC} = (P_N + P_M + P_S)/f_A = 1/0.37 = 2.7$$

We then have,

$$P_N = 2.7f_{AB} = 2.7(0.1) = 0.27$$

$$P_M = 2.7f_{AD} = 2.7(0.15) = 0.41$$

$$P_S = 2.7f_{AC} = 2.7(0.12) = 0.32$$

Therefore, an individual traveler leaving node A will find that he or she will take the North line 27% of the time, the Middle line 41% of the time, and the South line 32% of the time, if, as assumed, he or she takes the first vehicle arriving from either of these lines. This is represented in 3.15.

As a check, the expected total travel time from A to D may be estimated alternatively as

$$1.35 + 0.27(3 + 5) + 0.41(8) + 0.32(4 + 6) \approx 10$$

TABLE 3.8 Transit Network Assignment Computations

Iteration #1: $g_B = (\infty \cdot 0 + 5 \cdot \infty)/(0 + \infty) = (0.5 + 5 \cdot \infty)/\infty = 5$

$f_B = 0 + \infty = \infty$

2: $g_C = (\infty \cdot 0 + 6 \cdot \infty)/(0 + \infty) = (0.5 + 6 \cdot \infty)/\infty = 6$

$f_C = 0 + \infty = \infty$

3: $g_A = (\infty \cdot 0 + 8(0.1))/(0 + 0.1) = 13$

$f_B = 0 + 0.1 = 0.1$

4: $g_A = (13(0.1) + 8(0.15))/(0.1 + 0.15) = 10$

$f_A = 0.1 + 0.15 = 0.25$

5: $g_A = (10(0.25) + 10(0.12))/(0.25 + 0.12) = 10$

$f_A = 0.25 + 0.12 = 0.37$

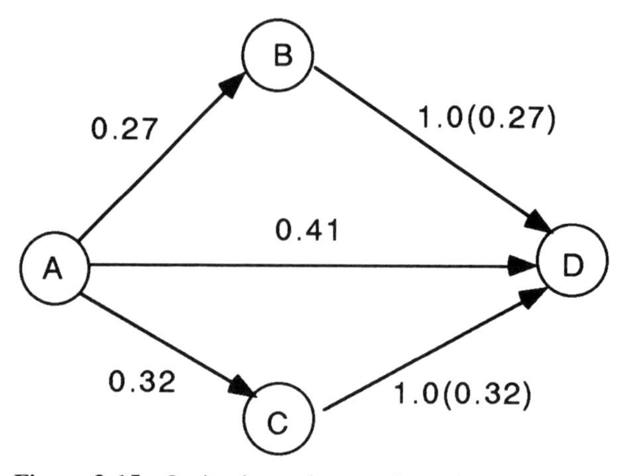

Figure 3.15 Optimal travel routes in probability terms.

In terms of the 200 travelers from A to D, the volumes on the network's links will be

$$v_{AB} = 0.27(200) = 200\, f_{AB}/f_A = 54/\text{hr.}$$

$$v_{AD} = 0.41(200) = 200\, f_{AD}/f_A = 81/\text{hr.}$$

$$v_{AC} = 0.33(200) = 200\, f_{AC}/f_A = 65/\text{hr.}$$

Consequently, the volume leaving B to go to D is equal to the sum of the incoming volumes to D, in this case

$$v_B = \sum_{a \in B+} v_a = 54/\text{hr.}$$

The volumes on the links leaving node B are similarly distributed, proportionally to the frequencies of the *optimal* links leaving B. In the present case, there is only one link leaving B, link BD, and it is necessarily optimal, as is confirmed in Table 3.7. Therefore,

$$v_{BD} = v_B(f_{BD}/f_B) = 54(\infty/\infty) = 54/\text{hr.}$$

In the same manner,

$$v_C = \sum_a v_a = 65/\text{hr.}$$

$$v_{CD} = v_C(f_{CD}/f_C) = 65(\infty/\infty) = 65/\text{hr.}$$

These link volumes are represented in Figure 3.16.

Other origin-destination volumes T_{ij} with a destination $j \neq D$ would be assigned to the network in a similar fashion, from an independent application

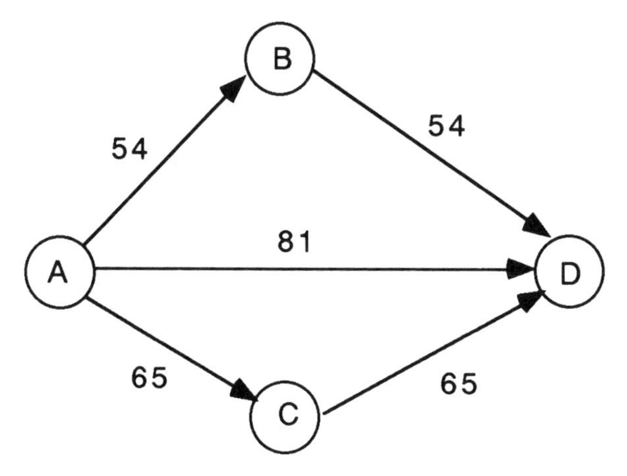

Figure 3.16 Transit link volumes.

of the algorithm. However, the assignment of volumes T_{iD} to the same destination from any other point would not require another application of the algorithm, since the optimal plans from all nodes to D, specifically from B and C to D, would have been identified, had the network not been simplified to include only those links used by travelers from A to D, as was done above.

3.4.3 Aggregate Demands: The R.T.'s Utility Maximization Problem

In this section, we develop the R.T.'s approach to the transit route demand problem, based on the mathematical program proposed by Spiess and Florian (1989).[19] In the present case, the R.T. is the traveler whose behavior averaged over time reproduces the expected aggregate route demands.

The problem formulation turns on the R.T.'s decisions of whether or not to use a given link *a to go to j*. The corresponding binary variable is then

$$x_{ij}^{a} = \begin{cases} 0 \text{ if, at node } i, \text{ link/line } a \text{ is not included in} \\ \quad \text{the optimal route to go to } j \\ 1 \text{ otherwise} \end{cases} ; \quad \forall\, i, j, a \quad (3.33)$$

In terms of these variables, the R.T.'s expected direct utility is equal to

$$\underset{(v_{aj}, x_{ij}^{a}, T_0)}{\text{Max}}\ U_R^t = -\left[\tau \sum_a t_a \sum_j v_{aj} + \overline{\omega} \sum_j \sum_i \frac{\sum_{a \in i^-} v_{aj}}{\sum_{a \in i^-} f_a x_{ij}^{a}} \right] + T_0 \quad (3.34)$$

[19]Other formulations of the transit network assignment problem are also presented and compared in De Cea and Fernandez (1989), and De Cea et al., (1989).

where the unknowns are the fractions v_{aj} of volumes v_a on a given link a going to destination j, the x_{ij}^a's defined in Formula (3.33), and as usual, the amount spent on other than transit travel, T_0.

The first term is the utility component corresponding to line haul time, in which parameter τ translates the importance of line-haul travel time with respect to waiting time. The second term is the total waiting time, in which the coefficient $\overline{\omega}$ reflects the nature of the p.d.f. of passengers and vehicle arrivals, as well as the importance of waiting time relative to travel time, and must be calibrated. (Its value was assumed to be equal to 1/2 in the above example.) The other symbols retain their previous meanings.

The constraints on the utility maximization translate as usual the budget limit and conservation of flow requirements in terms of *expected, or average* link volumes v_a, and are

$$\sum_a v_a c_a + T_0 = B \tag{3.35a}$$

$$v_{aj} = \frac{f_a x_{ij}^a}{\sum_{a \in i^-} f_a x_{ij}^a} \left[\sum_{a \in i^-} v_{aj} \right]; \qquad \forall\, i, j, \quad a \in i^- \tag{3.35b}$$

$$\sum_{a \in i^-} v_{aj} = \sum_{a \in i^+} v_{aj} + T_{ij}; \qquad \forall\, i, j \tag{3.35c}$$

$$v_a = \sum_j v_{aj}; \qquad \forall\, a, j \tag{3.35d}$$

$$x_{ij}^a = 0,\ 1; \qquad \forall\, i, j, a \tag{3.35e}$$

$$v_{aj} \geq 0; \qquad \forall\, a, j \tag{3.35f}$$

Constraints (3.35b) state that flows at a given node going *out* to a given destination are distributed to links leaving i proportionately to link frequencies f_a. This translates the assumption that the representative traveler always chooses the first available optimal link. Constraints (3.35c) state conservation of flow requirements at each node with respect to volumes to a given destination. Finally, Relationships (3.35d) define link volumes v_a as the sum of all destination-specific link volumes v_{aj}.

The flow variables are positive and continuous, but the x_{ij}^a's are binary (zero–one). Optimization problems in binary variables are typically harder to solve than problems in continuous ones. However, Problem (3.34)–(3.35) may be transformed into an equivalent linear program, by a change of variable from x_{ij}^a to the unknown expected waiting times at nodes w_{ij}:

$$w_{ij} = \frac{\sum_{a \in i^-} v_{aj}}{\sum_{a \in i^-} f_a x_{ij}^a}; \qquad \forall\, i, j \tag{3.36}$$

Consequently, Constraints (3.35b) are now written

$$v_{aj} = f_a x_{ij}^a w_{ij}; \qquad \forall\, i, j, \quad a \in i^- \tag{3.37a}$$

It may be shown (Spiess and Florian, 1989) that these constraints may be relaxed to a less restrictive form:

$$v_a^j \leq f_a w_{ij}; \qquad \forall\, i, j, \quad a \in i^- \tag{3.37b}$$

The transformed problem in the new variables, in which all variables are now continuous, is thus

$$\underset{(v_{aj}, w_{ij})}{\text{Max}}\; U_R^t = -\left[\tau \sum_a t_a \sum_j v_{aj} + \bar{\omega} \sum_i \sum_j w_{ij}\right] + T_0 \tag{3.38}$$

subject to

$$\sum_a v_a c_a + T_0 = B \tag{3.39a}$$

$$v_a^j \leq f_a w_{ij}; \qquad \forall\, i, j, \quad a \in i^- \tag{3.39b}$$

$$\sum_{a \in i^-} v_{aj} - \sum_{a \in i^+} v_{aj} = T_{ij}; \qquad \forall\, i, j \tag{3.39c}$$

$$v_{aj} \geq 0; \qquad \forall\, a, j \tag{3.39d}$$

This result may be summarized as follows.

PROPOSITION 3.5

The R.T.'s direct utility function in the deterministic transit route case is equal to

$$U_R^t = -\tau \sum_a t_a \sum_j v_{aj} - \bar{\omega} \sum_i \sum_j w_{ij} + T_0$$

The R.T.'s decisions are stated above in terms of *link* demands and waiting times. Using the relationships between link and route demands, the R.T.'s choices can alternatively be cast in terms of route demands. It may be noted that the constraints on the R.T.'s U.M. problem now include *feasibility* Constraints (3.39b), in addition to the previous budgetary and consistency constraints.

Because the R.T.'s utility function is linear in the demands, and thus not strictly concave, there may be multiple solutions to the above problem.[20] It is nevertheless possible to resolve this difficulty in the same manner, by introducing a random utility component in the specification of individual indirect utility Formula (3.28). (See Exercise 3.18.)

In any event, variable T_0 may again be eliminated, together with the budgetary constraint, resulting in a simpler and more familiar formulation. In this form, the problem is a standard linear programming problem, since now the objective function, as well as all the constraints, are linear, and in addition, all variables v_{aj} and w_{ij} are continuous. Problem (3.38)–(3.39) may thus be solved with standard linear programming codes. (See Exercise 3.9.) The same comments about the desirability of this approach as were made for the road networks apply, however.

It is useful to note that, as expected, when all line frequencies are infinite, and that consequently there is no wait at any of the nodes, the above problem formulation becomes the same as that of the MCR problem in the car case. (See Exercise 3.4.) In the same manner, it may be shown (Spiess and Florian, 1989) that the transit network assignment algorithm presented above is based on the optimality conditions for Problem (3.38)–(3.39).

Again, it is fairly obvious that

$$\tilde{W}_R^t = U_R^{t*} \tag{3.40}$$

Thus, we have

PROPOSITION 3.6

The traveler surplus from transit route choices is equal to

$$TS_R^t = -\sum_i \sum_j \sum_r T_{ijr}^*(\tau t_{ijr} + c_{ijr}) - \bar{\omega} \sum_i \sum_j w_{ij}^* \tag{3.41}$$

in which the T_{ijr}^*'s and the w_{ij}^*'s are the route demands and waiting times.

3.5 SUMMARY

In this chapter, we considered the basic problem of individual route choice, under uncongested conditions, for both car and transit passengers. This problem was solved at the individual level, in the deterministic case (when link

[20]A further theoretical difficulty in the present case is that the "utility function" is not directly cast in terms of the aggregate demands themselves, i.e., route flows, but rather, substitutes in the form of link volumes and waiting times. While, as we have seen in the car case, link volumes may easily be replaced by route volumes, using Formula (1.2b), the w_{ij}'s are independent of route volumes.

utilities are known with certainty), from application of "Minimum Cost Route" and its probabilistic version developed by Spiess. In the stochastic car case, the route choice problem was similarly solved from application of the "STOCH" algorithm. In all cases, the concept of "representative traveler" introduced in the previous chapter was applied to retrieve aggregate route demands providing network assignment as the solution to the representative traveler's utility maximization problem.

The stochastic transit case, both at the individual and R.T. (aggregate) levels, is left as a potential combination of these respective methods.

In the next chapter, the route choice problem will be combined, within the same framework of individual utility maximization, with the other traveler decisions, choice of destination, mode, and of whether to travel.

3.6 EXERCISES

3.1 For the network represented in Figure 3.17, where the numbers alongside the links represent the travel times, use the MCR algorithm to determine the minimum path from node 1 to node 7. All links are bidirectional.

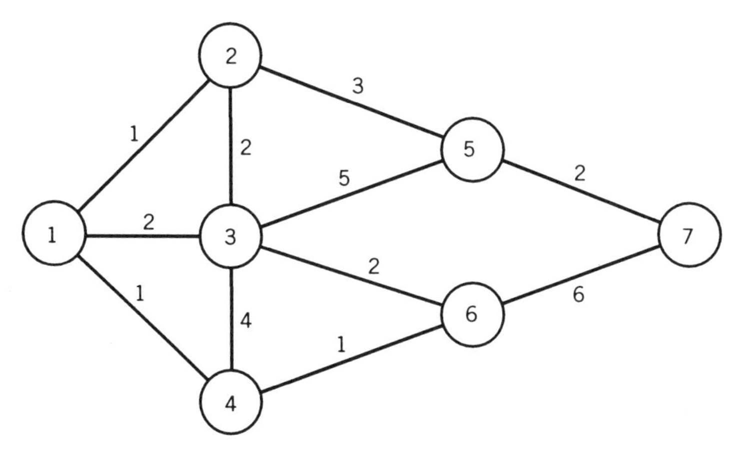

Figure 3.17 Network for Exercise 3.1.

3.2 Find the MCR from A to G for the network represented on Figure 3.18.

3.3 Formulate the MCR problem as a linear program of shipping a unit amount from a given origin to a given destination at the least possible cost, given the links' individual costs.

3.4 Show that when all link frequencies are infinite, i.e., when waiting times at all nodes are equal to zero, Problems (3.38)–(3.39) have the same formulation as that developed in Exercise 3.3.

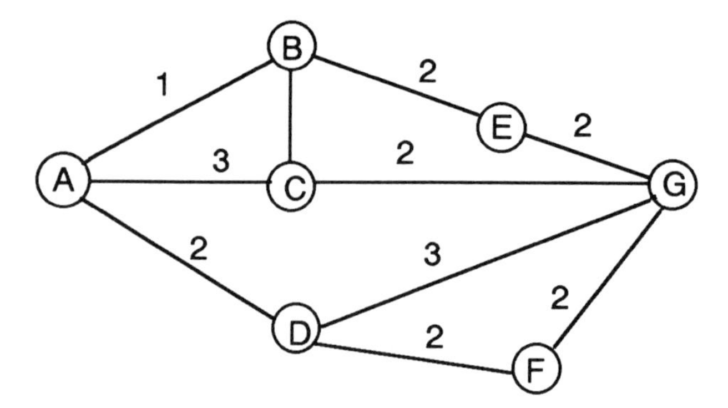

Figure 3.18 Network for Exercise 3.2.

3.5 For the car network represented in Figure 3.19 with link travel times as represented along the links, assign the demands $T_{16} = 100$, $T_{35} = 150$.

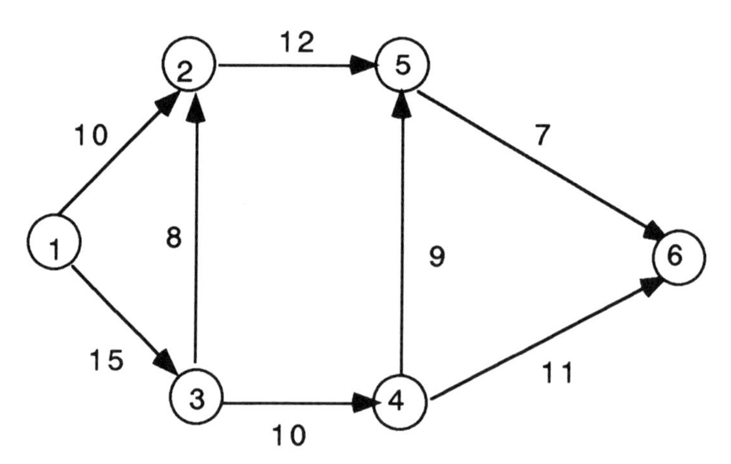

Figure 3.19 Network for Exercise 3.5.

3.6 For the transit network represented in Figure 3.20, with deterministic link travel times and frequencies (in vehicle/min) as represented along the links, in that order, assign the demands $T_{16} = 100$, $T_{35} = 150$. All access links have a two-minute walking time.

3.7 A transit network is given in Figure 3.21 (Spiess and Florian, 1989).
Link travel times, and the frequencies of the corresponding lines (in veh/min) are represented along them. Assume that the value of parameter $\overline{\omega}$ is equal to 1/2. The number of travelers from node A to node B is equal to 1000/hr.

a. Draw an appropriate network for the corresponding assignment problem, as simplified as possible.

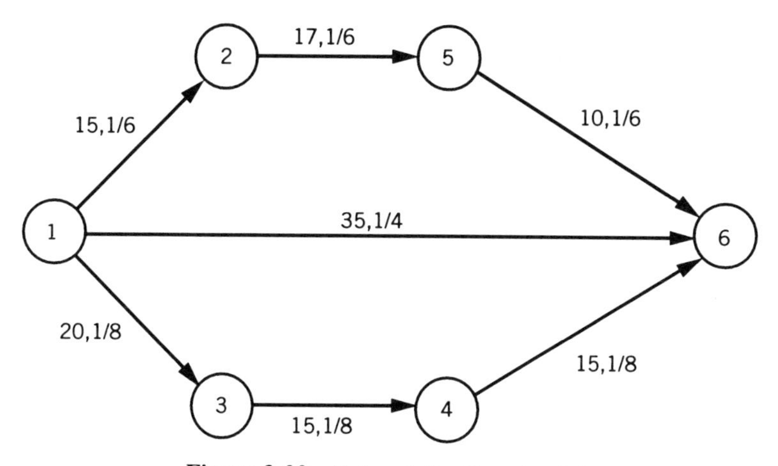

Figure 3.20 Network for Exercise 3.6.

b. Find the optimal routes from node A to node B.

c. Assign the 1000 travelers to the network's lines.

d. Find the optimal routes from X and Y to B.

3.8 Solve the example road network assignment problem in section 2.2 as linear program (3.8)–(3.9), using ready-made linear programming software, e.g., as found in "spreadsheets," or GAMS®, etc.

3.9 Solve the example transit network assignment problem in section 2.4 as linear program (3.38)–(3.39), using ready-made linear programming software, e.g., as found in "spreadsheets," or GAMS®, etc.

3.10 Apply the following alternative MCR algorithm to the example used in the text.

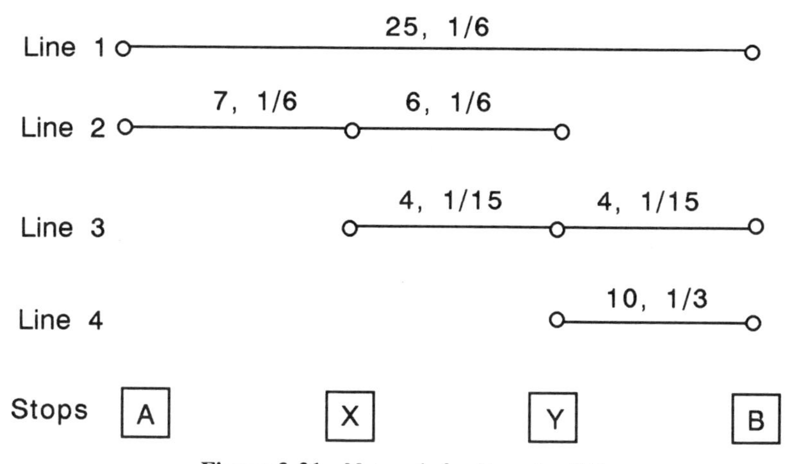

Figure 3.21 Network for Exercise 3.7.

1. Initialization

Set u_i equal to ∞ (in practice a very large number) for all nodes i other than the destination, and set $g_i = 0$ for the destination. g_i represents the current generalized cost to reach the destination j from node i.

Define the set $\{S\}$ of unexamined links as the whole network A.

Define the set A^+ of optimal links as empty.

3. Iterative step

2a. If $\{S\}$ is empty, stop. Otherwise, find the link (i, k) between nodes i and k such that the travel time from its *end* node plus the link's cost u_{ik} is the smallest of any link in $\{S\}$, i.e., such that

$$(t_a + g_{ik}) < (t_{a'} + g_{ik'}); \qquad \forall\, k'$$

Take that link out of $\{S\}$, i.e., set

$$\{S\} = \{S\} - a$$

Compare the current u_i for the *origin* node of the selected link a, to $g_k + t_a$. If the latter quantity is smaller than the former, i.e., if

$$(t_a + g_{ik}) < g_i$$

using link a from node i improves the present travel time from i to the destination. Then include link a in A_j^+, and go to 2b. Otherwise, go back to Step 2a.

2b. Update the current travel time from node i, by setting it to

$$g_i = (t_a + g_{ik})$$

Go back to 2a and iterate.

3.11 Perform the same stochastic route assignment as in the illustrative example in section 3.3 above, by maximizing utility function (3.26b) subject to standard constraints (3.9) with the use of nonlinear programming software, e.g., as found in "spreadsheets," or GAMS®, etc.

3.12* Show that generalizing a link's utility to the specification in Formula (3.22) does not affect Formula (3.21) for the route choice probability. (Hint: Repeat the same derivations, and use the fact that

$$\sum_a f_{kl}\delta^a_{ijr} = \sum_{ij}' f_{kl}\delta^a_{ijr} = f_{ij}' \tag{3.42}$$

so that this constant function of i and j may be included in K_{ij}.)

3.13 Using the STOCH algorithm, load the 250 travelers from 1 to 2 on the network used in the illustrative example.

3.14 Formulate the R.T.'s Utility Maximization Problem (3.8)–(3.9) in terms of link volumes instead of route volumes.

3.15* Demonstrate Formula (3.18).

3.16* Demonstrate Formula (3.26b).

3.17* Demonstrate Formula (3.18b).

3.18* Generalize the transit route choice problem to the probabilistic case, in the same manner as done for the car. (Hint: Do not confuse, or equate, the randomness in route utility specification with the randomness in travel times due to passenger/vehicle random arrivals.)

3.19* Show that the R.T.'s utility Formula (3.26) is concave, i.e., its negative is convex. (Hint: Evaluate the Hessian of the objective function; see section 1.7 of Appendix A, and show that it is negative definite.)

3.20 Show that choosing link *DA* instead of link *BA* in iteration 2 of the transit assignment algorithm in Table 3.7 leads to the same solution.

3.21 Assume in the illustrative example of section 2.4 that a user fee of $1.50 is charged on link 2. Assess its impact, in terms of the new link volumes, as well as on the value of the traveler surplus.

3.22 Assume in the illustrative example of section 3.3 that a user fee of $2 is charged on link 3. Assess its impact, in terms of the new link volumes, as well as on the value of the traveler surplus.

CHAPTER 4

COMBINED TRAVEL DEMAND MODELING UNDER UNCONGESTED CONDITIONS

4.1 INTRODUCTION: INDIVIDUAL AND COMBINED CHOICES, SINGLE AND JOINT DEMANDS

In this chapter, we integrate the route choice problem modeled in the previous chapter with the other choices facing the traveler, respectively choice of mode, of destination, and of whether to travel at all. As discussed in Chapter 1, all these choices are made jointly and must therefore be modeled together in a unified fashion.

To that effect, we shall retain the random utility framework which we have used in the route choice case. We will specifically use the results described in section 2.6 of Chapter 2, which concern sequential choices. This will lead to the application of the hierarchical, or "nested" logit model to represent the joint probabilities of combined choices.

The ability to integrate rigorously models of various dimensions of travel demand within the same methodology is one of the major advantages of the random utility approach. Perhaps even more advantageous is the fact that this ability is unaffected by the presence of demand externalities. This will be very useful when we begin to address the effects of network congestion, on the links and/or at the nodes, in the next chapter.

In the next section, we begin the process of integration in the case of the choices of mode and of route.

4.2 COMBINED MODE AND ROUTE CHOICES

We thus begin to backtrack toward the simultaneous consideration of all levels of choice, and consider the next higher level in the hierarchy of traveler choices, that of a mode. It may be remembered that the particular order of travel choices is to some extent arbitrary, as modal choice may precede destination choice, if this is more convenient analytically, or makes more sense conceptually. In any case, the chosen travel mode is no longer assumed given, as above, but is now unknown, and the very object of the individual traveler's choice. In other words, the modal demands T_{ijm} and T_{ijmr} are to be estimated *jointly*. However, the origin and destination of the traveler's trip are still fixed (given).

We must now formulate the utilities of the alternative choices facing the individual traveler. The difference from the previous case of route choice is that now the choice of a mode may be influenced by the attributes of the modal routes serving the given origin-destination, including travel time and cost. Consequently, utilities must be structured in the specific manner discussed in section 2.6 of Chapter 2, as follows.

The total utility received from a single trip on mode m *and* route r between the given locations i and j is specified as

$$\tilde{V}_{ijmr} = \tilde{U}_{ijm} + \tilde{U}_{ijmr} + \epsilon_{ijm} + \epsilon_{ijmr} \; ; \qquad \forall \; i, j, m, r \tag{4.1a}$$

The first utility is specific to modal choice m, while the second is specific to the combined mode-route choice. That is, the first term is the utility component which is common to all travelers using a given mode m, choices which include the same mode m. Similarly, the second term is the utility component which is common to the single choice consisting of a given mode and route; that is, the *conditional* utility received from a trip on route r, *given* that mode m has been selected, is equal to

$$\tilde{V}_{r/ijm} = U_{ijmr} + \epsilon_{ijmr} \tag{4.1b}$$

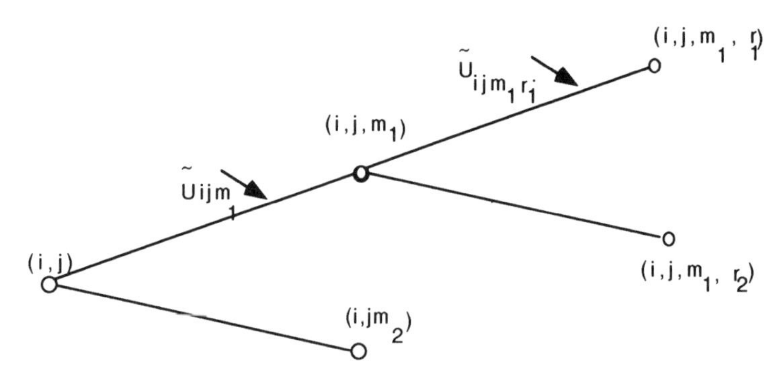

Figure 4.1 Structure of systematic utilities for mode-route choices.

The graphical representation of this utility structure is given in Figure 4.1. The utility of a given end node (i, j, m_x, r_y) is the sum of the link utilities along the path from node (i, j) to it.

In analogy to the specification of route utility in the previous chapter, we assume that the first term is specified as

$$\tilde{U}_{ijm} = b_{ij} + h_{ijm}; \qquad \forall\ i, j, m \tag{4.2}$$

b_{ij} is the average individual budget (income) for travelers between origin zone and destination j.[1] The term h_{ijm}, which may be interpreted as representing the "mode bias," or attractiveness, may be specified as a linear function of characteristics of travelers in zone i other than income and/or attributes of mode m (between i and j), other than travel cost and time.

$$h_{ijm} = a^0_{ijm} + \sum_k a^k_{ijm} X^k_{ijm}; \qquad \forall\ i, j, m \tag{4.3}$$

For further generality and flexibility, the X^k_{ijm} may themselves be prespecified as given, nonlinear transformations of other attributes:

$$X^k_{ijm} = \varphi_k(Y_{ijm}); \qquad \forall\ k, i, j, m \tag{4.4}$$

The second term, the conditional utility of a given modal route, is as specified in conformance with Formula (4.1):

$$\tilde{U}_{ijmr} = -\tau t_{ijmr} - c_{ijmr}; \qquad \forall\ i, j, m, r \tag{4.5}$$

According to the results discussed in section 2.6 of Chapter 2, the *expected* utility received from making modal choices after having decided to travel from

[1]If that is not available, b_i the average income of travelers in origin zones i, or b the overall average income may be used instead.

i to j is equal to

$$\tilde{W}_{m/ij} = \frac{1}{\beta_r} \ln \sum_r e^{\beta_r \tilde{U}_{ijmr}} = \frac{1}{\beta_r} \ln \sum_r e^{\beta_r(-c_{ijmr} - \tau t_{ijmr})};$$

$$\forall\ i, j, m \tag{4.6}$$

As in the case of a single-level choice, it is assumed that the random ϵ_{ijmr} are Gumbel-distributed, each with a mean zero and common variance σ_r^2. Also, in the same manner as above, it is assumed that the ϵ_{ijm}'s and the ϵ_{ijmr}'s are independent random variables. However, the utilities for two individual routes (i, j, m, r) and (i, j, m, r') on the same mode may be correlated, since they share the same error term ϵ_{ijm}. Consequently, according to Formula (2.32):

$$\text{Cor}(\epsilon_{ijmr}, \epsilon_{ijmr'}) = 1 - (\beta_m/\beta_r)^2: \qquad \forall\ i, j, m, r, r' \tag{4.7}$$

Thus, the ratio $1 - (\beta_m/\beta_r)^2$ measures the correlation between the utilities at the two levels of mode and route choice, and consequently, since both β's are positive by assumption.

$$\beta_m/\beta_r \leq 1 \tag{4.8}$$

This inequality should therefore in principle be reflected in the calibrated values of parameters β. It in turn implies that

$$\sigma_m^2 \geq \sigma_r^2 \tag{4.9}$$

This means that the uncertainty on modal utility would be expected to be greater than on route utility.

4.2.1 Individual Demands

With this utility specification, according to Formula (2.28), we may state the following result.

FACT 5.1

The marginal (unconditional) probability that an individual traveler between i and j will choose mode m is equal to

$$P_{m/ij} = \frac{e^{\beta_m(\tilde{U}_{ijm} + \tilde{W}_{m/ij})}}{\sum_m e^{\beta_m(\tilde{U}_{ijm} + \tilde{W}_{m/ij})}} = \frac{e^{\beta_m(h_{ijm} + \tilde{W}_{m/ij})}}{\sum_m e^{\beta_m(h_{ijm} + \tilde{W}_{m/ij})}}; \qquad \forall\ i, j, m \tag{4.10}$$

The value of parameters β_m and β_r in Formulas (4.10) and (4.6) respectively, are related to the variances of the random components ϵ_{ijm} and ϵ_{ijmr}

respectively. Specifically,

$$\beta_m = \pi / \sqrt{6} \sigma_{\epsilon_{ijm}} \tag{4.11}$$

$$\beta_r = \pi / \sqrt{6} \sigma_{\epsilon_{ijmr}} \tag{4.12}$$

These values, together with the values of the various parameters in the utility functions specified in Formulas (4.2) and (4.5) above, must be identified numerically as part of model calibration, or estimation. Calibration procedures for single-level, as well as multiple-level logit models are described in Chapter 7.

Since, according to conditional probabilities

$$P_{mr/ij} = P_{m/ij} P_{r/ijm}; \qquad \forall\ i, j, m, r \tag{4.13}$$

we have the next result:

FACT 5.2

The joint probability that an individual traveler between i and j will choose mode m and route r is equal to

$$P_{mr/ij} = \frac{e^{\beta_m(h_{ijm} + \tilde{W}_{m/ij})}}{\sum_m e^{\beta_m(h_{ijm} + \tilde{W}_{m/ij})}} \frac{e^{-\beta_r(c_{ijmr} + \tau t_{ijmr})}}{\sum_r e^{-\beta_r(c_{ijmr} + \tau t_{ijmr})}}; \qquad \forall\ i, j, m, r \tag{4.14}$$

The fact that this probability is not a multinomial (single-level) logit model defined on the combined choices (m, r) is due to the correlation between the random terms ϵ_{ijmr} and $\epsilon_{ijmr'}$. When this correlation is equal to zero, according to Formula (4.7), when $\beta_m = \beta_r = \beta$, then we might expect such a single-level model. In fact (see Exercise 4 2), in this case, Formula (4.14) becomes

$$P_{mr/ij} = \frac{e^{\beta(h_{ijm} - c_{ijmr} - \tau t_{ijmr})}}{\sum_m \sum_r e^{\beta(h_{ijm} - c_{ijmr} - \tau t_{ijmr})}}; \qquad \forall\ i, j, m, r \tag{4.15}$$

This is, indeed, a multinomial logit model, in which the argument is the sum of the systematic components of utility in Formulas (4.2) and (4.5) respectively

4.2.2 Aggregate Demands and the R.T.'s U.M. Problem

Having obtained the probabilities of combined choice of mode and route, given the numbers of travelers T_{ij} between each origin and destination, the aggregate travel demands for a given mode and modal route may then be obtained as

$$T_{ijm} = T_{ij} P_{m/ij}; \qquad \forall\ i, j, m \tag{4.16a}$$

$$T_{ijmr} = T_{ij} P_{mr/ij}; \qquad \forall\ i, j, m, r \tag{4.16b}$$

According to Formula (2.22) the expected received utility for each of the T_{ij} travelers from i to j from their combined choices of mode and route is equal to

$$\tilde{W}_{j/i} = b_{ij} + \frac{1}{\beta_m} \ln \sum_m \exp\left\{\beta_m\left(h_{ijm} + \frac{1}{\beta_r} \ln \sum_r e^{-\beta_r(\tau t_{ijmr} + c_{ijmr})}\right)\right\}; \qquad \forall\ i, j \tag{4.17a}$$

This result will be utilized in the next section, when combined choices of destination, mode, and route well be modeled. Another useful result in the same connection is that

$$\frac{\partial \tilde{W}_{j/i}}{\partial c_{ijmr}} = -P_{r/ijm}; \qquad \forall\ i, j, m, r \tag{4.17b}$$

Consequently, the total utility received by the R.T. from combined destination and route demands, on behalf of all travelers, is equal to

$$\tilde{W}_{MR} = B + \frac{1}{\beta_m} \sum_i \sum_j T_{ij} \ln \sum_m \exp\left\{\beta_m\left(h_{ijm} + \frac{1}{\beta_r} \ln \sum_r e^{\beta_r(-\tau t_{ijmr} - c_{ijmr})}\right)\right\} \tag{4.17c}$$

where, as usual, B is the total traveler budget/income:

$$B = \sum_{ij} T_{ij} b_{ij}$$

PROPOSITION 4.1

The traveler surplus corresponding to combined route and mode demands is equal to

$$TS_{MR} = \frac{1}{\beta_m} \sum_i \sum_j T_{ij} \ln \sum_m \exp\left\{\beta_m\left(h_{ijm} + \frac{1}{\beta} \ln \sum_r e^{-\beta_r(\tau t_{ijr} + c_{ijr})}\right)\right\} \tag{4.18a}$$

In the deterministic case when both β_m and β_r are infinite, the traveler surplus is equal to

$$TS_{MR} = \sum_i \sum_j T_{ij} \sum_m (h_{ijm} - \tau t_{ijm} - c_{ijm}) \tag{4.18b}$$

It is possible, in the same fashion as for the route choice problem in Chapter 3, to obtain the aggregate demands as specified in Formula (4.16) from the solution of the R.T.'s U.M. problem. (See Exercises 4.5 and 4.6.)

Finally, the aggregate route demands may be obtained from (see Exercise 4.10):

THEOREM 6.1

$$T_{ijmr} = -\frac{\partial \tilde{W}_{MR}}{\partial c_{ijmr}} \Big/ \frac{\partial \tilde{W}_{MR}}{\partial B}; \qquad \forall\, i, j, m, r \tag{4.19}$$

4.3 COMBINED DESTINATION, MODE, AND ROUTE CHOICES

Continuing the process of integration of individual travel choices and of the corresponding travel demands, we now consider the choice of destination. That is, the destination is no longer assumed given, as above, but is now unknown, as it is now being chosen by the individual traveler together with a mode and route. In other words, the modal demands T_{ij}, T_{ijm}, and T_{ijmr} are to be estimated *jointly*, given the T_i's.

We again specify the utilities facing the individual traveler from a given origin zone i. We expand on the same general hierarchical structure as above, by defining the total utility of a destination-mode-route combination as

$$\tilde{V}_{ijmr} = \tilde{U}_{ij} + \tilde{U}_{ijm} + \tilde{U}_{ijmr} + \epsilon_{ij} + \epsilon_{ijm} + \epsilon_{ijmr}; \qquad \forall\, i, j, m, r \tag{4.20a}$$

The first term, which represents the utility term common to all combined choices involving the same destination, received by all travelers going to destination j, is specified as

$$\tilde{U}_{ij} = b_i + h_{ij}; \qquad \forall\, i, j \tag{4.20b}$$

The term h_{ij}, which represents destination j's "attractiveness" from i, might itself be specified as a linear function of characteristics of travelers in zone i, other than income and/or attributes of destination j, including, importantly, cost of the activity to be conducted at the destination, for example, retail prices in the case of shopping travel.[2]

$$h_{ij} = a_{ij}^0 + \sum_k a_{ij}^k X_{ij}^k; \qquad \forall\ i, j \tag{4.20c}$$

The other two utility terms retain the same specification as in the two-level case in the previous section:[3]

$$\tilde{U}_{ijm} = h_{ijm}; \qquad \forall\ i, j, m \tag{4.20d}$$

In the same fashion, the third utility term is specified exactly as it was before, in the case of two levels of choice:

$$\tilde{U}_{ijmr} = -c_{ijmr} - \tau t_{ijmr}; \qquad \forall\ i, j, m, r \tag{4.20e}$$

The utility structure may be represented as in Figure 4.2.

4.3.1 Individual Demands

With this formulation, and under the same kind of assumptions about the characteristics of the ϵ_{ij}, ϵ_{ijm}, and ϵ_{ijmr} as were made above, but extended to an additional, third level, we have

$$P_{j/i} = \frac{e^{\beta_d(\tilde{U}_{ij} + \tilde{W}_{j/i})}}{\sum_j e^{\beta_d(\tilde{U}_{ij} + \tilde{W}_{j/i})}}; \qquad \forall\ i, j$$

[2]Note that if any of variables Y_{ij}^k represents the same factor as one of the X_{ijm}^k, it should then be defined as an average over all modes, and the corresponding X defined as the deviation from the average, as above.

[3]Note that the budget has been deleted, since it is now included in the first utility term in Formula (4.20b).

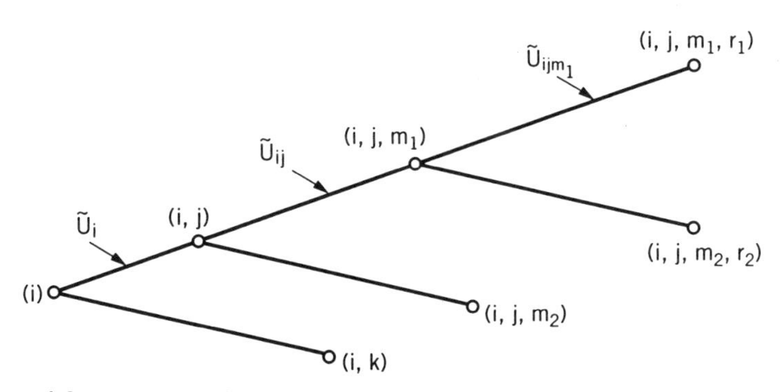

Figure 4.2 Structure of systematic utilities for destination-mode-route choices.

This result may be summarized as follows:

FACT 4.3

The marginal probability that an individual traveler traveling from zone i will choose to go to destination j is equal to

$$P_{j/i} = \frac{e^{\beta_d(h_{ij} + \tilde{W}_{j/i})}}{\sum_j e^{\beta_d(+\tilde{W}_{j/i})}}; \qquad \forall\ i, j \tag{4.21}$$

where the expected received utility $\tilde{W}_{j/i}$ from a given destination for a traveler in zone i is

$$\tilde{W}_{j/i} = \frac{1}{\beta_m} \ln \sum_m e^{\beta_m(\tilde{U}_{ijm} + \tilde{W}_{m/ij})}; \qquad \forall\ i, j \tag{4.22}$$

$\tilde{W}_{m/ij}$ itself is as defined in Formula (4.6b), so that

$$\tilde{W}_{j/i} = \frac{1}{\beta_m} \ln \sum_m \exp\left\{\beta_m\left(h_{ijm} + \frac{1}{\beta_r} \ln \sum_r e^{\beta_r(-\tau t_{ijmr} - c_{ijmr})}\right)\right\}; \qquad \forall\ i, j \tag{4.23}$$

which is Formula (4.17a). Consequently,

$$P_{jmr/i} = P_{j/i}P_{mr/ij} = P_{j/i}P_{m/ij}P_{r/mij}; \qquad \forall\ i, j, m, r$$

or, combining Formulas (4.21) and (4.15a), we have the following.

FACT 4.4

The joint probability that an individual traveler from zone i will choose to go to j on mode m and route r, is equal to

$$P_{jmr/i} = \frac{e^{\beta_d(h_{ij} + \tilde{W}_{j/i})}}{\sum_j e^{\beta_d(h_{ij} + \tilde{W}_{j/i})}} \frac{e^{\beta_m(h_{ijm} + \tilde{W}_{m/ij})}}{\sum_m e^{\beta_m(h_{ijm} + \tilde{W}_{m/ij})}} \frac{e^{-\beta_r(c_{ijmr} + \tau t_{ijmr})}}{\sum_r e^{-\beta_r(c_{ijmr} + \tau t_{ijmr})}}; \qquad \forall\ i, j, m, r \tag{4.24a}$$

As in the case of mode and destination choice above, when

$$\beta_d = \beta_m = \beta_r = \beta$$

Formula (4.24a) becomes (see Exercise 4.2):

$$P_{jmr/i} = \frac{e^{\beta(h_{ij} + h_{ijm} - \tau t_{ijmr} - c_{ijmr})}}{\sum\limits_j \sum\limits_m \sum\limits_r e^{\beta(h_{ij} + h_{ijm} - \tau t_{ijmr} - c_{ijmr})}}; \qquad \forall\ i, j, m, r \qquad (4.24b)$$

This is indeed a multinomial logit model, in which the argument is the sum of the systematic components of utility in Formulas (4.20) at the respective levels of destination, mode, and route choice.

4.3.2 Aggregate Demands and the R.T.'s U.M. Problem

Having obtained the probabilities of combined choice of destination, mode, and route, given the numbers of travelers T_i from each origin location, the aggregate travel demands for a given destination, mode, and route may then be obtained as

$$T_{ij} = T_i P_{j/i}; \qquad \forall\ i, j \qquad (4.25a)$$

$$T_{ijm} = T_{ij} P_{m/ij} = T_i P_{j/i} P_{m/ij}; \qquad \forall\ i, j, m \qquad (4.25b)$$

$$T_{ijmr} = T_{ijm} P_{jmr/i} = T_i P_{j/i} P_{m/ij} P_{r/ijm}; \qquad \forall\ i, j\ m, r \qquad (4.25c)$$

The R.T.'s indirect utility function, the total utility received by all travelers corresponding to the combined destination mode, and route demands, is then equal to

$$\tilde{W}_{DMR} = B + \frac{1}{\beta_d} \sum_i T_i \ln \sum_j \exp\left\{\beta_d\left(h_{ij} + \frac{1}{\beta_m} \ln \sum_m \exp\left\{\beta_m\left(h_{ijm} + \frac{1}{\beta_r} \ln \sum_r e^{-\beta_r(\tau t_{ijmr} + c_{ijmr})}\right)\right\}\right)\right\} \qquad (4.26a)$$

PROPOSITION 4.2

The traveler surplus from combined destination-mode-route demands is equal to

$$TS_{DMR} = \frac{1}{\beta_d} \sum_i T_i \ln \sum_j \exp\left\{\beta_d\left(h_{ij} + \frac{1}{\beta_m} \ln \sum_m \exp\left\{\beta_m\left(h_{ijm} + \frac{1}{\beta_r} \ln \sum_r e^{-\beta_r(\tau t_{ijmr} + c_{ijmr})}\right)\right\}\right)\right\} \qquad (4.26b)$$

In the deterministic case, this is equal to

$$TS_{DMR} = \sum_i \sum_j{}' \sum_m T_i(h_{ij} + h_{ijm} - \tau t_{ijm} - c_{ijm}) \qquad (4.26c)$$

In the same fashion as for the combined mode-route choice problem above, it is possible to obtain the aggregate demands as specified in Formulas (4.25) from the maximization of the R.T.'s utility representing the aggregate utility of all travelers. (See Exercise 4.7.) This is also shown in the more general case of variable costs, in Appendix 6.1 of Chapter 6.

Finally, it may be shown, as above, that aggregate demands may be retrieved from (see Exercise 4.11):

THEOREM 4.2

$$T_{ijmr} = -\frac{\partial \tilde{W}_{DMR}}{\partial c_{ijmr}} \Big/ \frac{\partial \tilde{W}_{DMR}}{\partial B}; \qquad \forall\, i, j, m, r \qquad (4.26d)$$

4.4 COMBINED TRAVEL, DESTINATION, MODE AND ROUTE CHOICES

Up till now, we have modeled the travel choices of *travelers* from a given zone i. In other words, the index i in all the probabilities estimated so far implied that they were conditional on the traveler's decision to travel (from location i) having already been made. To conclude the modeling of the traveler's choice process, we must now estimate the probability of the decision to travel in conjunction with all the others.

Given the present location i of a traveler (e.g., his or her residence, or place of work, etc.), and given a specific time period, the choice alternatives at the first level are:

1. To travel, that is, make *one* trip from zone i during that time period.[4]
2. Not to travel during that time period.[5]

To be consistent with the previous sections, index t (for travel) will refer to this level, with two values t and nt.

[4]It is assumed that the time period chosen is small enough (e.g., hour, day) that a resident only makes one such trip per period.

[5]This may be considered the "outside" alternative. Accordingly, its "cost" may also be set at the price of a substitute to travel (e.g., home delivery, etc.).

We now define the corresponding utilities. The total utility received by an individual in demand zone i from making one trip from area i to destination j, on mode m and route r (for the given purpose of interest), is specified as

$$\tilde{V}_{ijmr} = \tilde{U}_{it} + \tilde{U}_{ij} + \tilde{U}_{ijm} + \tilde{U}_{ijmr} + \epsilon_{it} + \epsilon_{ij} + \epsilon_{ijm} + \epsilon_{ijmr}; \qquad \forall\ i, j, m, r \tag{4.27}$$

The first utility term in Formula (4.27) is itself specified as

$$\tilde{U}_{it} = b_i + h_i; \qquad \forall\ i \tag{4.28a}$$

The utility of not making a trip from area i is specified as $\tilde{V}_{nt/i}$. Since logit probabilities are not affected by adding or subtracting a constant to all utilities this may arbitrarily be set to zero, for convenience:

$$\tilde{V}_{io} = \tilde{U}_{int} + \epsilon_{int} = 0 + \epsilon_{int}; \qquad \forall\ i \tag{4.28b}$$

The symbols in the formulas have the same meaning as previously. The term h_i, which may be interpreted as the attractiveness of traveling, may be specified further as a linear function of socioeconomic characteristics of travelers in zone i, such as age, and of the characteristics of demand area i, such as land use, zoning classification, distance from downtown, etc. The values of these factors are given (from surveys, secondary data such as the U.S. Census, or local government, etc.), as averages if the same variables are found in h_{ij} defined earlier.

$$h_i = a_i^0 + \sum_k a^k X_k^i; \qquad \forall\ i \tag{4.28c}$$

The other utility terms in formula (4.27) retrain their previous definitions, with the exception of the second utility, from which the traveler's budget/income is removed, since it is found in Formula (4.28a).

The graphical representation of this utility structure is given in Figure 4.3.

4.4.1 Individual Demands

Under these conditions, and again with the standard assumptions about the distribution of the various ϵ's, we have

$$P_{t/i} = \frac{e^{\beta_t(\tilde{U}_{it} + \tilde{W}_{t/i})}}{e^{\beta_t \tilde{U}_{int}} + e^{\beta_t(\tilde{U}_{it} + \tilde{W}_{t/i})}}; \qquad \forall\ i$$

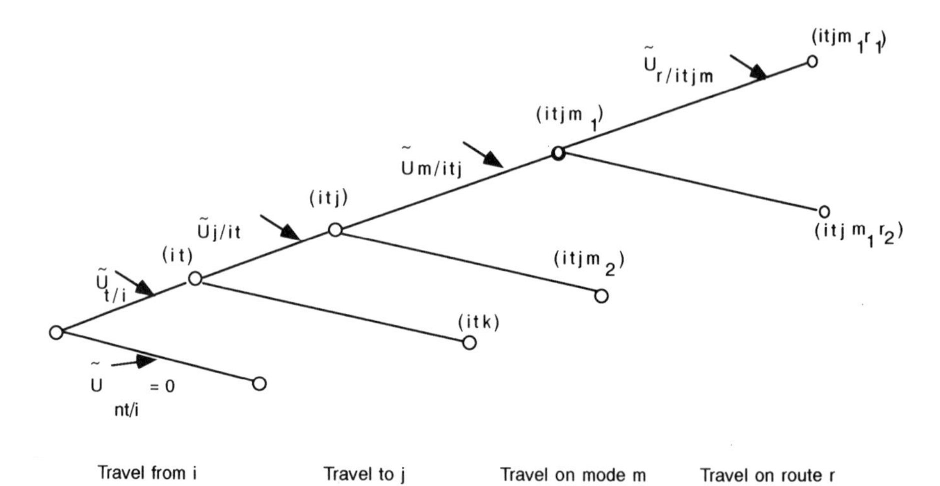

Figure 4.3 Utility structure for four-level travel choices.

that is, given Formula (4.28):

FACT 4.5

The marginal probability that an individual in zone i will decide to travel is equal to

$$P_{t/i} = \frac{e^{\beta_t(h_i + \tilde{W}_{t/i})}}{1 + e^{\beta_t(h_i + \tilde{W}_{t/i})}}; \qquad \forall\, i \tag{4.29}$$

where

$$\tilde{W}_{t/i} = \frac{1}{\beta_d} \ln \sum_j \exp\left\{\beta_d\left(h_{ij} + \frac{1}{\beta_m} \ln \sum_m \exp\left\{\beta_m\left(h_{ijm} + \frac{1}{\beta_r} \ln \sum_r e^{-\beta_r(\tau t_{ijmr} + c_{ijmr})}\right)\right\}\right)\right\} \tag{4.30}$$

Given that an individual is located in i, the probability $P_{jmr/i}$ may be expressed as the product of the respective conditional probabilities above:

$$P_{ijmr} = P_{t/i} P_{jmr/i} = P_{t/i} P_{j/i} P_{m/ij} P_{r/ijm}; \qquad \forall\, i, j, m, r \tag{4.31}$$

Combining Formulas (4.24a) and (4.29)[6] we have this result:

FACT 4.6

The probability that an individual located in i travels to destination j, on mode m, and following route r on that mode is equal to

$$P_{ijmr} = \frac{e^{\beta_t(h_i + \tilde{W}_{t/i})}}{1 + e^{\beta_t(h_i + \tilde{W}_{t/i})}} \frac{e^{\beta_d(h_{ij} + \tilde{W}_{j/i})}}{\sum_j e^{\beta_d(h_{ij} + \tilde{W}_{j/i})}} \frac{e^{\beta_m(h_{ijm} + \tilde{W}_{m/ij})}}{\sum_m e^{\beta_m(h_{ijm} + \tilde{W}_{m/ij})}} \frac{e^{-\beta_r(c_{ijmr} + \tau t_{ijmr})}}{\sum_r e^{-\beta_r(c_{ijmr} + \tau t_{ijmr})}};$$

$$\forall\ i, j, m, r \qquad (4.32a)$$

This nested, four-level logit model is the most general formulation of travel demand under our framework, for uncongested conditions. Overlooking the multiplicity of symbols, this formula, because of its particular hierarchical structure, should be considered a straightforward extension of the single-level logit model for car route choice with which we began. In fact, in practical, operational terms, the analyst will in effect work separately with logit models for the probabilities $P_{r/ijm}$, $P_{m/ij}$, $P_{j/i}$, and $P_{t/i}$, in that order, from their calibration to numerically estimating the corresponding probabilities.

In any case one of the advantages of this structured approach is its modularity. Consequently, one may add other levels of choice at any level of the above structure, and treat them in the same systematic manner as the others. For instance, another decision level, about the choice of time period for travel, may be incorporated.

In the special case when the respective values of parameters β_t, β_d, β_m, and β_r are all equal to β, the expression for P_{ijmr} in Formula (4.32a) becomes (see Exercise 4.2.):

$$P_{ijmr} = \frac{e^{\beta(h_i + h_{ij} + h_{ijm} - \tau t_{ijmr} - c_{ijmr})}}{[1 + e^{\beta h_i}] \sum_j \sum_m \sum_r e^{\beta(h_{ij} + h_{ijm} - \tau t_{ijmr} - c_{ijmr})}}; \qquad \forall\ i, j, m, r \qquad (4.32b)$$

This is indeed a multinomial logit model, in which the argument is the sum of the systematic components of utility in Formula (4.27) at the four respective levels of travel, mode, and route choice. Thus, all four levels of travel choice has collapsed into one with no dependency between them. This graphically represented in Figure 4.4, which may be compared with Figure 4.3.

[6]As mentioned earlier, P_{ijmr} now refers to the unconditional probability for *residents* of zone i of traveling to j on m and r.

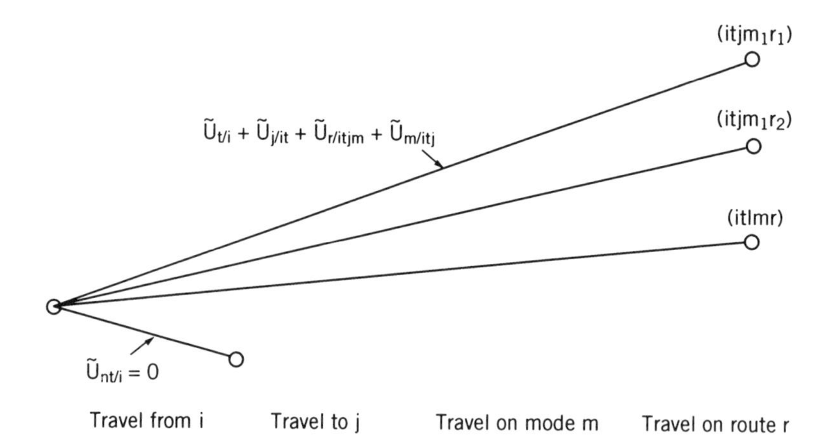

Figure 4.4 Utility structure in the absence of correlation.

4.4.2 Aggregate Demands and the R.T.'s U.M. Problem

In either case of independent or dependent utilities, all travel demands, measured in person-trips, may then be derived from the probability P_{ijmr} as expressed by Formulas (4.32). First, the demand for person-trips from i to j, on route r of mode m is

$$T_{ijmr} = N_i P_{ijmr}; \qquad \forall\ i, j, m, r \tag{4.33}$$

where N_i is the number of residents of zone i. Similarly, the demand for travel on mode m is equal to

$$T_m = \sum_{ijr} T_{ijmr} = \sum_{ijr} N_i P_{ijmr}; \qquad \forall\ m \tag{4.34}$$

The total demand for travel, in person-trips, is

$$T = \sum_{ijmr} T_{ijmr} = \sum_{i} N_i P_{t/i}; \tag{4.35}$$

Consequently, mode m's share of the overall ridership is

$$P_m = T_m / T; \tag{4.36}$$

Note that this is not equal to the probability that an individual travel taken at random from any demand zone will use mode m. Mode m's share of the ridership between i and j is, by definition, $P_{m/ij}$. Similarly, the demand for travel to j is equal to

$$T_j = \sum_{imr} T_{ijmr} = \sum_{imr} N_i P_{ijmr}; \qquad \forall\ j \tag{4.37}$$

and the demand for travel between i and j is equal to

$$T_{ij} = \sum_{mr} T_{ijmr} = \sum_{mr} N_i P_{ijmr}; \qquad \forall\ i, j \tag{4.38}$$

while the demand for travel between i and j on mode m is equal to

$$T_{ijm} = \sum_{r} T_{ijmr} = \sum_{r} N_i P_{ijmr}; \qquad \forall\ i, j, m \tag{4.39}$$

and so on.

Finally, the volume on a given modal link a_m may be estimated by using the link-route Relationship (1.2b)

$$v_{a_m} = \sum_i \sum_j \sum_r T_{ijm} \delta_{ijr}^{a_m}; \qquad \forall\ a_m \tag{4.40}$$

In summary, the framework presented above for the estimation of travel generation, distribution, modal and route choice is the most general, in the case of fixed attributes for the various choice alternatives, in particular travel times and costs. A major advantage of this hierarchical formulation is that, as we shall see in Chapter 6, it remains unchanged in the presence of congestion.

4.5 ILLUSTRATIVE EXAMPLE

We now illustrate this methodology with the following example of combined travel-destination-mode choice. In an hypothetical urban area, there are two main travel corridors, each served by two alternative modes, as represented in Figure 4.5. Thus, the only value for the index of the origin zone is $i = 1$, and the index of the destination zone takes the values A and B.

The route travel costs c_{ijm} for the two modes, measured in dollars, are as follows:

(i, j)	$1A$	$1B$
$m = 1$	9	12
2	11	8

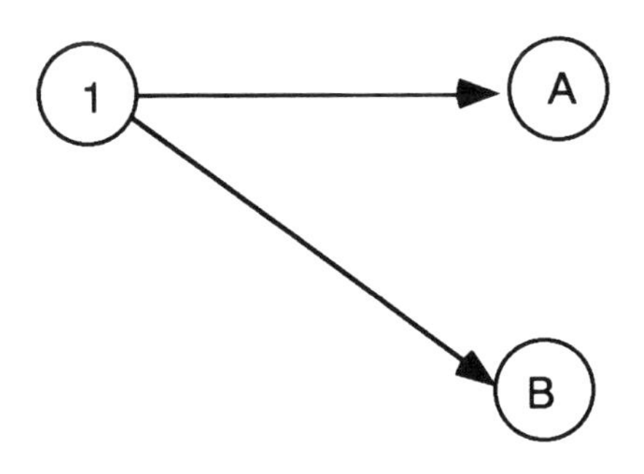

Figure 4.5 Network for illustrative example.

The values of the destination attractiveness terms, h_{ij} are estimated as follows:

$$h_{1A} = 4; \qquad h_{1B} = 5$$

The term h_1 is equal to 1.5. For simplicity, all other factors of utility in the above formulation, including the modal attractiveness h_{1jm}, are assumed equal to zero, and do not intervene. The number of residents in the single origin zone is $N_1 = 800$. Finally, the values of all parameters β may be assumed to be equal to 0.1.

Given this information, we want to estimate the numbers of travelers which choose each mode, the numbers of travelers which choose each destination, and the numbers of travelers which choose a given destination/mode combination.

The utility structure facing an individual resident of the origin zone may be represented as in Figure 4.6. The utilities of the various choices are represented above the nodes representing them. Note that we do not consider the choice of route, so that in effect, we are assuming that every traveler in a given corridor (i, j) uses the same route.This may in particular correspond to a situation in which information about routes (e.g., network topology, route costs, etc.), is not available, so that the fourth level of choice is dismissed, or "averaged out." Note also the negative sign for costs, and the positive signs for attractiveness.

We first estimate the probabilities of modal choice, at the last level of choice, given (i.e., conditional on) the preceding choice of destination. We have, in the first corridor

$$P_{m=1/1A} = \frac{e^{\beta_m \tilde{U}_{1A1}}}{\sum_m e^{\beta_m \tilde{U}_{1Am}}} = \frac{e^{\beta_m c_{1A1}}}{\sum_m e^{\beta_m c_{1Am}}} = \frac{e^{0.1(-9)}}{e^{0.1(-9)} + e^{0.1(-11)}} = 0.55$$

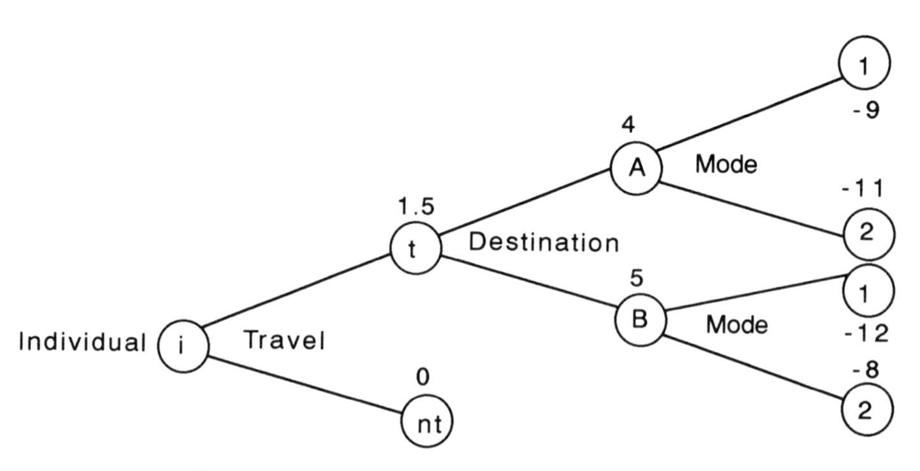

Figure 4.6 Utility structure for illustrative example.

Consequently,

$$P_{m=2/1A} = 1 - 0.55 = 0.45$$

Similarly, in the second corridor,

$$P_{m=1/1B} = \frac{e^{\beta_m \tilde{U}_{1B1}}}{\sum_m e^{\beta_m \tilde{U}_{1Bm}}} = \frac{e^{\beta_m c_{1B1}}}{\sum_m e^{\beta_m c_{1Bm}}} = \frac{e^{0.1(-12)}}{e^{0.1(-12)} + e^{0.1(-8)}} = 0.40$$

$$P_{m=2/1B} = 1 - 0.40 = 0.60$$

Next, we estimate the expected received utilities $\tilde{W}_{A/1}$ and $\tilde{W}_{B/1}$ for the respective destination:

$$\tilde{W}_{A/1} = \frac{1}{\beta_m} \ln \sum_m \exp(\beta_m \tilde{U}_{1Am})$$

$$= \frac{1}{0.1} \ln \sum_m e^{-\beta_m c_{1Am}} = \frac{1}{0.1} \ln (e^{0.1(-9)} + e^{0.1(-11)}) = -3.03$$

$$\tilde{W}_{B/1} = \frac{1}{\beta_m} \ln \sum_m \exp(\beta_m \tilde{U}_{1Bm})$$

$$= \frac{1}{0.1} \ln \sum_m e^{-\beta_m c_{1Bm}} = \frac{1}{0.1} \ln (e^{0.1(-12)} + e^{0.1(-8)}) = -2.87$$

Consequently, the conditional probabilities of choosing a given destination, given that an individual resident of zone 1 is traveling, are

$$P_{j=A/1} = \frac{e^{\beta_d(\tilde{U}_{1A} + \tilde{W}_{m/A})}}{\sum_j e^{\beta_d(\tilde{U}_{1j} + \tilde{W}_{m/j})}} = \frac{e^{\beta_d(h_{1A} + \tilde{W}_{m/A})}}{\sum_j e^{\beta_d(h_{1j} + \tilde{W}_{m/j})}} = \frac{e^{0.1(4-3.03)}}{e^{0.1(4-3.03)} + e^{0.1(5-2.87)}} = 0.45$$

so that consequently,

$$P_{j=B/1} = 1 - 0.45 = 0.55$$

Next, we estimate the expected received utilities $\tilde{W}_1$ of traveling from zone 1

$$\tilde{W}_1 = \frac{1}{\beta_d} \ln \sum_j e^{\beta_d(\tilde{U}_{1j} + \tilde{W}_{j/1})} = \frac{1}{\beta_d} \ln \sum_j e^{\beta_d(h_{1j} + \tilde{W}_{j/1})}$$

$$= \frac{1}{0.1} \ln (e^{0.1(4-3.01)} + e^{0.1(5-2.86)}) = 8.51$$

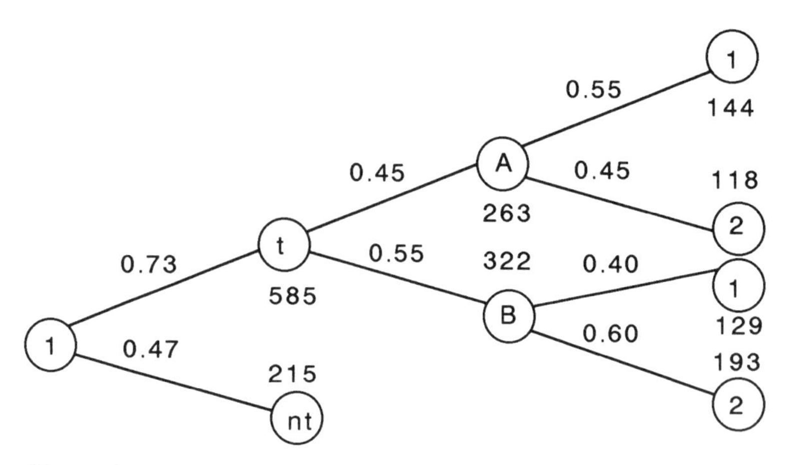

Figure 4.7 Probabilities of choice of travel choices and travel demands.

so that the probability of an individual resident of zone 1 deciding to travel is equal to

$$P_{t/1} = \frac{e^{\beta_t(\tilde{U}_1 + \tilde{W}_1)}}{1 + e^{\beta_t(\tilde{U}_1 + \tilde{W}_1)}} = \frac{e^{\beta_t(h_1 + \tilde{W}_1)}}{1 + e^{\beta_t(h_1 + \tilde{W}_1)}} = \frac{e^{0.1(1.5 + 8.51)}}{1 + e^{0.1(1.5 + 8.51)}} = 0.73$$

This means that 73% of the potential travelers will make a trip from the given area. Next 45% of them will choose destination A, and 55% destination B. The resulting choices, in terms of probability of choice (along the links) and numbers of travelers (at the nodes), are as represented in Figure 4.7.

The above calculations were performed to illustrate the application of the theory. However, it may be noticed that since all β's are equal, the nested logit model of combined travel, destination, and mode choice collapses into a single-level multinomial model, according to Formula (4.32b). The same travel demands could thus have been obtained in a simpler manner (see Exercise 4.9).

4.6 SUMMARY

In this chapter, we developed a model for the probabilistic outcome of the sequence of a traveler's decisions as to whether to travel or not, where to travel, on what mode, and following which itinerary. To that purpose, the probabilities of given choices at each of these respective levels were estimated, using the logit model, and combined appropriately. Specifically, we started from the lowest, or last level, the choice of route, which also corresponds to the phase of trip assignment in the sequential urban travel demand forecasting process. We then moved upward (or backward), to the level of the choice of mode, which corresponds to the phase of modal split, then to the level of destination choice, which corresponds to the phase of trip distribution. Finally, this was linked to the first level of choice, that of whether or not to travel, corresponding to the phase of trip generation, to obtain the number of trips over any modal route between any origin-destination pair.

In general, the respective travel demands are conditional on the demands at the preceding levels (as in the conventional approach discussed in section 1.4 of Chapter 1), but also dependent on the subsequent levels, and are thus *interactive*. For instance, travel generation, the first decision of whether to travel or not, is sensitive to the "composite" cost of travel over all destinations, modes and routes, through the expected received utility $\tilde{W}_{t/i}$. This quantity itself includes the expected received utility at the subsequent level, $\tilde{W}_{j/i}$, and so forth. Nevertheless, this multilevel structure allows for the individual examination of given phases in the urban travel demand forecasting process. For instance, interest might be focused on modal split, or on trip distribution.

The application of the random utility framework offers major advantages, compared to the conventional approach discussed in section 1.4 of Chapter 1. Among them is the total integration of all aspects of travel demand in a consistent microeconomic theory of travel demand. The hierarchical structure is very effective in representing individual alternatives which are correlated. For instance, individual transit modes, such as bus and subway, may be associated in the traveler's mind with an unmeasurable but common lack of privacy and/or comfort.

A major advantage of the nested logit formulation of travel demands is that it remains unchanged in the presence of demand externalities due to congestion, as we shall see in Chapter 6.

4.7 EXERCISES

4.1 The daily numbers of travelers from two given areas $i = 1$ and 2, are, on the average, $T_1 = 100$; $T_2 = 200$. It may be assumed that the utility of a travel destination $j = 1, 2$ is:

$$\tilde{V}_{ij} = \tilde{U}_{ij} + \epsilon_{ij} = -0.5\, t_{ij} - c_j + \epsilon_{ij} \tag{4.41}$$

where c_j is the parking charge at destination j, and t_{ij} is the travel time. The values of the t_{ij}'s and c_j's are given below

t_{ij}	To: $j = 1$	2
From $i = 1$	1	2
2	2	1

$c_1 = 1$; $c_2 = 0.5$

The value of β_d may be assumed to be equal to 1.0

a. Determine the origin-destination demands T_{ij}'s.

b. Estimate the total and average travel time and parking charge incurred by travelers in the respective origin zones. Estimate the traveler surplus.

c. Estimate the elasticities of T_{12} with respect to t_{12} and t_{21}, respectively.

d. What is the probability that, *on any given day*, 50 travelers will travel from zone 1 to destination 2?

4.2 Show the validity of Formulas (4.15b), (4.24b), and (4.32b) for the case when all β's are equal.

4.3 Given the conditional probabilities of choice at each of three levels i, j, m below, and the numbers of residents in the three zones T_i, estimate the numbers of travelers going to each destination T_j, and using each mode T_m. (Hint: Use Formula (4.32a) in the deterministic route choice case, and follow the steps in the illustrative example in section 4.5.)

$i =$	1	2	3
$P_{t/i} =$	0.2	0.6	0.2
$T_i =$	500	1000	350

$P_{j/i}$ $j =$	1	2	3
$i = 1$	0.2	0.3	0.5
2	0.3	0.4	0.3

$P_{m/j;i=1}$	$m = 1$	2
$j = 1$	0.3	0.7
2	0.4	0.6
3	0.5	0.5

$P_{m/j;i=2}$	$m = 1$	2
$j = 1$	0.4	0.6
2	0.2	0.8
3	0.7	0.3

4.4 A transportation system with two origins, two destinations, and two modes may be represented as in Figure 4.8.

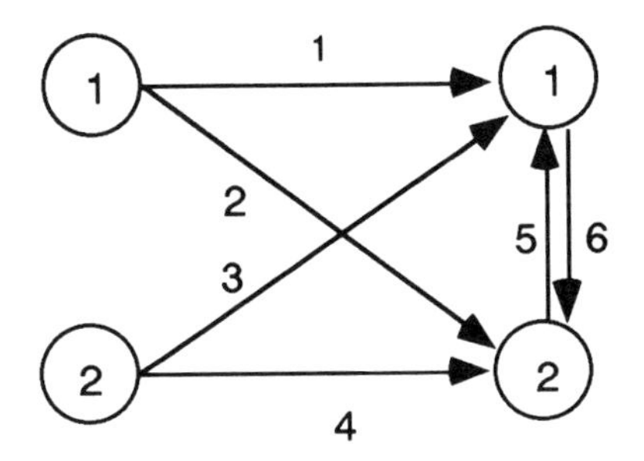

Figure 4.8 Network for Exercise 4.4.

The link costs for the two modes are as follows:

Link #	1	2	3	4	5	6
Cost $m = 1$	10	13	8	12	9	15
Cost $m = 2$	8	10	9	11	7	6

The values of the factors h_{it}, h_{ij} and h_{ijm} in Formulas (4.38a) are as follows:

$$h_1 = 7.94 \qquad h_2 = 6.10$$

$$h_{i1} = 3;\ \forall\ i \qquad h_2 = 5;\ \forall\ i$$

$$h_{ij1} = -3;\ \forall\ i, j \qquad h_{ij2} = -5;\ \forall\ i, j$$

The numbers of travelers in each area are:

$$N_1 = 350 \qquad N_2 = 200$$

Finally, the values of parameters β are:

$$\beta_t = 1.25; \qquad \beta_d = 0.75; \qquad \beta_m = 1.5; \qquad \beta_r = 0.5$$

a. Identify the routes from each origin to each destination and their costs.

b. Estimate the probabilities of choice $P_{m/ij}$ of each modal route between a given origin and destination, using Formula (4.10).

c. Estimate the probabilities of choice $P_{mr/ij}$ of each mode and route between a given origin and destination, using Formula (4.15a).

e. Estimate the probabilities $P_{j/i}$ of choice of each destination from a given origin, using Formula (4.21).

f. Estimate the probabilities $P_{t/i}$ of traveling from a given origin, using Formula (4.29).

g. Estimate the probabilities $P_{r/ijm}$ of traveling on a given modal route to a given destination from a given origin, using Formula (3.15a).

h. What would the answer to question (g) be if all β's are now equal to 1.25?

i. Estimate the demand for travel T_j to the various destinations j, using Formula (4.37).

j. Estimate the demand for travel T_{ij} between given origins i and destinations j using Formula (4.38).

k. Estimate the demands T_{ijm} for travel between i and j on mode m using Formula (4.39).

l. Estimate the demands T_i for travel from i.

m. Estimate the demands T_{im} for travel from i on mode m for all i and m.

n. Estimate the demands T_{jm} for travel to j on mode m for all j and m.

o. Estimate the demands T_r for travel on the various modal routes.

p. Estimate the volumes on the various links, using Formula (4.40).

4.5 Show that given origin demands T_i, the joint destination-route demands T_{ij} and T_{ijr} may be obtained as the solution of the R.T.'s utility maximization problem

$$\text{Max } U_{MR}(T_{ij}, T_{ijr}, T_0) = B - \left[\tau \sum_{ijr} T_{ijr} t_{ijr} + \frac{1}{\beta_r} \sum_{ijr} T_{ijr} \ln T_{ijr} + \frac{1}{\beta_d'} \sum_i \sum_j T_{ij} \ln T_{ij} + \sum_j h_{ij} \sum_i T_{ij} \right] + T_0 \tag{4.42}$$

such that

$$\sum_{ijr} T_{ijr} c_{ijr} + T_0 = B \tag{4.43a}$$

$$\sum_j T_{ij} = T_i; \qquad \forall\, i \tag{4.43b}$$

$$\sum_r T_{ijr} = T_{ij}; \qquad \forall\, i, j \tag{4.43c}$$

$$T_{ij} \geq 0,\ T_{ijr} \geq 0; \qquad \forall\, i, j \tag{4.43d}$$

where

$$\frac{1}{\beta_d'} = \frac{1}{\beta_d} - \frac{1}{\beta_r} \tag{4.44}$$

4.6 Show that the optimal value of the R.M. function in the above program is equal to the R.T.'s received utility, as specified in Formula (4.17c).

4.7 Show that given origin demands T_i, the joint destination-route demands T_{ij} and T_{ijr} may be obtained, in the case of deterministic route choice, as the solution of the (nonlinear) optimization problem

$$\text{Max } U_{MR}(T_{ij}, T_0) = B - \left[\tau \sum_{ij} T_{ij} t_{ij} + \frac{1}{\beta_d} \sum_i \sum_j T_{ij} \ln T_{ij} + \sum_j h_{ij} \sum_i T_{ij} \right] + T_0 \tag{4.45}$$

such that

$$\sum_{ij} T_{ij} c_{ij} + T_0 = B \tag{4.46a}$$

$$\sum_j T_{ij} = T_i; \qquad \forall\, i \tag{4.46b}$$

$$T_{ij} \geq 0,\ T_{ijr} \geq 0; \qquad \forall\, i, j \tag{4.46c}$$

where t_{ij} is the travel time between i and j (i.e., the minimum route time), and c_{ij} is the travel cost.

4.8 Show that given origin demands T_i, the joint destination-mode-route demands T_{ij}, T_{ijm}, and T_{ijmr} may be obtained, again in the case of deterministic route choice, as the solution of the (nonlinear) optimization problem

$$\underset{(T_{ij}, T_{ijm}, T_0)}{\text{Max}} \quad U_{DMR} = -\left[\sum_{ijm} T_{ijm} t_{ijm} + \frac{1}{\beta_m} \sum_m \sum_{ij} T_{ijm} \ln T_{ijm} + \frac{1}{\beta'_d} \sum_{ij} T_{ij} \ln T_{ij} + \sum_{ijm} T_{ijm} h_{ijm} + \sum_j h_j \sum_i T_{ij} \right] + T_0 \tag{4.47}$$

such that

$$\sum_{ijm} T_{ijm} c_{ijm} + T_0 = B \tag{4.48a}$$

$$\sum_m T_{ijm} = T_{ij}; \qquad \forall\, i, j \tag{4.48b}$$

$$\sum_j T_{ij} = T_i; \qquad \forall\, i \tag{4.48c}$$

$$T_i \geq 0;\ T_{ij} \geq 0,\ T_{ijm} \geq 0; \qquad \forall\, i, j, m \tag{4.48d}$$

Show that the optimal value of the objective function is equal to the R.T.'s received utility, as specified in Formula (4.26a).

4.9 Derive the probabilities of travel choice obtained in the illustrative example of section 4.5 by exploiting the fact that all β's are equal. Hint: Use Formula (4.32a), which in the present case is

$$P_{tjm/1} = \frac{e^{\beta \tilde{U}_1}}{1 + e^{\beta \tilde{U}_1}} \cdot \frac{e^{\beta(\tilde{U}_j + \tilde{U}_m)}}{\sum_{jm} e^{\beta(\tilde{U}_j + \tilde{U}_m)}} \tag{4.49}$$

4.10 Demonstrate the validity of Formulas (4.19).

4.11 Demonstrate the validity of Formulas (4.26d).

CHAPTER 5

ROUTE CHOICE MODELING UNDER CONGESTED CONDITIONS

5.1 INTRODUCTION: THE CONCEPTUAL AND METHODOLOGICAL IMPLICATIONS OF DEMAND EXTERNALITIES

In the preceding two chapters, travel demands were modeled under the assumption that all of their determinants were fixed. In particular, travel times and costs, which intervened in the utilities of the various travel choices at all levels, were assumed to be independent of travel demands. While this may be a reasonable assumption in some situations, in general it is not valid in most urban settings.

Specifically, the utility of a given car route may be a function of the demand on the route, since increasing levels of traffic results in decreasing speeds of travel, increased difficulty in passing, and thus in increasing travel times, as may readily be observed in many urban areas. The same is also true in the case of transit routes, where a higher number of passengers boarding and alighting may result in longer dwell times at the stations.

Similarly, the utility of a given destination may also be affected by the number of travelers going to it. For instance, the availability of parking at a given location, as measured by the average time it takes to find a space at any given arrival time, will decrease with the number of cars seeking parking there. In other situations, such as in shopping travel, the utility of a given destination may be an increasing function of its volume, reflecting a "bandwagon" effect.

In all cases, the fundamental difference between the absence and presence of congestion is that in the former case the utility of a given travel choice is fixed, at least in the short term (i.e., until changed by supplier's actions, such as, increasing the frequency of transit service). In the latter case, these utilities may now also be affected by the choices made by the travelers themselves. That is, any individual traveler's behavior is now dependent on and influenced by that of all other travelers. This phenomenon is known in microeconomics as a "demand externality."

The fundamental implication of this situation is that some utility factors, specifically modal route travel times and costs, and destination service times and costs, are now unknown, since they depend on the unknown demands for the routes, or for the destinations, and so forth. Consequently, the value of these factors must be determined internally, or endogenously, as part of the model's output, together, and consistently with, the various travel demands.

At the conceptual level, since the travel demand model is now in effect required to provide *more* information than previously, additional statements must be included in the model formulation. That is, the relationships between travel demands, which must still be driven by utility maximization, as previously, and the variable utility factors must be specified. From a methodological standpoint, it would be rather convenient if the random utility approach and the resultant formulation of demands as nested logit functions used in the uncongested case could be maintained.

It is clear that the only way to derive aggregate travel demands is at the aggregate level, from the solution of the R.T.'s utility maximization problem.

Specifically, so far we have been able to obtain the unknown travel demands by simply inserting given values into logit functions. These functions had the same specification for individual demands as for aggregate demands. As was systematically shown, an alternative, equivalent approach was to obtain the aggregate demands from the R.T.'s U.M. problem.

In the congested case, simply adding individual demands is no longer possible, so that consequently we shall have to obtain aggregate demands as the solution of the same problem. Because of the presence of congestion, the various R.T.'s problems will be somewhat more complex than they were in the uncongested case, and in particular will no longer be *linear* programs. This increase in analytical complexity is the price for modeling inherently more complex traveler behavior.

In the next section, we begin to examine the effects of congestion on the various levels of traveler choice and the corresponding travel demands. Just as in the uncongested case, all four levels may be considered at once, as we shall in fact see later. However, it will be convenient to consider them in the same order as we did in the uncongested case in the previous chapters. The pedagogical advantages of such an approach are a gradual increase in analytical complexity as well as complete parallelism between the uncongested and congested cases.

Thus, as in the uncongested case, we first examine the last choice level, that of a route, and its counterpart in terms of aggregate demand estimation, trip (network) assignment. We will then consider each preceding travel choice in turn, backtracking all the way to the decision of whether to travel. As in Chapter 3, we focus first on the car mode. The problem we thus (re)consider is how to determine the (conditional) probability $P_{r/ijc}$ that a traveler going from zone i to destination j with the car chooses route r. The equivalent problem is to determine the numbers of trips, T_{ijr}'s, on each route r from any origin zone i to any destination j, given the numbers T_{ij}.[1]

The basic difference, as discussed above, is that the values of the t_{ijr}'s are now unknown, that is, are no longer given externally to the model, but must be estimated internally as part of the model's solution. We assume that all other factors in the utility, including the values of the route charges, c_{ijr}, are fixed and given as in Chapter 3, at least in the short term. In the longer term, of course, some of these values may also vary. However, their values would then be set by the suppliers of travel, and thus come from the *supply* side, instead of the *demand* side, as in the case of the t_{ijr}'s.

In the present congested case, the determination of the unknown values of route utilities must, of course, remain consistent with the utility maximization principle which has been used in the uncongested route choice case, either in its deterministic or stochastic form. To reiterate, at equilibrium, no traveler can *unilaterally* increase his or her *received* utility by changing routes. This, in turn, implies that all *used* routes connecting a given origin with a given

[1]The implicit index c of the car mode will be omitted from now on for simplicity.

destination offer *equal*, maximum utility, and all *unused* routes offer a lower utility.

Route utilities, consistent with the uncongested case, are defined as

$$\tilde{V}_{ijr} = b_{ij} - c_{ijr} - \tau t^*_{ijr} + \epsilon_{ijr} = b_{ij} - g^*_{ijr} + \epsilon_{ijr}; \qquad \forall\, i, j, r \tag{5.1}$$

where now the values of t^*_{ijr} and g^*_{ijr} are unknown, and must have their equilibrium values. This utility may also be estimated in terms of the utilities of the individual links which compose the route, from Formula (3.9a):

$$\tilde{V}_{ijr} = b_{ij} - \sum_a (g_a(v^*_a) + \epsilon_a)\delta^a_{ijr}; \qquad \forall\, i, j, r \tag{5.2}$$

where

$$g_a(v_a) = \tau t_a(v_a) + c_a; \qquad \forall\, a \tag{5.3}$$

and where now a link's travel time t_a is a function of the link's demand v_a, as specified by the "link congestion function," "link performance function," or simply link cost function $t_a(\ )$. Consequently, Formula (5.1) may then be written in terms of a link's generalized cost function $g_a(\ )$.

In practical terms, there are several possible specifications which may be used in this context. Probably the most commonly used for the car mode is the B.P.R. function (U.S. Bureau of Public Roads, 1964):

$$t_a(x_a) = t^0_a \left[1 + 0.15 \left(\frac{v_a}{K_a} \right)^4 \right]; \qquad \forall\, a \tag{5.4}$$

where t^0_a, the "free flow" travel time, is the (minimum) travel cost at zero volume, and K_a is a given "level of service" of the link. This value may, for instance, be set equal to the flow at which travel time is 15% higher than the minimum travel time at zero flow.[2]

In addition, there are several other possible specifications for link travel time functions. (See, for instance, Branston, 1976, for a review.) It should also be emphasized that link demands are assumed to be continuous variables (i.e., measured by positive real numbers), and not integers, so that the link travel time functions are continuous.

The typical shape of a link performance function is represented in Figure 5.1. Link congestion functions are also available for the case of transit (Harris, 1989).

[2]In this regard, the link's "capacity" as defined in the Highway Capacity Manual (Transportation Research Board, 1985) is the flow corresponding to the lowest "level of service," or worst travel conditions, under prevailing conditions.

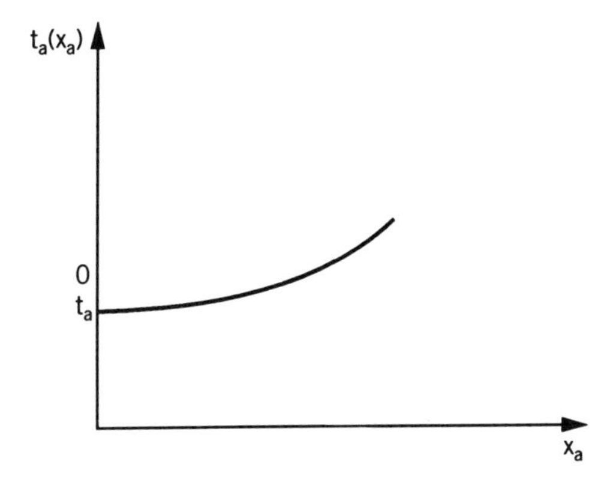

Figure 5.1 Link congestion function.

It is important to note that link-generalized cost functions must be strictly increasing in the link's demand, to guarantee uniqueness of the route demands (see section 5.2.2). Specifically, we require

$$\frac{\partial g_a(v_a)}{\partial v_a} > 0; \qquad \forall\ a,\ v_a \tag{5.5}$$

Since the term c_a, which represents the link's user charge, is constant, and since link cost functions, $t_a(\ )$, such as defined, for instance, by Formula (5.4), are strictly increasing in the link's volume, Condition (5.5) will be always be observed in practice.

It should also be noted that we have specified a link's travel time as a function of its own flow only. More generally, link performance functions may also be functions of demands on other links. For example, the speed of travel in one direction may be affected by the volume in the opposite direction. Furthermore, when individual modal networks interact, as when bus and cars use the same network's links, each modal travel time may be a function of other modes' demands. We will consider this case in section 3.3 of Chapter 6, in connection with modal split.

Following the presentation in the uncongested route choice case, we first examine the special case of deterministic utility.

5.2 CAR ROUTE CHOICE: DETERMINISTIC CASE

5.2.1 The R.T.'s Utility Maximization Problem and Traveler Surplus

In the deterministic case, the utility received by an individual traveling between i and j from taking route r is, as in the uncongested case, assumed to be primarily related to the route's travel time t_{ijmr} and the cost c_{ijmr}, respectively.

$$\tilde{V}_{ijr} = b_{ij} - \tau t^*_{ijr} - c_{ijr}; \qquad \forall\, i, j, r \tag{5.6}$$

where the endogenously determined, *unknown* term t^*_{ijr} is the equilibrium value of the route's travel time, as determined by equilibrium route demands T_{ijr}.[3]

Because of the presence of congestion on the network, the aggregate route demands cannot be obtained as the simple addition of individual route demands. That is, the route choices of individual travelers are interdependent, since routes' utilities are functions of travelers' choices, and thus of route volumes. Nevertheless, as in the uncongested case of Formula (3.4b), the aggregate route demands will be equal to:[4]

$$T^*_{ijr} = \begin{cases} T_{ij} & \text{if } g_{ijr} = g^*_{ijr} = \text{Min}_k\, g_{ijk} \\ 0 & \text{if } g_{ijr} > \text{Min}_k\, g_{ijk} \end{cases}; \qquad \forall\, i, j, r \tag{5.7a}$$

that is, the route between i and j with the maximum utility captures all the origin-destination demand.[5] In terms of probabilities of route choice, this means, again as in the uncongested case of Formula (3.4a):

$$P_{ijr} = \begin{cases} 1 & \text{if } g_{ijr} = g^*_{ijr} \\ 0 & \text{if } g_{ijr} > g^*_{ijr} \end{cases}; \qquad \forall\, i, j, r \tag{5.7b}$$

The major difference from the uncongested case, of course, is that the route utilities $\tilde{V}_{ijr}$ are unknown since the t_{ijr}'s are variable. In the congested case, aggregate demands must be derived at the aggregate level, as the solution of the R.T.'s U.M. problem.

[3]Intersection delays may also be incorporated (See Exercise 5.27).

[4]As in the uncongested case, however, route demands as specified in Formulas (5.7a) are not continuous, and furthermore, as we shall soon see, are not unique. However, we shall see that, again, these problems are resolved when route choice is considered probabilistic, and deterministic choice its limit case.

[5]This assumes that only one route from i to j will offer the maximum utility, as will generally be the case. If not, and as in the uncongested case, all such routes will capture an equal fraction of T_{ij}.

PROPOSITION 5.1

The R.T.'s direct utility from route choices in the presence of network externalities is equal to

$$U_R = -\tau \sum_a \int_0^{v_a} t_a(v)\, dv + T_0 \tag{5.8a}$$

where, for short

$$v_a = \sum_i \sum_j \sum_r (T_{ijr} + S_{ijr})\delta^a_{ijr}; \qquad \forall\, a \tag{5.8b}$$

and the S_{ijr}'s are the *given* route demands corresponding to all travel purposes other than the given one.

Functions $t_a(\;)$ are the link travel time functions, as defined in Formula (5.4).

The incorporation of the terms S_{ijr} is to take into account the fact that congestion is created by travelers for *all purposes*. However, from the standpoint of the R.T., these terms are *given*, and the result of known choices by another R.T. representing travelers for other purposes. If the route utilities of this second R.T. are the same as the first R.T.'s then they will make the same route choices, and (given) origin-destination demands S_{ij} may simply be added to the given T_{ij}, in the determination of the equilibrium utilities/travel times. If not, the route problem involves two different network users, and becomes somewhat more complex. The topic of multiple users is beyond the scope of this book, however. Several additional comments on this issue will be made later.

Consequently, the R.T.'s utility maximization problem is

$$\underset{(T_{ijr}, T_0)}{\text{Max}}\; U_R = -\tau \sum_a \int_0^{\Sigma_i\Sigma_j\Sigma_r(T_{ijr}+S_{ijr})\delta^a_{ijr}} t_a(v)\, dv + T_0 \tag{5.9a}$$

subject to the same Constraints (3.9a–c) as in the uncongested case, which are reproduced here.

$$\sum_i \sum_j \sum_r T_{ijr} c_{ijr} + T_0 = B \tag{5.9b}$$

$$\sum_r T_{ijr} = T_{ij}; \qquad \forall\, i, j \tag{5.9c}$$

$$T_{ijr} \geq 0;\; T_0 \geq 0; \qquad \forall\, i, j, r \tag{5.9d}$$

As usual, we may eliminate variable T_0 from the problem, so as to simplify it. This may be done by replacing the second term in the objective function by its expression in budgetary Constraint (5.9b), which then becomes unnecessary. Furthermore, we may change the problem into a minimization problem, by changing the sign of the objective function, which may thus be written as

$$\underset{(T_{ijr})}{\text{Min}}\ U'_R = \tau \sum_a \int_0^{\Sigma_i \Sigma_j \Sigma_r (T_{ijr} + S_{ijr})\delta^a_{ijr}} t_a(v)\, dv + \sum_i \sum_j \sum_r T_{ijr} c_{ijr}$$
$$= \sum_a \int_0^{v_a} g_a(v)\, dv \tag{5.10a}$$

after deletion of the constant term B. U'_R represents the R.T.'s disutility of travel. This is now to be minimized subject to the balance of demand requirements, which are that the route demands for the given travel purpose between a given origin and destination must add up to the given origin-destination demands T_{ij}:

$$\sum_r T_{ijr} = T_{ij}; \qquad \forall\ i, j \tag{5.10b}$$

In addition, all route demands must as always be nonnegative:[6]

$$T_{ijr} \geq 0; \qquad \forall\ i, j, r; \tag{5.10c}$$

To show that solving the R.T.'s U.M. Problem (5.10a–c) produces the aggregate demands of Formula (5.7a), the Karush-Kuhn-Tucker conditions (K.K.T) are now stated,[7] with respect to route demands T_{ijr}.[8,9] These conditions are

$$\frac{\partial U'_R}{\partial T_{ijr}} - \sum_i \sum_j \mu_{ij} \frac{\partial}{\partial T_{ijr}} \left\{ \sum_r T_{ijr} - T_{ij} \right\} \geq 0; \qquad \forall\ i, j, r \tag{5.11a}$$

$$T_{ijr} \left[\frac{\partial U'_R}{\partial T_{ijr}} - \sum_i \sum_j \mu_{ij} \frac{\partial}{\partial T_{ijr}} \left\{ \sum_r T_{ijr} - T_{ij} \right\} \right] = 0; \qquad \forall\ i, j, r \tag{5.11b}$$

in which the μ_{ij} are the dual variables for Constraints (5.10b), and the T_{ij}'s are given constants. Using the "chain rule" of partial derivatives, the derivatives of U'_R with respect to the route demands T_{ijr} are equal to:[10]

$$\frac{\partial U'_R}{\partial T_{ijr}} = \sum_a \frac{\partial U'_R}{\partial v_a} \cdot \frac{\partial v_a}{\partial T_{ijr}} = \sum_a g_a(v_a) \frac{\partial v_a}{\partial T_{ijr}} = \sum_a g_a(v_a) \delta^a_{ijr}; \qquad \forall\ i, j, r \tag{5.12}$$

Remembering that $t(\cdot)$ is link a's travel time function, and given the definition of δ^a_{ijr}, it is apparent that the term $\Sigma_a\, g_a(v_a)\delta^a_{ijr}$ in Formula (5.12) represents

[6]Route volumes are assumed to be continuous variables.

[7]These conditions are described in section 2.2 of Appendix A.

[8]Alternatively, the Lagrangean may be formed, and all partial derivatives with respect to the route flows and the Lagrangean coefficients set equal to zero, as in section 8 of Chapter 2.

[9]These conditions are *necessary* conditions which must hold at the minimum of the objective function. Note that we are assuming for the moment that there is a solution. The existence and uniqueness of the demands will be examined below, after we establish that the problem formulation above is indeed correct.

[10]The same expression may, of course, be obtained by differentiating directly with respect to T_{ijr} (see Exercise 5.19). However, we shall use this result later.

the sum of the travel costs on the links which compose route r between origin i and destination j, and is therefore equal to the travel cost g^*_{ijr}, on that route. Conditions (5.11) may then be written

$$g^*_{ijr} \geq \mu_{ij}; \qquad \forall\, i, j, r \tag{5.13a}$$

$$T^*_{ijr}(g^*_{ijr} - \mu_{ij}) = 0; \qquad \forall\, i, j, r \tag{5.13b}$$

Equations (5.13a) imply that μ_{ij} is smaller than the generalized cost on any route r between i and j. Equations (5.13b) implies that if T_{ijr} is positive, if route r between i and j is used, then the second factor must be equal to zero, so that μ_{ij} represents the generalized cost on that route, which is then the minimum cost between i and j. Conversely, if the route is not used, if T_{ijr} is equal to zero, then inequality (5.13a) is operative, and consequently, the generalized cost on the route is greater than, or equal to, the minimum generalized cost. These facts imply that

$$T^*_{ijr} = 0; \qquad \text{if } g^*_{ijr} > \operatorname*{Min}_k\,(g_{ijk}); \qquad \forall\, i, j, r \tag{5.14a}$$

$$T^*_{ijr} \geq 0; \qquad \text{if } g^*_{ijr} = \operatorname*{Min}_k\,(g_{ijk}); \qquad \forall\, i, j, r \tag{5.14b}$$

Relationships (5.14) are precisely the expression of the aggregate demands of Formula (5.7a), as desired. This result legitimizes the expression of function U_R in Formula (5.7a) as the R.T.'s direct utility.

The R.T.'s received utility from route choices in the presence of network externalities is equal to the optimal value of the direct utility function (5.9a),

$$\tilde{W}_R = B - \tau \sum_a \int_0^{v_a^*} t_a(v)\, dv - \sum_i \sum_j \sum_r T^*_{ijr} c_{ijr} = B - \sum_a \int_0^{v_a^*} g_a(v)\, dv \tag{5.15a}$$

where v^*_a and T^*_{ijr} are the equilibrium link volumes and route demands, respectively.

It may be noted that, due to externalities, the aggregate received utility is *not* equal to the simple addition of the individual received utilities, as specified in Formula (5.1). Comparing with the counterpart Formula (3.11c) for the uncongested case, it may be seen that the (negative) effect of congestion on the R.T.'s utility, or the community welfare function, is measured by the term:

$$-\tau \left[\sum_a \int_0^{v_a^*} t_a(v)\, dv - \sum_i \sum_j \sum_r T^*_{ijr} t^*_{ijr} \right] \tag{5.15b}$$

Finally, we may state the following result.

PROPOSITION 5.2

The traveler surplus from deterministic route choices in the presence of network externalities is equal to

$$TS_R = -\tau \sum_a \int_0^{v_a^*} t_a(v)\, dv - \sum_j \sum_r T_{ijr}^* c_{ijr} = - \sum_a \int_0^{v_a^*} g_a(v)\, dv \tag{5.16}$$

5.2.2 Existence and Uniqueness of Solutions[11]

Although we now know that the solution to the R.T.'s utility maximization program (5.10), *if* it exists, produces the correct aggregate demands, we still do not know whether such a solution does exist, nor do we know whether it is unique. In this section, we address these issues, which were set aside earlier. These are obviously relevant, not only in terms of application of this approach, but also from a theoretical point of view. We shall investigate these issues in terms of link demands, v_a, since it is easier analytically.

Some results from mathematical optimization theory may be invoked in this connection. (See section 2 of Appendix A.) If the set of feasible solutions to the mathematical optimization program is nonempty, compact (i.e., closed and bounded), and convex, then the program will have at least one solution (Rockafellar, 1970). In the present case, because the constraints on the R.T.'s utility maximization problem are linear, the feasible set is compact and convex. Also, it is clear that if there is at least one route r for all origins i and destinations j for which the given T_{ij} is nonzero, and if the total traveler's budget, or income, is sufficient to pay for the monetary costs of travel (two assumptions implicit in the formulation above), then it is always possible to assign the demand T_{ij} to the first such route, and zero demands to the others. Such an assignment, although not necessarily optimal, constitutes a feasible solution to the problem, and the feasible set will therefore be nonempty. Consequently, the R.T.'s U.M. problem will always have at least one solution.

Since the problem has a solution, if the objective function (5.10a) is everywhere convex in the variables v_a, then the solution will be unique (Rockafellar, 1970). The convexity of $U_R'(v_a)$ depends on whether its Hessian, that is, the matrix with general element

$$x_{ab} = \frac{\partial^2 U_R'(v)}{\partial v_a\, \partial v_b} \qquad \forall\ a, b \tag{5.17}$$

[11]This section may be skipped without loss of continuity, and the existence and uniqueness of the model's solution taken for granted.

is positive definite (Rockafellar, 1970). The first partial derivatives of $U'_R(v)$ in Formula (5.10a), with respect to the link demands v_a, are equal to

$$\frac{\partial U'_R(v)}{\partial v_a} = \frac{\partial \left[\sum_a \int_0^{v_a} g_a(v)\, dv \right]}{\partial v_a} = g_a(v_a); \qquad \forall\, a \tag{5.18a}$$

so that the second partial derivatives are

$$\frac{\partial^2 U'_R}{\partial v_a^2} = \frac{\partial g_a(v_a)}{\partial v_a}; \qquad \forall\, a \tag{5.18b}$$

$$\frac{\partial^2 U'_R}{\partial v_a\, \partial v_b} = \frac{\partial g_a(v_a)}{\partial v_b} = 0; \qquad \forall\, a, b \tag{5.18c}$$

Since we assume that Condition (5.5) is verified, all terms $(\partial g_a(v_a)/\partial v_a)$ are strictly positive, and the Hessian of U'_R is a diagonal matrix, where all diagonal terms are strictly positive, and all other terms are equal to zero. Such a matrix is positive definite (Rockafellar, 1970). Therefore, $U'_R(v_a)$ is strictly convex. Consequently, the solution to Problem (5.10a) will always be unique, *in terms of link demands*.

However, it is important to note that the solution may not be unique in terms of *route demands*, as may easily be verified analytically (See Exercise 5.2.) Thus, in practical terms, the R.T.'s utility maximization problem should be cast, and solved, in terms of link demands. Optimal route generalized costs $(\tau t_{ij}^* + c_{ij})$ may then be obtained from link demands through Formula (5.2). Having the route costs, route demands are finally obtained from Formulas (5.7a). However, since they are the solution of a MCR problem with given link costs, there may, *in principle*, be several minimum cost routes.

5.2.3 Solution Algorithm

Having formulated the R.T.'s utility maximization problem which produces the correct aggregate route demands (network assignment) in the presence of demand externalities, we now face the issue of how to solve it. Generally, an analytical model may never ultimately be disassociated from the procedure, or algorithm, which solves it numerically, as the model's features (e.g., nature of the objective function, or constraints) affect the solution method. Conversely, a model is only as good and efficient as the algorithm which solves it. In any case, it is clear that in practice, given the size of realistic networks, the R.T.'s utility maximization problem, and all others to come in subsequent sections and chapters, must be solved with a systematic procedure, or "algorithm," and with computers.

In the case of the R.T.'s utility maximization problem above, the objective function is convex, and the constraints are all linear, thus defining a convex feasible region. As such, it may be solved with standard techniques for problems of this type. While there are several possible methods available, a particularly efficient one in the present case is the "convex combination" method, also known as Frank-Wolfe algorithm, which is described below. This algorithm is based on a linear approximation of the objective function. It may be considered an example of the general "feasible descent" technique. (See, for instance, Bazaara, 1993 for details.) The principle of the method is as follows.

At iteration k of the algorithm, the value of the objective function $U'_R(\boldsymbol{v})$ in Formula (5.10a) is approximated as the linear function in the auxiliary variables y_a:[12]

$$U'_R(\boldsymbol{y}^k) = U'_R(\boldsymbol{v}^k) + \sum_a \left[\frac{U'_R(\boldsymbol{v})}{\partial v_a}\right]_k (y_a^k - v_a^k)$$

The original problem is then replaced with the auxiliary, or sub problem

$$\text{Min } U'_R(\boldsymbol{y}^k) = \text{Min}\left\{U'_R(\boldsymbol{v}^k) + \sum_a \left[\frac{\partial U'_R(\boldsymbol{v})}{\partial v_a}\right]_k (y_a^k - v_a^k)\right\} \quad (5.19)$$

in which the v_a^k's are known, subject to Constraints (5.10b,c) in the auxiliary variables Z. In this objective function, since the unknowns are the y_j's, the terms which are functions only of the v_a^k's are constant, and may be dropped from the linearized objective function. Therefore, the first term $U'_M(\boldsymbol{v}^k)$ may be dropped from the linearized objective function, as well as all terms of the form

$$\left[\frac{\partial U'_R(\boldsymbol{v})}{\partial v_a}\right]_k v_a^k$$

Consequently, the auxiliary objective function may be rewritten

$$\underset{y}{\text{Min}}\, f(\boldsymbol{y}) = \left\{\sum_a \left[\frac{\partial U'_R(\boldsymbol{v})}{\partial v_a}\right]_k y_a^k\right\} \quad (5.20)$$

subject to linear constraints

$$\sum_r Z_{ijr} = T_{ij}; \qquad \forall\, i, j \quad (5.21a)$$

$$Z_{ijr} \geq 0; \qquad \forall\, i, j, r; \quad (5.21b)$$

[12]Bold script refers to the matrix of variables, i.e., $\boldsymbol{x} = (x_1, \cdots, x_j, \cdots, x_n)$.

Since the value of the terms $[\partial U'_R(\boldsymbol{v})/\partial v_a]$ is known at the kth iteration, objective function $f(y)$ in Formula (5.20) is a linear combination of the unknown values y_j, and since constraints (5.21a) are linear, the nonlinear problem has been approximated by a linear program. Its solution, $\boldsymbol{y}^k$ may be shown to define the vector from current solution point $\boldsymbol{v}^k$ to "auxiliary" solution point $\boldsymbol{y}^k$, along which there is the greatest rate of decrease in the value of the objective function $U'_R(\boldsymbol{v}^k)$ from its current value. (See Exercise 5.8.)

Once this direction is identified by solving the linear problem above, the best estimate of the solution to the original problem is found between the points $\boldsymbol{v}^k$ and $\boldsymbol{y}^k$. This is because $\boldsymbol{y}^k$, being the solution to a linear program, automatically lies on the boundary of the feasible region, at the intersection of two or more of the constraints. The position of the solution to the original, nonlinear problem (i.e., the size of the move along that direction from the current solution $\boldsymbol{v}^k$) may then be determined by solving the problem:

$$\operatorname*{Min}_{\theta} \; U'_R(\boldsymbol{v}^k + \theta(\boldsymbol{y}^k - \boldsymbol{v}^k)) \tag{5.22a}$$

such that

$$0 \leq \theta \leq 1 \tag{5.22b}$$

This problem in the unknown θ is a one-dimensional optimization problem for which the feasible region is a given range of values (i.e., from zero to one). This may be solved using the bisection method, since $U'_R(\theta)$ is a continuous, single-valued function, which is known to have only one maximum in the interval [0, 1]. This problem is illustrated in Figure 5.2.

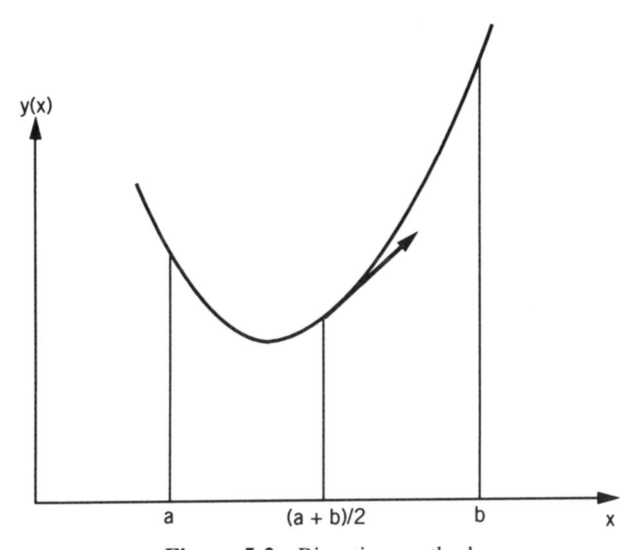

Figure 5.2 Bisection method.

Its principle is very simple. As can be seen in Figure 5.2, if at a given point θ the derivative of $U_R'(\theta)$ is negative, this implies that the minimum value of $U_R'(\theta)$ lies to the right of the point, so that the interval to the left of the point may be discarded. Conversely, of course, if the derivative of $U_R'(\theta)$ is positive, this implies that the minimum value of $U_R'(\theta)$ lies to the left of the point, so that the interval to the right of the point may be discarded.

The bisection method (for minimization) may then be described in the following way.

ALGORITHM 5.1. BISECTION METHOD

Update iteration counter by one.[13]

Set $\theta = (a + b)/2$ (i.e., the midpoint of the *current* interval $[a, b]$).

Compute the value of the derivative $F(\theta) = (dU_R'/d\theta)$

If $F(\theta) \leq 0$ set $a = \theta$ (i.e., the updated lower bound of the interval is the current midpoint).

If $F(\theta) \geq 0$ set $b = \theta$ (i.e., the updated upper bound of the interval is the current midpoint).

If $b - a \leq \epsilon$ stop.[14] An estimate of the solution is $\theta = (a + b)/2$ (i.e., the midpoint of the current interval $[a, b]$).

If $b - a \leq \epsilon$ iterate, i.e., go back to the top of the iterative step.

It is easy to see that after n iterations, the width of the current interval in which the estimated solution lies is equal to

$$(b_0 - a_0)/2^n$$

This relationship may be used to determine the number of iterations required, given the tolerance ϵ.

It is easy to show (see Exercise 5.24) that

$$F(\theta) = \frac{dU_R'}{d\theta} = \sum_a (y_a - v_a) g_a(v_a + \theta(y_a - v_a)) \tag{5.23}$$

This expression may then be used directly in the application of the bisection algorithm, thus bypassing the numerical estimation of the value of $U_R'(\theta)$ and then its θ derivative at each iteration.

Once the value of θ is found, the new, updated solution to the nonlinear problem is estimated as

$$v_a^{k+1} = v_a^k + \theta(y_a^k - v_a^k); \qquad \forall\, a \tag{5.24}$$

[13] The counter is initially set at zero, for the first iteration.

[14] ϵ is a tolerance level, whose value is equal to *twice* the width of the final interval in which the solution is desired to lie.

Since the value of the objective function is improved at each iteration, and the auxiliary's problem optimality conditions are the same as the original problem's, the algorithm is guaranteed to eventually converge to the exact solution.

Throughout the iterations, progress of convergence toward the exact solution and the "distance" from it may be assessed. Because of the convexity of the objective function, we have (see section 1.7 of Appendix A):

$$U_R'(\boldsymbol{v}^*) \geq U_R'(\boldsymbol{v}^k) + (\boldsymbol{v}^* - \boldsymbol{v}^k) \cdot \nabla U_R'(\boldsymbol{v}^k)$$

Because at iteration k the auxiliary problem's solution minimizes $\boldsymbol{y}^k \cdot \nabla U_R'(\boldsymbol{y}^k)$ for all feasible $\boldsymbol{y}$, we also have

$$U_R'(\boldsymbol{v}^k) + (\boldsymbol{v}^* - \boldsymbol{v}^k) \cdot \nabla U_R'(\boldsymbol{v}^k) \geq U_R'(\boldsymbol{v}^k) + (\boldsymbol{y}^k - \boldsymbol{v}^k) \cdot \nabla U_R'(\boldsymbol{v}^k)$$

Consequently, the r.h.s. value, which may be estimated as part of the iteration's computations, is a lower bound for the value of the objective function;

$$|U_R'(\boldsymbol{v}^k) - U_R'(\boldsymbol{v}^*)| \leq |(\boldsymbol{y}^k - \boldsymbol{v}^k) \cdot \nabla U_R'(\boldsymbol{v}^k)|; \qquad \forall\, k \qquad (5.25)$$

The algorithm may then be stopped whenever the maximum difference between the current solution and the optimal solution, the r.h.s. of Formula (5.25), is less than a prespecified tolerance.

The great advantage of using the linearization method for solving network equilibrium problems is that the auxiliary, linear Problem (5.20)–(5.21) becomes a simple minimum cost route problem. Indeed, since the route demands T_{ijr} may be used as alternative variables instead of the v_a's, we may express the objective function of the linearized problem in terms of auxiliary route demands Z_{ijr} instead of auxiliary link demands y_a. As was seen in Formula 5.12, the derivatives of the objective function (5.10a) in terms of the route demands are equal to

$$\frac{U_R'}{\partial T_{ijr}} = \sum_a (\tau t_a(v_a) + c_a)\delta_{ijr}^a = \sum_a g_a(v_a)\delta_{ijr}^a = g_{ijr}; \qquad \forall\, i, j, r \quad (5.26)$$

It may be readily recognized that this quantity, as was pointed out then is equal to the generalized travel cost between i and j on route r. Consequently, the linearized objective function for this problem at iteration k may be written as

$$\operatorname*{Min}_{Z} \left\{ \sum_i \sum_j \sum_r \left[\frac{\partial U_R'(T)}{\partial T_{ijr}} \right]_k Z_{ijr}^k \right\} = \sum_i \sum_j \sum_r g_{ijr}^k Z_{ijr}^k \qquad (5.27)$$

Since Z_{ijr} represents the (auxiliary) demand on route r between i and j, it is clear that objective function (5.27) represents the total generalized cost incurred by all travelers between all points i and j. This total cost will then obviously be minimized when all travelers between any two given points i and

j are assigned to the minimum (generalized) cost route between these two points. Thus, the solution of the linear problem may be obtained by identifying the minimum cost routes between any two points on the network. This may be performed with the use of the MCR algorithm described in section 3.2.3.

The iterative step of the solution algorithm for Problem (5.10) may then be summarized as follows:

ALGORITHM 5.2. LINEARIZATION OF DETERMINISTIC ROUTE CHOICE PROBLEM

Initialization

Set iteration counter k = zero.

Set all links demands at zero: $v_a = 0$.

Determine the corresponding link travel costs $g_a(0)$ from the link performance functions and the given link costs.

Assign each of the given T_{ij}'s to the corresponding minimum generalized cost route between i and j, using the MCR algorithm with these initial link costs, resulting in link demands v_a^0.

Iterative step

A. Update the link travel costs: $g_a = g_a(v_a^k)$.

B. Assign the given T_{ij} to the minimum cost route between i and j, using the MCR algorithm with these current link times, resulting in auxiliary link demands y_a^k.

C. Determine the value θ which minimizes

$$\underset{(0 \leq \theta \leq 1)}{\text{Min}} \quad U_R' = \sum_a \int_0^{v_a^k + \theta(y_a^k - v_a^k)} g_a(v)\, dv \tag{5.28}$$

where $g_a(v)$ is defined by Formula (5.4), using the bisection method.

D. Update the current solution $\boldsymbol{v}$ as

$$v_a^{k+1} = v_a^k + \theta(y_a^k - v_a^k); \qquad \forall\, a \tag{5.29}$$

E. Check whether

$$|(\boldsymbol{y}^k - \boldsymbol{v}^k) \cdot \nabla U_R'(\boldsymbol{v}^k)| = \sum_a (y_a^k - v_a^k) \left[\frac{\partial U_R'}{\partial v_a}\right]_k \leq \epsilon$$

where ϵ is a given tolerance.

If so, stop; if not, iterate.

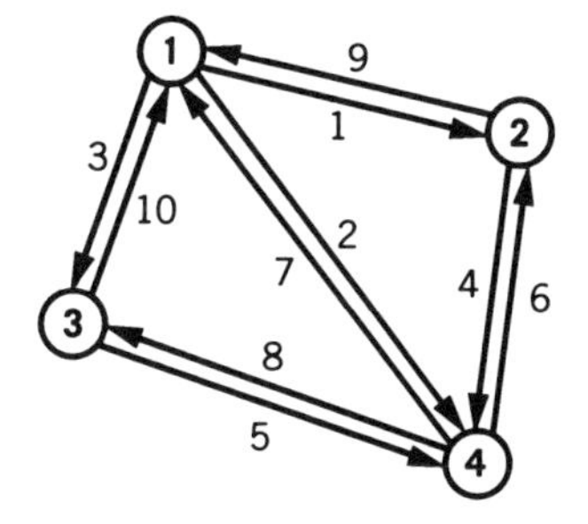

Figure 5.3 Example network.

5.2.4 Illustrative Example: Application of Linearization Algorithm

We shall now illustrate the application of the Frank-Wolfe algorithm to the hypothetical network used in the example of route choice/network assignment in Figure 5.3. The network's configuration is replicated in Figure 5.3 for convenience.

The major difference is that link travel times, instead of having the fixed values we have used so far, are now variable, a function of the link's demand. We shall use the BPR's specification given in Formula 5.4, where the values of the "free flow" travel times are the constant values used in the uncongested case, as listed in Table 1.3, and the levels of service K_a are as given in Table 5.1.

We again assume that there are no link charges, $c_a = 0$; $\forall\ a$. The numbers of travelers T_{14} and T_{12} from node 1 to node 4 and from node 1 to node 2 respectively, are given, and equal to

$$T_{14} = 150$$

$$T_{12} = 250$$

TABLE 5.1 Link Characteristics

Link	t_a^0	K_a
1	5	55
2	15	50
3	6	60
4	8	50
5	7	55
6	8	60
7	15	55
8	7	50
9	5	50
10	6	55

as in the uncongested case. For simplification, we assume that there are no other travelers (i.e., for other purposes than the one under analysis) using the network, so that $S_{ij} = 0$; $\forall\ i, j$. This might be the case, for example, for travel to work during the rush hour period.

Given these data, we would like to estimate the equilibrium demands on the respective alternative routes between these two origin-destination pairs, as well as on the individual network links, applying the Frank-Wolfe algorithm. Table 5.2 presents the results of the first two iterations of the algorithm.

At the algorithm's termination, the estimated equilibrium link demands, route demands, and route costs are as respectively represented in Tables 5.3 and 5.4 on page 156. (See Exercise 5.11.)

Graphically, equilibrium link demands may be represented as in Figure 5.4a on page 157.

All unrepresented links have a zero demand. Nonzero route demands may be represented graphically as in Figure 5.4b. It may be verified that, as expected, the utility maximization principle is observed. That is, all used routes between a given origin and a given destination cost the same, although they do not necessarily carry the same demand. Routes which are not used cost more. These loadings may be compared with the uncongested loadings. In particular, the total travel time for all travelers may be computed from the link demands and costs in Table 5.3, and is equal to 27,128 minutes, or an average travel time of 68 minutes per traveler. It should be noted that since there is congestion, the negative of this value does not measure the traveler surplus. (See Formulas (5.16a and b).) In the present case, the traveler surplus is equal to $-9{,}120$.

If, in the same manner as in the deterministic case, a user fee of \$10 is charged on links 4 and 6, and if, in reaction to road pricing, the value of τ is now equal to 0.1, reflecting increased sensitivity to monetary costs, link volumes are as in Figure 5.4c.

The total travel time is now equal to 37,480 minutes. The traveler surplus is now equal to $-1{,}543$. Traveler surplus when there were no user charges must be valued at $0.1(-9.120) = -912$, to take into account the change in value of τ from 1 to 0.1. Consequently, charging a user fee results in a net decrease in community welfare of 631, measured in dollars.

The convex combinations method is very efficient in solving the congested route choice problem since the linear optimization subproblems which must be solved at each iteration turn out to be precisely the same route choice problem in the *uncongested* case, that is, the problem of assigning given demands T_{ij} to the minimum cost route between i and j, which was solved in section 2.3 of Chapter 3. It should also be noted that the δ^a_{ijr} coefficients, although appearing in the objective Function (5.10a) above, are in fact not used in the algorithm, which, as described above, essentially entails repeatedly loading the given T_{ij}'s on minimum cost routes. Thus, in the deterministic case, route enumeration is not required. This will no longer be true in the stochastic case, as we shall soon see.

TABLE 5.2 Frank Wolfe Algorithm for Deterministic Route Choice Problem

A. Tabular summary of the first two iterations

Link flows

Iteration k = 0					1			2	
Link a	v_a	t_a	v_a	t_a	y_a	v_a	t_a	y_a	v_a
1	0	5	325	919.42	0	214.175	177.46	0	170.483
2	0	15	0	15.00	400	136.400	139.61	0	108.574
3	0	6	75	8.20	0	49.425	6.41	400	120.942
4	0	8	75	14.08	0	49.425	9.15	0	39.342
5	0	7	75	10.63	0	49.425	7.68	400	120.942
6	0	8	0	8.00	250	85.250	12.89	250	118.859
					$\theta_1 = 0.341$			$\theta_2 = 0.204$	

Route flows

Iteration k =				0	1		2	
O–D	Route	Links	t_r	T_r	t_r	Z_r	t_r	Z_r
1 to 2	1	1	5	250	919.42		177.46	
T_{12} = 250	2	2,6	23		23.00	250	152.50	
	3	3,5,6	21		26.83		26.99	250
1 to 4	1	2	15		15.00	150	139.61	
T_{14} = 150	2	1,4	13	75	933.49		186.60	
	3	3,5	13	75	18.83		14.10	150

B. Detailed computations for first iteration

Initialization

1. Set iteration counter $k = 0$
2. Set link flows $v_a = 0$
3. Determine the corresponding link costs $t_a(0)$ from the link performance functions.

$$t_a = t_a^0\left(1 + 0.15\left(\frac{0}{K_a}\right)^4\right) = t_a^0; \qquad \forall\, a$$

4. Assign each of the given T_{ij}'s to the corresponding minimum cost route between i and j, using the MCR algorithm with t_a^0, resulting in link demands v_a^0.

$T_{12} = 250 \rightarrow$ Load 250 onto link 1, since $t_{ijr} = 5$ is minimum cost.

$T_{14} = 150 \rightarrow$ Load ($150/2 = 75$) onto links 1 and 4, and links 3 and 5,

since routes 2 and 3 have equal minimum costs of $t_{ijr} = 13$.

TABLE 5.2 (*Continued*)

Iterative Step

A. Update link travel costs:

$$t_a^1 = t_1^0 \left[1 + 0.15 \left(\frac{v_a^1}{K_a} \right)^4 \right]; \qquad \forall\, a$$

$$t_1^1 = 5 \left[1 + 0.15 \left(\frac{325}{55} \right)^4 \right] = 919.42$$

$$t_2^1 = 5 \left[1 + 0.15 \left(\frac{0}{50} \right)^4 \right] = 15.0$$

$$t_3^1 = 6 \left[1 + 0.15 \left(\frac{75}{60} \right)^4 \right] = 8.20$$

$$t_4^1 = 8 \left[1 + 0.15 \left(\frac{75}{50} \right)^4 \right] = 14.07$$

$$t_5^1 = 7 \left[1 + 0.15 \left(\frac{75}{55} \right)^4 \right] = 10.63$$

$$t_6^1 = 8 \left[1 + 0.15 \left(\frac{0}{60} \right)^4 \right] = 8.0$$

B. Assign the given T_{ij} to the minimum cost route between i and j, using the MCR algorithm with these current link times, resulting in auxiliary link demands Z_a^K

$R_\#$	t_{ijr}	$R_\#$	t_{ijr}
1	919.42	7	15
2	15 + 8 = 23	8	919.42 + 14.07 = 933.49
3	8.20 + 10.63 + 8 = 26.83	9	8.20 + 10.63 = 18.83

$$T_{12} = 250 \rightarrow \text{Load 250 onto route 2, (Links 2,6)}$$

$$T_{14} = 150 \rightarrow \text{Load 150 onto route 1, (Link 2)}$$

C. Solve:

$$\underset{(0 \leqslant \theta \leqslant 1)}{\text{Min}} \quad U'^k = \sum_a \int_a^{v_a^k + \theta(y_a^k - v_a^k)} t_a(x)\, dx$$

$$= \int_0^{325 + \theta(0 - 3.25)} 5 \left(1 + 0.15 \left(\frac{x}{55} \right)^4 \right) dx + \int_0^{0 + \theta(400 - 0)} 15 \left(1 + 0.15 \left(\frac{x}{50} \right)^4 \right) dx$$

$$+ \int_0^{75 + \theta(0 - 75)} 6 \left(1 + 0.15 \left(\frac{x}{60} \right)^4 \right) dx + \int_0^{75 + \theta(0 - 75)} 8 \left(1 + 0.15 \left(\frac{x}{50} \right)^4 \right) dx$$

$$+ \int_0^{75 + \theta(0 - 75)} 7 \left(1 + 0.15 \left(\frac{x}{55} \right)^4 \right) dx + \int_0^{0 + \theta(250 - 0)} 8 \left(1 + 0.15 \left(\frac{x}{60} \right)^4 \right) dx$$

TABLE 5.2 (*Continued*)

Estimate

$$\frac{dU'^k}{d\theta} = \sum_a t_a^0 \left(1 + 0.15 \left(\frac{v_a^k}{K_a}\right)^4\right) (y_a^k - v_a^k)$$

$$= -1625[1 + 182.88(1 - \theta)^4] + 6000[1 + 614.4\theta^4] - 450[1 + 0.37(1 - \theta)^4]$$
$$- 600[1 + 0.76(1 - \theta)^4] - 525[1 + 0.52(1 - \theta)^4] + 2000[1 + 45.21\theta^4]$$

Apply bisection method

Iteration #	$dU'_k(\theta)/d\theta$	a	b	θ	$dU'k(\theta)/d\theta$
0	—	0	1	0.5	−222,221.53
1	−222,221.53	0	0.5	0.25	+75,759.75
2	+75,759.75	0.25	0.5	0.375	−34,005.38
3	−34,005.38	0.25	0.375	0.3125	+25,772.68
4	+25,772.68	0.3125	0.375	0.34375	−2250.26
5	−2250.26			0.3414	26.77

$$\text{Therefore } \theta_1^* = 0.3414$$

D. Update the current solution v as $v_a^{k+1} = v_a^k + \theta^*(y_a^k - v_a^k)$.
From E, $\theta = 0.3414$, therefore $v_a^2 = v_a^1 + 0.3414\,(y_a^1 - v_a^1)$.

$$v_1^2 = 325 + 0.341(0 - 325) = 214.175$$

$$v_2^2 = 0 + 0.341(400 - 0) = 136.4$$

$$v_3^2 = 75 + 0.341(0 - 75) = 49.425$$

$$v_4^2 = 75 + 0.341(0 - 75) = 49.425$$

$$v_5^2 = 75 + 0.341(0 - 75) = 49.425$$

$$v_6^2 = 0 + 0.341(250 - 0) = 85.25$$

E. Estimate the value of the convergence criterion

$$|(y^k - v^k) \cdot \nabla U(v^k)| = \sum_a (y_a^k - v_a^k) \left[\frac{\partial U}{\partial v_a}\right]_k$$

$$(0 - 214.05)(5)\left[1 + 0.15\left(\frac{214.05}{55}\right)^4\right] = -37{,}898.97$$

$$(400 - 136.56)(15)\left[1 + 0.15\left(\frac{136.56}{50}\right)^4\right] = 36{,}933.70$$

$$(0 - 49.40)(6)\left[1 + 0.15\left(\frac{49.40}{60}\right)^4\right] = -316.83$$

TABLE 5.2 (*Continued*)

$$(0 - 49.40)(8)\left[1 + 0.15\left(\frac{49.40}{50}\right)^4\right] = -451.69$$

$$(0 - 49.40)(7)\left[1 + 0.15\left(\frac{49.40}{55}\right)^4\right] = -379.56$$

$$(250 - 85.35)(8)\left[1 + 0.15\left(\frac{85.35}{60}\right)^4\right] = \underline{2,126.21}$$

$$12.86$$

Moreover, it may be seen that the convex combination algorithm is in essence similar to the heuristic "capacity restraint" method described in section 4.4 of Chapter 1. Both methods rely on performing a series of minimum cost route assignments. Starting from an initial estimate of equilibrium loadings, each new, improved estimate is obtained as a weighted average ("convex combination"), of the previous estimate and of current one, with weights θ and $(1 - \theta)$, respectively. Specifically, Formula (5.29) implies

$$\boldsymbol{v}^{k+1} = \theta y^k + \boldsymbol{v}^k(1 - \theta) \tag{5.30}$$

The major improvement over the heuristic, however, which incidentally results in the only additional difficulty in this "exact" approach, lies in the determination of the optimal weights to be used in the averaging of successive solutions, that is, of the value of θ. This in turn only requires the minimization

TABLE 5.3 Link Demands and Costs

Link	Demand	Time
1	169	71
2	106	59
3	126	23
5	126	36
6	81	12

TABLE 5.4 Route Costs

Route #	Demand	Cost
1	169	71
2	81	71
3	0	71
7	106	59
8	0	79
9	44	59

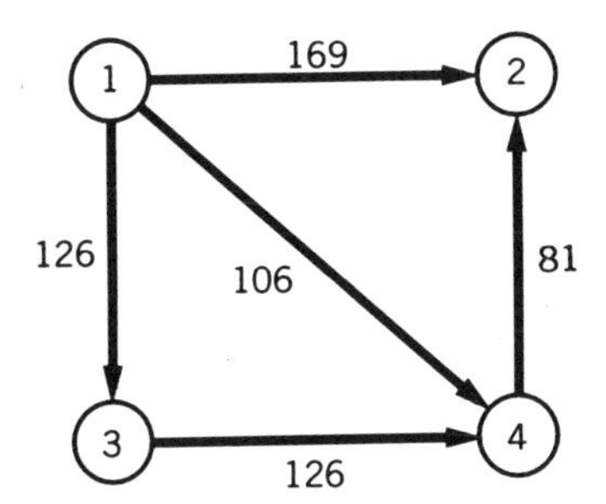

Figure 5.4a Equilibrium link demands.

of a function of a single variable in a given interval of values, which is solved with the rather simple "bisection" method. The advantage is that as already noted, the convex combination algorithm is *guaranteed* to approximate the *exact* solution within a given tolerance in a finite number of iterations. Thus, the convex combination method provides a better estimate of the exact equilibrium demands with less computational effort.

Before we move on to the case of random utility, it should be mentioned that the same deterministic R.T. utility maximization problem in the presence of congestion externalities above may also equivalently be formulated solely in terms of link demands (see Exercise 5.1):

$$\underset{(v_{aj})}{\text{Min}}\ U'_M = \sum_a \int_0^{\Sigma_j v_{aj}} g_a(v)\, dv \tag{5.31}$$

subject to

$$\sum_{a \in i^-} v_{aj} - \sum_{a \in i^+} v_{aj} = T_{ij}; \qquad \forall\, i, j \tag{5.32a}$$

$$v_{aj} \geq 0; \qquad \forall\, j, a \tag{5.32b}$$

where i^- and i^+ mean the set of links leaving and entering node i respectively, and

$$T_{ii} = -\sum_k T_{ki}; \qquad \forall\, i \tag{5.33}$$

as already defined in Formula 3.27.

This version of the R.T.'s problem may be solved with the Frank-Wolfe algorithm in the same manner as Problem (5.10a–c), by performing and optimally combining a series of minimum cost route assignments with fixed link costs.

Finally, the deterministic route choice problem on congested networks may also equivalently be formulated as a "variational inequality" problem, to which specialized algorithms may be applied. One advantage of this approach is that

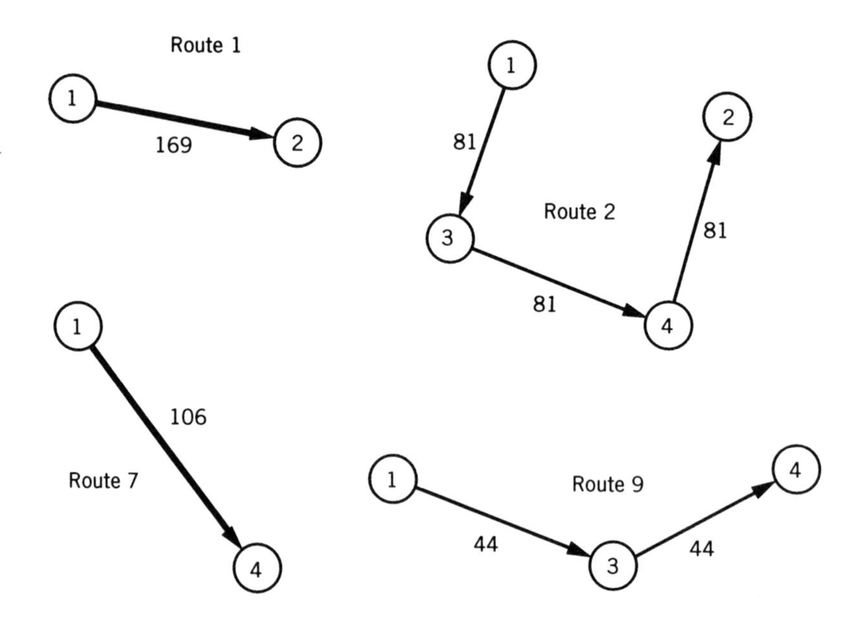

Alternative equilibrium solution

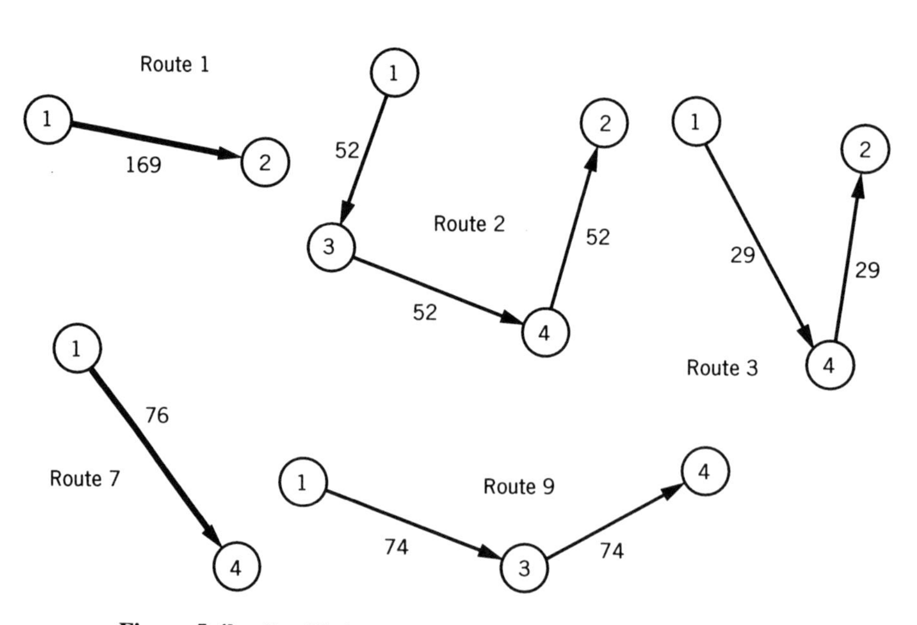

Figure 5.4b Equilibrium route demands for illustrative example.

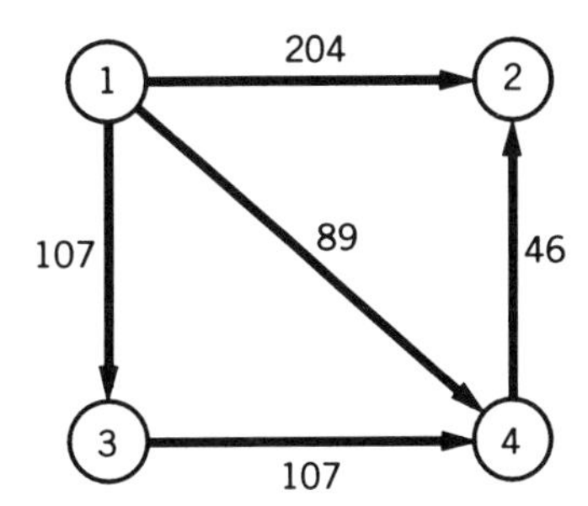

Figure 5.4c Equilibrium link volumes with user charges.

it may be applied to generalized formulations of the problem, including where link costs are functions of demands on other links, and when several R.T.'s with different utility functions use the same network concurrently. This is particularly convenient to model the interactions between various different classes of travelers using the network simultaneously, as when travelers for different purposes have different utilities from a given route. (See, for instance, Nagurney, 1993 for a detailed discussion.)

5.3 CAR ROUTE CHOICE: PROBABILISTIC CASE

The model developed above for the route choice problem under congestion was based on the assumption of deterministic route utilities. In this section, we generalize the formulation to the case of random route utility. Specifically, we now assume that the individual utility received by an individual traveling between i and j from taking route r has the same specification as in Formula (3.14) for the probabilistic uncongested case,

$$\begin{aligned}\tilde{V}_{ijr} &= \tilde{U}^*_{ijr} + \epsilon_{ijr} = b_{ij} - \tau t^*_{ijr} - c_{ijr} + \epsilon_{ijr} \\ &= b_{ij} - g^*_{ijr} + \epsilon_{ijr}; \qquad \forall\, i, j, r \end{aligned} \tag{5.34}$$

in which the random terms ϵ_{ijr}'s are assumed to be i.i.d. Gumbel variables. As in the deterministic congested case, the link travel times t_a are variable (unknown).

5.3.1 Network Assignment: The Fixed Point Method

Under the same utility maximization principle, the probability that a given individual traveler between i and j selected at random will choose route r, as in the uncongested case of Formula (3.15a), is equal to

$$P_{r/ij} = \frac{e^{\beta_r \tilde{U}^*_{ijr}}}{\sum_r e^{\beta_r \tilde{U}^*_{ijr}}} = \frac{e^{-\beta_r(\tau t^*_{ijr} + c_{ijr})}}{\sum_r e^{-\beta_r(\tau t^*_{ijr} + c_{ijr})}} = \frac{e^{-\beta_r g^*_{ijr}}}{\sum_r e^{-\beta_r g^*_{ijr}}}; \qquad \forall\, i, j, r \tag{5.35a}$$

In terms of route demands, this is equivalent to

$$T_{ijr} = T_{ij} \frac{e^{-\beta_r g_{ijr}^*}}{\sum_r e^{-\beta_r g_{ijr}^*}}; \qquad \forall\ i, j, r \tag{5.35b}$$

These individual probabilities depend on route times, on individual link demands, and thus ultimately on aggregate route demands. The unknown route demands are then the solution of the system of equations obtained by eliminating the route costs in Equations (5.35a), as follows.[15]

First, using Formula (1.2a), the route demands are equal to

$$T_{ijr}^* = T_{ij} P_{r/ij} = \frac{T_{ij} \exp\left(-\beta_r \sum_a g_a(v_a^*)\delta_{ijr}^a\right)}{\sum_r \exp\left(-\beta_r \sum_a g_a(v_a^*)\delta_{ijr}^a\right)}; \qquad \forall\ i, j, r \tag{5.36}$$

Then, replacing the link demands by their expression in terms of the route demands through Formula (1.2b), the T_{ijr}'s are solution of the system

$$T_{ijr}^* = \frac{T_{ij} \exp\left(-\beta_r \sum_a \delta_{ijr}^a g_a\left(\sum_i \sum_j \sum_r (T_{ijr}^* + S_{ijr})\delta_{a,r}^{i,j}\right)\right)}{\sum_r \exp\left(-\beta_r \sum_a \delta_{ijr}^a g_a\left(\sum_i \sum_j \sum_r (T_{ijr}^* + S_{ijr})\delta_{a,r}^{i,j}\right)\right)}; \qquad \forall\ i, j, r \tag{5.37}$$

In matrix notation,[16] this square system of equations is of the form

$$\boldsymbol{T} = \boldsymbol{F}(\boldsymbol{T}) \tag{5.38}$$

This particular form of equation system is sometimes called a "fixed point" problem. It should be noted that the specification of the values of the "link-route" incidence coefficients δ_{ijr}^a requires route enumeration. As already noted in the case of fixed utility in section 3.3 of Chapter 3, this may pose practical problems with realistic networks, because of the large number of routes.

As in the deterministic case, the value of the S_{ijr}'s are assumed given. If

[15]The reason why this analytical approach is not possible in the deterministic case is that route volumes may not be expressed by equalities, but by Inequalities (5.7).

[16]Bold symbols refer to arrays of values (vectors and matrices).

not, they may be evaluated concurrently with the T_{ijr}'s by *simultaneously* solving the other problem.[17]

$$S^*_{ijr} = \frac{S_{ij} \exp\left(-\beta_r \sum_a \delta^a_{ijr} g_a \left(\sum_i \sum_j \sum_r (T_{ijr} + S^*_{ijr}) \delta^{i,j}_{a,r}\right)\right)}{\sum_r \exp\left(-\beta_r \sum_a \delta^a_{ijr} g_a \left(\sum_i \sum_j \sum_r (T_{ijr} + S^*_{ijr}) \delta^{i,j}_{a,r}\right)\right)}; \qquad \forall\, i, j, r$$

(5.39)

Once the aggregate route demands are estimated, the probabilities of an individual choice of route are equal to

$$P_{r/ij} = T^*_{ijr}/T_{ij}; \qquad \forall\, i, j, r$$

5.3.2 Illustrative Example

Let us illustrate this approach in the case of the same example network we used earlier. The data is the same. However, we now assume that instead of being equal to ∞, as in the deterministic case, the value of parameter β is equal to 0.05. As already noted, since only travel times are used to represent disutility, we may set the value of τ at 1.0. The system of Equations (5.37) then takes the specific form shown on p. 163.

All other route demands are equal to zero. Also, all S_{ijr}'s are assumed equal to zero. After expressing the functions t_a according to Formula (5.4) in which the values of t^0_a and K_a are the given values in Table 5.1, this becomes a system of six nonlinear equations in six unknowns, the route demands T_{121}, T_{122}, T_{123}, T_{141}, T_{142}, and T_{145}.

Fixed point problems such as this may be solved with simple, "quasi-balancing" algorithms. The general idea behind the method, as graphically represented in Figure 5.5, is to start from an initial value for the left-hand side of Formula (5.37), compute the corresponding value of the right-hand side, assign this value to the left-hand side, and repeat this "recycling" of values for the unknowns until there is only a negligible change in them from one iteration to the next.

[17]The joint solution then constitutes a "Cournot-Nash" equilibrium, where the two R.T.'s representing the two types of travelers each maximize their utility, each taking for given the route choices of the other. In other words, at equilibrium

$$\tilde{W}(\boldsymbol{T}^*, \boldsymbol{S}) \geq \tilde{W}(\boldsymbol{T}, \boldsymbol{S}); \qquad \forall\, \boldsymbol{T}, \boldsymbol{S}$$

$$\tilde{Y}(\boldsymbol{T}, \boldsymbol{S}^*) \geq \tilde{Y}(\boldsymbol{T}, \boldsymbol{S}); \qquad \forall\, \boldsymbol{T}, \boldsymbol{S}$$

where $\tilde{Y}$ is the received utility of the R.T. representing the S_{ij} travelers.

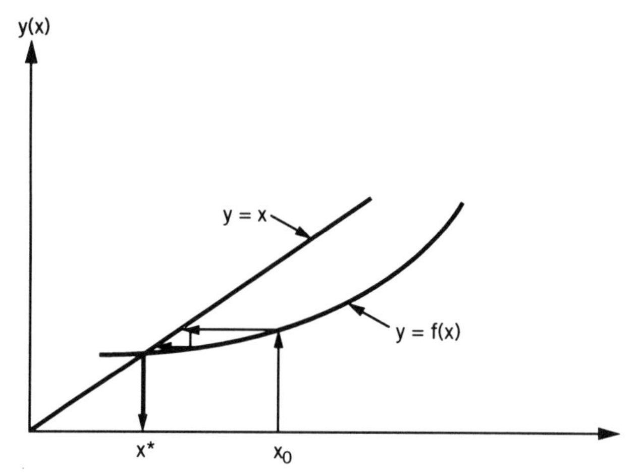

Figure 5.5 "Quasi-balancing" solution to the Fixed Point problem in one variable.

This procedure may be formalized as follows, in the case of Equations (5.29).

ALGORITHM 5.3. FIXED POINT ALGORITHM

Iterative step

Set iteration counter at $k + 1$ (or initially at 1). Given the current (estimated) values for the route demands $\boldsymbol{T}^k$ (or initial, guessed values $\boldsymbol{T}^1$), estimate the next value $\boldsymbol{T}^{k+1}$ from Equations (5.37):

$$\boldsymbol{T}^{k+1} = \boldsymbol{F}(\boldsymbol{T}^k) \tag{5.40}$$

If $k > 1$ check for convergence:

$$\|\boldsymbol{T}^{k+1} - \boldsymbol{T}^k\| \leq \epsilon \tag{5.41}$$

where ϵ represents a predetermined convergence tolerance. If (5.41) holds, stop. If not, iterate.

For algorithms of this type, however, there is a necessary condition for values of the unknown variables, T_{ijr}, in this case, to converge to a finite solution. (See Exercise 5.9.) A major advantage of this type of method is that it is easy to translate into code for computer implementation. Alternatively, "spreadsheet" computer applications may be used. (See Exercise 5.6.)

$$T_{121} = 250 \frac{\exp(-0.05t_1(T_{121} + T_{133} + T_{142}))}{\exp(-0.05t_1(T_{121} + T_{133} + T_{142})) + \exp(-0.05\{t_3(T_{122} + T_{131} + T_{143}) + t_5(T_{122} + T_{143}) + t_6(T_{122} + T_{123})\} + \exp(-0.05\{t_2(T_{123} + T_{132} + T_{141}) + t_6(T_{122} + T_{123})\}}$$

$$T_{122} = 250 \frac{\exp(-0.05\{t_3(T_{122} + T_{131} + T_{143}) + t_5(T_{122} + T_{143}) + t_6(T_{122} + T_{123})\})}{\exp(-0.05t_1(T_{121} + T_{133} + T_{142})) + \exp(-0.05\{t_3(T_{122} + T_{131} + T_{143}) + t_5(T_{122} + T_{143}) + t_6(T_{122} + T_{123})\} + \exp(-0.05\{t_2(T_{123} + T_{132} + T_{141}) + t_6(T_{122} + T_{123})\}}$$

$$T_{123} = 250 \frac{\exp(-0.05\{t_2(T_{123} + T_{132} + T_{141}) + t_6(T_{122} + T_{123})\})}{\exp(-0.05t_1(T_{121} + T_{133} + T_{142})) + \exp(-0.05\{t_3(T_{122} + T_{131} + T_{143}) + t_5(T_{122} + T_{143}) + t_6(T_{122} + T_{123})\} + \exp(-0.05\{t_2(T_{123} + T_{132} + T_{141}) + t_6(T_{122} + T_{123})\}}$$

$$T_{141} = 150 \frac{\exp(-0.05t_2(T_{123} + T_{132} + T_{141}))}{\exp(-0.05t_2(T_{123} + T_{132} + T_{141})) + \exp(-0.05\{t_1(T_{121} + T_{133} + T_{142}) + t_4(T_{133} + T_{141})\} + \exp(-0.05\{t_3(T_{122} + T_{143}) + t_5(T_{122} + T_{143})\}}$$

$$T_{142} = 150 \frac{\exp(-0.05t_1(T_{121} + T_{133} + T_{142}) + t_4(T_{133} + T_{142}))}{\exp(-0.05t_2(T_{123} + T_{132} + T_{141})) + \exp(-0.05\{t_1(T_{121} + T_{133} + T_{142}) + t_4(T_{133} + T_{142})\} + \exp(-0.05\{t_3(T_{122} + T_{143}) + t_5(T_{122} + T_{143})\}}$$

$$T_{143} = 150 \frac{\exp(-0.05\{t_3(T_{122} + T_{131} + T_{143}) + t_5(T_{122} + T_{143})\})}{\exp(-0.05t_2(T_{123} + T_{132} + T_{141})) + \exp(-0.05\{t_1(T_{121} + T_{133} + T_{142}) + t_4(T_{133} + T_{142})\} + \exp(-0.05\{t_3(T_{122} + T_{143}) + t_5(T_{122} + T_{143})\}}$$

More generally, a standard method for solving square systems of nonlinear equations

$$F_1(x_1, x_2, \ldots x_j, \ldots, x_n) = 0$$

$$\vdots$$

$$F_i(x_1, x_2, \ldots x_j, \ldots, x_n) = 0$$

$$\vdots$$

$$F_n(x_1, x_2, \ldots x_j, \ldots, x_n) = 0 \quad (5.42)$$

such as Equations (5.37) in particular, is the Newton-Raphson method. The essence of the procedure may be conveyed graphically in the case of a single unknown, as in Figure 5.6.

The algorithm's basic iterative step is

$$x^{k+1} = x^k - f^k/f'^k; \qquad \forall\, k \quad (5.43)$$

In the general, multivariate case, the algorithm may be stated in the following way.

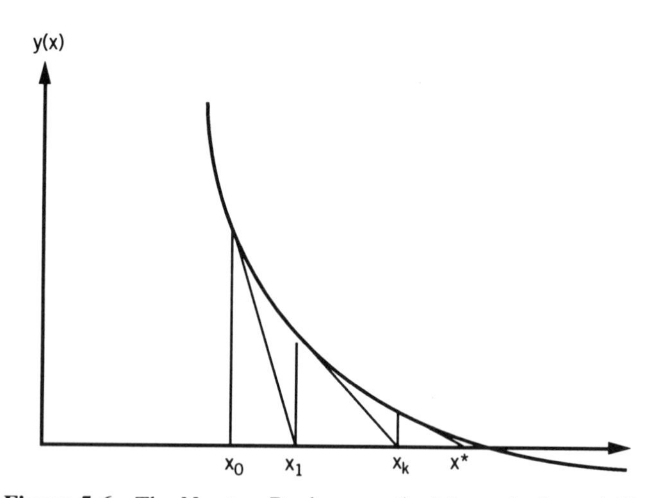

Figure 5.6 The Newton-Raphson method for a single variable.

ALGORITHM 5.4. NEWTON'S METHOD

Initialization

Set counter at zero.
Assume initial solution $\boldsymbol{X}^0$.

Iteration

Set $k = k + 1$.
Evaluate

$$\boldsymbol{X}^{k+1} = \boldsymbol{X}^k - \boldsymbol{J}^{-1}\boldsymbol{F} \tag{5.44a}$$

where J is the Jacobian of functions F_i in the equation system to be solved, that is the matrix with general element

$$a_{ij} = \frac{\partial F_i(x_1, \ldots, x_j, \ldots, x_n)}{\partial x_j}; \qquad \forall\ i, j \tag{5.44b}$$

Check convergence

$$\|\boldsymbol{X}^{k+1} - \boldsymbol{X}^k\| \leq \epsilon$$

If so, stop. If not, iterate.

The advantage of this method over the quasi-balancing method is that it is guaranteed to approximate a solution in a finite number of iterations.[18] While this method is significantly more intricate to code than the quasi-balancing method, it is readily available as part of most commercially available computer applications for mathematical analysis. (See Exercise 5.7.)

5.3.3 Aggregate Demands: The R.T.'s Utility Maximization Problem and Traveler Surplus

In this section, and in keeping with the practice we have followed so far, we formulate the R.T.'s utility maximization problem as an alternative approach to the derivation of the aggregate demands.

[18]The tradeoff is that this method requires the analytical expression of the partial derivatives of the above functions. This may be difficult to obtain. In addition, the method requires the inversion of the Jacobian at each iteration. There may be values of the unknowns which make this impossible. The performance of the method is also rather sensitive to how good an estimate of the final solution the starting point represents.

PROPOSITION 5.3

The R.T.'s direct utility function in the congested probabilistic route choice case is equal to

$$U_R = -\tau \sum_a \int_0^{\Sigma_i \Sigma_j \Sigma_r (T_{ijr} + S_{ijr})\delta_{ijr}^a} t_a(v)\, dv - \frac{1}{\beta_r} \sum_i \sum_j \sum_r T_{ijr} \ln T_{ijr} + T_0 \tag{5.45}$$

This is the same as the utility Function (5.10a) in the deterministic case, with the addition of the second, "entropy" term. Consequently, the R.T.'s utility maximization problem in this case is

$$\text{Max } U_R(T_{ijr}, T_0) = -\tau \sum_a \int_0^{\Sigma_i \Sigma_j \Sigma_r (T_{ijr} + S_{ijr})\delta_{ijr}^a} t_a(v)\, dv - \frac{1}{\beta_r} \sum_i \sum_j \sum_r T_{ijr} \ln T_{ijr} + T_0 \tag{5.46}$$

subject to the same constraints as in the deterministic, congested (and also uncongested) case, the budgetary Constraint (5.9b), and the conservation of demand Constraints (5.9c). Nonnegativity Constraints (5.10d) are of course always present.

As usual, the R.T.'s problem's formulation may be simplified, through elimination of variable T_0 and the budgetary constraint. The problem is then

$$\begin{aligned} \text{Min } U'_R(T_{ijr}) &= \sum_a \int_0^{\Sigma_i \Sigma_j \Sigma_r (T_{ijr} + S_{ijr})\delta_{ijr}^a} g_a(v)\, dv + \frac{1}{\beta_r} \sum_i \sum_j \sum_r T_{ijr} \ln T_{ijr} \\ &= \sum_a \int_0^{v_a(T_{ijr})} g_a(v)\, dv + \frac{1}{\beta_r} \sum_i \sum_j \sum_r T_{ijr} \ln T_{ijr} \end{aligned} \tag{5.47}$$

subject to

$$\sum_r T_{ijr} = T_{ij}; \qquad \forall\, i, j \tag{5.48a}$$

$$v_a \geq 0; \; T_{ijr} > 0; \qquad \forall\, i, j, r \tag{5.48b}$$

As before, the S_{ijr}'s are assumed known if the corresponding travelers do not have the same route utilities. If they do, the determination of the S_{ijr}'s given the values of the S_{ij}'s is achieved simply through adding the constraints

$$\sum_r S_{ijr} = S_{ij}; \qquad \forall\, i, j \tag{5.48c}$$

$$S_{ijr} > 0; \qquad \forall\, i, j, r \tag{5.48d}$$

to the above problem. This R.T.'s U.M. problem generalizes the formulation of Fisk (1980). As expected, when the route utilities are deterministic ($\beta_r = \infty$), the second utility term vanishes. The formulation is then the same as in the deterministic case above.

It is straightforward to show, in the same manner as in the deterministic case, that the solution to Problem (5.47)–(5.48a–d) produces aggregate route demands conforming to Formulas (5.35b). This is demonstrated in Appendix 5.1, at the end of this chapter.

It is important to note that, as in the uncongested case, at any given time the generalized costs on all used routes in the probabilistic case are *not* equal, as they were in the deterministic case.

The R.T.'s received utility in the probabilistic case is equal to (see Exercise 5.21):

$$\tilde{W}_R = B - \sum_a \int_0^{v_a^*} g_a(v)\, dv - \frac{1}{\beta_r} \sum_i \sum_j T_{ij} \ln T_{ij}$$
$$+ \frac{1}{\beta_r} \sum_i \sum_j T_{ij} \ln \sum_r e^{-\beta_r g_{ijr}^*} + \sum_i \sum_j \sum_r T_{ijr}^* g_{ijr}^* \tag{5.49a}$$

where, as usual, the v_a^* may be expressed in terms of the route demands T_{ijr}^* (and if applicable, S_{ijr}^*). Consequently, we have the following result.

PROPOSITION 5.4

The traveler surplus in the probabilistic, congested route choice case is equal to

$$TS_R = -\tau \sum_a \int_0^{v_a^*} t_a(v)\, dv - \frac{1}{\beta_r} \sum_i \sum_j T_{ij} \ln T_{ij}$$
$$+ \frac{1}{\beta_r} \sum_i \sum_j T_{ij} \ln \sum_r e^{-\beta_r g_{ijr}^*} + \sum_i \sum_j \sum_r T_{ijr}^* t_{ijr}^*$$

As expected from the uncongested case of Formula (3.18b), the equilibrium aggregate route demands may be retrieved using the following theorem

THEOREM 5.1

$$T^*_{ijr} = -\frac{\partial \tilde{W}_R}{\partial c_{ijr}} \bigg/ \frac{\partial \tilde{W}_R}{\partial B} \qquad \forall\ i, j, r \tag{5.50}$$

This important result, which is demonstrated in Appendix 5.2, shows that in the general congested stochastic case, the R.T.'s utility function in Formula (5.46) is legitimate from a microeconomic standpoint. The use of nonlinear optimization programs of the type (5.47)–(5.48) has been standard practice in network analysis and transportation planning, but without being rigorously justified from a "behavioral" standpoint. (See for instance Boyce et al., 1988.) Other justifications have been offered, including "entropy maximization" (see Wilson, 1970, for instance), or "cost efficiency" (Smith, 1987). However, it was commonly accepted that the objective functions in such programs did not represent utility functions in the strict sense, but rather, analytical devices to obtain the desired aggregate demands. The above results show that, when properly formulated, such optimization programs are theoretically sound.

The same results apply to the uncongested case, as is also shown in Appendix 5.1. Finally, using a continuity argument, the proof is also valid in the limit, deterministic case, either congested or uncongested.

As noted above, the R.T.'s utility maximization problem in the stochastic case constitutes a generalization of the deterministic case, Problem (5.10). Specifically, when the value of parameter β becomes infinite the latter problem reverts to the former, since the term

$$\frac{1}{\beta_r} \sum_i \sum_j \sum_r T_{ijr} \ln T_{ijr} \tag{5.51}$$

disappears from the objective function. Since an infinite value for β corresponds to a zero variance in the random terms of route utilities, in this case these errors are equal to their expected values (zero). Individual travelers are then assigned deterministically to their maximum utility route.

From this perspective, Equation (5.50) holds in all cases, and the technical problems noted in connection with the deterministic case are obviated. That is to say, by considering the deterministic case as a special limit case, which is attained in a continuous manner, of the general stochastic case, the R.T.'s approach to the derivation of the aggregate demands is justified theoretically.

The other limit case $\beta = 0$ corresponds to a situation where the variance of these errors is infinite. Consequently, *all* routes between a given origin and destination offer the same probabilistic utility at equilibrium, and consequently, all route demands should also be equal. In that case, the optimal solution to the R.T.'s problem is

$$T^*_{ijr} = (1/R_{ij})T_{ij}; \qquad \forall\ i, j, r$$

in which R_{ij} is the number of routes between i and j.

5.3.4 Existence and Uniqueness of Model Solutions[19]

We now investigate the existence and uniqueness of the solutions to the R.T.'s U.M. Problem (5.47)–(5.48). Again, this will be done with respect to objective function (5.47). We first show that objective function (5.47) is convex. We can, without loss of generality, write $U'_R(v, T)$ as

$$U'_R(\boldsymbol{v}, \boldsymbol{T}) = U_1(\boldsymbol{v}) + U_2(\boldsymbol{T}) \tag{5.52a}$$

with

$$U_1 = \sum_a \int_0^{v_a} g_a(v)\, dv \tag{5.52b}$$

$$U_2 = \frac{1}{\beta_r} \sum_i \sum_j \sum_r T_{ijr} \ln T_{ijr} \tag{5.52c}$$

The convexity of U_1 with respect to the route demands T_{ijr} was established in section 5.3.4, when we investigated the uniqueness of the solutions to the deterministic route choice problem. We now examine similarly the Hessian of $U_2(\boldsymbol{T})$, or the matrix with general element

$$\frac{\partial^2 U_2(\boldsymbol{T})}{\partial T_{ijr}\, \partial T_{kln}}$$

The second partial derivatives of $U_2(T_{ijr})$ are respectively:

$$\frac{\partial^2 U_2}{\partial T^2_{ijr}} = \frac{1}{\beta_r T_{ijr}}; \qquad \forall\ i, j, r \tag{5.53a}$$

$$\frac{\partial^2 U_2}{\partial T_{ijr}\, \partial T_{kln}} = 0; \qquad \forall\ i, j, r, k, l, n \tag{5.53b}$$

Since β is strictly positive, and T_{ijr} is strictly positive, being given by the logit function (5.35b), the Hessian of U_2 is a diagonal matrix with strictly

[19]This section may be skipped without loss of continuity.

positive elements, and consequently, U_2 is a strictly convex function. Consequently, U'_R being the sum of a convex function and a strictly convex function, is a strictly convex function. Since the constraints are linear, and the feasible region is therefore convex, the KKT conditions are necessary *and* sufficient. (See section 2.2 of Appendix A.) It is shown in Appendix A that the solution to the KKT conditions is Formula (5.35b). This therefore is also the solution to Problem (5.47)–(5.48). The fact that U' is a strictly convex function guarantees that this is the unique solution to the problem.

5.3.5 Solving the Model

The R.T.'s problem (5.47)–(5.48) in the stochastic case, as noted, constitutes an extension of the deterministic problem. In the former problem, the "entropy" term (5.52c) is added to the objective function. This extended version of the problem may be solved with an algorithm derived from the convex combination algorithm, and which is based on a *partial* rather than full, linearization of the objective function, applied only to the original first term (Evans, 1976). This algorithm, which was originally developed as an adaptation of the Frank-Wolfe algorithm to combined route and destination choice problems, may also be used for stochastic route choice problems. Like the Frank-Wolfe algorithm, it is based on a linearization of the objective function, but in this case, the linearization is limited to the link cost integral terms.

At iteration k, the auxiliary problem is

$$\text{Min } U'^k_R(y_a, Z_{ijr}) = \left[\sum_a y_a g_a^{k-1} + \frac{1}{\beta_r} \sum_{ijr} Z^k_{ijr} \ln Z^k_{ijr} \right] \tag{5.54}$$

subject to Constraints (5.48) in the Z variables, in which the g_a^{k-1} are the current link-generalized costs, computed from the previous solution, and auxiliary variables y_a^k and Z^k_{ijr}, respectively, replace variables v_a and T_{ijr}. This is precisely the stochastic route choice problem with fixed-link costs g_a^{k-1}, as stated in the *uncongested* case.

It may be shown that the solution to Problem (5.54), (5.48) represents a "descent" solution to the main (original) problem, that is, the value of the function $U'_R(\boldsymbol{v})$ decreases from its present value $U'_R(\boldsymbol{v}^k)$ along the direction represented by the vector $(\boldsymbol{y}^k - \boldsymbol{v}^k)$. Furthermore, it may be shown that this auxiliary problem is a convex problem (i.e., in which the objective function is a convex function and the feasible region is convex). (See Exercise 5.14.) Therefore, it admits only one solution, which may then be identified from solving directly the problem's first-order conditions. These may be shown to be

$$Z^k_{ijr} = Z_{ij} \frac{e^{-\beta_r(\tau t^k_{ijr} + c_{ijr})}}{\sum_r e^{-\beta_r(\tau t^k_{ijr} + c_{ijr})}}; \qquad \forall\, i, j, r \tag{5.55}$$

where the travel times t_{ijr}^k at iteration k are given by

$$t_{ijr}^k = \sum_a \delta_{ijr}^a t_a^{k-1}; \qquad \forall\, i, j, r \tag{5.56}$$

that is, the route costs corresponding to the given link costs at the current iteration. (See Exercise 5.22.)

Thus, it is apparent that solving the subproblem amounts to distributing the given T_{ij}'s from a given origin i to destinations j according to logit Formula (5.55), as in the uncongested case. It may be remembered from section 3 of Chapter 3 that this requires route enumeration, which may be unfeasible for realistic networks. In this case, application of the STOCH algorithm described in section 3.2 of Chapter 3, or any other appropriate method of obviating this requirement, would be required.

Once the solution to the auxiliary problem is obtained in whatever manner is chosen, the remainder of the iterative step is the same as in the Frank-Wolfe algorithm. The iterative step of Evans' algorithm in the case of the stochastic congested route choice problem is as follows:

ALGORITHM 5.5. PARTIAL LINEARIZATION FOR THE STOCHASTIC CONGESTED ROUTE CHOICE PROBLEM

Iterative step

Given a feasible solution (v_a, T_{ijr}) the corresponding link travel costs, and the given T_{ij}'s:

A. Determine the route costs between given origins i and destinations j, according to Formula (1.2a).

B. Estimate the auxiliary Z_{ijr}'s, using logit Formulas (5.55).

C. Determine the corresponding auxiliary link demands y_a^k

E. Solve for θ which minimizes the function:

$$U'_R(\boldsymbol{v}^k + \theta(\boldsymbol{y}^k - \boldsymbol{v}^k), \boldsymbol{T}^k + \theta(\boldsymbol{Z}^k - \boldsymbol{T}^k)) \tag{5.57}$$

in the interval [0, 1].

F. Once the optimal value of θ is found, the new, improved solution is

$$v_a^{k+1} = v_a^k + \theta(y_a^k - v_a^k); \qquad \forall\, a \tag{5.58a}$$

$$T_{ijr}^{k+1} = T_{ijr}^k + \theta(Z_{ijr}^k - T_{ijr}^k); \qquad \forall\, i, j, r \tag{5.58b}$$

(Continued)

G. Check whether

$$|(\boldsymbol{y}^k - \boldsymbol{v}^k) \cdot \nabla_v U'_R(\boldsymbol{Z}^k, \boldsymbol{v}^k) + (\boldsymbol{Z}^k - \boldsymbol{T}^k) \cdot \nabla_Z U'_R(\boldsymbol{Z}^k, \boldsymbol{v}^k)| \leq \epsilon \tag{5.59}$$

where ϵ is a given tolerance. If so stop; if not, iterate.

A simplified version of the algorithm may alternatively be applied, in which instead of determining the optimal value of the "step-size" θ, a fixed value, equal to

$$\theta_k = 1/(k + 1) \tag{5.60}$$

at iteration k is used. The advantage is that Step E does not have to be performed, while the procedure is still guaranteed to converge (Powell and Sheffi, 1982, Chen and Alfa, 1991). The drawback, however, is that this procedure, which is sometimes called the MSA method, is less efficient, and results in a larger number of iterations. (See Exercise 5.10.)

5.3.6 Illustrative Example

We now illustrate the application of the linearization, fixed step-size algorithm to the same example problem we have already solved. Table 5.5a presents the computations for the first iteration of the Evans algorithm. Tables 5.5b and 5.5c present the equilibrium solution.

Graphically, link loadings may be represented as in Figure 5.7.

All unrepresented links have a zero demand. Nonzero route demands may be represented graphically as in Figure 5.8.

It may be verified that, as already announced, the costs on all used routes are *not* equal, in contrast with the results of the deterministic route assignment of the previous section. It may be seen from comparing these results with those of the deterministic UE solution of the previous section that a lower value of β_r would result in a greater dispersion of origin-destination demands across routes. Conversely, the larger the value of β_r, the closer the solution would come to the deterministic solution of the previous section.

In the present case, the observed total travel time for all travelers,

$$T = \sum_a t_a^* v_a^* \tag{5.61}$$

is equal to 28,580 minutes, corresponding to an average individual travel time of 71 minutes. This value may be compared to its counterpart in the deterministic case, which was 67 minutes. The traveler surplus is equal to −44,300. In the same manner as in the deterministic case of section 2.3, the impacts of user charges on the link volumes, total travel time, and traveler surplus may be assessed. (See Exercise 5.26.)

TABLE 5.5a Frank Wolfe Algorithm for Probabilistic Route Choice Problem

A. Determine the route costs at zero volumes between given origins i and destinations j, according to $t^0_{ijr} = \Sigma_a t^0_a \delta^a_{ijr}$

Iteration	1	2	3	4	5	6
v^0_a	0	0	0	0	0	0
t^0_a	5	15	6	8	7	8
K_a	55	50	60	50	55	60
Estimate the T_{ijr}'s using Formula (5.36)						
T^1_{121}	134.70					
T^1_{122}		54.77				54.77
T^1_{123}			60.53		60.53	60.53
T^1_{141}		46.72				
T^1_{142}	51.64			51.64		
T^1_{143}			51.64		51.64	
v^1_a	186.34	101.49	112.17	51.64	112.17	115.30
t^1_a	103.82	53.19	16.99	9.37	25.17	24.36

$$T_{12} = 250; \quad T_{14} = 150$$

$$T^1_{121} = 250 \frac{e^{-0.05(5)}}{e^{-0.05(5)} + e^{-0.05(23)} + e^{-0.05(21)}}$$

$$T^1_{121} = 134.70; \quad T^1_{122} = 54.77; \quad T^1_{123} = 60.53$$

$$T^1_{141} = 150 \frac{e^{-0.05(15)}}{e^{-0.05(15)} + e^{-0.05(13)} + e^{-0.05(13)}}$$

$$T^1_{141} = 46.72; \quad T^1_{142} = 51.64; \quad T^1_{143} = 51.64$$

TABLE 5.5a (*Continued*)

B. Estimate the auxiliary Z_{ijr}'s, using logit formulas (5.55)

	1	2	3	4	5	6
t_a^1	103.82	53.19	16.99	9.37	25.17	24.36
Z_{121}^1	22.37					
Z_{122}^1		83.20				83.20
Z_{123}^1			144.43		144.43	144.43
Z_{141}^1		53.85				
Z_{142}^1	2.68			2.68		
Z_{143}^1			93.47		93.47	
y_a^1	25.05	137.05	237.90	2.68	237.90	227.63

$$Z_{121}^1 = 250 \frac{e^{-0.05(103.82)}}{e^{-0.05(103.82)} + e^{-0.05(53.19 + 24.36)} + e^{-0.05(16.99 + 25.17 + 24.36)}}$$

$$Z_{121}^1 = 22.37; \quad Z_{122}^1 = 83.20; \quad Z_{123}^1 = 144.43$$

$$Z_{141}^1 = 150 \frac{e^{-0.05(53.19)}}{e^{-0.05(53.19) - 0.05(103.82 + 9.37)} + e^{-0.05(16.99 + 25.17)}}$$

$$Z_{141}^1 = 53.85; \quad Z_{142}^1 = 2.68; \quad Z_{143}^1 = 93.47$$

C. Solve for θ which minimizes the function: $U'(\boldsymbol{v}^1 + \theta(\boldsymbol{y}^1 - \boldsymbol{v}^1), \boldsymbol{T}^1 + \theta(\boldsymbol{Z}^1 - \boldsymbol{T}^1))$ in the interval (0, 1).

$$\underset{0 \le \theta \le 1}{\text{Min}} \; U'_{(\theta)} = \sum_a \int^{(v_a^1 + \theta(y_a^1 - v_a^1))} t_a(v)\, dv + \frac{1}{\beta_r} \sum_i \sum_j \sum_r (T_{ijr}^1 + \theta(Z_{ijr}^1 - T_{ijr}^1)) \ln (T_{ijr}^1 + \theta\,(Z_{ijr}^1 - T_{ijr}^1))$$

$$\frac{dU'_{(\theta)}}{d\theta} = \sum_a t_a\,[v_a + \theta(y_a - v_a)](y_a - v_a) + \frac{1}{\beta_r} \sum_{ijr} [1 + \ln (T_{ijr} + \theta\,(Z - T_{ijr}))](Z_{ijr} - T_{ijr})$$

$$\begin{aligned}
&= 5[1 + 0.15((186.34 - 161.29\theta)/55)^4](-161.29) + 15[1 + 0.15((101.49 - 35.56\theta)/50)^4](-35.56) \\
&\quad + 6[1 + 0.15((112.17 + 125.73\theta/60)^4](125.73) + 8[1 + 0.15((51.64 + 48.96\theta)/50)^4](-48.96) \\
&\quad + 7\,[1 + 0.15((112.17 + 125.73\theta)/55)^4](125.73) + 8[1 + 0.15((115.30 + 112.33\theta)/60)^4](112.33) \\
&\quad + (1/0.05)[1 + \ln (134.70 - 112.33\theta)](-112.33) + (1/0.05)[1 + \ln (54.77 + 28.43\theta)](28.43) \\
&\quad + (1/0.05)\,[1 + \ln (60.53 + 83.90\theta)](83.90) + (1/0.05)[1 + \ln (46.72 + 7.13\theta)](7.13) \\
&\quad + (1/0.05)[1 + \ln (51.64 - 48.96\theta)](-48.96) + (1/0.05)[1 + \ln (51.64 + 41.83\theta)](41.83)
\end{aligned}$$

D. Apply Bisection Method

Iteration #	$dU'(\theta)/d\theta$	a	b	New θ	$dU'(\theta)/d\theta$
0		0	0.5	0.25	−7737.90
1	−7737.90	0	0.25	0.125	+2460.39
2	+2460.39	0.125	0.25	0.1875	−2614.75
3	−2614.75	0.125	0.1875	0.1563	−83.58

Therefore $\theta_1^* = 0.1563$

TABLE 5.5a (*Continued*)

E. Performing the move, the new improved solution is, in terms of route flows,

$$T^2_{ijr} = T^1_{ijr} + \theta(Z^2_{ijr} - T^2_{ijr}); \qquad \forall\, i, j, r$$

$$T^2_{121} = 134.70 + 0.1563(22.37 - 134.70) = 117.14$$

$$T^2_{122} = 54.77 + 0.1563(83.20 - 54.77) = 59.21$$

$$T^2_{123} = 60.53 + 0.1563(144.43 - 60.53) = 73.64$$

$$T^2_{141} = 46.72 + 0.1563(53.85 - 46.72) = 47.83$$

$$T^2_{142} = 51.64 + 0.1563(2.68 - 51.64) = 43.99$$

$$T^2_{143} = 51.64 + 0.1563(93.47 - 51.64) = 58.18$$

Link volumes are then:

link a =	1	2	3	4	5	6
T^2_{121}	117.14					
T^2_{122}		59.21				59.21
T^2_{123}			73.64		73.64	73.64
T^2_{147}		47.83				
T^2_{148}	43.99			43.99		
T^2_{149}			58.18		58.18	
v^2_a	161.13	107.05	131.82	43.99	131.82	132.86
t^2_a						

Based on these volumes, link travel times are updated to t^2_a, and the next iteration performed in the same manner.

TABLE 5.5b Equilibrium Link Demands and Costs

Link #	Demand	Cost
1	166	68
2	108	63
3	126	24
7	30	8
8	126	36
9	114	24

TABLE 5.5c Equilibrium Route Demands and Costs

Route #	Demand	Cost
1	136	68
2	61	84
3	53	87
7	55	63
8	30	76
9	65	60

Thus, under either assumption, the utility maximizing behavior of individual travelers (or of the R.T. representing them), has rather similar effects in terms of the aggregate, total travel cost. In the next section, we shall examine what the behavior of the R.T. should be in order to optimize not individual travel times, but aggregate travel times.

In closing this section, it is useful to note that it turns out that for highly congested networks, when link demands are close to capacity, route demands estimated under the stochastic assumption are close to those under the deterministic one (Sheffi and Powell, 1981). (See Exercise 5.11.) This observation is important, since on the one hand, realistic contemporary urban conditions are often highly congested, and on the other, the practical application of the stochastic model is more difficult than that of the deterministic, for the reasons discussed earlier. Thus, the deterministic model may be a satisfactory approximation to the more accurate probabilistic one in many specific situations.

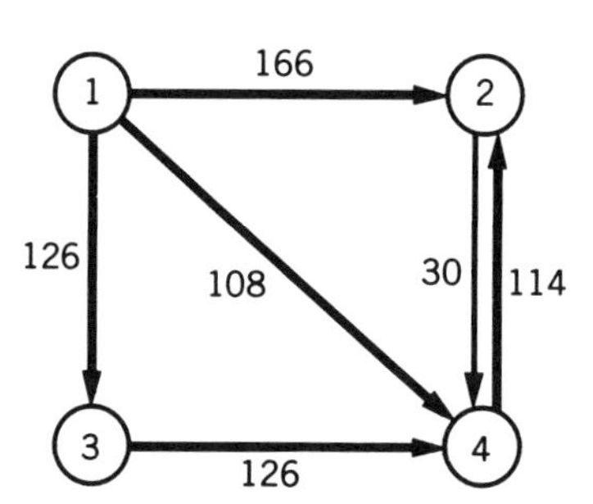

Figure 5.7 Stochastic equilibrium link demands.

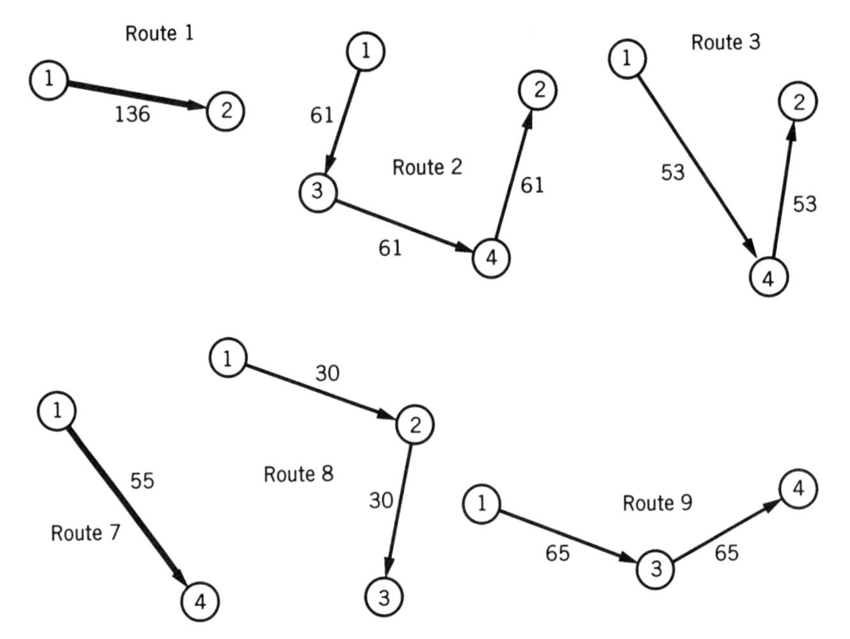

Figure 5.8 Stochastic equilibrium route demands.

5.4 SOCIALLY OPTIMAL CAR ROUTE ASSIGNMENT

In the preceding sections, route choice was modeled under the assumption of individual utility maximizing behavior, reflecting the individual traveler's own self-interest. This "behavioral" principle is a reasonable one when using the route choice model for *predictive* purposes. However, the analyst may alternatively be interested in a network equilibrium assignment of given origin-destination volumes T_{ij} which would correspond not to individual interest, but rather public, community interest, for instance, minimum aggregate travel cost, including time.

In such a case, the network assignment problem would then be to

$$\text{Max } S_{(v_a)} = -\sum_a v_a g_a(v_a) \tag{5.62}$$

subject to

$$\sum_r T_{ijr} = T_{ij}; \qquad \forall\ i, j \tag{5.63a}$$

$$v_a = \sum_i \sum_j \sum_r T_{ijr} \delta^a_{ijr}; \qquad \forall\ a \tag{5.63b}$$

$$T_{ijr} \geq 0,\ v_a \geq 0; \qquad \forall\ i, j, r, a \tag{5.63c}$$

The solution to this problem is sometimes referred to as the "System Equilibrium" (SE) route assignment. It may be used in a *normative*, as opposed to predictive, mode, for instance, to provide a description of ideal, as opposed to expected, network conditions.

5.4.1 The R.T.'s U.M. Problem

It should be fairly obvious that when link travel costs are constant, in the absence of demand externalities created by congestion,

$$\text{Max } S_{(v_a)} = -\sum_a v_a g_a(v_a) = -\sum_a v_a(\tau t_a + c_a) \tag{5.64}$$

and the individually and socially optimal solutions are one and the same. (See Exercise 5.23.) In the presence of congestion, however, this will no longer be true.

This raises the question of what would be socially optimal behavior. In this connection, it may be noted that the objective function of the SE problem Formula (5.62) may be cast in the same form as the R.T.'s utility function, Formula (5.8d), if one could write

$$\sum_a v_a g_a(v_a) = \sum_a \int_0^{v_a} h_a(v)\, dv \tag{5.65}$$

That is (see Exercise 5.23), since the R.T. problem and the S.E. problem have the same conservation of demand constraints, we have the following.

FACT 5.1

If the link functions h_a are defined as

$$h_a(v_a) = g_a(v_a) + v_a \frac{\partial g_a(v_a)}{\partial v_a}; \qquad \forall\, a \tag{5.66}$$

the R.T's U.M. problem solution will be socially optimal.

This result may not seem so unexpected when it is seen that the socially optimal link cost function is equal to the original one, plus an additional cost, the second term in the r.h.s. of Formula (5.66), which is equal to the product of the link's volume and the marginal link cost, that is, the increase in total (i.e., social) link cost created by one additional traveler using the link at that volume. In other words, the socially optimal pricing of road travel under congested conditions is to charge the marginal cost.

The traveler surplus for the SE route assignment may be estimated from Formula (5.16) where functions g_a are replaced by functions h_a, for all links a. (See Exercise 5.20.)

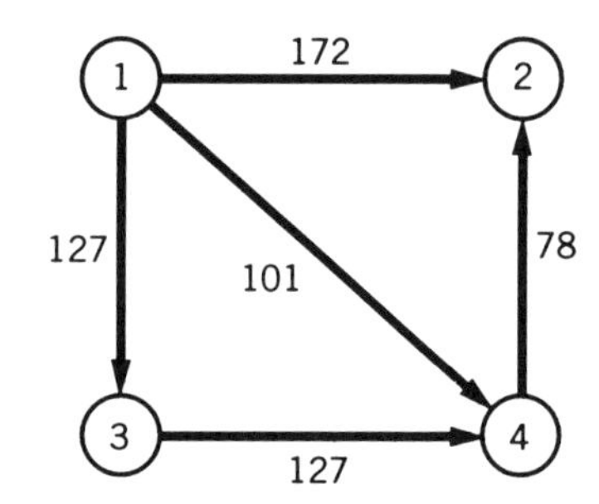

Figure 5.9 Socially optimal equilibrium link demands.

5.4.2 Illustrative Example: Socially Optimal Network Assignment

To illustrate, in the same hypothetical situation as in the previous two examples, the system optimal solution is graphically represented in Figures 5.9 and 5.10.

It may be seen that this solution is rather close to the UE solution. This is due to the high level of congestion on the network. For purposes of comparison with the UE values, link and route travel times are represented in Tables 5.6a and b.

The total travel time is 26,560 minutes, or 66 minutes per traveler, as opposed to 67 for the UE assignment. As noted, the UE and SE solutions will become closer as the level of network congestion increases. Also, the two solutions are one and the same when there is no congestion (when travel times are constant). Thus, in this sense, there is a silver lining in high levels of congestion. Consequently, also, the efficiency of the SE network assignment relative to the UE assignment is highest at middle levels of congestion.

In closing this section, mention should be made of "Braess' Paradox." This refers to the (apparently) counterintuitive observation that the addition of a link to a given transportation system may in some cases actually increase the total travel time. This phenomenon, which was first examined by Braess (1968) loses its mystery when it is remembered that travelers minimize their own individual travel costs, and not the average cost, which in general is different. The two are only equal in the uncongested case.

Thus, unless the link's integration into the existing network is specifically designed with respect to the SE objective, there is no reason to expect improving average conditions. On the other hand, some individual travelers, between certain origins and destinations, may benefit. Issues of optimal network design are examined briefly in Chapter 9.

5.5 DETERMINISTIC TRANSIT ROUTE CHOICE

In this section, we now consider the route choice problem in the specific case of the transit mode. It may be remembered from section 3.5.2 that, in the case of transit networks and fixed travel times, an individual traveler's choice of a

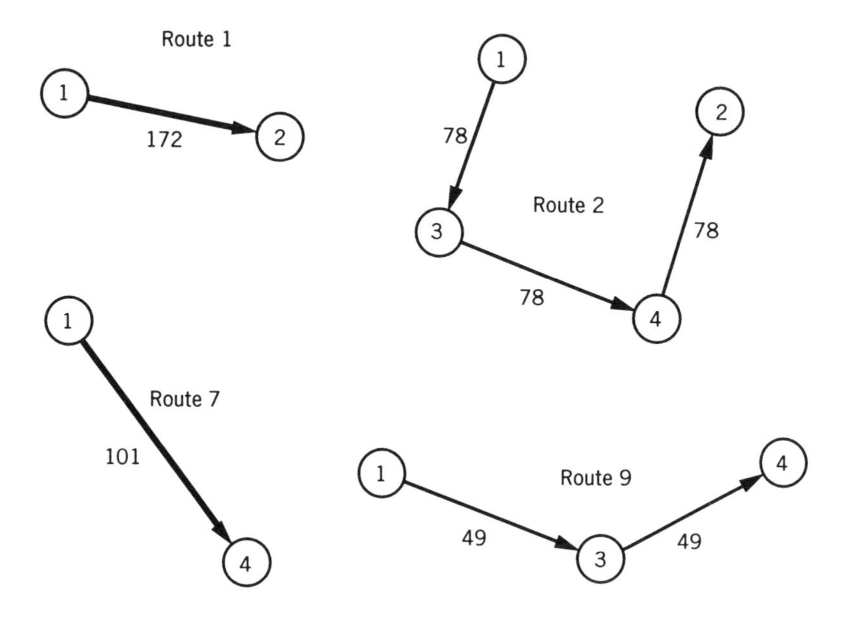

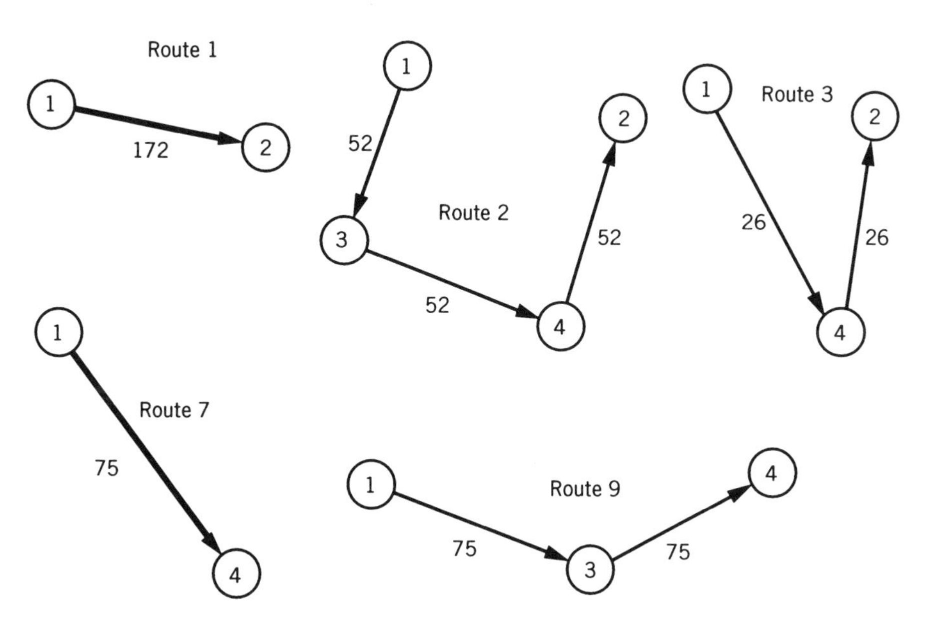

Figure 5.10 Socially optimal equilibrium route demands.

TABLE 5.6a Link Costs

Link #	Cost
1	77
2	53
3	24
4	8
5	37
6	11

TABLE 5.6b Route Costs

Route #	**Cost**
1	77
2	72
3	64
7	53
8	85
9	60

route was based on the principle of minimization of expected generalized travel costs, including expected waiting times at the network's nodes. In this section, we consider the case when the average travel cost for one traveler depends on the route choices of all the other *transit* travelers, but not car travelers. That is, we assume that there is no interaction between the transit network and the car network, that, in effect, transit links are separate from, and independent of, highway links. This would in particular be true for subway networks, as well as buses using reserved lanes. The more general case (e.g., when both transit and cars use the same links), is examined in Section 6.3.3 of Chapter 6.

In keeping with the notation used in section 5.1, transit link travel times t_a are now a function of transit demands,

$$t_a = t_a(v_a); \qquad \forall\ a$$

where functions $t_a(\ )$ are continuous, nondecreasing functions. These functions may approximate several types of congestion effects. First, in particular on the "walk" links, this relationship of cost to demand may reflect increased boarding and alighting time. On travel links themselves, it may represent traveler "discomfort," including a decreased average probability of finding a seat, which may be estimated from results of queuing theory.[20] Finally, in particular in subway systems, it may also represent increased vehicle travel time due to increased vehicular density on the system's links.

[20]This approach would require ignoring the order of boarding.

Given link cost functions, we want to determine as assignment of individual travelers to the transit network in which, in accordance with the UE principle, no traveler may reduce further his or her expected travel time by changing behavior.

5.5.1 The R.T.'s U.M. Problem

In analogy with the car case described earlier, as well as in extension of the uncongested transit case in section 4.3 of Chapter 3, we have the following.

> **FACT 5.3**
>
> The R.T.'s direct utility in the congested transit route choice case is equal to
>
> $$U_R^t = -\tau \sum_a \int_0^{\Sigma_j v_{aj}} t_a(x)\, dx - \overline{\omega} \sum_i \sum_j w_{ij} + T_0$$

Consequently, the R.T.'s utility maximization problem is then

$$\underset{(v_{aj}, w_{ij})}{\text{Max}}\ U_R^t = \left[-\tau \sum_a \int_0^{\Sigma_j v_{aj}} t_a(x)\, dx + \overline{\omega} \sum_i \sum_j w_{ij} \right] + T_0 \tag{5.67}$$

subject to

$$\sum_a v_a c_a + T_0 = B \tag{5.68a}$$

$$v_{aj} \leq f_a w_{ij}; \qquad \forall\, i, j, a \in i^- \tag{5.68b}$$

$$\sum_{a \in i^-} v_{aj} - \sum_{a \in i^+} v_{aj} = T_{ij}; \qquad \forall\, i, j \tag{5.68c}$$

$$v_{aj}, w_{ij}, T_0 \geq 0, \qquad \forall\, i, j, a \tag{5.68d}$$

It may be shown, using the KKT conditions, that the solution to Problem (5.67)–(5.68) conforms to the maximum expected utility principle at the individual level, in the same manner as in the car case. That is, at equilibrium, only travel *strategies* (see below) which offer the minimum *expected* travel cost will be used, and those which offer a higher expected cost will not. It may also be shown, using the same approach as for Problem (5.10) for street networks, that the solution always exists, and is always unique, provided that the link cost functions are strictly increasing.

The similarity of the problem's formulation with that of the car case suggests that this problem may also be solved using the same Frank-Wolfe linearization algorithm as is used for the solution of the user equilibrium car route assignment problem. However, the auxiliary subproblem at each iteration now consists of assigning the given origin-destination demands to the transit network, given the current, fixed-link costs. The minimum expected route cost algorithm described in section 3.4.1 of Chapter 3 for uncongested conditions may then be used to that effect.

Just as in the case of constant travel times, it is worthwhile noting that the above model is a generalization of the model presented in section 5.2 for the car mode. Indeed, if all link frequencies f_a are infinite (i.e., transit vehicles are now, like the private car, instantly available), Problem (5.67)–(5.68) becomes equivalent to Problem (5.10).

5.5.2 Illustrative Example

As an illustration of the application of the congested transit network assignment model, let us revisit the prototype transit network we have used in the preceding chapter, and assume now that the link cost functions reflect the effects of congestion discussed earlier. The congested travel cost c_a for link a is assumed to be given by the congestion function

$$t_a(v_a) = t_a^0 + 0.06(v_a/K_a); \qquad \forall\, a \tag{5.69}$$

where t_a^0 is the uncongested travel time (in minutes) used in the illustrative example in section 4.2 of Chapter 3, and k_a is a given coefficient, whose values are given in Table 5.7.

We assume that the given T_{ij}'s are equal to $T_{12} = 250$; $T_{14} = 150$. Also, we assume that there are no "return" links, numbers 6 through 10.

Parameter $\overline{\omega}$ is equal to 0.5. With these data, the demands on the network links are represented in Figure 5.11.

The optimal links from node 1 to node 4 and their probabilities of being used by an individual traveler are represented in Figure 5.12.

TABLE 5.7

Link a	$K(a)$
1	5
2	15
3	5
4	10
5	10
6	5
7	15
8	5
9	10
10	10

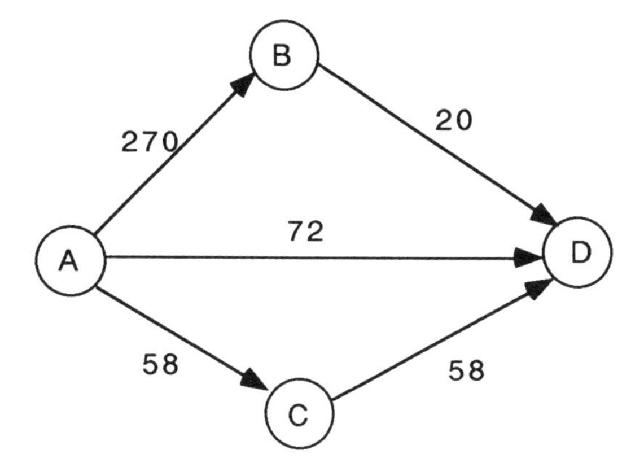

Figure 5.11 Congested transit network assignment.

The average waiting time per passenger at node 1 is 5 minutes going to 2, and 1.6 minutes going to 4. The link travel times are presented in Table 5.8.

It should be noted that individual destinations cannot be addressed separately, as in the uncongested case; when there are several concurrent origin-destination pairs it will thus not be possible to identify the optimal links for individual travelers from solving the R.T.'s problem, in terms of individual origin-destinations. Also, by reformulating the R.T.'s problem in terms of possible *strategies* s to go to a given destination j, rather than link demands as above, optimal *strategies* may be identified (Spiess and Florian, 1989).

A strategy s from a given node i to a given destination j is defined as a *set* of links, any one of which may be taken, in order of *first* being available. To illustrate, in the case of the present network, there are seven strategies from node 1 to node 4. These are described in Table 5.9 on page 186.

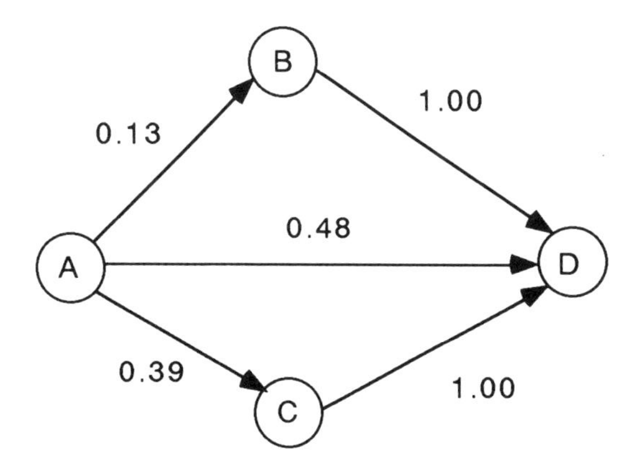

Figure 5.12 Equilibrium transit route demands.

TABLE 5.8 Congestion Travel Times on Transit Links

Link #	Cost
1	6.24
2	8.29
3	4.59
4	5.12
5	6.37

Thus, using strategy 5 implies using whichever of links 2 or 3 is first available at node A, and then link 5. The single strategy (#8) available from 1 to 2 consists of using route 1, i.e., link 1.

The alternative formulation of the R.T.'s problem is following (Spiess, 1984):

$$\underset{(v_a^s, w_i^s, T_{is}, T_0)}{\text{Max}} \quad U_R^t = -\left[\sum_a \tau \int_0^{\Sigma_s v_a^s} t_a(x)\, dx + \overline{\omega} \sum_i \sum_s w_i^s + \sum_a \sum_s v_a^s c_a \right] \tag{5.70}$$

subject to

$$v_a^s \le \delta_a^s f_a w_i^s; \qquad \forall\, a \in i^-, s \tag{5.71a}$$

$$\sum_{a \in i^-} v_a^s - \sum_{a \in i^+} v_a^s = T_{is}; \qquad \forall\, i, s \tag{5.71b}$$

$$\sum_s T_{is} = T_{ij}; \qquad \forall\, i, j \tag{5.71c}$$

$$v_i^s \ge 0; \; w_i^s \ge 0, \; T_0 \ge 0; \; \forall\, i, s \tag{5.71d}$$

TABLE 5.9 Strategies from 1 to 4

Strategy number	Description	Component links
1	A→D	2
2	A→B→D	1,4
3	A→C→D	3,5
4	A→D *or* A→B→D	1,2,4
5	A→D *or* A→C→D	2,3,5
6	A→B→D *or* A→C→D	1,3,4,5
7	A→D *or* A→B→D *or* A→C→D	1,2,3,4,5

in which index s is specifically understood to refer to strategies for travel from a specific origin i to a specific destination j, and would thus be written in full as s_{ij}, or $s \in S_j$. v_a^s is the number of passengers using strategy s from node i. c_a is the given cost of travel on link a. The given link-strategy incidence binary variables δ_a^s have a value of 0 if link a is not part of strategy s and 1 if it is, and play a role similar to that of the link-route incidence variables δ_{ijr}^a in the car case. T_{is} is the volume using strategy s from i.

It may be shown, using the KKT conditions, that at the solution, only strategies which offer the minimum expected cost are used, in accordance with the U.M. principle.

Using the strategy-based formulation (5.70)–(5.71), the transit network assignment in the case above is represented as in Tables 5.10 and 5.11 below, corresponding to Figures 5.11 and 5.12 above.

It may be verified that the U.M. principle, in terms of expected strategy costs, is observed. These costs are equal to the values of the dual variables for constraints (5.71b). Costs g_{ij} are the dual variables for constraints (5.71c).

It should be pointed out that, as described in Exercise 3.23 in Chapter 3, the uncongested link-based formulation may be retrieved from the congested case above by setting the link cost functions as the constants t_a^0. Solving the illustrative example in section 3.4.2 on this basis would result in all 200 travelers using strategy 7 as defined in Table 5.9 above, corresponding to the link-based solution represented in Figures 3.15 and 3.16.

Finally, the strategy-based formulation may be generalized to a probabilistic version. (See Exercises 5.28 to 5.30.)

Table 5.10 Volume on Link *a* Following Strategy *s*, v_a^s

	Strategy	
Link #	8	7
1	250	20
2		72
3		58
4		20
5		58

TABLE 5.11 Volume From *i* on Strategy *s*, T_{is}

	Strategy	
	8	7
1	250	150
2		20
3		58

5.6 SUMMARY

In this chapter, the results obtained for the route choice problem under uncongested conditions were systematically extended to take into account the presence of demand externalities, both in the car and the transit case. Because of the fact that individual traveler choices are affected by the choices of all other travelers, the analytical role of the R.T. in generating aggregate demands is now central, as congestion effects must be modeled at the aggregate level.

Solving the R.T.'s utility maximization problem required the application of various numerical procedures. In particular, the ''linearization'' and ''partial linearization'' algorithms were described. These procedures are guaranteed to approximate the exact travel demands within a given tolerance in a finite number of iterations, and are thus far preferable to heuristic, informal procedures.

The concepts of the R.T.'s received utility and traveler surplus were seen to provide, as in the uncongested case, a measure for community welfare. A model for prescriptive assignment of given origin-destination demands to a network so as to optimize community utility, as opposed to individual utility, was also formulated.

In the next chapter, route choice under congested conditions will be integrated with the other traveler choices, following the parallel with the uncongested case in Chapter 4.

5.7 EXERCISES

5.1 Show that the solution of Problem (5.31)–(5.32) produces the demands in Formula (5.7a). (Hint: Show that this problem is equivalent to Problem (5.10).)

5.2 Show that objective function (5.10a) is not strictly convex in the variables T_{ijr}, and that consequently, the solution of the deterministic route demand Problem (5.10) in terms of the route demands may not be unique.

5.3 For the simple network represented in Figure 5.13, the link travel time functions are

$$t_{12}(v) = 15 + 0.05v_{12}$$

$$t_{23}(v) = 25 + 0.07v_{23}$$

$$t_{24}(v) = 20 + 0.04v_{24}$$

where t_{ij} is the travel cost from node i to node j, and v_{ij} is the demand on link ij. There are no user charges. Also, the travel demands are as

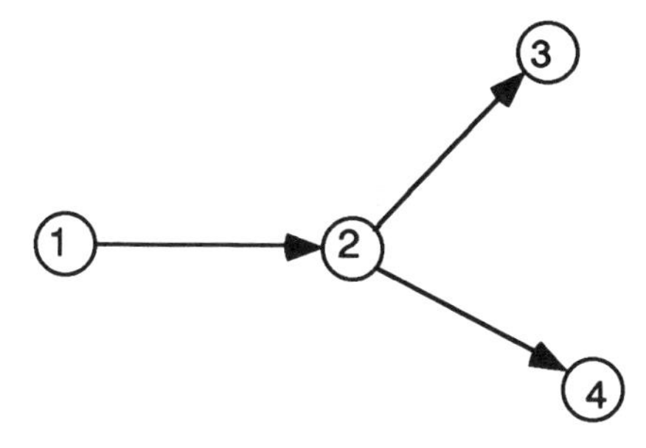

Figure 5.13 Network for Exercise 5.3.

follows

$$T_{13} = 10000$$

$$T_{14} = 5000$$

Determine the equilibrium traffic assignment for the network under these conditions. Discuss the potential problems which linear link travel time functions may bring.

5.4 A network consists of three parallel links connecting two nodes 1 and 2, as represented in Figure 5.14. The individual generalized link cost functions for the three links are respectively:

$$g_1(v) = 10 + 15v$$

$$g_2(v) = 5 + 20v$$

$$g_3(v) = 15 + 5v$$

where g_i is the cost of travel on link i, and v is the demand on the link. If these three links were to be replaced by a single link, for instance, to simplify network representation, what should be the cost function for the link so as to make it analytically equivalent to the three links it replaces?

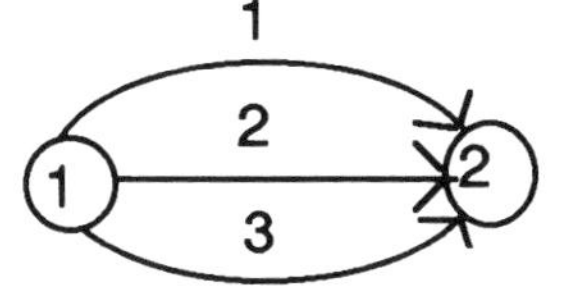

Figure 5.14 Network for Exercise 5.14.

5.5 If the iterative relationship in the quasi-balancing method is expressed as

$$X_j = \left[\Phi \left(X_j^{1/\gamma_1} \right) \right]^{-\gamma_2} \tag{5.72}$$

the sufficient condition for convergence is (Phiri, 1980):

$$|\gamma_1| > |\gamma_2|.$$

Show that the condition is met in the case of the system of Equations (5.37) for the SUE.

5.6 Write a macro implementing the fixed point Algorithm 5.3 for implementation on a spreadsheet. (Note that the principle of the method is to introduce a circular relationship between the unknowns, i.e., precisely of the kind which is illegal in spreadsheets. However, all spreadsheets provide a special override option to that effect.) Apply it to solve the SUE problem in the case of the network in Exercise 5.4. Use a value for $\beta = 0.5$.

5.7 Use mathematical analysis software (e.g., Mathematica®, Mathcad®, etc.), to solve the system of Equations (5.37). Compare with the application of the fixed point method in the previous exercise, both in terms of solution values, and in terms of computational efficiency.

5.8 Show that the solution to auxiliary Problem (5.20)–(5.21) represents a descent direction for the original (main) problem. (Hint: show that the value of the function $U(\boldsymbol{v})$ decreases from its present value $U(\boldsymbol{v}^k)$ along the direction represented by the vector $(\boldsymbol{y}^k - \boldsymbol{v}^k)$, i.e., $U'(\theta)$ in Formula (5.23) is negative for all values of θ.)

5.9 Apply the MSA algorithm to the illustrative SUE route assignment in section 5.3.2 using predetermined step size, i.e., set at iteration k

$$\theta^k = 1/(k + 1) \tag{5.73}$$

to replace step E of Algorithm 5.5. Compare the results and efficiency of the two approaches.

5.10 Complete Table 5.2 by continuing the convex combinations iterations until flows are within 1 unit of the solution, to obtain the results in Tables 5.3 and 5.4.

5.11 Solve the SUE problem in the illustrative example of section 5.3.2 with levels of service equal to one-tenth of the present values. Verify that the solution is very close to that obtained in the deterministic case in the illustrative route assignment in section 5.2.4.

5.12 Show that the S_{ij} travelers for "general" travel purposes (i.e., other than the one of interest) are also assumed to choose routes according to the UE principle. (Hint: Write the KKT conditions for the S_{ijr} variables.)

5.13 Show that at convergence of the linearization algorithm, the KKT conditions for the original problem are the same as those for the auxiliary problem. (Hint: Show that $U'(\theta)$ in Formula (5.23) tends to 0^- when θ tends to 0^+.)

5.14 Show that auxiliary Problem (5.54) subject to Constraints (5.48) is convex.

5.15 Going back to the spatial system of Exercise 4.1, the travel costs and destination costs are now variable, and depend on the link demands and trip ends, respectively, as follows.

$$t_{11} = 1 + 2T_{11};\ t_{12} = 2 + T_{12};\ t_{21} = 2 + 0.5T_{21};\ t_{22} = 1 + T_{22};$$

$$s_3 = 1 + 2T_3;\ s_4 = 0.5 + 3T_4$$

Obtain the equilibrium travel demands T_{ij}'s.

5.16 Consider the network represented in Figure 5.15, where link numbers are as usual represented alongside them. The link generalized cost functions are respectively equal to

$$t_1(v) = 1 + v_1$$

$$t_2(v) = 0.5 + 2v_2$$

$$t_3(v) = 1 + 0.5v_3$$

where the volumes are in hundreds. The total number of travelers from node 1 is equal to 100 in the same units. Given this information, formulate the R.T.'s U.M. problem which produces the equilibrium assignment of these travelers on the network. Verify that the equilibrium conditions hold.

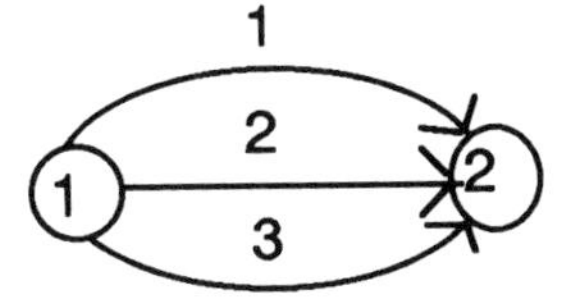

Figure 5.15 Network for Exercise 5.16.

5.17 Derive Formula (5.15a). (Hint: The R.T.'s received utility may be estimated as the optimal value of the R.T.'s direct utility function, i.e., at the solution of the R.T.'s optimization problem.)

5.18 Derive Formula (5.50).

5.19 Obtain directly the derivatives of Function (5.10a) with respect to the route demands.

5.20 Formulate the traveler surplus *TS* in the socially optimal case of section 5.4.1. Show that its value is indeed optimal, i.e. greater than that for any possible network assignment.

5.21 Demonstrate Formula (5.49a).

5.22 Derive the optimality conditions for Problem (5.54)–(5.48). (Hint: Use KKT conditions.)

5.23 Show that when the link cost functions have the expression in Formula (5.66), the solution to R.T.'s Problem (5.65)–(5.63) is a socially optimal one.

5.24 Demonstrate Formula (5.23).

5.25 Show that the equilibrium link travel costs, either in the deterministic or the stochastic user equilibrium route choice case, are equal to the dual variables of the definitional Constraints (5.8b).

5.26 With reference to the illustrative example in section 5.3.6, assume that a \$3 toll is now charged on link 1. Evaluate its impacts on the link volumes, total travel time and traveler surplus.

5.27 Refine the formulation of the R.T.'s U.M. Problem (5.9), to incorporate time delays at intersections. (Hint: Use the link-based formulation, Problem (5.31)–(5.32). Introduce additional variables $v_{a_1a_2}$ representing the flow going from link a_1 to link a_2. Define intersection delay functions of these variables and add them to the utility function. State conservation of flow constraints at intersections.)

5.28 Assume that the expected cost of a given strategy in the transit assignment problem is not known, e.g. due to uncertainty about link travel times, but is a random variable. Assuming these to be i.i.d. Gumbel, generalize Model (5.70)–(5.71) to represent probabilistic, congested transit assignment.

Show, using the KKT conditions, that at the R.T.U.M. problem's solution, individual travelers choose the strategy offering minimum *random* expected cost.

5.29 Assuming a value $\beta = 0.1$ in the above formulation, determine the assignments of travelers to strategies, as well as link volumes in the illustrative example of section 5.5.2. Compare the respective results and discuss the usefulness of the stochastic formulation.

5.30 Devise an algorithm to solve the stochastic version of the transit assignment problem. (Hint: combine the Frank-Wolfe algorithm with the MECR and STOCH algorithms.)

APPENDIX 5.1 OBTAINING THE AGGREGATE ROUTE DEMANDS IN THE STOCHASTIC CASE FROM THE R.T.'S UTILITY MAXIMIZATION PROBLEM

The same approach as in the deterministic case will be utilized again, stating the KKT conditions. In terms of the route demand variables T_{ijr}, these conditions are expressed by Relationships (5.11a,b). Referring to objective function (5.10a), we have, for objective Function (5.47):

$$\frac{\partial U'_R}{\partial T^*_{ijr}} = \sum_a g_a(v^*_a)\delta^a_{ijr} + \frac{\partial}{\partial T^*_{ijr}}\left\{\frac{1}{\beta_r}\sum_i\sum_j\sum_r T^*_{ijr}\ln T^*_{ijr}\right\} \tag{5.74}$$

Since the only difference between the two problems is the addition of the term

$$\frac{1}{\beta_r}\sum_i\sum_j\sum_r T^*_{ijr}\ln T^*_{ijr}$$

in objective Function (5.47), then

$$\frac{\partial U'_R}{\partial T^*_{ijr}} = g^*_{ijr} + \frac{1}{\beta_r}(\ln T^*_{ijr} + 1) \tag{5.75}$$

so that the KKT conditions are

$$g^*_{ijr} + \frac{1}{\beta_r}(\ln T^*_{ijr} + 1) - \mu_{ij} \geq 0 \qquad \forall\, i, j, r \tag{5.76a}$$

$$T^*_{ijr}\left(g^*_{ijr} + \frac{1}{\beta_r}(\ln T^*_{ijr} + 1) - \mu_{ij}\right) = 0 \qquad \forall\, i, j, r \tag{5.76b}$$

For $\ln T^*_{ijr}$ to be defined, T^*_{ijr} must be strictly positive, so that the second factor in the left-hand side of (5.76b) must be equal to zero. This implies that Inequation (5.76a) may be dismissed, and also that

$$\ln T^*_{ijr} = -\beta_r(g^*_{ijr} - \mu_{ij}) - 1$$

or

$$T_{ijr} = e^{-\beta_r g^*_{ijr}} e^{\beta_r \mu_{ij} - 1}$$

From Condition (5.48a) we have

$$\sum_r e^{-\beta_r g^*_{ijr}} e^{\beta_r \mu_{ij} - 1} = e^{\beta_r \mu_{ij} - 1}\sum_r e^{-\beta_r g^*_{ijr}} = T_{ij}; \qquad \forall\, i, j$$

and

$$e^{\beta_r \mu_{ij}} - 1 = \frac{T_{ij}}{\sum_r e^{-\beta_r g^*_{ijr}}}; \qquad \forall\, i, j \tag{5.76c}$$

so that

$$T_{ijr} = T_{ij} \frac{e^{-\beta_r g^*_{ijr}}}{\sum_r e^{-\beta_r g^*_{ijr}}}; \qquad \forall\, i, j, r \tag{5.77}$$

which is precisely Formula (5.35b), as expected, and in which g_{ijr} is the equilibrium generalized cost of route r between i and j. Since T_{ij} is strictly positive, so is T^*_{ijr} and the requirement that $\ln T^*_{ijr} \neq 0$ is met, so that Formula (5.77) does indeed constitute the solution to the problem. If any T_{ij} is equal to zero, then the corresponding T^*_{ijr} are equal to zero for all r, and Formula (5.77) is not necessary.

APPENDIX 5.2 OBTAINING AGGREGATE DEMANDS FROM THE R.T.'S INDIRECT UTILITY FUNCTION IN THE PRESENCE OF DEMAND EXTERNALITIES

The R.T.'s indirect utility is given by Formula (5.49a):

$$\tilde{W}_R = B - \tau \sum_a \int_0^{v^*_a} t_a(v)\, dv + \frac{1}{\beta} \sum_i \sum_j T_{ij} \ln \sum_r e^{-\beta g^*_{ijr}} + \tau \sum_i \sum_j \sum_r T^*_{ijr} t^*_{ijr}$$
$$- \frac{1}{\beta} \sum_i \sum_j T_{ij} \ln T_{ij}$$

Since the T_{ij}'s and B are given, we have

$$\frac{\partial \tilde{W}_R}{\partial c_{kln}} = -\tau \frac{\partial}{\partial c_{kln}} \sum_a \int_0^{v^*_a} t_a(v)\, dv + \frac{1}{\beta} \sum_i \sum_j T_{ij} \frac{\partial}{\partial c_{kln}} \ln \sum_r e^{-\beta g^*_{ijr}} + \tau \frac{\partial}{\partial c_{kln}} \sum_i \sum_j \sum_r T^*_{ijr} t^*_{ijr} \tag{5.78}$$

From the "chain rule," we have

$$\frac{\partial \sum_a \int_0^{\Sigma_i \Sigma_j \Sigma_r (T^*_{ijr} + S_{ijr})\delta^a_{ijr}} t_a(v)\, dv}{\partial c_{kln}} = \sum_i \sum_j \sum_r \frac{\partial}{\partial T^*_{ijr}} \left[\sum_a \int_0^{\Sigma_i \Sigma_j \Sigma_r (T^*_{ijr} + S_{ijr})\delta^a_{ijr}} t_a(v)\, dv \right] \frac{\partial T^*_{ijr}}{\partial c_{kln}}$$

Each term in this sum may be estimated as

$$\frac{\partial \sum_a \int_0^{\Sigma_i \Sigma_j \Sigma_r (T_{ijr} + S_{ijr})\delta^a_{ijr}} g_a(v)\, dv}{\partial T_{ijr}} = \sum_a \frac{\partial \int_0^{\Sigma_i \Sigma_j \Sigma_r (T_{ijr} + S_{ijr})\delta^a_{ijr}} g_a(v)\, dv}{\partial T_{ijr}}$$

$$= \sum_a \left| g_a\left(\sum_i \sum_j \sum_r (T_{ijr} + S_{ijr})\, \delta^a_{ijr} \right) \right| \frac{\partial \sum_i \sum_j \sum_r (T_{ijr} + S_{ijr})\, \delta^a_{ijr}}{\partial T_{ijr}} = \sum_a g_a(v_a)\, \delta^a_{ijr} = g_{ijr}$$

so that the first term in Formula (5.78) is equal to

$$-\tau \sum_i \sum_j \sum_r t^*_{ijr} \frac{\partial T^*_{ijr}}{\partial c_{kln}}$$

Next, the derivative of the second term in Formula (5.78) is equal to

$$\frac{\partial}{\partial c_{kln}} \left\{ \frac{1}{\beta} \sum_i \sum_j T_{ij} \ln \sum_r e^{-\beta g^*_{ijr}} \right\} = \frac{1}{\beta} \sum_i \sum_j T_{ij} \frac{1}{\sum_r e^{-\beta g^*_{ijr}}} \sum_r \frac{\partial e^{-\beta g^*_{ijr}}}{\partial c_{kln}}$$

$$= \frac{1}{\beta} \sum_i \sum_j T_{ij} \frac{e^{-\beta g^*_{ijr}}}{\sum_r e^{-\beta g^*_{ijr}}} \frac{\partial e^{-\beta g^*_{ijr}}}{\partial c_{kln}} = -\sum_i \sum_j T_{ij} \sum_r P^*_{r/ij} \left(\frac{\partial c_{ijr} + \tau \partial t^*_{ijr}}{\partial c_{kln}} \right)$$

$$= T_{kl} P^*_{n/kl} - \tau \sum_i \sum_j \sum_r T_{ij} P^*_{ij/r} \frac{\partial t^*_{ijr}}{\partial c_{kln}} = -T^*_{kln} - \tau \sum_i \sum_j \sum_r T^*_{ijr} \frac{\partial t^*_{ijr}}{\partial c_{kln}}$$

Finally, the derivative of the third term in Formula (5.78) is equal to

$$\tau \frac{\partial}{\partial c_{kln}} \sum_i \sum_j \sum_r T^*_{ijr} t^*_{ijr} = \tau \sum_i \sum_j \sum_r t^*_{ijr} \frac{\partial T^*_{ijr}}{\partial c_{kln}} + \tau \sum_i \sum_j \sum_r T^*_{ijr} \frac{\partial t^*_{ijr}}{\partial c_{kln}}$$

Adding all terms, we have

$$\frac{\partial \tilde{W}_R}{\partial c_{kln}} = -T^*_{kln}; \qquad \forall\, k, l, n \tag{5.79}$$

and since

$$\frac{\partial \tilde{W}_R}{\partial B} = 1 \tag{5.80}$$

Formula (5.50) is established. It should be noted that the proof is also valid for the uncongested probabilistic case, as then the first term in the indirect utility function is the negative of the third term, so that they cancel each other out, even before taking the derivatives. This may also be verified directly from Formula (3.16b).

Using a continuity argument, the proof is also valid in the limit, deterministic congested case. Indeed, in this case, the relationship

$$T^*_{ijr} = T_{ij} P_{r/ij}; \qquad \forall\, i, j, r$$

is still valid, and is in fact Formula (5.7a).

Finally, the validity of the proof in the deterministic uncongested case may be verified directly from Formula (3.11c). Thus, in all cases Roy's identity applies to route demands.

CHAPTER 6

COMBINED TRAVEL DEMAND MODELING UNDER CONGESTED CONDITIONS

6.1 INTRODUCTION: MODELING THE EFFECTS OF CONGESTION ON JOINT TRAVEL DEMANDS

Having modeled route choice in the presence of congestion, in this chapter we combine it with the other choices facing an individual traveler, namely destination, route, and decision to travel, by ''backtracking'' one decision level at a time, from the present lowest level of route choice to the highest level of decision to travel, as we did in the uncongested case. In so doing, we will retain the same methodological approach as was used to perform congested network assignment, that is, obtain the respective demands T_{ijmr}, T_{ijm}, T_{ij}, and T_i as the solution of utility maximization problems on the part of the R.T. As in the case of route choice, this approach is necessary because of the presence of congestion on the network, and/or at the destinations. That is, the respective demands, modal, by destination, and so on, cannot be obtained simply as the addition of individual demands. Rather, the interaction between individual travel choices induced by demand externalities must be taken into account at the aggregate level.

The benefits of using the R.T. approach are twofold. First, the same nested logit models as were developed in the uncongested case will still obtain. The formulation of the congested case will then truly generalize that of the uncongested case. Those factors of choice, including link, destination, and modal costs, which are variable, that is, unknown in the congested case, will now be determined as part of the solution of the respective R.T.'s utility maximization problems. Thus, there will be complete parallelism between the uncongested and congested case.

Second, and perhaps more importantly, the four dimensions of travel demand will be totally and rigorously integrated, without the need to resort to heuristic ''feedbacks'' between them. That is, the models are solved in a rigorous manner, through the application of algorithms which are guaranteed to converge to an internally consistent solution. In particular, the values of the respective travel costs will be compatible with one another, with the demands they underlie, and with congestion effects. Thus, the travel demands will be in equilibrium not only internally, in terms of individual demands (e.g., for route, or modes), but jointly, in terms of *combined* demands (e.g., for modal routes to destinations, etc.). Joint demands are also consistent conceptually, as they represent the choices of a *single* decision maker, the R.T., under common constraints.

6.2 COMBINED DESTINATION AND ROUTE CHOICE

6.2.1 Specification of Variable Destination Utilities

In this section, we consider the *combined* choices of destination and route, *for a given mode*, when the factors of these respective choices, for example, route travel times and destination costs, have variable and thus unknown values. This may be due to congestion, both on the network and at the facilities visited by travelers at the destinations. To illustrate the flexibility of the framework, we shall change from the uncongested case and place in the present case the choice of mode at a higher hierarchical level than the choice of destination.[1] Since index m is fixed, it will not be carried throughout the derivations, but will implicitly be present.

Thus, we assume that the origin-destination demanded T_{ij} are no longer given, as was assumed in the previous chapter, but are now unknown. Thus, we assume in this first version of the model that the trip origins by mode T_i are given, and that we are to determine the aggregate route demands T_{ijr}'s, *together* with the aggregate origin-destination demands T_{ij}'s. In addition we assume that route choice conforms to the stochastic UE principle, that all travelers choose the itinerary which minimizes their random route utilities. Destination utilities are assumed to be random, as in the uncongested case in section 2 of Chapter 4, but are now variable, a function of destination volumes, reflecting externalities at facilities (e.g., parking lots, stores, etc.) at the destination.

This may translate two major kinds of effects. First, variable destination utilities may reflect congestion at the facility visited at the destination, such as parking lot, stores, etc. Facility congestion may in turn be caused by a limited capacity of the facility to serve travelers, for example, if the facility's parking lot is so small that customers must either wait for a parking space to become available, or park at some distance from the facility. Variable destination utilities may also reflect supply-demand relationships which determine the price of the activity conducted at the destination (e.g., parking charges, price of retail goods, etc.).[2]

Thus, we assume that the main effect of facility congestion is reflected in a "service cost" s_j, reflecting, for instance, the average time performing the travel purpose (e.g., shopping, getting service, etc.) at destination j. We assume that the relationship of s_j to the volume T_j at the destination is specified in the form of a destination congestion function $s_j(T_j)$. Since the volumes T_j at the various facilities are unknown, the service costs s_j are also unknown, and must consequently be determined as part of the model's solution.

The specific expression of functions $s_j()$ may, for instance, be derived from

[1] As discussed in Chapter 1, the validity of this ordering should be confirmed during model calibration (see the next chapter). If it is not, it is straightforward to place modal choice lower than destination choice in the hierarchical structure. (See next section.)

[2] While we shall not pursue this aspect in depth, the reader may refer to Oppenheim (1994).

queueing theory, and will reflect the nature of the consumer arrival process, the number of "servers," and other characteristics of the queueing process. In particular, when destination demands are Poisson distributed and travelers are served by a single server, the destination cost function may be represented by a service time expressed by the "Pollaczek-Kintchine" formula (Gross and Harris, 1974).

$$s_j(T_j) = \begin{cases} \gamma_j T_j / 2(1 - \zeta_j T_j); & \text{for } \zeta_j T_j < 1 \\ \infty & \text{otherwise} \end{cases} \tag{6.1}$$

where ζ_j and γ_j are respectively the first and second moments of the distribution of random service times at j. For simplification and generality, we will use in the illustrative examples a simple function of the type

$$s_j(T_j) = s_j^0 + k_j T_j^{\omega_j}; \qquad \forall\, j \tag{6.2}$$

where s_j^0 is the "fixed," or base destination cost at zero demand. Parameter k_j, which has the dimension of a unit, or marginal, cost, has a positive value and acts as a scaling and conversion factor between demand levels and service cost. Parameter ω_j is a dimensionless parameter, whose (positive) value translates the severity of congestion effects, that is, the rate at which the cost increases with demand, and implies that functions s_j are strictly increasing in their argument T_j.[3]

The values of parameters k_j and ω_j must be calibrated, but we assume for the purpose of exposition that they are known.[4] Graphically, functions s_j may be represented as in Figure 6.1.

Service cost functions s_j thus play roles similar to those of link cost functions t_a.

In analogy with the specification of individual traveler utilities in the uncontested Case (4.20), the utility received by an individual traveler from making a single trip to a given destination j, for the travel purpose at hand, using a given mode m from origin i, is equal to

$$\begin{aligned} \tilde{V}_{ijr} &= \tilde{U}_{ij} + \tilde{U}_{ijr} + \epsilon_{ij} + \epsilon_{ijr} \\ &= b_i + h_{ij} - \pi s_j^* - \tau t_{ijr}^* - c_{ijr} + \epsilon_{ij} + \epsilon_{ijr}; \qquad \forall\, i, j, r \end{aligned} \tag{6.3}$$

[3]Conversely, a negative value for ω indicates positive externalities, i.e., functions are strictly decreasing in their argument T_j. While congestion externalities will usually be negative, positive externalities might be present if any other destination attribute becomes more attractive with increasing destination volumes, translating a "bandwagon" effect. It is not necessary to specify which type of externality effects are present, as it turns out that it does not affect the properties of the model, in particular, the uniqueness of its solution, as long as *only one* type of externality is present. (See Oppenheim, 1993a and Exercise 6.1.)

[4]It should be noted that in this particular case calibration may be performed *outside* the model (i.e., independently of the calibration of the other model parameters), for instance, on the basis of observations of the values of the service time, as a function of demand.

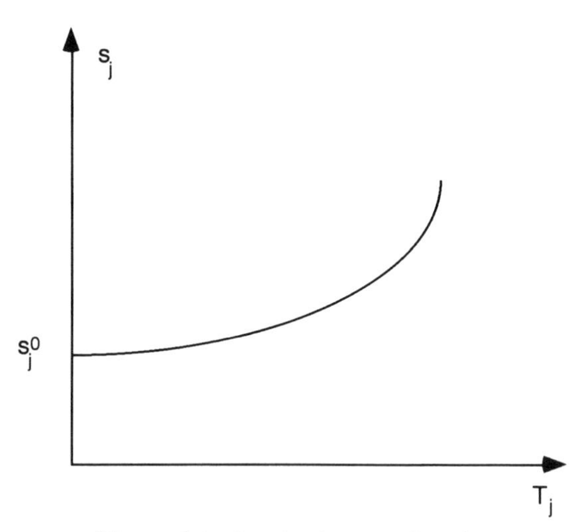

Figure 6.1 Destination cost function.

s_j^* is the equilibrium destination cost, whose value is given by functions $s_j(T_j)$, where T_j is to be determined. t_{ijr}^* is the equilibrium travel time on route r between i and j, and is to be determined, as in the previous chapter. c_{ijr} is the route's cost, and is given. The other symbols retain the meaning they had in the previous chapter. In particular, b_i is the income/budget of travelers from zone i.

6.2.2 Individual and Aggregate Demand Functions

With the standard assumptions about the nature of the random terms ϵ_{ij}, the probability $P_{j/i}$ that a randomly chosen individual traveler from i on mode m is observed to go to destination j has the same specification as in the uncongested case, Formula (4.21):

$$P_{j/i} = \frac{e^{\beta_d(h_{ij} - \pi s_j^* + \tilde{W}_{j/i}^*)}}{\sum_j e^{\beta_d(h_{ij} - \pi s_j^* + \tilde{W}_{j/i}^*)}}; \qquad \forall\ i, j \tag{6.4a}$$

where $\tilde{W}_{j/i}$ is the expected received utility from a choice of route to j from i, net of the budget, and is specified by Formula (4.22);

$$\tilde{W}_{j/i} = \frac{1}{\beta_r} \ln \sum_r e^{-\beta g_{ijr}^*}; \qquad \forall\ i, j \tag{6.4b}$$

and similarly, the probability $P_{r/ij}$ of choosing a given route between i and j has the same form as in the route choice case, as given by Formula (3.15a)

$$P_{r/ij} = \frac{e^{-\beta_r(\overset{*}{t}_{ijr} + c_{ijr})}}{\sum_r e^{-\beta_r(\overset{*}{t}_{ijr} + c_{ijr})}} = \frac{e^{-\beta_r \overset{*}{g}_{ijr}}}{\sum_r e^{-\beta_r \overset{*}{g}_{ijr}}}; \qquad \forall\, i, j, r \tag{6.5}$$

in which the t_{ijr}^*'s are the unknown, equilibrium values reflecting network externalities, and the s_j^*'s are the unknown, equilibrium values reflecting destination externalities. These values conform to the "user equilibrium" principle, that is, reflect the choice by every individual traveler of the maximum utility route-destination combination available to him or her. As usual

$$P_{jr/i} = P_{r/ij} P_{j/i}; \qquad \forall\, i, j, r \tag{6.6a}$$

so that given the origin volumes by mode T_i, the corresponding origin-destination volumes are then equal to

$$T_{ij} = T_i P_{j/i}; \qquad \forall\, i, j \tag{6.6b}$$

6.2.3 The R.T.'s Utility Maximization Problem

PROPOSITION 6.1

The R.T.'s direct utility function for joint destination and route choices in the congested case is equal to

$$U_{DR} = -\tau \sum_a \int_0^{v_a} t_a(v)\, dv - \frac{1}{\beta_r} \sum_i \sum_j \sum_r T_{ijr} \ln T_{ijr}$$
$$- \frac{1}{\beta_d'} \sum_i \sum_j T_{ij} \ln T_{ij} + \sum_i \sum_j h_{ij} T_{ij}$$
$$- \pi \sum_j \int_0^{\Sigma_i T_{ij}} s_j(x)\, dx + T_0 \tag{6.7a}$$

where as usual

$$v_a = \sum_i \sum_j \sum_r (T_{ijr} + S_{ijr}) \delta_{ijr}^a; \qquad \forall\, a \tag{6.7b}$$

and where

$$\beta_d' = \frac{\beta_d \beta_r}{\beta_r - \beta_d} \rightarrow \frac{1}{\beta_d'} = \frac{1}{\beta_d} - \frac{1}{\beta_r} \tag{6.7c}$$

Consequently, the aggregate destination/route volumes T_{ij}, together with the unknown values of the destination and route times, s_j and t_{ij}, may be obtained as solutions to the R.T.'s utility maximization problem:

$$\begin{aligned}\text{Max } U_{DR}\,(T_{ij},\, T_{ijr},\, T_0) \\ = -\tau \sum_a \int_0^{\Sigma_i \Sigma_j \Sigma_r (T_{ijr} + S_{ijr})\delta^a_{ijr}} t_a(v)\, dv - \frac{1}{\beta_r} \sum_i \sum_j \sum_r T_{ijr} \ln T_{ijr} \\ - \frac{1}{\beta'_d} \sum_i \sum_j T_{ij} \ln T_{ij} + \sum_i \sum_j h_{ij} T_{ij} - \pi \sum_j \int_0^{\Sigma_i T_{ij}} s_j(x)\, dx + T_0\end{aligned} \tag{6.8}$$

The familiar constraints are

$$\sum_i \sum_j \sum_r T_{ijr} c_{ijr} + T_0 = B \tag{6.9a}$$

$$\sum_j T_{ij} = T_i; \qquad \forall\, i \tag{6.9b}$$

$$\sum_r T_{ijr} = T_{ij}; \qquad \forall\, i, j \tag{6.9c}$$

Finally, all variables must be nonnegative-valued, as indicated:

$$v_a \geq 0;\; T_{ij} > 0,\; T_{ijr} > 0,\; T_0; \qquad \forall\, i, j, r \tag{6.9d}$$

Variables T_{ijr} represent the demand on route r between i and j for the travel purpose of interest, and are unknown. Variables S_{ijr} represent interzonal (origin-destination) demands for all other purposes, and are assumed known and given. The incorporation of these latter variables in the problem is required, as in the route assignment problem in the previous chapter, since link travel times on a network shared by multiple types of users are dependent on the travel decisions of all users. Functions $t_a(v)$ represent the link travel time functions for given mode m, as defined in Formula (5.4). The modal link-route incidence variables δ^a_{ijr} are defined as they were for a single mode in Formula (1.1):

$$\delta^a_{ijr} = \begin{cases} 1 \text{ if link } a \text{ is part of the mode's route } r \text{ between } i \text{ and } j \\ 0 \text{ otherwise} \end{cases}; \qquad \forall\, i, j, r, a$$

We now elaborate briefly on the meaning of the respective constraints on the R.T.'s problem. Constraints (6.9b) state that the number of trip origins from a given zone i, on the *given* implicit mode m, is equal to the given level T_i. Constraint (6.9c) requires the sum of modal trips for the travel purpose of interest on all routes r between locations i and j to be equal to the number of modal trips between i and j.

Since β'_d must be positive for the R.T.'s utility function to have a meaning, Relationship (6.7b) implies that

$$\frac{1}{\beta'_d} = \frac{1}{\beta_d} - \frac{1}{\beta_r} \geq 0$$

or

$$\beta_d \leq \beta_d \rightarrow \sigma_d \geq \sigma_r$$

If this condition does not hold in a specific situation, the orders of these levels should be inverted. In the limit case when $\beta_d = \beta_r = \beta$, the term in the origin-destination volumes T_{ij} disappears from the R.T.'s utility function, and the solution conforms to the multinomial logit

$$T_{ijr} = T_i \frac{e^{\beta(h_{ij} - \pi s_j^* - g_{ijr}^*)}}{\sum_j \sum_r e^{\beta(h_{ij} - \pi s_j^* - g_{ijr}^*)}}; \qquad \forall\, i, j, r \tag{6.10a}$$

where $s_j^* = s_j(T_j^*)$, and g_{ijr}^* is the equilibrium route cost. When $\beta_r = \infty$, the solution conforms to the multinomial logit

$$T_{ijr} = T_i \frac{e^{\beta_d(h_{ij} - \pi s_j^* - g_{ij}^*)}}{\sum_j e^{\beta_d(h_{ij} - \pi s_j^* - g_{ij}^*)}} \text{ if } r = r^* \text{ and 0 otherwise}; \qquad \forall\, i, j, r \tag{6.10b}$$

where g_{ij}^* is the minimum route cost between i and j, and $s_j^* = s_j(T_j^*)$.

The R.T.'s received utility is equal to

$$\begin{aligned} \tilde{W}_{DR} = {} & B + \frac{1}{\beta_d} E_i - \sum_a \int_0^{v_a^*} g_a(v)\, dv + \sum_i \sum_j \sum_r T_{ijr}^* g_{ijr}^* \\ & - \pi \left[\sum_j \int_0^{\Sigma_i T_{ij}^*} s_j(x)\, dx - \sum_i \sum_j T_{ij}^* s_j^* \right] \\ & + \frac{1}{\beta_d} \sum_i T_i \ln \sum_j e^{\beta_d g_{ij}^*} \end{aligned} \tag{6.11a}$$

where, for compactness,

$$g_{ij} = h_{ij} - \pi s_j \left(\sum_i T_{ij} \right) + \tilde{W}_{j/i}; \ \forall\, i, j \tag{6.11b}$$

and

$$E_i = -\sum_i T_i \ln T_i$$

which is a constant in the present case where the T_i's are given.

Consequently, we have the following result.

PROPOSITION 6.2

The traveler surplus from combined route and destination demands in the congested case is equal to

$$TS_{DR} = -\sum_a \int_0^{v_a^*} g_a(v)\, dv + \sum_i \sum_j \sum_r T_{ijr}^* g_{ijr}^*$$
$$-\pi \left[\sum_j \int_0^{\Sigma_i T_{ij}^*} s_j(x)\, dx - \sum_i \sum_j T_{ij}^* s_j^* \right] + \frac{1}{\beta_d} \sum_i T_i \ln \sum_j e^{\beta_d g_{ij}^*} \tag{6.11c}$$

It is shown in Appendix 6.1 that the solution to the R.T.'s utility maximization Problem (6.8)–(6.9) retrieves the aggregate Demands (6.4a), (6.5), in which the unknown origin-destination and destination costs have their equilibrium values, as discussed earlier.[5]

In summary, the solution of the R.T.'s utility maximization problem does retrieve the *combined equilibrium* destination-route demands, in which the variable route travel times and the variables destination costs have endogenously determined values which are compatible with the principle of *individual* random utility maximization. This important result justifies the R.T.'s approach, not only in the case of a single travel choice, such as route, as seen in the previous chapter, but in the case of joint travel choices.

The legitimacy of the R.T.'s utility maximization approach may be further confirmed by showing that the same relationship between the R.T.'s received utility and the aggregate travel demands in the uncongested case, Formula (4.19), remains valid in this joint congested demand case in the presence of network and destination externalities. Specifically, the following result is shown in Appendix 6.5

[5]Note that we are, for the moment, assuming that the program does indeed have a solution. This is not entirely obvious, but is examined below.

THEOREM 6.1

$$T^*_{ijr} = -\frac{\partial \tilde{W}_{DR}}{\partial c_{ijr}} \Big/ \frac{\partial \tilde{W}_{DR}}{\partial B}; \qquad \forall\, i, j, r \tag{6.11d}$$

The power of the R.T. approach should begin to be apparent. In any event, we might ask at this point whether Program (6.8)–(6.9) does indeed have a solution, whether given the inputs T_i, S_{ij}, and so on, one can always and uniquely obtain the origin-destination and the route demands. The answer to that question, as demonstrated in Appendix 6.2, is yes. Furthermore, the useful fact that such a solution is unique, in terms of the link demands, is also demonstrated there. Finally, it may be shown that under appropriate conditions, the solution to the R.T.'s utility maximization problem is also unique in the case of positive destination externalities, when parameter π in utility Function (6.3) is negative. (See Exercise 6.1.)

6.2.4 Solution Algorithm

An algorithm, directly based on the same Evans' partial linearization method (Evans, 1972) which was used in the previous chapter in connection with the SUE problem, may be used to solve the combined route/destination choice problem. At iteration k, the subproblem to solve for identifying a descent direction is

$$\text{Min } U'^{k}_{DR} = \sum_a y^k_a g^{k-1}_a + \frac{1}{\beta'_d} \sum_{ij} Z^k_{ij} \ln Z^k_{ij} + \sum_j \left((h_{ij} + \pi s^{k-1}_j) \sum_i Z^k_{ij} \right)$$
$$+ \frac{1}{\beta_r} \sum_{ijr} Z^k_{ijr} \ln Z^k_{ijr} \tag{6.12}$$

where U'_{DR} is the negative of the utility U_{DR}, that is, the disutility,[6] subject to Constraints (6.9a–d) where link demands v are replaced by auxiliary link demands y, and travel demands T are replaced by auxiliary demands Z. g^{k-1}_a and s^{k-1}_j are respectively the current (updated) values of the link and destination costs functions, computed on the basis of the latest values for the link and destination demands respectively. It may be shown that this problem is a convex problem (see Exercise 6.3), which therefore may be solved directly from its first-order conditions. Leaving the demonstrations of these respective statements to the interested reader, as they follow closely the demonstrations above,

[6]Minimizing the disutility is more logical, since there is a predominance of negative utility factors in the expression of U_{DR} in Formula (6.7a).

the first-order conditions are

$$Z_{ij}^{k} = T_i \frac{e^{-\beta_d g_{ij}^{k-1}}}{\sum_j e^{-\beta_d g_{ij}^{k-1}}}; \qquad \forall\ i, j \tag{6.13a}$$

and

$$Z_{ijr}^{k} = \frac{Z_{ij}^{k} e^{-\beta_r g_{ijr}^{k-1}}}{\sum_r e^{-\beta_r g_{ijr}^{k-1}}}; \qquad \forall\ i, j, r \tag{6.13b}$$

where

$$g_{ij}^{k-1} = h_{ij} - \pi s_j\left(\sum_i T_{ij}^{k-1}\right) + \tilde{W}_{j/i}^{k-1}; \qquad \forall\ i, j \tag{6.14a}$$

$$\tilde{W}_{j/i}^{k-1} = \frac{1}{\beta_r} \ln \sum_r e^{-\beta_r g_{ijr}^{k-1}}; \qquad \forall\ i, j \tag{6.14b}$$

In all these computations, the given S_{ijr} representing the general route demands must be added to the Z_{ijr}. If only the S_{ij}'s are given *and* if the general-purpose travelers have the same route utilities as the T_{ij} travelers, the S_{ij}'s may then simply be added to the T_{ij}'s and both assigned to destinations and routes together. Thus, solving the subproblem amounts to distributing the given T_i's from a given origin i to destinations j according to Formula (6.13a), and the resulting Z_{ij} to routes according to Formula (6.13b).

In the special case when route choice is deterministic, the optimality conditions for the subproblem are the deterministic "user equilibrium" conditions. This stage of the algorithm is then modified accordingly. (See Exercise 6.12.)

Once the auxiliary solution is obtained, the optimal size of the move along the descent direction is then obtained by minimizing, with respect to $0 \le \theta \le 1$, the function of θ,[7]

$$U_{DR}^{\prime k}(\theta) = U_{DR}^{\prime}(\boldsymbol{v}^k + \theta(\boldsymbol{y}^k - \boldsymbol{v}^k), \boldsymbol{T}^k + \theta(\boldsymbol{Z}^k - \boldsymbol{T}^k)) \tag{6.15a}$$

It may be shown in this connection that (see Exercise 6.21):

$$\begin{aligned}\frac{dU_{DR}^{\prime k}(\theta)}{d\theta} &= \sum_a (y_a - v_a) g_a(v_a + \theta(y_a - v_a)) \\ &+ \frac{1}{\beta_d^{\prime}} \sum_i \sum_j (Z_{ij} - T_{ij}) \ln (T_{ij} + \theta(Z_{ij} - T_{ij}) \\ &- \sum_i \sum_j h_{ij}(Z_{ij} - T_{ij}) + \pi \sum_j (Z_j - T_j) s_j(T_j + \theta(Z_j - T_j)) \\ &+ \frac{1}{\beta_r} \sum_i \sum_j \sum_r (Z_{ijr} - T_{ijr}) \ln (T_{ijr} + \theta(Z_{ijr} - T_{ijr}))\end{aligned} \tag{6.15b}$$

[7] $U_{DR}^{\prime k}(\theta)$ is a function of θ only, as v_a^k, y_a^k, T_{ij}^k and Z_{ij}^k are all constants, that is, the last specified values for the primary and auxiliary variables.

Once the optimal value of θ is found, the new, improved solution is

$$T_{ijr}^{k+1} = T_{ijr}^k + \theta(Z_{ijr}^k - T_{ijr}^k); \qquad \forall\, i, j, r \tag{6.16a}$$

$$T_{ij}^{k+1} = T_{ij}^k + \theta(Z_{ij}^k - T_{ij}^k); \qquad \forall\, i, j \tag{6.16b}$$

This basic algorithmic step is then iterated until stopped with an appropriate termination criterion.

The algorithm may then be formulated as follows.

ALGORITHM 6.1 PARTIAL LINEARIZATION FOR COMBINED DESTINATION-ROUTE DEMANDS

An initial feasible solution may be obtained by using travel costs corresponding to zero link and origin-destination demands.

Iterative step

Given a current feasible solution, (T_{ij}^k, T_{ijr}^k), the current link travel costs, g_a^{k-1}, and the given T_i's:

A. Determine route costs g_{ijr} between origins i and destinations j from Formula (1.2a), destination costs s_j from Formula (6.2), origin-destination utilities g_{ij} from Formula (6.14a), and the corresponding $\tilde{W}_{j/i}$'s from Formula (6.14b).

B. Estimate the auxiliary Z_{ij}'s, using the nested logit Formulas (6.13a).

C. Assign Z_{ij} to the routes between i and j on mode m, according to Formula (6.13b), or using the STOCH algorithm to obviate route enumeration.

D. Solve

$$\begin{aligned} \underset{(0 \le \theta \le 1)}{\text{Min}} \quad U_{DR}'^k(\theta) &= \sum_a \int_0^{v_a^k + \theta(y_a^k - v_a^k)} g_a(v_a^k)\, dv \\ &+ \frac{1}{\beta_r} \sum_i \sum_j \sum_r (T_{ijr}^k + \theta(Z_{ijr}^k - T_{ijr}^k)) \ln (T_{ijr}^k + \theta(Z_{ijr}^k - T_{ijr}^k)) \\ &+ \frac{1}{\beta_d'} \sum_{ij} (T_{ij}^k + \theta(Z_{ij}^k - T_{ij}^k)) \ln (T_{ij}^k + \theta(Z_{ij}^k - T_{ij}^k)) \\ &- \sum_{ij} h_{ij}(T_{ij}^k + \theta(Z_{ij}^k - T_{ij}^k)) + \pi \sum_j \int_0^{T_j^k + \theta(Z_j^k - T_j^k)} s_j(x)\, dx \end{aligned} \tag{6.17}$$

(Continued)

where, for short

$$v_a = \sum_i \sum_j \sum_r (T_{ijr} + S_{ijr})\delta^a_{i,j,r}; \qquad \forall a$$

$$y_a = \sum_i \sum_j \sum_r (Z_{ijr} + S_{ijr})\delta^a_{i,j,r}$$

$$T_j = \sum_i T_{ij}$$

$$; \qquad \forall j$$

$$Z_j = \sum_i Z_{ij}$$

E. Update the solution as

$$T^{k+1}_{ij} = T^k_{ij} + \theta(Z^k_{ij} - T^k_{ij}); \qquad \forall i, j$$

$$T^{k+1}_{ijr} = T^k_{ijr} + \theta(Z^k_{ijr} - T^k_{ijr}); \qquad \forall i, j, r$$

F. Check convergence, i.e., whether

$$|(\boldsymbol{y}^k - \boldsymbol{v}^k) \cdot \nabla_v U'_{DR}(\boldsymbol{T}^k, \boldsymbol{v}^k) + (\boldsymbol{Z}^k - \boldsymbol{T}^k) \cdot \nabla_T U'_{DR}(\boldsymbol{T}^k, \boldsymbol{v}^k)| \leq \epsilon$$

where ϵ is a given tolerance. If so, stop; if not, iterate.

6.2.5 Illustrative Example: Application of Evans' Algorithm

We now illustrate the application of the model to the same network as we have been using all along, the last time in connection with route assignment in section 5.3.6. The difference is that now the origin destination demands T_{ij} are no longer given (i.e., $T_{12} = 250$; $T_{14} = 150$), but must be determined as the result of travelers' choice of destination. The total number of travelers T_1 out of node 1 is assumed to be equal to 150. All other T_i's are assumed equal to zero (the only travel taking place is from node 1). In addition, and to generalize with respect to the illustrative example for the route choice problem, we now assume that there are "general" travelers, for whom the destination is *known*, in the amounts

$$S_{12} = 100$$

$$S_{14} = 150$$

We assume that the destination attractiveness terms h_{ij} (in minutes) are equal to

$$h_{12} = -2$$

$$h_{13} = -1000$$

$$h_{14} = -4$$

These may, for instance, be interpreted as the time required to park in the absence of congestion at the destination. The very large value for destination 3 translates the fact that there is no parking there, thus prohibiting its choice as a destination.

Further, we assume that the destination cost functions s_j are specified as

$$s_j = k_j \left(\sum_i T_{ij} \right)^{\omega_j}; \qquad j = 2, 4$$

with

$$k_2 = 0.1; \quad \omega_2 = 0.6$$

$$k_4 = 0.2; \quad \omega_4 = 0.9$$

These may, for instance, be interpreted as the effect on parking time of congestion at the destination. The link cost functions remain the same as previously. We assume that the values of the model parameters, as obtained from calibration, are

$$\tau = 1$$

$$\pi = 1$$

$$\beta_d = 0.05$$

Finally, we assume that route choice is deterministic. Given these data, we would like to estimate the T_{ij}'s (i.e., in effect T_{12}, T_{14}, since we expect all other origin-destination demands to be zero), as well as the route demands T_{12r} and T_{14r}. The first three iterations of the Frank-Wolfe algorithm applied to this problem are represented in Table 6.1.

The equilibrium solution is represented in Figures 6.2–6.5.

Link travel costs are represented in Table 6.2.

Route travel costs for the routes of interest are represented in Table 6.3.

Thus, it can be verified that, as expected, the routes which carry a demand are those which offer the minimum cost, specifically routes 1 and 2 between nodes 1 and 2 and routes 7 and 9 between nodes 1 and 4. Note that route 3 between 1 and 2 is not used, even though it offers the same travel cost as routes 1 and 2. This is not inconsistent with the principle of user equilibrium. Also, note that even though general travelers S_{ij} are assigned to different routes than specific travelers T_{ij}, in reality it would not be possible to distinguish between the two, as obviously, interchanging them one for one on these routes would affect neither the travel costs nor the destination costs, which are only a function of destination demands.

The equilibrium destination costs at destinations 2 and 4 are equal to $s_2 = 1.4$ and $s_4 = 9.3$ respectively, so that the total destination costs are $(2 + 1.4)$ and $(4 + 9.3)$, respectively. Destination costs for the given-purpose travelers amount to 780 minutes or 5.2 minutes on the average for the 150 travelers.

TABLE 6.1 Evans' Algorithm for the Combined Destination/Route Choice Problem

Initialization	Links						Destinations	
	1	2	3	4	5	6	2	4
v_a^0	0	0	0	0	0	0		
t_a^0	5	15	6	8	7	8	$t_{12}^0 = 5$	$t_{14}^0 = 13$
T_{12}	93	0	0	0	0	0	$s_2^0 = 0$	$s_4^0 = 0$
S_{12}	100	0	0	0	0	0		
T_{14}	29	0	29	29	29	0	$T_{12}^0 = 150 \dfrac{e^{-0.05(5+2)}}{e^{-0.05(5+2)} + e^{-0.05(13+4)}}$ $= 92$	
S_{14}	75	0	75	75	75	0	$T_{14}^0 = 58$	
v_a^1	296	0	104	104	104	0		

Iteration 1	Links						Destinations	
	1	2	3	4	5	6	1	2
Update t_a^1	638	15	14	30	20	8	$t_{12}^1 = 23$	$t_{14}^1 = 15$
T_{12}	0	103	0	0	0	103	$s_2^1 = 1.61$	$s_4^1 = 24.2$
S_{12}	0	100	0	0	0	100	$Z_{12}^1 = 150 \frac{e^{-0.05(27.35)}}{e^{-0.05(27.35)} + e^{-0.05(43.2)}}$ $= 103$	
T_{14}	0	47	0	0	0	0	$Z_{14}^1 = 47$	
S_{14}	0	150	0	0	0	0		
y_a^1	0	400	0	0	0	203		
							$\theta = 0.318$	
Move v_a^2	202	127	71	71	71	65	$T_{12}^1 = 96$	$T_{14}^1 = 54$

	Links						Destinations		θ
Iteration # 2	1	2	3	4	5	6	1	2	
Update t_a^1	142	109	7.7	12.8	9.8	9.6	$t_{12}^2 = 27.1$	$t_{14}^2 = 17.5$	
T_{12}	0	0	100	0	100	100	$s_2^2 = 2.37$	$s_4^2 = 23.8$	
S_{12}	0	0	100	0	100	100			
T_{14}	0	0	50	0	50	0			
S_{14}	0	0	150	0	150	0	$Z_{12}^2 = 100$	$Z_{14}^2 = 50$	
y_a^2	0	0	400	0	400	200			
							$\theta = 0.173$		
Move v_a^3	167	105	128	58	128	88	$T_{12}^2 = 97$	$T_{14}^1 = 53$	

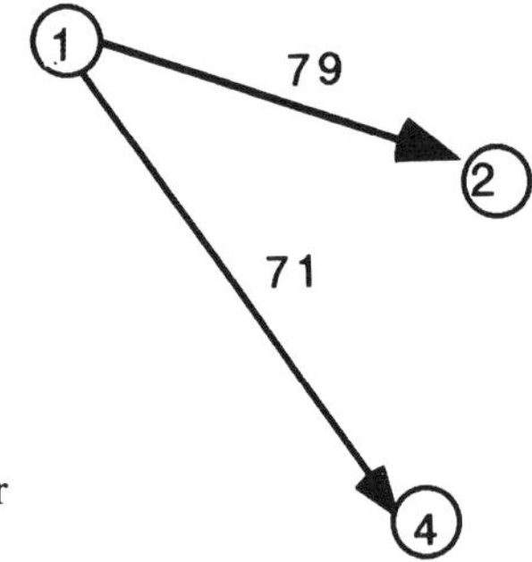

Figure 6.2 Equilibrium origin-destination demands for given-purpose travel.

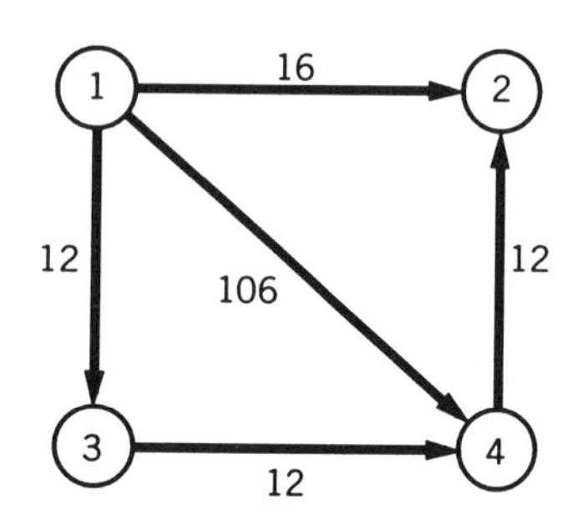

Figure 6.3 Equilibrium link demands.

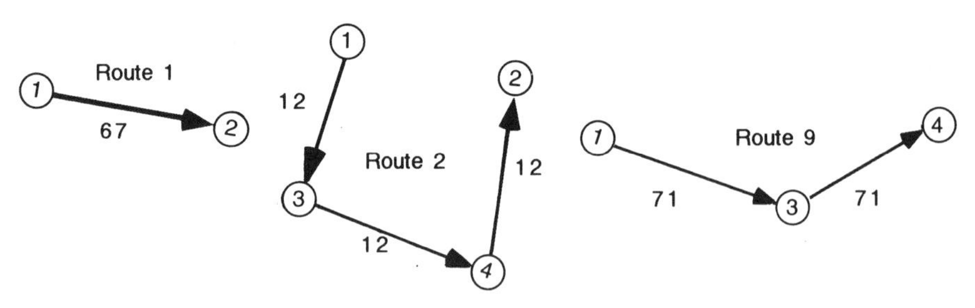

Figure 6.4 Equilibrium route demands for given-purpose travel.

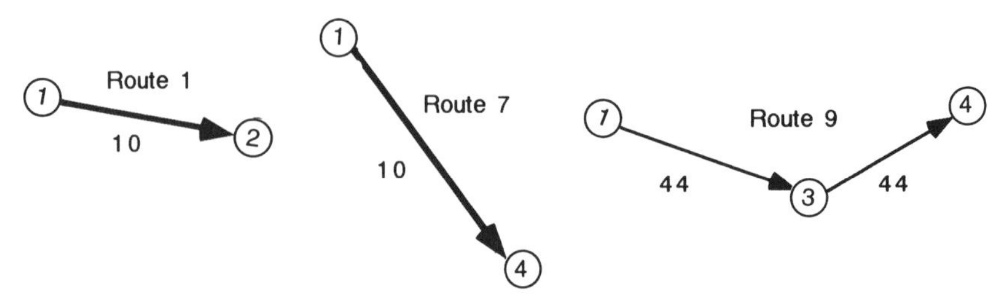

Figure 6.5 Equilibrium route demands for general travel.

TABLE 6.2 Link Travel Costs

Link #	Travel Cost
1	69
2	61
3	24
4	8
5	37
6	8

The total travel cost for all travelers is equal to 25,743 minutes, or 66.35 minutes per traveler on the average. The traveler's surplus is equal to −22,270.

Finally, it is easy to verify that these respective solution values for route travel times and demands satisfy the logit model in Formula (6.4), as required (see Exercise 6.5). The impacts of user charges on travel demands, total travel time, and traveler surplus may be assessed as in the route choice case (see Exercise 6.22).

Thus, the solution to Problem (6.8)–(6.9) reflects all the qualitative properties of individual utility maximizing choices, as well as those of congested networks and destinations. Specifically, for this solution, no individual traveler may further reduce deterministic route travel costs by unilaterally changing choice of route, nor increase further total utility (including destination-related) by unilaterally changing destination choice. Considering the richness of these properties, and the relative ease with which the R.T.'s problem may be solved with Evans' algorithm, this approach to combined destination and route demand estimation is remarkably powerful.

6.3 COMBINED MODE, DESTINATION, AND ROUTE CHOICES

Having modeled destination and route choice, we now backtrack one more step in our hierarchical framework to include modal choice, together with the subsequent choices of destination and route considered earlier. Whereas pre-

TABLE 6.3 Route Travel Costs

Route #	Travel Cost
1	69
2	69
3	69
4	24
5	68
6	84
7	61
8	77
9	61

viously there was only one mode of travel available (i.e., the single value of index m for travel mode was fixed), there are now several alternative modes available for travel between any given origin i and destination j. Consequently, whereas previously the givens were T_i and the unknowns were T_{ij} and T_{ijr} (or equivalently v_a), now the givens are T_i and the unknowns are T_{ijm} and T_{ijmr}. The choice of destination is assumed to precede that of a mode.

6.3.1 Nonscheduled, Independent Modes

6.3.1.1 Individual and aggregate demand functions We make several simplifying assumptions in the first version of the model. First, that all modes have independent networks. In other words, that the demand for travel on one mode's links does not affect the travel times/costs for any of the other modes. This might be the case, for instance, when considering a car network and a subway network, or, in general, any individual networks with separate links. This assumption will be relaxed in the next section. We also assume, without loss of generality, that destination costs s_j are fixed.

Thus, in accordance with the general model represented in Formula (4.21), the (marginal) probability that an individual traveler in demand zone i chooses destination j (irrespective of mode) has the same specification as in the uncongested case, and is accordingly equal to

$$P_{j/i} = \frac{e^{\beta_d(h_{ij} + \tilde{W}^*_{j/i})}}{\sum_j e^{\beta_d(h_{ij} + \tilde{W}^*_{j/i})}}; \qquad \forall\ i, j \tag{6.18a}$$

in which $\tilde{W}^*_{j/i}$ is the equilibrium expected value of the utility received from one trip from location i to location j, and is equal to

$$\tilde{W}^*_{j/i} = \frac{1}{\beta_m} \ln \sum_m e^{\beta_m(h_{ijm} + \tilde{W}^*_{m/ij})}; \qquad \forall\ i, j \tag{6.18b}$$

in which, in turn, $\tilde{W}^*_{m/ij}$ is equal to

$$\tilde{W}^*_{m/ij} = \frac{1}{\beta_r} \ln \sum_r e^{-\beta_r g^*_{ijmr}}; \qquad \forall\ i, j, m \tag{6.18c}$$

For simplification, the term h_{ij} includes the fixed destination costs s_j. Furthermore, the conditional probability $P_{m/ij}$ that an individual traveler having chosen to go to destination j now chooses mode m also has the same specification as in the previous section, and is equal to

$$P_{m/ij} = \frac{e^{\beta_m(h_{ijm} + W^*_{m/ij})}}{\sum_m e^{\beta_m(h_{ijm} + \tilde{W}^*_{m/ij})}}; \qquad \forall\ i, j, m \tag{6.19}$$

Finally, again as in the previous section

$$P_{r/ijm} = \frac{e^{-\beta_r g^*_{ijmr}}}{\sum_r e^{-\beta_r g^*_{ijmr}}}; \qquad \forall\, i, j, m, r \tag{6.20}$$

Given $P_{j/i}$ and $P_{m/ij}$ and $P_{r/ijm}$ as specified above, the joint conditional probability that an individual traveler chooses destination j *and* access mode m, given that he or she is located in zone i, is

$$P_{jmr/i} = P_{j/i} P_{m/ij} P_{r/ijm}; \qquad \forall\, i, j, m, r \tag{6.21a}$$

Consequently, the demand for travel from origin i to destination j, using mode m and route r, is equal to

$$T_{ijmr} = T_i P_{jmr/i} = T_i P_{j/i} P_{m/ij} P_{r/ijm}; \qquad \forall\, i, j, m, r \tag{6.21b}$$

As always in the case of variable travel costs, the values of the t^*_{ijmr}'s are unknown, as they depend on the modal origin-destination and route demands, both of which are unknown. Consequently, $\tilde{W}^*_{j/i}$ and $\tilde{W}^*_{m/ij}$ are also unknown.

6.3.1.2 *The R.T.'s Utility Maximization Problem*

PROPOSITION 6.3

The R.T.'s utility function for joint destination, mode, and route choices is

$$U_{DMR} = -\tau \sum_m \sum_{a_m} \int_0^{v_a} t_a^m(x)\, dx - \frac{1}{\beta_r} \sum_{ijmr} T_{ijmr} \ln T_{ijmr}$$

$$- \frac{1}{\beta'_m} \sum_{ijm} T_{ijm} \ln T_{ijm} - \frac{1}{\beta'_d} \sum_{ij} T_{ij} \ln T_{ij} + \sum_{ijm} T_{ijm} h_{ijm} + \sum_{ij} h_{ij} T_{ij} + T_0 \tag{6.22a}$$

where

$$\beta'_m = \frac{\beta_m \beta_r}{\beta_r - \beta_m} \rightarrow \frac{1}{\beta'_m} = \frac{1}{\beta_m} - \frac{1}{\beta_r}$$

$$\beta'_d = \frac{\beta_m \beta_d}{\beta_m - \beta_d} \rightarrow \frac{1}{\beta'_d} = \frac{1}{\beta_d} - \frac{1}{\beta_m} \tag{6.22b}$$

The R.T.'s U.M. problem is consequently

$$\text{Max } U'_{DMR}(T_{ij}, T_{ijm}, T_{ijmr}, T_0)$$

$$= -\sum_m \sum_{a_m} \int_0^{\Sigma_i \Sigma_j \Sigma_r (T_{ijmr} + S_{ijmr})\delta^a_{ijr}} t^m_a(x)\, dx - \frac{1}{\beta_r} \sum_{ijmr} T_{ijmr} \ln T_{ijmr}$$

$$- \frac{1}{\beta'_m} \sum_{ijm} T_{ijm} \ln T_{ijm} - \frac{1}{\beta'_d} \sum_{ij} T_{ij} \ln T_{ij} + \sum_{ijm} T_{ijm} h_{ijm} \tag{6.23}$$

$$+ \sum_{ij} h_{ij} T_{ij} + T_0$$

The variables, or unknowns, are the route demands T_{ijr}, the origin-destination demands T_{ij} and their components by mode, T_{ijm}, and the budgetary slack T_0. The constraints on the R.T.'s problem are

$$\sum_{ijmr} T_{ijmr} c_{ijmr} + T_0 = B \tag{6.24a}$$

$$\sum_r T_{ijmr} = T_{ijm}; \qquad \forall\, i, j, m \tag{6.24b}$$

$$\sum_m T_{ijm} = T_{ij}; \qquad \forall\, i, j \tag{6.24c}$$

$$\sum_j T_{ij} = T_i; \qquad \forall\, i \tag{6.24d}$$

In addition, as usual, all variables must be positive or nonnegative.

$$T_{ij} > 0,\ T_{ijm} > 0,\ T_{ijmr} > 0,\ T_0 \geq 0; \qquad \forall\, i, j, m, r \tag{6.24e}$$

Because of definitional relationship (6.22b), the variances of the random utilities at the respective levels should be such that

$$\sigma_d^2 \geq \sigma_m^2 \geq \sigma_r^2$$

Note the similarity with the statement of the R.T.'s problem in the case of route/destination choice, Problem (6.8)–(6.9). The constraints are, as usual, demand conservation requirements. The first set of Constraints (6.24b) translates the requirements that the sum of route demands for a given mode between a given origin and destination be equal to the unknown modal origin-destination demand. The second set of Constraints (6.24c) translates the requirements that the sums of modal origin-destination demands be equal to the unknown total origin-destination demand. The third set of Constraints (6.24d), translates the requirements that the sums of origin-destination demands be equal to the given (known) trip origins.

It is worth noting that this formulation accommodates "nontravel" modes, for example, telecommuting. In this case, in the first term of the objective function, the functions $g_a^m(v)$ for nontravel modes will be equal to constants, reflecting the fact that for such modes there is no congestion, so that travel costs are fixed. Also, in the case of nontravel modes, as already mentioned, we may assume by way of simplification (since we are not concerned about the problem of route choice for these modes) that there is only one route consisting of a single link between any origin and any destination.

In the same manner as above, and under the same general conditions, it may be shown that the solution to the R.T.'s U.M. Problem (6.23)–(6.24) always exists and is unique. (See Exercise 6.9.) The demonstration that the solution to the R.T.'s Problem retrieves the aggregate travel demands specified in Formulas (6.18) to (6.20) is presented in Appendix 6.3 at the end of the chapter.

In the same manner as in the previous case, the received utility of the R.T. may be estimated as the optimal value of the R.T.'s utility function (see Exercise 6.13), and is equal to

$$\tilde{W}_{DMR} = B - \frac{1}{\beta_d} \sum_i T_i \ln T_i - \sum_m \sum_a \int_0^{v_a^{m*}} g_a(v)\, dv + \sum_{ijmr} T^*_{ijmr} g^*_{ijmr} + \frac{1}{\beta_d} \sum_i T_i \ln \sum_j e^{\beta_d g^*_{ij}} \tag{6.25a}$$

where g_{ij} is specified in Formula (6.11b). This leads to the next result.

PROPOSITION 6.4

The traveler surplus for combined destination, mode, and route choices in the congested case is equal to

$$TS_{DMR} = -\sum_m \sum_a \int_0^{v_a^{m*}} g_a(v)\, dv + \sum_{ijmr} T^*_{ijmr} g^*_{ijmr} + \frac{1}{\beta_d} \sum_i T_i \ln \sum_j e^{\beta_d g^*_{ij}} \tag{6.25b}$$

The same relationship still holds between this function and the aggregate modal travel demands (see Exercise 6.14):

THEOREM 6.2

$$T^*_{ijmr} = -\frac{\partial \tilde{W}_{DMR}}{\partial c_{ijmr}} \Big/ \frac{\partial \tilde{W}_{DMR}}{\partial B}; \qquad \forall\ i, j, m, r \tag{6.26}$$

6.3.1.3 Solution Algorithm Problem (6.23)–(6.24) for the combined mode-destination-route problem may be solved with the same "partial linearization" algorithm as was used for the destination-route problem. This feature, incidentally, is one of the great advantages of using the R.T.'s approach to travel demand modeling. The general principle underlying the algorithm remains the same as in the previous cases with fewer travel choices. That is, at each iteration the subproblem to solve is the R.T.'s problem in the absence of demand externalities, that is, with fixed travel costs set at their last estimated value. As we have seen, the only difference between the present case and the preceding one is the addition of an "entropy term."

In the present case of the destination, mode, route choice Problem (6.23)–(6.24), the auxiliary problem at iteration k is then

$$\text{Min } U'^{k}_{DMR}(Z_{ij}, Z_{ijm}, Z_{ijmr}) = \sum_{ijmr} Z_{ijmr} g^{k-1}_{ijmr} + \frac{1}{\beta_r} \sum_{ijmr} Z_{ijmr} \ln Z_{ijmr}$$

$$+ \frac{1}{\beta'_d} \sum_{ijm} Z^k_{ijm} \ln Z^k_{ijm} + \frac{1}{\beta'_m} \sum_{ij} Z^k_{ij} \ln Z^k_{ij} - \sum_{ijm} h_{ijm} Z^k_{ijm} - \sum_{ij} Z^k_{ij} h_{ij} \qquad (6.27)$$

subject to the same Constraints (6.24) in terms of the auxiliary variables Z^k_{ijmr}, Z^k_{ijm}, and Z^k_{ij} which respectively replace the variables T_{ijmr}, T_{ijm}, T_{ij}. In this expression, the *fixed* link travel costs (the only variable costs) are estimated as

$$g^{m,k-1}_a = g^m_a \left(\sum_i \sum_j \sum_r (T^{k-1}_{ijmr} + S_{ijmr}) \delta^a_{ijr} \right); \qquad \forall\, a, m \qquad (6.28)$$

where S_{ijmr} are the given "general" demands on mode m's route r between i and j. It may be shown that the solution to Problem (6.27)–(6.24) represents a descent solution to the main (original) problem. In other words, along the direction represented by the vector $(\boldsymbol{Z}^k - \boldsymbol{T}^k)$, the value of the function $U(T)$ decreases from its present value $U(\boldsymbol{T}^k)$.

Furthermore, it may be shown that this auxiliary problem is a convex problem (i.e., in which the objective function is a convex function and the feasible region is convex). Therefore, it may be solved directly from its first-order optimality conditions. These may be shown to be, with standard notation

$$Z^k_{ijmr} = T_i \frac{e^{\beta_d(h_{ij} + \tilde{W}^{k-1}_{j/i})}}{\sum_j e^{\beta_d(h_{ij} + \tilde{W}^{k-1}_{j/i})}} \cdot \frac{e^{\beta_m(\tilde{W}^{k-1}_{m/ij} + h_{ijm})}}{\sum_m e^{\beta_m(\tilde{W}^{k-1}_{m/ij} + h_{ijm})}} \cdot \frac{e^{-\beta_r g_{ijmr}}}{\sum_r e^{-\beta_r g_{ijmr}}};$$

$$\forall\, i, j, m, r \qquad (6.29)$$

with the variable utilities fixed at the value corresponding to the preceding solution.

It is then apparent from Relationships (6.29) that solving the subproblem at each iteration amounts to distributing the given T_i's from a given origin i to destination j according to Formula (6.18a), and then distributing the resulting T_{ij}'s between a given origin i and a given destination j to modes m according to Formula (6.19). Finally, the resulting route demands are obtained from a stochastic network assignment with fixed link costs, which may be performed in practice with the STOCH algorithm discussed in section 3.2 of Chapter 3.

The optimal size of the move along the descent direction is then obtained in the same manner as in the Frank-Wolfe algorithm, by minimizing the function of θ

$$U'^{k}_{DMR}(T^k_{ij} + \theta(Z^k_{ij} - T^k_{ij}), T^k_{ijm} + \theta(Z^k_{ijm} - T^k_{ijm}),$$
$$T^k_{ijmr} + \theta(Z^k_{ijmr} - T^k_{ijmr})) \tag{6.30a}$$

All values in this function, except for θ, are known, so that this problem is a constrained optimization for a function of a single variable θ. This may be achieved with the bisection method. In the same manner as in the previous case of route choice only, Formula (5.23), it may be shown that (see Exercise 6.17):

$$\begin{aligned}
\frac{dU'^{k}_{DMR}(\theta)}{d\theta} &= \sum_m \sum_a (y_a - v_a) g_a(v_a + \theta(y_a - v_a)) \\
&+ \frac{1}{\beta_r} \sum_{ijmr} (Z_{ijmr} - T_{ijmr}) \ln (T_{ijmr} + \theta(Z_{ijmr} - T_{ijmr})) \\
&+ \frac{1}{\beta'_d} \sum_i \sum_j (Z_{ij} - T_{ij}) \ln (T_{ij} + \theta(Z_{ij} - T_{ij})) \\
&+ \frac{1}{\beta'_m} \sum_{ijm} (Z_{ijm} - T_{ijm}) \ln (T_{ijm} + \theta(Z_{ijm} - T_{ijm})) \\
&- \sum_{ijm} h_{ijm}(Z_{ijm} - T_{ijm}) - \sum_i \sum_j h_{ij}(Z_{ij} - T_{ij}) \\
&+ \pi \sum_j (Z_j - T_j) s_j(T_j + \theta(Z_j - T_j))
\end{aligned} \tag{6.30b}$$

Once the optimal value of θ is found, the new, improved solution is:

$$T^{k+1}_{ij} = T^k_{ij} + \theta(Z^k_{ij} - T^k_{ij}); \qquad \forall\, i, j \tag{6.31a}$$

$$T^{k+1}_{ijm} = T^k_{ijm} + \theta(Z^k_{ijm} - T^k_{ijm}); \qquad \forall\, i, j, m \tag{6.31b}$$

$$T^{k+1}_{ijmr} = T^k_{ijmr} + \theta(Z^k_{ijmr} - T^k_{ijmr}); \qquad \forall\, i, j, m, r \tag{6.31c}$$

ALGORITHM 6.2 PARTIAL LINEARIZATION FOR COMBINED DESTINATION, MODE, AND ROUTE DEMANDS

An initial feasible solution may be obtained by using travel costs corresponding to zero modal link and origin-destination demands.

Given a current feasible solution, $(T^k_{ij}, T^k_{ijm}, T^k_{ijmr})$, the current link travel costs, $g_a^{m,k-1}$, and the given T_i's:

A. Determine the routes' costs g_{ijmr} between origins i and destinations j, and the corresponding $\tilde{W}_{m/ij}$ and $\tilde{W}_{j/i}$'s.

B. Estimate the auxiliary Z_{ij}'s and Z_{ijm}'s according to Formulas (6.18)–(6.20).

C. Assign Z_{ijm} to the routes between i and j on mode m, using the STOCH algorithm on each modal network to obviate route enumeration.

D. Solve

$$\begin{aligned}\underset{(0 \le \theta \le 1)}{\text{Min}} \quad U'^{k}_{DMR} &= \sum_m \sum_a \int^{v^k_{a_m} + \theta(y^k_{a_m} - v^k_{a_m})} g^m_a(v)\, dv \\ &+ \frac{1}{\beta_r} \sum_{ijmr} (T^k_{ijmr} + \theta(Z^k_{ijmr} - T^k_{ijmr})) \ln (T^k_{ijmr} + \theta(Z^k_{ijmr} - T^k_{ijmr})) \\ &+ \frac{1}{\beta'_m} \sum_{ijm} (T^k_{ijm} + \theta(Z^k_{ijm} - T^k_{ijm})) \ln (T^k_{ijm} + \theta(Z^k_{ijm} - T^k_{ijm})) \\ &+ \frac{1}{\beta_d} \sum_{ij} (T^k_{ij} + \theta(Z^k_{ij} - T^k_{ij})) \ln (T^k_{ij} + \theta(Z^k_{ij} - T^k_{ij})) \\ &- \sum_{ijm} h_{ijm}(T^k_{ijm} + \theta(Z^k_{ijm} - T^k_{ijm})) - \sum_{ij} h_{ij}(T^k_{ij} + \theta(Z^k_{ij} - T^k_{ij}))\end{aligned} \tag{6.32}$$

where, for short

$$\left.\begin{aligned} v^m_a &= \sum_i \sum_j \sum_r (T_{ijmr} + S_{ijmr})\, \delta^{a_m}_{ijr} \\ y^m_a &= \sum_i \sum_j \sum_r (Z_{ijmr} + S_{ijmr})\, \delta^{a_m}_{ijr} \end{aligned}\right\} ; \quad \forall\, a, m \tag{6.33}$$

E. Update the solution as

$$T^{k+1}_{ij} = T^k_{ij} + \theta(Z^k_{ij} - T^k_{ij}); \qquad \forall\, i, j \tag{6.34a}$$

$$T^{k+1}_{ijm} = T^k_{ijm} + \theta(Z^k_{ijm} - T^k_{ijm}); \qquad \forall\, i, j, m \tag{6.34b}$$

$$T^{k+1}_{ijmr} = T^k_{ijmr} + \theta(Z^k_{ijmr} - T^k_{ijmr}); \qquad \forall\, i, j, m, r \tag{6.34c}$$

F. Check convergence, i.e., whether

$$|(\boldsymbol{y}^k - \boldsymbol{v}^k) \cdot \nabla_v U'_{DMR}(\boldsymbol{T}^k, \boldsymbol{v}^k) + (\boldsymbol{Z}^k - \boldsymbol{T}^k) \cdot \nabla_T U'_{DMR}(\boldsymbol{T}^k, \boldsymbol{v}^k)| \leq \epsilon \tag{6.35}$$

where ϵ is a given tolerance. If so, stop; if not, iterate.

6.3.2 Scheduled, Independent Modes

In the above formulation of the combined mode-destination and route choice, it was assumed that both modes were instantly available, and that consequently the criterion of route choice in both cases was the minimization of (line-haul) travel time. However, in the case of scheduled modes such as public transportation, travelers must incur a wait, so that the criterion is minimization of expected travel time, line haul plus average waiting time, as in the transit route choice model developed in section 4 of Chapter 3.

6.3.2.1 The R.T.'s Utility Maximization Problem The objective function for the transit route choice problem thus includes an extra term: $\overline{\omega}\Sigma_i \Sigma_s w_{is}$ where the terms are the average waiting times at node i for travelers with destination j using strategy s. Consequently, the R.T.'s utility function for the combined route, destination, and mode choice problem, when one of the modes is scheduled, and in the deterministic route choice case, is then, using the "strategy" based formulation in Section 5.5 of Chapter 5.

$$\begin{aligned}
\text{Max } U_{MR}(T_0, T_{ij}, T_{ijm}, T_{ijr}, T_{ist}, v_a^c, v_{as}^t, v_a^m, w_{is}) \\
= -\tau \sum_m \sum_{a_m} \int_0^{v_a^m} t_a^m(x)\, dx - \overline{\omega} \sum_i \sum_s w_{is} \\
- \frac{1}{\beta'_m} \sum_{ijm} T_{ijm} \ln T_{ijm} - \frac{1}{\beta'_d} \sum_{ij} T_{ij} \ln T_{ij} \\
+ \sum_{ijm} T_{ijm} h_{ijm} + \sum_j h_{ij} T_{ij} + T_0
\end{aligned} \tag{6.36}$$

subject to

$$v_{as}^t = \delta_s^a f_a w_{is}; \qquad \forall\, a \in i^-, s \tag{6.37a}$$

$$\sum_r T_{ijr} = T_{ijc}; \qquad \forall\, i, j \tag{6.37b}$$

$$\sum_{a \in i^-} v_{as}^t - \sum_{a \in i^+} v_{as}^t = T_{ist}; \qquad \forall\, a, i, s \tag{6.37c}$$

$$\sum_m T_{ijm} = T_{ij}; \qquad \forall\, i, j \tag{6.37d}$$

$$\sum_{s \in S_j} T_{ist} = T_{ijt}; \qquad \forall\, i, j \qquad (6.37e)$$

$$\sum_j T_{ij} = T_i; \qquad \forall\, i \qquad (6.37f)$$

$$v_a^c,\, w_{is},\, T_{ij},\, T_{ijm},\, T_{ijr},\, T_{ist},\, v_{as}^t,\, T_0 \geq 0, \qquad \forall\, i, j, a, m, s \qquad (6.37g)$$

where

$$v_a^c = \sum_i \sum_j \sum_r (T_{ijmr} + S_{ijmr})\,\delta_{ijr}^{a_c}; \qquad \forall\, a_c \qquad (6.38a)$$

$$v_a^t = \sum_s v_{as}^t; \qquad \forall\, a_s \qquad (6.38b)$$

6.3.2.2 Solution Algorithm The solution to the problem may be obtained with Evans' algorithm. In this case, the terms $\Sigma_i\, \Sigma_s\, w_{is}$, which are already linear in the w_{is} variables, are included "as is" in the linearized objective function for the subproblem. The only other difference from the algorithm above is that the transit network is loaded with the MECR algorithm, described in section 3.4.1 of Chapter 3, instead of the MCR algorithm.

6.3.2.3 Illustrative Example We now illustrate the application of the combined mode-destination-route choice model. Using again the same prototypical spatial structure as in section 6.2.5, we now assume that, in addition to the car network as above, the same transit network as in Example 3.4.2 serves the given locations. Transit travel times and frequencies are as represented in Tables 3.6a and 3.6b respectively. We assume that both networks are congested. The respective modal links' congestion functions are the same as have been used in Chapter 5. The number of travelers in zone 1 is equal to 400.[8] To accommodate the transit demand from 1 to 3, a single strategy (for simplification), number 15, is defined as taking the South line, link 3. The modal costs c_{ijm} of interzonal travel are specified in Table 6.4.

We assume for simplification, but without loss of generality, that all parameters h_{ij} and $h_{ijm} = 0$. As in the previous example, the high car cost to zone 3 reflects the absence of parking. The values of the parameters are

$$\overline{\omega} = 0.5$$

$$\tau = 1$$

$$\beta_m = 0.12 \rightarrow \beta'_m = 0.12$$

$$\beta_d = 0.92 \rightarrow \beta'_d = 0.39$$

[8]We assume for computational simplification, but without loss of generality, that there are no other given T_i's. Other T_i's would be *simultaneously* assigned in the same manner.

TABLE 6.4 Modal Costs of Interzonal Travel c_{ijm}

$i = 1; j = 2$	3	4
$c_{ijc} =$ 2	50	3
$c_{ijt} =$ 6	4	8

With these values, the equilibrium link volumes for the two modes are as represented in Figure 6.6, while the equilibrium demands and travel costs are described in Tables 6.5 to 6.7.

It may be verified that all these values together with the given travel and destination costs are in equilibrium, that is, are compatible with one another when inserted into Formulas (6.18–6.20). (See Exercise 6.19).

The impacts of user charges on travel demands, total travel time, and traveler surplus may be assessed as in the route choice case. (See Exercise 6.23.)

6.3.3 Interacting Modes

In the previous two sections, the respective modal networks were assumed to be independent, and not to interact in any other way than each receiving a fraction of the given total origin-destination demands. However, this assumption may not be valid in some cases. That is, a given link on either modal network may (or may not) be shared by both types of vehicles. For instance, buses in general share the same road network as cars.

In such a case, we may assume by way of simplification that any vehicle on the bus link incurs the same travel time. Link travel times may then in turn be estimated by expressing in the link cost function the total demand as the sum of the respective demands for each type of vehicular traffic,

$$t_a(v_a) = t_a(v_{a_c} + e_b v_{a_t}); \qquad \forall\, a \in A^+ \tag{6.39}$$

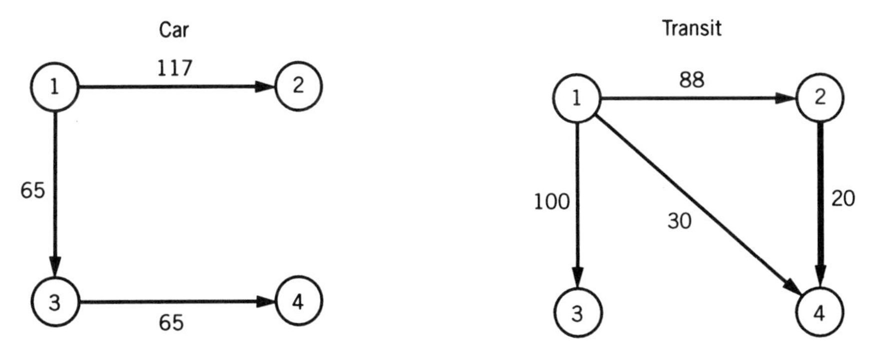

Figure 6.6 Equilibrium link volumes.

TABLE 6.5 Equilibrium Origin-Destination Demands

$T_{12} = 185$
$T_{13} = 100$
$T_{14} = 115$

TABLE 6.6 Equilibrium Origin-Destination Modal Demands

$T_{12c} = 117$, all using route 1; $T_{12t} = 68$, all using strategy 1
$T_{13c} = 0$; $T_{13t} = 100$, all using strategy 15
$T_{14c} = 65$, all using route 9; $T_{14t} = 50$, all using strategy 14

TABLE 6.7 Equilibrium Modal Travel and Waiting Times

$t_{1c} = 8.2$, $t_{2c} = 15$; $t_{3c} = 6.2$; $t_{4c} = 8$; $t_{4c} = 7.3$
$t_{1t} = 4.0$, $t_{2t} = 8.1$; $t_{3t} = 5.2$; $t_{4t} = 5.1$; $t_{4t} = 6$
$w_{11} = 5$; $w_{115} = 4.12$; $w_{111} = 2$
$t_{12c} = 8.2$, $t_{13c} = 6.2$; $t_{14c} = 13.5$
$t_{12t} = 9.0$; $t_{13t} = 9.3$; $t_{14t} = 10.6$

where A^+ is the set of links used both by the car and the bus modes, and e_b is the bus-car equivalency, the number of cars a bus represents in terms of using link capacity.[9]

More generally, however, interacting modes may have different travel time functions, reflecting the mode's own characteristics. For instance, in the example above, bus speeds may be better specified by their own bus travel time functions. Finally, it is also possible that vehicular traffic on separate modal links influence one another. For instance, the intensity of traffic on general lanes may, for various reasons, affect the speed on a reserved HOV lane.

In such cases, each mode may then be considered to have its own network, whether this corresponds to physical reality or not. Each link will then be characterized by given *modal* link cost functions. To illustrate the concept, let a_c and a_t be the two links bearing car and transit traffic from a given node i to a given node j, as represented in Figure 6.7.

The respective link costs functions are specified as

$$t_{a_c} = g_a^c(v_a^c, v_a^t); \qquad \forall\, a_c \tag{6.40a}$$

$$t_{a_t} = g_a^t(v_a^c, v_a^t); \qquad \forall\, a_t \tag{6.40b}$$

[9]This coefficient depends on the type of arterial traffic conditions, etc. (See Institute of Traffic Engineers, 1990).

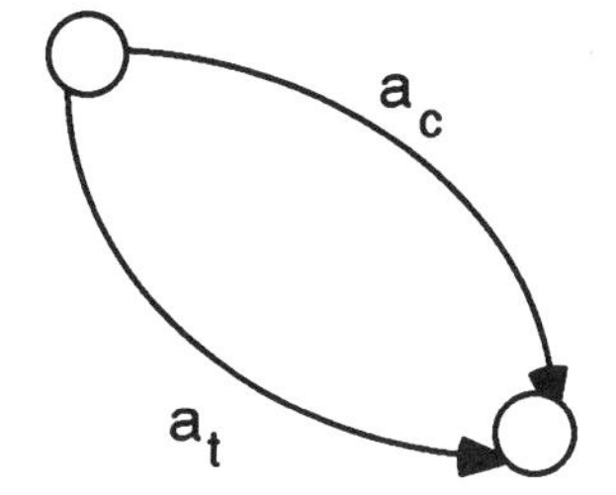

Figure 6.7 Interacting modal links.

If these functions are such that the following condition is met

$$\frac{\partial g_a^c(v_a^c, v_a^t)}{\partial v_a^t} = \frac{\partial g_a^t(v_a^c, v_a^t)}{\partial v_a^c}; \qquad \forall\, a_c;\, a_t \tag{6.41a}$$

that is, when the effects of an additional unit of modal demand on the other mode's link travel times are symmetric, the R.T.'s utility optimization problem whose solution provides the equilibrium modal, destinations, and route demands, is (see Exercise 6.7):

$$\begin{aligned}\text{Min } U'_{MR}(v_a^m, T_{ij}, T_{ijm}) &= \tfrac{1}{2}\sum_a \int_0^{v_a^c} (g_a^c(x, v_a^t) + g_a^c(x, 0))\, dx \\ &+ \tfrac{1}{2}\sum_a \int_0^{v_a^t} (g_a^t(v_a^c, x) + g_a^t(0, x))\, dx \\ &+ \frac{1}{\beta'_m}\sum_{ijm} T_{ijm} \ln T_{ijm} \\ &+ \frac{1}{\beta'_d}\sum_{ij} T_{ij} \ln T_{ij} - \sum_{ijm} T_{ijm} h_{ijm} - \sum_{ij} h_{ij} T_{ij}\end{aligned} \tag{6.42}$$

where modal link demands v_a^c and v_a^t are replaced by their expression in terms of route demands, Formulas (1.2b), subject to the same Constraints (6.9). This problem may be solved with the same partial linearization algorithm as repeatedly used earlier.

If Condition (6.41a) does not hold, but nevertheless

$$\frac{\partial g_a^c}{\partial v_a^c}\frac{\partial g_a^t}{\partial v_b^t} - \frac{\partial g_a^c}{\partial v_a^t}\frac{\partial g_a^t}{\partial v_b^c} > 0 \qquad \forall\, a_c;\, a_t \tag{6.41b}$$

even though it is not possible to formally specify the R.T.'s utility optimization problem, in practice the equilibrium travel demands may be estimated from the application of a "diagonalized" version of the partial linearization algorithm. In the main, "outer loop" each individual mode's link demands are "freed" one at a time, while all the other modes' link demands are "frozen" at their last estimated level. In the secondary, "inner loop," the partial linearization algorithm for the combined destination and route problem as described above is applied to estimate the next values of the "free" mode's demands. It is useful to note that it turns out that it is sufficient in practice to apply only the first iteration of Evans' algorithm at each inner iteration. When modal route choices are deterministic, the same problems may be formulated as "variational inequality" problems (Nagurney, 1993), which does not require meeting Conditions (6.41a) or (6.41b).

In conclusion of Section 6.3, it should be pointed out that the case of combined mode and route choice, both for scheduled, as well as unscheduled modes, may be retrieved from the present case of combined mode, destination, and route choice, for all types of modes, i.e., scheduled, non-scheduled and interacting, simply by treating the unknowns T_{ij} as given constants. (See Exercises 6.25–6.29.)

6.4 COMBINED TRAVEL, MODE, DESTINATION, AND ROUTE CHOICE

We now backtrack one final level to the top of the hierarchical structure of travel choices, to incorporate the choice of whether to travel with all the other choices we have considered so far, namely the choices of route, destination, and mode. In other words, the numbers of trips T_i originating from a given origin i are now variable, and not given, as previously assumed in the congested case. Of course, travel costs are still assumed variable. Furthermore, we would like the resulting expression for the unknown demands T_i's to conform to the same formulation as when travel times are fixed, that is, Formula (4.29)

$$T_i = N_i \frac{e^{\beta_t(h_i + \tilde{W}^*_{t/i})}}{1 + e^{\beta_t(h_i + \tilde{W}^*_{t/i})}}; \qquad \forall\, i \tag{6.43a}$$

where N_i is the number of residents in zone i, and h_i is the given constant characteristic of zone i, which may be a linear function of characteristics of the zone and of its average resident relevant to trip making, as defined in Formula (4.28c), and $\tilde{W}^*_{t/i}$ is the expected utility received by an individual traveling from zone i, which is defined in Formula (4.30)

$$\tilde{W}^*_{t/i} = b_i + \frac{1}{\beta_d} \ln \sum_j e^{\beta_d(h_{ij} + \tilde{W}^*_{j/i})}; \qquad \forall\, i \tag{6.43b}$$

in which $\tilde{W}^*_{j/i}$ is the expected received utility from a trip made from i to j, which is defined in Formula (4.17a)

$$\tilde{W}^*_{j/i} = \frac{1}{\beta_m} \ln \sum_m e^{\beta_m(h_{ijm} + \tilde{W}^*_{m/ij})}; \qquad \forall\, i, j \tag{6.44a}$$

in which $\tilde{W}_{m/ij}$ is the expected utility received from a trip made on mode m between i and j, which is defined in Formula (4.6b)

$$\tilde{W}^*_{m/ij} = \frac{1}{\beta_r} \ln \sum_r e^{-\beta_r g^*_{ijmr}}; \qquad \forall\, i, j, m \tag{6.44b}$$

The other demands, T_{ij}, T_{ijm}, and T_{ijmr} should also have the same formulations as in the preceding sections (as in the uncongested case). The respective equilibrium values of the utilities $\tilde{W}^*_{m/ij}$, $\tilde{W}^*_{j/i}$ and $\tilde{W}^*_{t/i}$, as well as g^*_{ijmr}, must be consistent with the demands T_i, T_{ij}, T_{ijm}, and T_{ijmr} which they generate, and vice versa; these demands must be consistent with the equilibrium utilities to which they give rise.

6.4.1 The R.T.'s Utility Maximization Problem

It is again possible to formulate the R.T.'s utility maximization problem which produces such aggregate demands, in the same fashion as previously. The systematic mechanism by which such problems are constructed as we keep going up the hierarchical structure of traveler choices should by now be clear. It essentially consists of adding an "entropy"-type term in the objective function, as well as an additional constraint. Both expressions link the highest level demands so far with the new, higher ones.

PROPOSITION 6.5

The R.T.'s utility function for combined choice of travel, destination, mode, and route is equal to

$$\begin{aligned} U_{TJMR} = & -\tau \sum_m \sum_{a_m} \int_0^{v_a} t_a^m(x)\, dx - \frac{1}{\beta_r} \sum_{ijmr} T_{ijmr} \ln T_{ijmr} \\ & - \frac{1}{\beta_t} \sum_i (T_i \ln T_i + T_{i0} \ln T_{i0}) - \frac{1}{\beta'_d} \sum_{ij} T_{ij} \ln T_{ij} \\ & - \frac{1}{\beta'_m} \sum_{ijm} T_{ijm} \ln T_{ijm} + \sum_{ijm} h_{ijm} T_{ijm} + \sum_{ij} h_{ij} T_{ij} \\ & + \sum_i h_i T_i + T_0 \end{aligned} \tag{6.45a}$$

where v_a, β'_m and β'_d have the expressions in Formula (6.23) and

$$\beta'_t = \frac{\beta_t \beta_d}{\beta_d - \beta_t} \rightarrow \frac{1}{\beta'_t} = \frac{1}{\beta_t} - \frac{1}{\beta_d} \tag{6.45b}$$

Consequently, the R.T.'s problem in the present case is

$$\text{Min } U'_{TJMR}\,(T_0, T_i, T_{i0}, T_{ijm}, T_{ijmr}) = \tau \sum_m \sum_{a_m} \int_0^{\Sigma_i\Sigma_j\Sigma_r(T_{ijmr}+S_{ijmr})\delta^{a_m}{}_{ijr}} t_a^m(x)\,dx$$

$$+ \frac{1}{\beta_r} \sum_{ijmr} T_{ijmr} \ln T_{ijmr} + \frac{1}{\beta'_t} \sum_i (T_i \ln T_i + T_{i0} \ln T_{i0})$$

$$+ \frac{1}{\beta'_d} \sum_{ij} T_{ij} \ln T_{ij} + \frac{1}{\beta'_m} \sum_{ijm} T_{ijm} \ln T_{ijm} - \sum_{ijm} h_{ijm} T_{ijm}$$

$$- \sum_{ij} h_{ij} T_{ij} - \sum_i h_i T_i + T_0 \tag{6.46}$$

subject to Constraints

$$\sum_{ijmr} T_{ijmr} c_{ijmr} + T_0 = B \tag{6.47a}$$

$$\sum_{ijm} T_{ijmr} = T_{ijm}; \qquad \forall\, i, j, m \tag{6.47b}$$

$$\sum_m T_{ijm} = T_{ij}; \qquad \forall\, i, j \tag{6.47c}$$

$$\sum_j T_{ij} = T_i; \qquad \forall\, i \tag{6.47d}$$

$$T_i + T_{i0} = N_i; \qquad \forall\, i \tag{6.47e}$$

In addition, as usual, all variables must be nonnegative.

$$T_0 \geq 0;\; T_{i0} > 0,\; T_{ij} > 0,\; T_{ijm} > 0,\; T_{ijmr} > 0; \qquad \forall\, i, j, r, m \tag{6.47f}$$

In the same manner as above, and under the same general conditions, it may be shown that the solution to this problem always exists and is unique. It is shown in Appendix 6.4 that its solution retrieves the required equilibrium aggregate joint travel demands.

In the same manner as in the previous case, the received utility of the R.T. may be estimated as the optimal value of the R.T.'s utility function and is equal to

$$\tilde{W}_{TDMR} = B - \sum_m \sum_a \int_0^{v_a^{m*}} g_a(v)\,dv$$

$$+ \frac{1}{\beta_t} \sum_i N_i \ln \sum_j e^{-\beta_t g_i^*} + \sum_{ijmr} T^*_{ijmr} g^*_{ijmr} \tag{6.48a}$$

where

$$g_i = h_i + \tilde{W}_{t/i}; \qquad \forall i \tag{6.48b}$$

Consequently:

PROPOSITION 6.6

The traveler surplus for combined equilibrium travel demands by mode, destination, and route is equal to

$$TS_{TDMR} = -\sum_m \sum_a \int_0^{v_a^{m*}} g_a(v)\, dv + \sum_{ijmr} T^*_{ijmr} g^*_{ijmr} + \frac{1}{\beta_t} \sum_i N_i \ln \sum_j e^{-\beta_t g_i^*} \tag{6.48c}$$

If there is congestion at the destinations, the term

$$-\pi \left[\sum_j \int_0^{\Sigma_i T^*_{ij}} s_j(x)\, dx - \sum_i \sum_j T^*_{ij} s^*_j \right]$$

would be added to the welfare and surplus functions, respectively. Similarly, if there are demand externalities at the level of mode choice, the term

$$-\psi \left[\sum_m \int_0^{T^*_m} q_m(x)\, dx - T^*_m q^*_m \right]$$

would similarly be added, where $q_m(\ )$ is the modal congestion function. Parameters π and ψ may be either positive or negative, corresponding respectively to negative and positive externalities. (See Exercise 6.4.)

Under either specification, the same relationship again holds between the R.T.'s received utility and the aggregate modal travel demands (see Exercise 6.16):

THEOREM 6.3

$$T^*_{ijmr} = -\frac{\partial \tilde{W}_{TDMR}}{\partial c_{ijmr}} \Big/ \frac{\partial \tilde{W}_{TDMR}}{\partial B}; \qquad \forall\ i, j, m, r \tag{6.49}$$

This particular R.T. problem represents the culmination of the development of travel demand models in this text. Indeed, it is the most general formulation of travel demands we have developed, and in fact, includes them all. That is, all cases, individual or aggregate, uncongested or congested, single, or combined choices may be retrieved from this formulation. Specifically, uncongested demands may be obtained simply by replacing the link and/or destination time functions by the constants representing the corresponding times at zero volumes.

Also, all of the previous travel demands models T_{ij}, T_{ijm}, T_{ijmr} may be obtained by setting the corresponding variables T_i, T_{ij}, and so on, equal to constants. Demands at the individual level, the probabilities of choice of the respective travel alternatives, may be obtained by setting the given volumes N_i to one. Finally, these various actions may be compounded to cover all of the situations we have examined in this text. In that sense, we could have started the description of the demand side directly with this formulation.

6.4.2 Solution Algorithm

Problem (6.46)–(6.47) for the combined travel-mode-destination-route problem may be solved with the same "partial linearization" algorithm as was used for all of the combined problems above. The auxiliary problem at iteration k is then

$$\begin{aligned}\text{Min } U'^{k}_{TDMR}(Z_i^k, Z_{i0}^k, Z_{ij}^k, Z_{ijm}^k, Z_{ijmr}^k) &= \sum_{ijmr} Z_{ijmr} g_{ijmr}^k \\ &+ \frac{1}{\beta_r} \sum_{ijmr} Z_{ijmr}^k \ln Z_{ijmr}^k + \frac{1}{\beta'_d} \sum_{ijm} Z_{ijm}^k \ln Z_{ijm}^k \\ &+ \frac{1}{\beta'_m} \sum_{ij} Z_{ij}^k \ln Z_{ij}^k + \sum_{ijm} h_{ijm} Z_{ijm}^k + \sum_{ij} Z_{ij}^k h_{ij} \\ &+ \sum_i Z_i^k h_i + \frac{1}{\beta'_t} \sum_i (Z_i^k \ln Z_i^k + Z_{i0} \ln Z_{i0}^k) \end{aligned} \qquad (6.50)$$

subject to the same constraints (6.47) in terms of the auxiliary variables, Z_0^k, Z_i^k, Z_{ij}^k, Z_{ijm}^k, Z_{ijmr}^k, which respectively replace variables T_{i0}, T_i, T_{ij}, T_{ijm}, T_{ijmr}. In this expression, the *fixed* link travel costs (the only variable costs) are estimated as

$$g_a^{m,k} = g_a^m \left(\sum_{ijr} (T_{ijmr}^{k-1} + S_{ijmr}) \delta_{ijr}^a \right); \qquad \forall\, a, m \qquad (6.51)$$

It may be shown that the solution to Problem (6.47)–(6.50) represents a descent solution to the main (original) problem. In other words, along the direction represented by the vector $(\boldsymbol{Z}^k - \boldsymbol{T}^k)$, the value of the function $U'_{TDMR}(\boldsymbol{T})$ decreases from its present value $U'_{TDMR}(\boldsymbol{T}^k)$.

Furthermore, it may be shown that this auxiliary problem is a convex problem (i.e., in which the objective function is a convex function and the feasible region is convex). Therefore, it may be solved directly from its first-order conditions. These may be shown to be

$$Z_{ijmr}^{k} = N_i \frac{e^{\beta_t(\tilde{W}_{t/i}^{k-1} + h_i)}}{(1 + e^{\beta_t(\tilde{W}_{t/i}^{k-1} + h_i)})} \frac{e^{\beta_d(h_{ij} + \tilde{W}_{j/i}^{k-1})}}{\sum_j e^{\beta_d(h_{ij} + \tilde{W}_{j/i}^{k-1})}} \cdot \frac{e^{\beta_m(h_{ijm} + \tilde{W}_{m/ij}^{k-1})}}{\sum_m e^{\beta_m(h_{ijm} + \tilde{W}_{m/ij}^{k-1})}} \frac{e^{-\beta_r g_{ijmr}}}{\sum_r e^{-\beta_r g_{ijmr}}}; \qquad \forall\ i, j, m, r \qquad (6.52)$$

where the link costs, $\tilde{W}_{m/ij}^{k}$ and $\tilde{W}_{j/i}^{k}$ are specified as in the auxiliary problem in the previous case, and

$$\tilde{W}_{t/i}^{k} = \frac{1}{\beta_t} \ln \sum_j e^{\beta_t(h_{ij} + \tilde{W}_{j/i}^{k})}; \qquad \forall\ i, k \qquad (6.53)$$

As in the previous case also, solving this subproblem requires distributing the given N_i's from a given origin i into travelers T_i to destinations j, modes m and routes r according to nested logit Formula (6.52).

The optimal size of the move along the descent direction is then obtained in the same manner as in the Frank-Wolfe algorithm, by minimizing the function of θ

$$U_{TDMR}^{\prime k}(\theta) = U(T_i^k + \theta(Z_i^k - T_i^k), T_{i0}^k + \theta(Z_{i0}^k - T_{i0}^k), T_{ij}^k + \theta(Z_{ij}^k - T_{ij}^k), T_{ijm}^k + \theta(Z_{ijm}^k - T_{ijm}^k), T_{ijmr}^k + \theta(Z_{ijmr}^k - T_{ijmr}^k)) \qquad (6.54)$$

ALGORITHM 6.3 FOR THE COMBINED TRAVEL, DESTINATION, MODE, AND ROUTE PROBLEM

Initialization

An initial feasible solution may be obtained by distributing the given N_i's as in the general iteration below, but on the basis of travel costs corresponding to zero values for all v_a's.

Iterative step

Given a current feasible solution, $(T_i^k, T_{i0}^k, T_{ijm}^k, T_{ijm}^k, T_{ijmr}^k)$, the current link travel costs, $g_a^{m,k-1}$, and the given N_i's:

A. Determine the modal routes' costs between origins i and destinations j, and the corresponding g_{ijmr}^{k-1}, $\tilde{W}_{m/ij}^{k-1}$, $\tilde{W}_{j/i}^{k-1}$ and $\tilde{W}_{t/i}^{k-1}$'s.

(*Continued*)

B. Estimate the auxiliary Z_{ijmr}'s, using nested logit Formula (6.52).
C. Determine the corresponding auxiliary link demands $y^k_{a_m}$.
Update the travel times on this basis.
D. Solve

$$\begin{aligned}
\min_{(0 \le \theta \le 1)} \quad U'^k_{TDMR}(\theta) = & \sum_m \sum_a \int_0^{v^k_{a_m} + \theta(y^k_{a_m} - v^k_{a_m})} g^m_a(x)\, dx \\
& + \frac{1}{\beta_r} \sum_{ijmr} (T^k_{ijmr} + \theta(Z^k_{ijmr} - T^k_{ijmr})) \ln (T^k_{ijmr} + \theta(Z^k_{ijmr} - T^k_{ijmr})) \\
& + \frac{1}{\beta'_m} \sum_{ijm} (T^k_{ijm} + \theta(Z^k_{ijm} - T^k_{ijm})) \ln (T^k_{ijm} + \theta(Z^k_{ijm} - T^k_{ijm})) \\
& + \frac{1}{\beta'_d} \sum_{ij} (T^k_{ij} + \theta(Z^k_{ij} - T^k_{ij})) \ln (T^k_{ij} + \theta(Z^k_{ij} - T^k_{ij})) \\
& + \frac{1}{\beta'_t} \sum_{i} (T^k_{i} + \theta(Z^k_{i} - T^k_{i})) \ln (T^k_{i} + \theta(Z^k_{i} - T^k_{i})) \\
& + \frac{1}{\beta'_t} \sum_{i} (T^k_{i0} + \theta(Z^k_{i0} - T^k_{i0})) \ln (T^k_{i0} + \theta(Z^k_{i0} - T^k_{i0})) \\
& - \sum_{ijm} h_{ijm}(T^k_{ijm} + \theta(Z^k_{ijm} - T^k_{ijm})) \\
& - \sum_{ij} h_{ij}(T^k_{ij} + \theta(Z^k_{ij} - T^k_{ij})) - \sum_i h_i(T^k_i + \theta(Z^k_i - T^k_i))
\end{aligned} \tag{6.55}$$

E. Update the solution as

$$T^{k+1}_i = T^k_i + \theta(Z^k_i - T^k_i); \qquad \forall\, i \tag{6.56a}$$

$$T^{k+1}_{i0} = T^k_{i0} + \theta(Z^k_{i0} - T^k_{i0}); \qquad \forall\, i \tag{6.56b}$$

$$T^{k+1}_{ij} = T^k_{ij} + \theta(Z^k_{ij} - T^k_{ij}); \qquad \forall\, i, j \tag{6.56c}$$

$$T^{k+1}_{ijm} = T^k_{ijm} + \theta(Z^k_{ijm} - T^k_{ijm}); \qquad \forall\, i, j, m \tag{6.56d}$$

$$T^{k+1}_{ijmr} = T^k_{ijmr} + \theta(Z^k_{ijmr} - T^k_{ijmr}); \qquad \forall\, i, j, m, r \tag{6.56e}$$

F. Check convergence, i.e., whether

$$|(\boldsymbol{y}^k - \boldsymbol{v}^k)\nabla_v U'_{TDMR}(\boldsymbol{T}^k, \boldsymbol{v}^k) + (\boldsymbol{Z}^k - \boldsymbol{T}^k)\nabla_T U'_{TDMR}(\boldsymbol{T}^k, \boldsymbol{v}^k)| \le \epsilon \tag{6.57}$$

where ϵ is a given tolerance. If so, stop; if not, iterate.

When one or more of the modes is scheduled, the same (car) route and (transit) strategy-based formulation (6.36)–(6.37) may be used. This does not affect the terms in the variables T_{i0} and T_i.

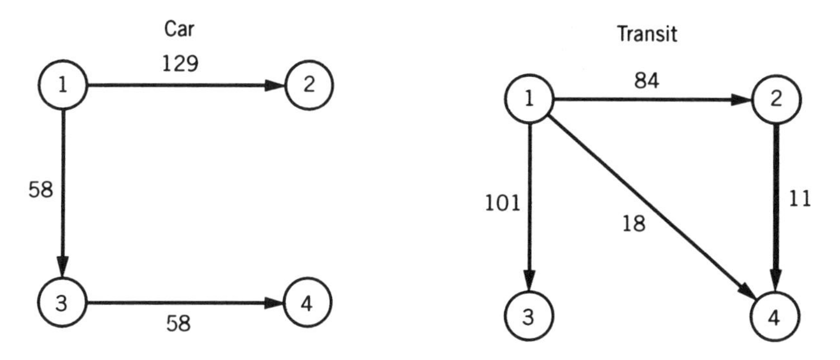

Figure 6.8 Equilibrium link demands.

6.4.3 Illustrative Example

As an example of the application of this model, we now revisit the previous example, and assume that the number of travelers out of zone 1, which was assumed to be equal to 400, is now unknown, and a function of the generalized travel cost from that zone.[10] There are 1000 residents (i.e., potential travelers) in zone 1. The costs of travel are as before. Parameter h_1 is equal to 60. Finally, parameter $\beta_i' = 0.1$. The values of the other parameters remain the same as previously. In addition, we assume that there are variable destination costs for the car mode at nodes 2 and 4, as specified in the illustrative example of section 6.2.5. The value of parameter π is the same, i.e., 1.

In addition, the values of the fixed destination attractiveness terms h_{1j} are equal to

$$h_{12} = 12;\ h_{13} = 10;\ h_{14} = 8.$$

The total number of travelers is equal to 390. The equilibrium travel demands corresponding to these data are represented in Figure 6.8 and Tables 6.8 to 6.10.

[10]Again for the sake of simplicity we assume, without loss of generality, that there are no travelers from the other three zones. If there were, the procedure would be exactly the same, but of course computationally more demanding.

TABLE 6.8 Equilibrium Origin-Destination Demands

$T_{12} = 202$
$T_{13} = 101$
$T_{14} = 87$

TABLE 6.9 Equilibrium Origin-Destination Modal Demands

$T_{12c} = 129$, all using route 1; $T_{12t} = 73$, all using strategy 1
$T_{13c} = 0$; $T_{13t} = 101$, all using strategy 15
$T_{14c} = 58$, all using route 9; $T_{14t} = 29$, all using strategy 11

TABLE 6.10 Equilibrium Modal Travel and Destination Times

$t_1^c = 9.5,\ t_2^c = 15;\ t_3^c = 6.1;\ t_4^c = 8;\quad t_4^c = 7.2$
$t_1^t = 4.0,\ t_2^t = 8.1;\ t_3^t = 5.2;\ t_4^t = 5.1;\ t_4^t = 6$

$t_{12c} = 9.5;\ t_{13c} = 6.1;\ t_{14c} = 13.3$
$t_{12t} = 9.0;\ t_{13t} = 9.4;\ t_{14t} = 10.5$
$s_2 = 1.3;\ s_4 = 4.0$

It may be verified that all these values, together with the given travel costs, are in equilibrium, that is, they are compatible with one another when inserted into the formulas for the aggregate demands. (See Exercise 6.20.)

The impact of user charges on travel demands, total travel time, and traveler surplus may be assessed as in the route choice case. (See Exercise 6.24.)

6.5 SUMMARY

In this chapter, we have formulated the R.T.'s utility maximization problems for combined traveler choices, under congested conditions. These problems, while being functionally similar to their counterparts for the uncongested case, determine the equilibrium values for the unknown, variable factors of choice, including travel times and the various expected received utilities. These equilibrium values are precisely those which motivate the individual travel choices of all travelers. These choices seek to maximize the overall utility of travel. At the same time, the travel demands resulting from these choices conversely give rise to precisely these utilities, through modal link and destination costs functions.

Thus, the R.T. approach which we have systematically used is quite powerful, as its produces forecasts of travel demands which are consistent both from the point of view of individual behavior (i.e., microeconomic principles) and physical phenomena (network congestion). Moreover, there is a strong conceptual methodological unity between the uncongested and the congested case, and between the specific models for various combinations of travel choices.

Just as remarkable is the fact that these quantitatively and qualitatively rigorous forecasts may be obtained in a relatively straightforward and systematic manner. Thus, there is no need to resort to "trial and error," "feedbacks," or ''incremental'' solution techniques, as for each model an exact algorithm is available which is guaranteed to approximate the unique solution within a known precision in a finite number of iterations. Specifically, it should be clear that the determination of the optimal step size θ at each iteration of the exact algorithm implements rigorously the basic intuition underlying these informal schemes. Both the ad-hoc and the exact approach to dealing with variable factors of demand are based on repeatedly applying procedures which are applicable with fixed utilities. The difference is that in the former approach this is done in a ''bootstrap'' fashion, while in the latter it is justified and guided by optimization theory.

Finally, the apparent complexity of the respective R.T. problems should not hide their essentially simple, and systematic structure. There is a strong similarity between all these models. Each is obtained from the route choice problem through the consecutive addition at each higher level of choice of ''entropy'' terms of the form $T_{ij} \ln T_{ij}$. Accordingly, the various versions of the partial linearization algorithm are all essentially similar. The only differences between them are in the number of logit distributions of demands during each iteration, and the specification of the function $U'(\theta)$, reflecting the changes above.

6.6 EXERCISES

6.1 Derive a sufficient condition for the solution to the combined route choice/destination choice problem to be unique, in the case of positive destination externalities, i.e., when parameter π in utility specification is negative. (Hint: Compute the Hessian of the objective function, i.e., the matrix with general element

$$\frac{\partial^2 U(T_{ijm})}{\partial T_{ijm} \partial T_{klm}}$$

and establish the sufficient condition for its being positive definite.)

6.2 Show that Subproblem (6.27)–(6.24) in connection with Algorithm 6.2 for the combined mode/route/destination problem is a convex problem.

6.3 Show that Subproblem (6.12)–(6.8) in connection with Algorithm 6.1 for the combined route/destination problem is a convex problem.

6.4 Generalize Problem (6.23)–(6.24) to include a variable destination cost instead of the fixed cost s_j (Hint: Add a term similar to the second integral in the objective function of Problem (6.8)–(6.9) for the combined destination-route choice problem). Show that the problem's solution retrieves the desired equilibrium conditions.

6.5 Verify that the solution values for the travel times and demands in the illustrative example of section 6.2.5 satisfy the logit model in Formula (6.4)–(6.5) with equilibrium costs, as they must.

6.6 Formulate the R.T.'s utility optimization problem whose solution produces the joint mode-destination-route demands when mode choice is assumed to precede destination choice in the hierarchical structure.[11]

6.7 Assume that the modal costs functions in Problem (6.42)–(6.9) for interacting modes, are defined as

$$g_a^c(v_a^c, v_a^t) \equiv g_a^t(v_a^c, v_a^t) = t_a^0\left[1 + 0.15\left(\frac{v_a^c + e_c v_a^t + v_a^0}{K_a}\right)\right]^4$$

where e_t is the car equivalent of one truck. This common link travel time function might be applicable, for instance, when links are used by two kinds of users (e.g., cars and trucks). Show that the solution produces equilibrium travel times on both modes.

6.8* Assume that for a given mode, the unknown number of trips originating in demand zone i, T_i, is a function of the expected utility of travel from that zone, of the form[12]

$$T_i = a_i \tilde{W}_i^{*\gamma}; \qquad \forall\, i \tag{6.58}$$

where

$$\tilde{W}_i^{*\gamma} = \frac{1}{\beta_d} \ln \sum_j e^{\beta_d(h_{ij} - \overset{*}{g}_{ij} - \pi s_j(\overset{*}{T}_j))}; \qquad \forall\, i \tag{6.59}$$

Consequently, the combined travel-destination and route demands are specified as

$$T_{ij}^* = a_i \tilde{W}_i^{*\gamma}(t, s) \frac{e^{\beta_d(h_{ij} - \overset{*}{g}_{ij} - \pi \overset{*}{s_j}(T_j))}}{\sum_j e^{\beta_d(h_{ij} - \overset{*}{g}_{ij} - \pi \overset{*}{s_j}(T_j))}}; \qquad \forall\, i, j \tag{6.60}$$

[11]It may be useful to remember in this connection that the ordering of the levels of choice is for analytical purposes only, as ultimately, individuals make all their travel choices simultaneously.
[12]Such a formulation, in the special case $\gamma = 1$, was proposed by Safwat and Magnanti (1988).

Show that the demands may be retrieved as the solution of the nonlinear optimization problem

$$\text{Min } F(\tilde{W}_i, T_{ij}, T_{ijr}) = \sum_i \int_0^{\tilde{W}_i} q_i(x)\, dx + \sum_a \int_0^{v_a} g_a(x)\, dx$$

$$+ \sum_i \sum_j T_{ij} \ln T_{ij} + \sum_j \left\{ -h_{ij} \sum_i T_{ij} + \pi \int_0^{\Sigma_i T_{ij}} s_j(x)\, dx \right\} \quad (6.61)$$

in which the functions $q_i(x)$ are defined as

$$q_i(x) = a_i \gamma x^{\gamma - 1}(x + \ln (a_i x^{\gamma})); \qquad \forall\, i \qquad (6.62a)$$

and subject to constraints

$$\sum_j T_{ij} = a_i \tilde{W}_i^{\gamma}; \qquad \forall\, i \qquad (6.62b)$$

$$\sum_i T_{ij} = T_j; \qquad \forall\, j \qquad (6.62c)$$

$$\sum_r S_{ijr} = S_{ij}; \qquad \forall\, i, j \qquad (6.62d)$$

and where, as usual,

$$v_a = \sum_i \sum_j \sum_r (T_{ijr} + S_{ijr})\, \delta_{ijr}^{a}; \qquad \forall\, a \qquad (6.62e)$$

Compare this formulation with that in section 6.4.1. In particular, determine whether Formula (6.49) still holds.

6.9 Show that under appropriate conditions the solution to the combined travel-mode-destination and route choice problem (6.23)–(6.24) always exists and is unique.

6.10 For the network represented in Figure 6.9, the travel costs and destination costs functions are respectively

$$t_1 = 1 + 0.02x; \qquad t_2 = 2 + 0.01x; \qquad t_3 = 2 + 0.05x$$

$$s_2 = 1 + 0.02x; \qquad s_3 = 0.5 + 0.03x$$

a. Given that 100 travelers travel from node 1, determine the equilibrium assignments on the network. Assume a value of 1 for all β parameters.

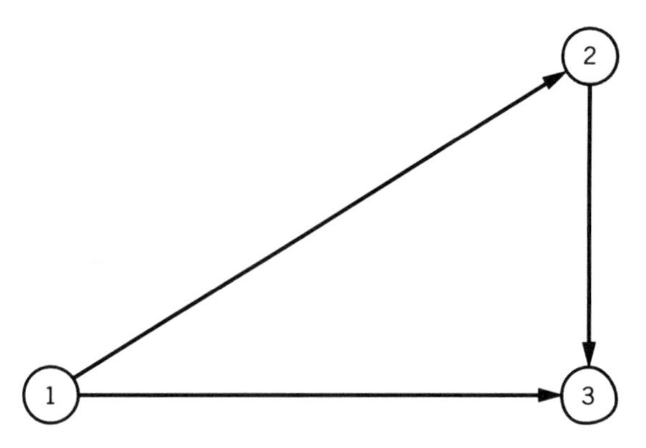

Figure 6.9 Network for Exercise 6.10.

b. Another mode is now available for travel to the two alternative destinations. The two modal networks are topologically similar and independent of one another. The fixed travel costs on the links of the second mode are

$$t_1 = 1.5; \qquad t_2 = 1.5; \qquad t_3 = 2$$

c. Determine the equilibrium assignments on the two networks.

6.11 Establish Formula (6.11a) for the R.T.'s received utility in the combined route-destination case.

6.12 Develop the optimality conditions for the subproblem in Algorithm 6.1 for solving the combined destination route choice, when route choice is deterministic.

6.13 Demonstrate the validity of Formula (6.25a) in the combined destination, mode and route choice case. (Hint: Use the relationship

$$\tilde{W}_{DMR} = \tilde{W}_{MR} + \tilde{W}_R + \frac{1}{\beta'_d} E_i - \sum_i \sum_j T^*_{ij} \tilde{W}^*_{j/i}$$

$$+ \frac{1}{\beta_d} \sum_i T_i \ln \sum_j e^{\beta_d g^*_{ij}} \tag{6.63}$$

where the first two terms are the expected received utilities with respect to the first two levels, whose derivatives have been estimated in the previous two sections. Also, remember that the modal networks are independent.) g_{ij} is defined in Formula (6.11b).

6.14 Establish Formula (6.26).

6.15 Assume that on the example network used in section 6.2.5, reserved lanes for car pools are opened, thus in effect creating a second, independent network corresponding to the original one. The capacities of the original links are only 66% of their previous values, assuming that one out of three lanes is now allocated to car pools. The travel time functions t_a for the new car pool links are the same as for the corresponding links, but with a capacity equal to 33% of the original capacity. As an incentive, car poolers get priority parking at the destination, which is estimated to save 6 minutes at node 2 and 8 minutes at node 6.

Assume that the "general" origin-destination demands all use the private car, so that

$$S_{121} = 100; \qquad S_{122} = 0$$

$$S_{141} = 150; \qquad S_{142} = 0$$

Determine all travel demands, T_{ij}, T_{ijm}, T_{ijmr}.

6.16 Demonstrate Formula (6.49).

6.17 Derive the expression for $U'(\theta)$ in Formula (6.15b).

6.18 Consider the spatial system represented in Figure 6.10. Link numbers are alongside them. The data are as follows, using the notation in the exposition. (All volumes are in hundreds)

$$T_1 = 10$$

$$t_1(v) = 15 + v_1$$

$$t_2(v) = 20 + 0.8v_2$$

$$s_2(T) = 10T_2$$

$$s_3(T) = 15T_3$$

$$\beta_d = 0.01; \qquad \pi = 1; \qquad \tau = 1$$

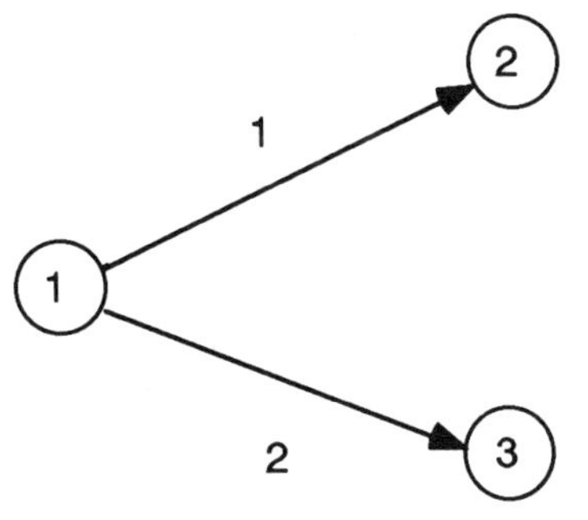

Figure 6.10 Network Exercise 6.18.

a. Write the R.T.'s U.M. problem which produces the equilibrium combined route and destination travel demands from node 1 to the two alternative destinations.

b. Solve the problem to estimate numerically the demands. (Hint: Use the fact that there are no routes and that link volumes are related, so that the program may be formulated as an unconstrained minimization problem in a single variable.)

c. Verify that the solution possess all of the expected equilibrium properties, i.e., the variable utilities (generalized travel costs) are equal for the two alternatives, and that the demand distribution to the respective destinations conforms to a logit model corresponding to these utilities.

6.19 Verify that the solution to the illustrative example in section 6.3.2.3 conforms numerically to Formulas (6.18–6.20).

6.20 Verify that the solution to illustrative example in section 6.4.3 conforms numerically to Formulas (6.43a), (6.18)–(6.20).

6.21 Demonstrate Formula (6.30b).

6.22 Assume in connection with the illustrative example in section 6.2.5 that tolls of $3 and $2 are respectively charged on links 1 and 5. Evaluate the impacts of these charges on the respective travel demands, total travel time and traveler surplus.

6.23 Assume in connection with the illustrative example in section 6.3.2.3 that tolls of $3 and $2 are respectively charged on links 1 and 5. Evaluate the impacts of these charges on the respective travel demands, total travel time and traveler surplus.

6.24 Assume in connection with the illustrative example in section 6.4.3 that tolls of $3 and $2 are respectively charged on links 1 and 5. Evaluate the impacts of these charges on the respective travel demands, total travel time and traveler surplus.

6.25. Specify the utility structure, as well as the probabilities of choice, in the case of combined mode and route choice for unscheduled modes.

6.26. Specify the utility structure, as well as the probabilities of choice, in the case of combined mode and route choice, when one or more of the modes is scheduled. (Hint: Use the strategy-based formulation of transit route choice described in section 5.5 of Chapter 5).

6.27. Formulate the R.T.'s problem corresponding to combined mode and route choice, for unscheduled as well as scheduled modes. (Hint: Use formulations (6.23)–(6.24)), or (6.36)–(6.37), replacing the unknowns T_{ij} by given constants).

6.28. Adapt Algorithm 6.2 to the combined mode and route assignment, for unscheduled, as well as for scheduled modes.

6.29. With reference to the illustrative example in section 6.3.2.2., we now assume that the origin-destination volumes T_{12} and T_{14} are no longer unknown, but are equal to

$$T_{12} = 250; \; T_{14} = 150.$$

Assume a deterministic route/strategy choice, and $\beta_{\mathrm{m}} = \beta'_{\mathrm{m}} = 0.1$. The other data remain the same, i.e., the problem is now only a problem of combined mode and route choice. Estimate the link and route car volumes, and the link and strategy transit volumes.

APPENDIX 6.1 THE SOLUTION TO THE R.T.'S UTILITY MAXIMIZATION PROBLEM (6.8)–(6.9) RETRIEVES THE AGGREGATE COMBINED ROUTE AND DESTINATION TRAVEL DEMANDS SPECIFIED IN FORMULAS (6.4A)–(6.5)

We shall make use of the Karush-Kuhn-Tucker conditions for optimality (KKT) for the above program, following the same approach as already used. Note that in all the derivations below, the variable times t_{ijr} and s_j by definition have their equilibrium values t^*_{ijr} and s^*_j. To lighten the notation, the asterisks will be deleted in this appendix.

First, as usual, we eliminate the budgetary slack variable T_0 from the utility function by replacing it by its expression from the budgetary constraint, and transforming the problem into a minimization problem

$$\text{Min } U'_{DR}(T_{ij}, T_{ijr}) = \sum_a \int_0^{\Sigma_i \Sigma_j \Sigma_r (T_{ijr} + S_{ijr})\delta^a_{ijr}} g_a(v)\, dv + \frac{1}{\beta_r} \sum_i T_{ijr} \ln T_{ijr}$$

$$+ \frac{1}{\beta_d} \sum_{ij} T_{ij} \ln T_{ij} - \sum_{ij} h_{ij} T_{ij} + \pi \sum_j \int_0^{\Sigma_i T_{ij}} s_j(x)\, dx \tag{6.64}$$

With respect to its first variables, the route demands T_{ijr}, it is apparent that the KKT conditions are exactly the same as for Problem (5.47) for the stochastic route choice, since the two problems are exactly similar with respect to these variables, both in terms of the first two terms in objective Function (6.8) and Constraint (6.9c). Consequently, using the results in Appendix 5.I of Chapter 5, the KKT conditions in the present case would also lead to Formula (5.76c), which is replicated here

$$e^{\beta_r \mu_{ij} - 1} = \frac{T_{ij}}{\sum_r e^{-\beta_r g_{ijr}}}; \qquad \forall\, i, j \tag{6.65}$$

where the μ_{ij}'s are the dual variables for Constraints (6.9c), and

$$T_{ijr} = \frac{T_{ij} e^{-\beta_r g_{ijr}}}{\sum_r e^{-\beta_r g_{ijr}}}; \qquad \forall\, i, j, r$$

Next, the KKT conditions with respect to the origin-destination demands T_{ij} are

$$\frac{\partial U'_{DR}}{\partial T_{ij}} - \sum_i \lambda_i \frac{\partial}{\partial T_{ij}} \left\{ \sum_j T_{ij} - T_{ij} \right\} - \sum_i \sum_j \mu_{ij} \frac{\partial}{\partial T_{ij}} \left\{ \sum_r T_{ijr} - T_{ij} \right\} \geq 0; \qquad \forall\, i, j$$

$$T_{ij} \left[\frac{\partial U'_{DR}}{\partial T_{ij}} - \sum_i \lambda_i \frac{\partial}{\partial T_{ij}} \left\{ \sum_j T_{ij} - T_i \right\} - \sum_i \sum_j \mu_{ij} \frac{\partial}{\partial T_{ij}} \left\{ \sum_r T_{ijr} - T_{ij} \right\} \right] = 0; \qquad \forall\, i, j$$

Using the fact that

$$\frac{\partial}{\partial T_{ij}} \sum_j \int_0^{\Sigma_i T_{ij}} s_j(x)\, dx = s_j \left(\sum_i T_{ij} \right) = s_j(T_j); \qquad \forall\, i, j$$

The derivatives of the objective function U'_{DR} with respect to the origin-destination demands T_{ij} are equal to

$$\frac{\partial U'_{DR}}{\partial T_{ij}} = \frac{1}{\beta'_d} (\ln T_{ij} + 1) - h_{ij} + \pi s_j(T_j); \qquad \forall\, i, j;$$

The above conditions may then be written

$$\frac{1}{\beta'_d} (\ln T_{ij} + 1) - h_{ij} + \pi s_j(T_j) - \lambda_i + \mu_{ij} \geq 0 \qquad \forall\, i, j,$$

$$T_{ij} \left[\frac{1}{\beta'_d} (\ln T_{ij} + 1) - h_{ij} + \pi s_j(T_j) - \lambda_i + \mu_{ij} \right] = 0 \qquad \forall\, i, j,$$

For $\ln T_{ij}$ to be defined, T_{ij} must be strictly positive, so that the second factor in the left-hand side of the last equality must be equal to zero, which implies

$$\ln T_{ij} = \beta'_d (h_{ij} - \pi s_j(T_i) - \mu_{ij} + \lambda_i) - 1; \qquad \forall\, i, j$$

From Formula (6.65):

$$\mu_{ij} = \frac{1}{\beta_r}\left(\ln T_{ij} - \ln \sum_r e^{-\beta_r g_{ijr}} + 1\right)$$

$$= \frac{1}{\beta_r}\left(\ln T_{ij} + 1\right) - \tilde{W}_{j/i}; \qquad \forall\, i, j \tag{6.66b}$$

Since

$$\tilde{W}_{j/i} = \frac{1}{\beta_r} \ln \sum_r e^{-\beta_r g_{ijr}}; \qquad \forall\, i, j$$

and given the definition of β'_d in Formula (6.7b), Formula (6.66) may also be written, after eliminating μ_{ij} from the previous equation

$$\ln T_{ij} = \beta_d g_{ij} + \beta_d \lambda_i - 1; \qquad \forall\, i, j$$

where, as defined in Formula (6.11b)

$$g_{ij} = h_{ij} - \pi s_j + \tilde{W}_{j/i}; \qquad \forall\, i, j$$

This implies

$$T_{ij} = e^{\beta_d g_{ij}} e^{\beta_d \lambda_i - 1}; \qquad \forall\, i, j \tag{6.67}$$

From Constraint (6.9b), we have

$$\sum_j e^{\beta_d g_{ij}} e^{\beta_d \lambda_i - 1} = e^{\beta_d \lambda_i - 1} \sum_j e^{\beta_d g_{ij}} = T_i; \qquad \forall\, i$$

and therefore,[13]

$$e^{\beta_d \lambda_i - 1} = \frac{T_i}{\sum_j e^{\beta_d g_{ij}}}; \qquad \forall\, i \tag{6.68}$$

Consequently,

$$T_{ij} = T_i \frac{e^{\beta_d g_{ij}}}{\sum_j e^{\beta_d g_{ij}}} = \frac{e^{\beta_d (h_{ij} - \pi s_j + \tilde{W}_{j/i})}}{\sum_j e^{\beta_d (h_{ij} - \pi s_j + \tilde{W}_{j/i})}}; \qquad \forall\, i, j$$

[13]Note the exact similarity with Formula (6.65), where β_d has replaced β_r and λ_i has replaced μ_{ij}. This relationship will be useful in the next section, when we include the next higher level of choice, that of mode. The fact that this relationship may then be considered the ''link'' between successive levels of choice is not surprising, since it involves the individual's received utilities, in the present case $\tilde{W}_{j/i}$.

Equivalently,

$$P_{j/i} = \frac{T_{ij}}{T_i} = \frac{e^{\beta_d g_{ij}}}{\sum_j e^{\beta_d g_{ij}}} = \frac{e^{\beta_d(h_{ij} - \pi s_j + \tilde{W}_{j/i})}}{\sum_j e^{\beta_d(h_{ij} - \pi s_j + \tilde{W}_{j/i})}}; \qquad \forall\, i, j$$

The last formula is precisely the specification for the demands which we expected, as specified in Formulas (6.4a), in which the travel times t_{ijr} are the "user equilibrium" travel times, and the destination costs s_j have their equilibrium congested values.

APPENDIX 6.2 EXISTENCE AND UNIQUENESS OF SOLUTION TO THE R.T.'S UTILITY MAXIMIZATION PROBLEM IN THE CONGESTED, JOINT ROUTE AND DESTINATION CHOICE CASE

In order to demonstrate the existence of solutions to the R.T.'s disutility minimization problem above, we must show that the objective function in Expression (6.8) does have a minimum in the "feasible" region represented by the set of Constraints (6.9). A basic result in calculus is that if the "feasible" region, R, is convex, closed, bounded, and nonempty, then any continuous function will have at least one minimum in R (Rockafellar, 1970).

First, it is clear that since the feasible region R in the multidimensional space $(T_{ijr}, S_{ijr}, T_{ij})$ is defined by linear equality Constraints (6.9), R is convex, closed and bounded. Also, it is clear that for any given set of given values T_i, R is nonempty, as it contains, for instance, the point

$$T_{ij} = \begin{cases} T_i; & j = i \\ 0; & j \neq i \end{cases} \qquad \forall\, i, j$$

$$T_{ijr} = \begin{cases} T_{ij}; & \forall\, i, j, r = 1 \\ 0; & \forall\, i, j, r \neq i \end{cases}$$

$$S_{ijr} = \begin{cases} S_{ij}; & \forall\, i, j, r = 1 \\ 0; & \forall\, i, j, r \neq i \end{cases}$$

Therefore, R is not empty, in terms of the variables T_{ijr}, S_{ijr}, T_{ij}. Since the correspondence between the T_{ijr}'s and S_{ijr}'s and the link demands v_a is linear, Formula (6.7b), the feasible region in terms of the variables and v_a is also convex, compact and nonempty. Consequently, Program (6.8)–(6.9) possesses at least one solution.

Although the fact that the above problem does have a solution is reassuring, we still do not know whether that solution is unique, or multiple. To examine this next issue, we make use of another result in calculus, which states that if the feasible region R is compact, convex, and nonempty, and if the objective function is everywhere strictly convex, then the solution to a nonlinear mathematical minimization program will be unique (Rockafellar, 1970). Since we have already shown that the feasible region R defined by Equations (6.9) is compact, convex, and nonempty, we must show that utility Function (6.8) is strictly, concave, or that its negative U' is strictly convex.

We can, without loss of generality, write $U'_{DR}(\boldsymbol{v}, \boldsymbol{T})$ as

$$U'_{DR}(\boldsymbol{v}, \boldsymbol{T}) = U'_1(\boldsymbol{T}) + U'_2(\boldsymbol{T}) \tag{6.69a}$$

with

$$U'_1 = \sum_a \int_0^{\Sigma_i \Sigma_j \Sigma_r (T_{ijr} + S_{ijr}) \delta^a_{ijr}} g_a(x)\, dx + \frac{1}{\beta_d} \sum_i \sum_j T_{ij} \ln T_{ij}; \tag{6.69b}$$

$$U'_2 = -\sum_{ij} h_{ij} T_{ij} + \sum_j \pi \int_0^{\Sigma_i T_{ij}} s_j(x)\, dx \tag{6.69c}$$

It was demonstrated in section 5.2.2 in the previous chapter, in connection with the route choice problem, that since the functions $g_a(v_a)$ which define the link travel times are assumed to be strictly increasing, $U'_1(\boldsymbol{T})$ is strictly convex. We must then examine whether $U'_2(\boldsymbol{T})$ is convex. We begin with the case of negative externalities, when the functions $s_j(T_j)$ defined in Equation (6.2) are increasing in the T_j's. In this case, each of the terms

$$\int_0^{\Sigma_i T_{ij}} s_j(x)\, dx$$

in Function U_2 above is a function $f_j(\Sigma_i T_{ij})$ of variables T_{ij}'s. Because the functions $s_j(v)$ are positive, and strictly increasing, functions $f_j(v)$ are increasing, and strictly convex. Furthermore, since the argument $\Sigma_i T_{ij}$ of these latter functions is a linear, and therefore also convex function of the T_{ij}'s, the function $f_j(\Sigma_i T_{ij})$ is convex in the T_{ij}'s. (See for instance, Berge, 1966.)

In addition, each of the terms $-h_{ij} \Sigma_i T_{ij}$ is obviously convex, being a linear function; hence the function

$$\sum_j \left\{ -h_{ij} \sum_i T_{ij} + \pi \int_0^{\Sigma_i T_{ij}} s_j(x)\, dx \right\}$$

is itself convex. Thus, Function U'_2 is the sum of two convex functions and is therefore convex. Therefore, U'_{DR} is strictly convex, and thus, the solution to Problem (6.8)–(6.9) is unique. That is, the combined destination route-demand functions are uniquely defined.

APPENDIX 6.3 THE SOLUTION TO THE R.T'S UTILITY MAXIMIZATION PROBLEM (6.23)–(6.24) RETRIEVES THE AGGREGATE COMBINED ROUTE, DESTINATION, AND MODE TRAVEL DEMANDS SPECIFIED IN FORMULAS (6.18)–(6.20)

We again state the necessary KKT conditions at the problem's solution. First, with respect to the modal route demands T_{ijmr}, these conditions mean that there exists quantities μ_{ij} which are such that

$$\frac{\partial U'_{DMR}}{\partial T_{ijmr}} - \sum_{ijm} \mu_{ijm} \frac{\partial}{\partial T_{ijmr}} \left(\sum_{r} T_{ijmr} - T_{ijmr} \right) \geq 0; \qquad \forall\, i, j, m, r$$

$$T_{ijmr} \left[\frac{\partial U'_{DMR}}{\partial T_{ijmr}} - \sum_{ijm} \mu_{ijm} \frac{\partial}{\partial T_{ijmr}} \left(\sum_{r} T_{ijmr} - T_{ijm} \right) \right] = 0; \qquad \forall\, i, j, m, r$$

It is apparent that these conditions constitute the generalization to the multimodal case of conditions (5.11) for the case of a single mode. Therefore, we have

$$T_{ijmr} = \frac{T_{ijm} e^{-\beta_r g_{ijmr}}}{\sum_{r} e^{-\beta_r g_{ijmr}}}; \qquad \forall\, i, j, m, r$$

For nontravel modes (e.g. telecommuting), the cost is fixed, and there is only one available route. This implies that $P_{r/ijm} = T_{ijmr}/T_{ijm}$ conforms exactly to Formula (6.20), as required.

Furthermore, the value of μ_{ijm} is equal to

$$\mu_{ijm} = \frac{1}{\beta_r} \left(\ln T_{ijm} - \ln \sum_{r} e^{-\beta_r g_{ijmr}} + 1 \right) = \frac{1}{\beta_r} (\ln T_{ijm} + 1) - \tilde{W}_{j/i}; \qquad \forall\, i, j, m \quad (6.70)$$

Similarly, the KKT conditions with respect to the T_{ijm} variables imply that there are values λ_{ij} such that

$$\frac{\partial U'_{DMR}}{\partial T_{ijm}} - \sum_{ij} \lambda_{ij} \frac{\partial}{T_{ijm}} \left(\sum_{m} T_{ijm} - T_{ij} \right) - \sum_{ijm} \mu_{ijm} \frac{\partial}{\partial T_{ijm}} \left(\sum_{r} T_{ijmr} - T_{ijm} \right) \geq 0; \qquad \forall\, i, j, m$$

$$T_{ijm} \left[\frac{\partial U'_{DMR}}{\partial T_{ijm}} - \sum_{ij} \lambda_{ij} \frac{\partial}{T_{ijm}} \left(\sum_{m} T_{ijm} - T_{ij} \right) - \sum_{ijm} \mu_{ijm} \frac{\partial}{\partial T_{ijm}} \left(\sum_{r} T_{ijmr} - T_{ijm} \right) \right] \geq 0; \qquad \forall\, i, j, m$$

These conditions are similar to those with respect to T_{ij} in the Appendix 6.1, so that we have

$$T_{ijm} = T_{ij} \frac{e^{\beta_d(h_{ijm} + \tilde{W}_{m/ij})}}{\sum_j e^{\beta_d(h_{ijm} + \tilde{W}_{m/ij})}}; \qquad \forall\ i, j, m \tag{6.71}$$

and

$$\lambda_{ij} = \frac{1}{\beta_m} (\ln T_{ij} + 1) - \tilde{W}_{j/i}; \qquad \forall\ i, j$$

This implies that $P_{m/ij} = T_{ijm}/T_{ij}$ conforms exactly to Formula (6.10), as required.

Next, the KKT conditions with respect to the origin-destination (total) demands T_{ij} state that there exist values α_i such that

$$\frac{\partial U'_{DMR}}{\partial T_{ij}} - \sum_i \alpha_i \frac{\partial}{T_{ij}} \left(\sum_j T_{ij} - T_i \right) - \sum_{ij} \mu_{ij} \frac{\partial}{\partial T_{ij}} \left(\sum_m T_{ijm} - T_{ij} \right) \geq 0; \qquad \forall\ i, j$$

$$T_{ij} \left[\frac{\partial U'_{DMR}}{\partial T_{ij}} - \sum_i \alpha_i \frac{\partial}{T_{ij}} \left(\sum_j T_{ij} - T_i \right) - \sum_{ij} \mu_{ij} \frac{\partial}{\partial T_{ij}} \left(\sum_m T_{ijm} - T_{ij} \right) \right] \geq 0; \qquad \forall\ i, j$$

Carrying through the derivations, in the same manner as above, and using the value of λ_{ij} estimated above, as well as the Constraint

$$\sum_j T_{ij} = T_i; \qquad \forall\ i$$

leads to the result that

$$\alpha_i = \frac{1}{\beta_d} (\ln T_i + 1) - \tilde{W}_{t/i}; \qquad \forall\ i \tag{6.72}$$

where

$$\tilde{W}_{t/i} = \frac{1}{\beta_d} \ln \sum_j e^{\beta_d g_{ij}}; \qquad \forall\ i$$

and

$$g_{ij} = h_{ij} + \tilde{W}_{j/i}; \qquad \forall\, i, j$$

as defined in Formula (6.11b) and

$$T_{ij} = T_i \frac{e^{\beta_d(\tilde{W}_{j/i} + h_{ij})}}{\sum_j e^{\beta_d(\tilde{W}_{j/i} + h_{ij})}}; \qquad \forall\, i, j$$

The similarity between the results at the successive hierarchical levels should be apparent, and will again be seen in the next Appendix 6.4.

In any case, $P_{j/i} = T_{ij}/T_i$ conforms exactly to Formula (6.18a), as required. Thus, the solution to the R.T. utility maximization problem (6.23)–(6.26) with given T_i's does indeed retrieve the appropriate aggregate combined route, mode, and destination demands, in which the generalized modal travel costs have their equilibrium values. In the next appendix, this will be generalized to include variable T_i's as well.

APPENDIX 6.4 THE SOLUTION TO THE R.T.'S UTILITY MAXIMIZATION PROBLEM (6.46), (6.47) RETRIEVES THE AGGREGATE COMBINED ROUTE, DESTINATION, AND MODE TRAVEL DEMANDS WITH ENDOGENOUS T_i'S

This will again be done by stating the KKT conditions. In fact, it may be noticed that the utility function, as well as the constraints for the combined travel, mode, destination, and route problem are the same as for the preceding combined mode, destination, and route problem, in terms of their common variables T_{ijr}, T_{ij}, and T_{ij}. The change from the latter to the former problems is in the addition of the terms

$$-\frac{1}{\beta_t}\sum_i (T_i \ln T_i + T_{i0} \ln T_{i0}) + \sum_i h_i T_i$$

in the objective function, and of the Constraint

$$T_i + T_{i0} = N_i; \qquad \forall\, i$$

Therefore, the KKT conditions in terms of variables T_{ijr}, T_{ijm}, and T_{ij} remain the same, leading to the same Formula

$$T_{ijmr} = T_i \frac{e^{\beta_d(h_{ij} + \tilde{W}_{j/i})}}{\sum_j e^{\beta_d(h_{ij} + \tilde{W}_{j/i})}} \cdot \frac{e^{\beta_m(+\tilde{\omega}_{m/ij} + h_{ijm})}}{\sum_m e^{\beta_m(+\tilde{\omega}_{m/ij} + h_{ijm})}} \cdot \frac{e^{-\beta_r g_{ijmr}}}{\sum_r e^{-\beta_r g_{ijmr}}}$$

The model's properties in terms of the first level of traveler choice will then come from the statement of the KKT conditions with respect to variables T_i and T_{i0}. The KKT conditions in terms of the variables T_i may be written

$$\frac{\partial U'_{TDMR}}{\partial T_i} - \sum_i \alpha_i \frac{\partial}{T_{ij}} \left(\sum_j T_{ij} - T_i \right)$$

$$- \sum_i \overline{\omega}_i \frac{\partial}{\partial T_i} (T_i + T_{i0} - N_i) \geq 0; \qquad \forall\, i$$

$$T_i \left[\frac{\partial U'_{TDMR}}{\partial T_i} - \sum_i \alpha_i \frac{\partial}{T_{ij}} \left(\sum_j T_{ij} - T_i \right) \right.$$

$$\left. - \sum_i \overline{\omega}_i \frac{\partial}{\partial T_i} (T_i + T_{i0} - N_i) \right] = 0; \qquad \forall\, i$$

We have

$$\frac{\partial U'_{TDMR}}{\partial T_i} = -\frac{1}{\beta'_t} (\ln T_i + 1) + h_i; \qquad \forall\, i$$

remembering the specification of β'_t in Formula (6.45b) and the expression for α_i in Formula (6.72) in the previous Appendix 6.3, these conditions may then be written, after some standard derivations, as

$$T_i = e^{\beta_t(\tilde{W}_{t/i} + h_i)} e^{\beta_t \overline{\omega}_i - 1}; \qquad \forall\, i$$

Similarly, the KKT derivations with respect to variables T_{i0} would show that

$$T_{i0} = e^{\beta_t \overline{\omega}_i - 1}; \qquad \forall\, i$$

Finally, making use of the Constraint

$$T_i + T_{i0} = N_i; \qquad \forall\, i$$

we have

$$e^{\beta_t \overline{\omega}_i - 1}(1 + e^{\beta_t(\tilde{W}_{t/i} + h_i)}) = N_i \qquad \forall\, i$$

and consequently

$$e^{\beta_t \overline{\omega}_i - 1} = \frac{N_i}{1 + e^{\beta_t(\tilde{W}_{t/i} + h_i)}} = T_{i0}; \qquad \forall\, i$$

so that

$$T_i = N_i \frac{e^{\beta_t(\tilde{W}_{t/i} + h_i)}}{1 + e^{\beta_t(\tilde{W}_{t/i} + h_i)}}; \qquad \forall\, i$$

There expressions are exactly Formulas (6.43a). Thus, the solution to Problem (6.45)–(6.46) produces the appropriate travel demands T_{ijmr}, T_{ijm}, T_{ij}, T_i, and T_{i0}, with the equilibrium values of their factors, $\tilde{W}_{t/i}$, $\tilde{W}_{j/i}$, $\tilde{W}_{m/ij}$ and g_{ijmr}.

APPENDIX 6.5 DEMONSTRATING ROY'S IDENTITY FOR THE COMBINED DESTINATION-ROUTE CASE

The R.T.'s received utility is equal to the optimal value of the R.T.'s utility at the solution of Problem (6.8)–(6.9). (See Exercise 6.11)

$$\begin{aligned} \tilde{W}_{DR} = & -\sum_a \int_0^{v_a^*} g_a(v)\, dv + \sum_i \sum_j \sum_r T_{ijr}^* g_{ijr}^* \\ & - \pi \left[\sum_j \int_0^{\Sigma_i T_{ij}^*} s_j(x)\, dx - \sum_i \sum_j T_{ij}^* s_j^* \right] \\ & + \frac{1}{\beta_d} \sum_i T_i \ln \sum_j e^{\beta_d g_{ij}^*} + B + \frac{1}{\beta_d} E_i \end{aligned} \qquad (6.11a)$$

where, for compactness,

$$g_{ij} = h_{ij} - \pi s_j \left(\sum_i T_{ij} \right) + \tilde{W}_{j/i}; \qquad \forall\, i, j$$

First, it was seen in Appendix 5.2 that the derivatives with respect to c_{kln} of the first and second terms in the above expression add up to

$$\tau \sum_i \sum_j \sum_r T_{ijr}^* \frac{\partial t_{ijr}^*}{\partial c_{kln}}$$

Next, we have,

$$\frac{\partial}{\partial c_{kln}} \left[\int_0^{T_j^*} s_j(x)\, dx \right] = \sum_p \left\{ \frac{\partial}{\partial T_p^*} \left[\int_0^{T_j^*} s_j(x)\, dx \right] \frac{\partial T_p^*}{\partial c_{kln}} \right\}$$

$$= \frac{\partial}{\partial T_{ij}^*} \left[\int_0^{\Sigma_i T_{ij}^*} s_j(x)\, dx \right] \frac{\partial T_j^*}{\partial c_{kln}} = \sum_i s_j\, (T_j^*) \frac{\partial T_{ij}^*}{\partial c_{kln}}; \qquad \forall\, i, j, r$$

so that the derivative of the third term in the expression above is equal to

$$-\pi \sum_j \sum_i s_j^* \frac{\partial T_{ij}^*}{\partial c_{kln}}$$

Next, the derivative of the fourth term is

$$\pi \left[\sum_i \sum_j s_j^* \frac{\partial T_{ij}^*}{\partial c_{kln}} + \sum_i \sum_j T_{ij}^* \frac{\partial s_j^*}{\partial c_{kln}} \right]$$

Next, the derivative of the fifth term is equal to

$$\frac{1}{\beta_d} \sum_i T_i \frac{1}{\sum_j e^{\beta_d g_{ij}^*}} \sum_j \frac{\partial e^{\beta_d g_{ij}^*}}{\partial c_{kln}} = \frac{\beta}{\beta} \sum_i T_i \sum_j P_{j/i}^* \frac{\partial g_{ij}^*}{\partial c_{kln}}$$

$$= \sum_i \sum_j T_{ij}^* \frac{\partial}{\partial c_{kln}} (h_{ij} - \pi s_j^* + \tilde{\omega}_{j/i})$$

$$= -\pi \sum_i \sum_j T_{ij}^* \frac{\partial s_j^*}{\partial c_{kln}} + \sum_i \sum_j T_{ij}^* \frac{\partial \tilde{\omega}_{j/i}}{\partial c_{kln}}$$

Since

$$\tilde{W}_{j/i} = \frac{1}{\beta_r} \ln \sum_r e^{-\beta_r g_{ijr}}; \qquad \forall\, i, j$$

we may utilize the result shown in Appendix 5.2 that the last term is equal to

$$-T_{kln} - \tau \sum_i \sum_j \sum_r T_{ijr}^* \frac{\partial t_{ijr}}{\partial c_{kln}}$$

Collating all these terms, we have finally

$$\frac{\partial \tilde{W}_{DR}}{\partial c_{kln}} = -T_{kln}; \qquad \forall\, k, l, n$$

from which again

$$-\frac{\partial \tilde{W}_{DR}}{\partial c_{kln}} \bigg/ \frac{\partial \tilde{W}_{DR}}{\partial B} = T_{kln}^*; \qquad \forall\, k, l, n$$

The same approach would establish Formulas (6.26) and (6.49).

CHAPTER 7

MODEL PARAMETER ESTIMATION

7.1 INTRODUCTION: THE PROCESS OF MODEL ESTIMATION

So far in this text, we have exclusively concentrated on the theoretical development of travel demand models. In particular, all parameters appearing in these models were left unspecified, as symbols without a numerical value. Actual model application to forecast travel demand requires that these parameters be given specific values. In this chapter, we address the issue of parameter estimation as a prerequisite to practical application. Let us begin by clarifying the difference between variables, coefficients, and parameters.

The models of travel demand which have been developed in the previous chapters incorporate various inputs whose values reflect prevailing conditions under which the model is applied. These inputs may be classified into two main categories, respectively model coefficients and model parameters. For instance, the quantity τ in utility specification Formula (3.1) is a parameter. It has two roles. First, its value reflects the sensitivity of travelers to travel time, as a component of the disutility or generalized travel cost. Second, it acts as a conversion coefficient between units of travel time and units of money. On the other hand, the value of t_a^0 in Formula (5.4) is a coefficient. Its value, as explained then, measures the travel time on a given link at zero volume.

The basic difference between coefficients and parameters is that in general, the value of coefficients will be estimated from direct observations or measurements, or through application of specific formulas or special techniques, but in any event *exogenously* to the model itself. For instance, the value of N_i, the number of potential travelers in a given demand zone at a future date may be the result of a demographic projection; the estimation of the value of t_a^0 may come from surveys external to the travel demand model. Accordingly, the estimation of model coefficients will not be addressed, since it is either relatively straightforward, as in the latter case, or it involves relatively specialized knowledge, as in the former.

On the other hand, parameter values cannot be measured directly, but must be estimated *endogenously* to the model, specifically from the predicted values for the model's variables. In general, the parameters which must be estimated in our travel demand models intervene in the specification of the utilities facing the traveler at the various levels of choice. Additionally, they also specify the level of variability in random utility, in the form of the β values in the resulting logit models.

As seen in the previous chapters, a given model must be supplemented with a solution algorithm, which may then be considered, at least for practical purposes, an integral part of the model. In the same fashion, a given model must also be supplemented with a parameter/model estimation procedure. The purpose of this chapter is to present techniques for parameter estimation.

It may be useful to articulate at the outset the general principle underlying model estimation, even though it may appear somewhat obvious. Specifically, the unknown parameters should be given values such that the travel demands predicted with the model thus "calibrated" conform, to the greatest extent possible, to travel demands as observed in reality. In other words, a model should be so specified as to be maximally consistent with empirical observations.

Having stated the qualitative principle underlying parameter estimation, we need to translate it into operational, mathematical terms, and specify how we measure the "conformity" between predicted and observed values for the model's variables. Depending on whether the utility components are fixed or variable, that is, on whether demand externalities are present, two main approaches to calibration, disaggregate and aggregate, respectively, are applicable, which in turn correspond to the maximum likelihood and maximum entropy principles, respectively.

In the former case, because no externalities exist, individual travelers' choices are independent of one another, and consequently, parameters may be calibrated on the basis of observations of individual travel behavior and personal characteristics. This obviously leads to the most detailed level of model specification, as it is consistent with the fact that the travel demand models are based on *individual* traveler choices, and should therefore be identified at that level. When externalities exist, travel demands may be characterized only at their equilibrium, aggregate levels. Consequently, parameter values in this case are, of necessity, based on less detailed information, since aggregate values (e.g., total travel volumes) may be obtained from individual observations of trip making, but not vice versa. This is the compromise one has to make for the ability to deal with the more general congested case.

7.2 MAXIMUM LIKELIHOOD METHOD WITH INDIVIDUAL UTILITY SPECIFICATION

In this section, we consider the simplest case, that of fixed values for all factors of travel choices, including in particular destination costs s_j and travel costs c_{ijm}. In this case, the joint probability distributions P_{ijmr}, which constitute the models of individual travel demands, are numerically specified up to the values of the parameters. Thus, the problem of travel demand model calibration is in fact the problem of identifying the parameters in probability distribution functions.

In this form, the problem is a standard, classical problem in statistics and econometrics. In this particular instance, the above criterion for conformity of model predictions and "real world" observations may be based on probability theory. It states that parameter values should maximize the probability, or likelihood, *as predicted by the model*, of occurrence of an observed sample of individual travel demands. This principle is called the "maximum likelihood" principle (M.L.) (Wonnacott and Wonnacott, 1977).

The M.L. method may be applied at the disaggregate level, on the basis of individual observations of traveler characteristics and choices. Alternatively, it may be performed at the aggregate level, on the basis of observations of zonal characteristics and travel volumes. In the former case, the utility of a given choice may then be a function of individual traveler characteristics, such as income and age. In the latter case, utility functions may only be specified in terms of average factors, such as average income in demand zone i.

In the next section, we examine the disaggregate case first.

7.2.1 Maximum Likelihood Equations

In the disaggregate formulation, the probability of a given travel choice, for example, destination j, by a given individual traveler, is specified as a function of the individual's characteristics as well as those of the alternatives. Let us illustrate the former situation first, in the case of a multinomial model of combined route and destination choice, when the former is deterministic. In that case, as seen in Formula (4.15) in Chapter 4, the two levels of choice collapse into one. The procedure for estimating models for multiple choice levels is discussed in section 7.32.

For each individual traveler, the conditional probability that he or she chooses destination j given that the decision to travel and the mode have already been made is, according to Formula (4.4):

$$P_{j/n} = \frac{e^{\beta(\alpha_j' - c_{nj} - \tau' t_{nj})}}{\sum_j e^{\beta(\alpha_j' - c_{nj} - \tau' t_{nj})}}; \qquad \forall\, n, j \tag{7.1a}$$

t_{nj} is the given, minimum travel cost from traveler n's location to destination j. c_{nj} is the given travel cost. α_j' is a parameter characteristic of each destination, which corresponds to h_{ij} in our general formulation.

Since only the products $\beta\alpha_j'$ and $\beta\tau'$ appear in the model, the parameters to calibrate are in effect β and

$$\alpha_j = \beta\alpha_j'$$

$$\tau = \beta\tau'$$

and we may write the above model as

$$P_{j/n} = \frac{e^{(\alpha_j - \beta c_{nj} - \tau t_{nj})}}{\sum_j e^{(\alpha_j - \beta c_{nj} - \tau t_{nj})}} = \frac{e^{\tilde{U}_{nj}}}{\sum_j e^{\tilde{U}_{nj}}}; \qquad \forall\, n, j \tag{7.1b}$$

where

$$\tilde{U}_{nj} = \alpha_j - \beta c_{nj} - \tau t_{nj}; \qquad \forall\, n, j$$

We assume that we have a set of observations for the choices of destination for a sample of N individuals, numbered $1, 2, \ldots, n, \ldots, N$. Such data might, for instance, come from "focus groups," or panels. We represent individual n's choice of destination, j_n, through the following (binary) indicator:

$$\delta_{nj} = \begin{cases} 1 & \text{if traveler } n \text{ has chosen destination } j \\ 0 & \text{if not} \end{cases} \quad ; \qquad \forall\, n, j \qquad (7.2)$$

Since the route chosen to a given destination is *always* the minimum cost route, and thus does not depend on the individual, it is not necessary to keep a record of the chosen route. In this manner, the choices of the n individuals are completely represented by the matrix of δ_{nj}'s. Given this information, the probability that the first traveler in the sample chooses the destination he or she has chosen, j_1, is equal to

$$P_{1/1}^{\delta_{11}} P_{2/1}^{\delta_{12}}, \ldots, P_{J/1}^{\delta_{1J}} = \prod_{j=1}^{J} P_{j/1}^{\delta_{1j}}$$

Similarly, the probability that the nth traveler chooses destination j_n is

$$P_{1/n}^{\delta_{n1}} P_{2/n}^{\delta_{n2}}, \ldots, P_{J/n}^{\delta_{nJ}} = \prod_{j=1}^{J} P_{j/n}^{\delta_{nj}}$$

If we assume that the n travelers have been selected at random and that their choices are independent of one another, the overall probability, that traveler 1 chooses destination j_1, *and* that traveler 2 chooses destination j_2, and so on, until the last traveler, is equal to the product of the respective individual probabilities above (see section 3.2 of Appendix A):

$$L = \prod_{n=1}^{N} \prod_{j=1}^{J} P_{j/n}^{\delta_{nj}} \qquad (7.3a)$$

This expression represents the "likelihood" of the observed sample of destination choices. Each of the probabilities $P_{n/j}$ is a function of the unknown parameter values β, α_j and τ, through Formula (7.1b). Therefore, L is also a function of the unknown parameters. Application of the M.L. principle stated above requires that L be maximized with respect to the unknown parameter values. Given the fact that L is a product of many terms, it is clearly easier to maximize its logarithm Λ, rather than L itself. This is because the ln function is strictly monotone; maximizing the logarithm of a function $f(x)$ is equivalent to maximizing the function.

$$\Lambda = \ln \prod_{n=1}^{N} \prod_{j=1}^{J} P_{j/n}(\alpha_j, \beta, \tau)^{\delta_{nj}} \qquad (7.3b)$$

Thus, the unknown parameter values are a solution of the optimization problem:

$$\underset{(\alpha_j,\beta,\tau)}{\text{Max}}\ \Lambda = \underset{(\alpha_j,\beta,\tau)}{\text{Max}} \left[\ln \prod_{n=1}^{N} \prod_{j=1}^{J} P_{j/n}(\alpha_j, \beta, \tau)^{\delta_{nj}} \right]$$

Using Formula (A.13) in Appendix A, the above problem is

$$\underset{(\alpha_j,\beta,\tau)}{\text{Max}} \left[\sum_{n=1}^{N} \sum_{j=1}^{J} \delta_{nj} \ln P_{j/n}(\alpha_j, \beta, \tau) \right] \tag{7.4}$$

in which $P_{j/n}$ has the expression in Formula (7.1b) above, and the terms δ_{nj} are given, representing observed travelers' choices, as described earlier.

We must then find the unconstrained optimum of a function of several variables, β, τ, and α_j. Thus, at the solution of Problem (7.4), all partial derivatives of the function to maximize with respect to the unknowns must be equal to zero. (See section 2.1 in Appendix A.) The unknown parameter values are therefore the solutions to Equations

$$\frac{\partial \Lambda}{\partial \beta} = 0 \tag{7.5a}$$

$$\frac{\partial \Lambda}{\partial \tau} = 0 \tag{7.5b}$$

$$\frac{\partial \Lambda}{\partial \alpha_j} = 0; \qquad \forall j \tag{7.5c}$$

Since the δ_{jn} are constant with respect to the parameters, Equation (7.5a) may be formulated as

$$\frac{\partial \Lambda}{\partial \beta} = \frac{\partial}{\partial \beta} \sum_{n=1}^{N} \sum_{j=1}^{J} \delta_{nj} \ln P_{j/n} = \sum_{n} \sum_{j} \delta_{nj} \frac{\partial}{\partial \beta} \ln P_{j/n} = \sum_{n} \sum_{j} \frac{\delta_{nj}}{P_{j/n}} \frac{\partial P_{j/n}}{\partial \beta}$$

Furthermore, we have

$$\frac{\partial P_{j/n}}{\partial \beta} = \frac{-c_{nj} e^{\tilde{U}_{nj}} \sum_{k=1}^{J} e^{\tilde{U}_{nk}} + e^{\tilde{U}_{nj}} \sum_{k} c_{nk} e^{\tilde{U}_{nk}}}{\left[\sum_{k} e^{\tilde{U}_{nk}} \right]^2}$$

$$= -c_{nj} P_{j/n} + P_{j/n} \sum_{k} c_{nk} P_{k/n} = P_{j/n} \left(-c_{nj} + \sum_{k} c_{nk} P_{k/n} \right)$$

Consequently, Condition (7.5a) may be written

$$\frac{\partial \Lambda}{\partial \beta} = \sum_n \sum_j \frac{\delta_{nj}}{P_{j/n}} P_{j/n} \left(-c_{nj} + \sum_k c_{nk} P_{k/n} \right)$$

$$= \sum_n \sum_j \delta_{nj} \left(-c_{nj} + \sum_k c_{nk} P_{k/n} \right) = 0$$

so that finally:

$$\sum_n \sum_j c_{nj} P_{j/n} = \sum_n \sum_j c_{nj} \delta_{nj} \tag{7.6}$$

The left-hand side of Equation (7.6) represents the total travel costs for the travelers in the sample, as *predicted* by the model. Similarly, the right-hand side of Equation (7.6) represents the total *observed* travel costs for the travelers in the sample. Therefore, the first M.L. condition states that these two quantities are equal.

Following the same kind of derivations, Condition (7.5b), with respect to the second parameter τ, may similarly be written (see Exercise 7.1):

$$\sum_n \sum_j t_{nj} P_{j/n} = \sum_n \sum_j t_{nj} \delta_{nj} \tag{7.7}$$

This condition states that the total *observed* travel time incurred by the travelers in the sample is equal to the expected value of the same quantity, as *predicted* by the model.

Finally, we have

$$\frac{\partial \Lambda}{\partial \alpha_j} = \sum_j \frac{\delta_{nj}}{P_{j/n}} \frac{\partial P_{j/n}}{\alpha_j}; \qquad \forall j$$

with

$$\frac{\partial P_{j/n}}{\partial \alpha_j} = P_{j/n}(1 - P_{j/n}); \qquad \forall j$$

so that, following similar derivations (see Exercise 7.1), Conditions (7.5c) may be expressed as

$$\sum_n P_{j/n} = \sum_n \delta_{nj} = \hat{T}_j; \qquad \forall j \tag{7.8a}$$

These last conditions may be rewritten as

$$\frac{1}{T} \sum_n 1 P_{j/n} = \frac{\hat{T}_j}{T}; \qquad \forall j \tag{7.8b}$$

or

$$P_j = \hat{P}_j; \qquad \forall\, j \tag{7.8c}$$

The predicted share of destination j's demands is equal to the observed share.

It is important to note that any one of the J Equations (7.8a) follows from the $(J - 1)$ others. Indeed, the sum of the respective right-hand sides is always equal to the sample size

$$\sum_j \hat{T}_j = N$$

so that only $(J - 1)$ equations are independent. Consequently, only $(J - 1)$ of the α_j's may be identified, and (any) one of them has to be fixed arbitrarily. This is consistent with the fact that utilities of given alternatives are not absolute, but relative to one another; that is, adding or subtracting a constant from all utilities, replacing U_{nj} by $U_{nj} + K$ for all individuals n, would not change the probabilities of their choices. Fixing one of the α_j's then amounts to choosing a reference point for the utilities.

In any case, in this form it may be seen that, in general, each of the calibration conditions above states that the observed value of traveler's expenditures of a given factor of travel choice (i.e., c, t, or 1) is equal to its expected value as predicted by the model. Thus, if additional choice factors had intervened *linearly* in the utility function, for example, through a more elaborate form for the h_{ij}'s, and the choice probability consequently had been

$$P_{j/n} = \frac{\exp\left(\alpha_j + \sum_k \alpha_j^k X_{nj}^k - \beta c_{nj} - \tau t_{nj}\right)}{\sum_j \exp\left(\alpha_j + \sum_k \alpha_j^k X_{nj}^k - \beta c_{nj} - \tau t_{nj}\right)}; \qquad \forall\, n, j \tag{7.9a}$$

the equations which would have been added to the above system to identify the values of parameters α_j^k would have been similar to those above. This may be summarized in the following result.

FACT 7.1

The M.L. equations for fixed travel choice factors X^k are of the general form

$$\sum_j \sum_n X_{nj}^k P_{j/n} = \sum_j \sum_n X_{nj}^k \delta_{nj}; \qquad \forall\, k \tag{7.9b}$$

This provides great flexibility, as X_{nj} may itself be specified as a predetermined function of the characteristics of the traveler and the destination. For instance, in the case of shopping travel, X_{nj} might be defined as

$$X_{nj} = \ln R_j^{\gamma}/I_n^{\eta}; \qquad \forall\ n, j$$

where R_j and I_n are respectively the destination's average activity price and average traveler's income.

The $(J + 2)$ unknowns β, τ and the α_j's intervene in the system of $(J + 2)$ equations (7.5), through the expression of the $P_{j/n}$'s in Formula (7.1b). The solution will be unique, *up to* one value of the α_j's, as discussed above. That is, it may easily be shown that once the "base" utility has been set, the objective function for unconstrained Problem (7.4) is everywhere strictly convex. (See Exercise 7.1.) This system of nonlinear equations may be solved with Algorithm 5.4 in Chapter 5.

7.2.2 Illustrative Example

We shall now illustrate the M.L. method to calibrate a combined destination-route model of the form below, in connection with the same hypothetical network we have been using all along.

$$P_{j/n} = \frac{e^{\beta(\alpha_j - c_{nj} - \tau t_{nj})}}{\sum_j e^{\beta(\alpha_j - c_{nj} - \tau t_{nj})}}; \qquad \forall\ n, j \qquad (7.10)$$

A sample of ten travelers have been surveyed, and their place of residence and choice of destination in connection with shopping, in the form of variables δ_{nj} as defined by Formula (7.2) above, have been recorded. The observed values are represented in Tables 7.1 and 7.2. All other values are equal to zero. Thus, traveler 1 has gone to destination 3, traveler 2 has gone to destination 1, and so on. Given their residence zone, and the origin-destination travel times on the given network, as represented in Table 1.8, the individual travel times of each of these travelers to each of the alternative destinations are represented in Table 7.3.

Finally, tolls paid by travelers are represented in Table 7.4. All other travel costs are assumed to be equal to zero.

With these data, the first M.L. equation for the parameter of travel time τ

TABLE 7.1 Residence Zone of Travelers

$n =$	1	2	3	4	5	6	7	8	9	10
$i =$	3	1	3	1	4	4	4	1	2	1

TABLE 7.2 Observed Destination Choices

(n, j)	δ_{nj}
1.3	1
2.1	1
3.3	1
4.1	1
5.4	1
6.4	1
7.4	1
8.1	1
9.2	1
10.1	1

is then, setting $\alpha_0 = 0$

$$7 \frac{e^{(0-0.75\beta-7\tau)}}{e^{(0-0.75\beta-7\tau)} + e^{(\alpha_2-0\beta-8\tau)} + e^{(\alpha_3-0\beta-2\tau)} + e^{(\alpha_4-0\beta-9\tau)}}$$

$$+ 8 \frac{e^{(\alpha_2-0\beta-8\tau)}}{e^{(0-0.75\beta-7\tau)} + e^{(\alpha_2-0\beta-8\tau)} + e^{(\alpha_3-0\beta-2\tau)} + e^{(\alpha_4-0\beta-9\tau)}}$$

$$+ 2 \frac{e^{(\alpha_3-0\beta-2\tau)}}{e^{(0-0.75\beta-7\tau)} + e^{(\alpha_2-0\beta-8\tau)} + e^{(\alpha_3-0\beta-2\tau)} + e^{(\alpha_4-0\beta-9\tau)}}$$

$$+ 9 \frac{e^{(\alpha_4-0\beta-9\tau)}}{e^{(0-0.75\beta-7\tau)} + e^{(\alpha_2-0\beta-8\tau)} + e^{(\alpha_3-0\beta-2\tau)} + e^{(\alpha_4-0\beta-9\tau)}} + \text{etc.}$$

$$= 2 + 2 + 1 + 2 + 11 + 2 + 3 + 2 + 1 + 2 = 28$$

where only the terms for the first traveler have been represented. The equation for parameter β is similar. The equation for parameter α_2, the "attractiveness" of destination 2 is

$$\frac{e^{(\alpha_2-0\beta-8\tau)}}{e^{(0-0.75\beta-7\tau)} + e^{(\alpha_2-0\beta-8\tau)} + e^{(\alpha_3-0\beta-2\tau)} + e^{(\alpha_4-0\beta-9\tau)}}$$

$$+ \frac{e^{(\alpha_2-1\beta-6\tau)}}{e^{(0-0\beta-2\tau)} + e^{(\alpha_2-1\beta-6\tau)} + e^{(\alpha_3-0\beta-8\tau)} + e^{(\alpha_4-0\beta-11\tau)}}$$

$$+ \text{etc} = 1$$

where only the terms for the first two travelers have been represented. The equation for the other α's are similar. Note that there are only five equations, since α_1 was set equal to zero.

The resulting system may then be solved with Algorithm 5.4 in Chapter 5

TABLE 7.3 Individual Travel Times

(n, j)	$t(n, j)$	(n, j)	$t(n, j)$
1.1	7	6.1	9
1.2	8	6.2	8
1.3	2	6.3	14
1.4	9	6.4	2
2.1	2	7.1	10
2.2	6	7.2	8
2.3	8	7.3	9
2.4	11	7.4	3
3.1	9	8.1	2
3.2	6	8.2	6
3.3	1	8.3	6
3.4	8	8.4	11
4.1	2	9.1	8
4.2	8	9.2	1
4.3	8	9.3	12
4.4	10	9.4	7
5.1	9	10.1	2
5.2	9	10.2	6
5.3	3	10.3	7
5.4	11	10.4	14

for the solution of square nonlinear systems. The solution is

$$\hat{\beta} = 7.467 \qquad \hat{\alpha}_3 = -2.012$$

$$\hat{\tau} = 0.770 \qquad \hat{\alpha}_4 = +2.843$$

$$\hat{\alpha}_2 = -1.740$$

where the caret symbol (^) signifies estimated (observed) values. The calibrated destination-route choice model may then be specified as shown on page 263.

TABLE 7.4 Individual Travel Costs

(n, j)	$c(n, j)$
1.1	0.75
2.2	1
3.1	0.75
4.2	1
5.2	2
6.2	2
7.2	2
8.2	1
9.4	1.5
10.2	1

$$P_{1/n} = \frac{e^{(-7.467c_{nj} - 0.77t_{nj})}}{e^{(-7.467c_{nj} - 0.77t_{nj})} + e^{(-1.740 - 7.467c_{nj} - 0.77t_{nj})} + e^{(-2.012 - 7.467c_{nj} - 0.77t_{nj})} + e^{(2.843 - 7.467c_{nj} - 0.77t_{nj})}}; \qquad \forall\, n$$

$$P_{2/n} = \frac{e^{(1.740 - 7.467c_{nj} - 0.77t_{nj})}}{e^{(-7.467c_{nj} - 0.77t_{nj})} + e^{(-1.740 - 7.467c_{nj} - 0.77t_{nj})} + e^{(-2.012 - 7.467c_{nj} - 0.77t_{nj})} + e^{(2.843 - 7.467c_{nj} - 0.77t_{nj})}}; \qquad \forall\, n$$

$$P_{3/n} = \frac{e^{(-2.012 - 2.304c_{nj} - 0.77t_{nj})}}{e^{(-7.467c_{nj} - 0.77t_{nj})} + e^{(-1.740 - 7.467c_{nj} - 0.77t_{nj})} + e^{(-2.012 - 7.467c_{nj} - 0.77t_{nj})} + e^{(2.843 - 7.467c_{nj} - 0.77t_{nj})}}; \qquad \forall\, n$$

$$P_{4/n} = \frac{e^{(2.843 - 7.467c_{nj} - 0.77t_{nj})}}{e^{(-7.467c_{nj} - 0.77t_{nj})} + e^{(-1.740 - 7.467c_{nj} - 0.77t_{nj})} + e^{(-2.012 - 7.467c_{nj} - 0.77t_{nj})} + e^{(2.843 - 7.467c_{nj} - 0.77t_{nj})}}; \qquad \forall\, n$$

TABLE 7.5 Predicted Probabilities of Destination Choice $P_{j/n}$

$j =$	1	2	3	4
$n = 1$		0.008	0.626	0.366
2	0.982		0.001	0.016
3		0.017	0.620	0.363
4	0.964		0.001	0.035
5	0.055		0.744	0.202
6				1.000
7				1.000
8	0.978		0.006	0.016
9	0.025	0.975		
10	0.996		0.003	0.001

With these values, the predicted probabilities of destination choice for the ten individuals are in Table 7.5.

As expected, the estimated values $\hat{\beta}$ and $\hat{\tau}$ are positive. The ratio of their respective values provides some indication about the monetary value of time to the sample of travelers. (See Exercise 7.9.) The large value of $\hat{\beta}$ relative to that of $\hat{\tau}$ corresponds to the fact that the observed origin-destination pattern was "one-to-one," that is, travelers in a given origin zone were systematically observed to choose the same zone as a destination. As a result, the estimated value of $\hat{\beta}$ is likely to be imprecise. This indicates a deterministic destination choice, at least on the basis of this very limited sample, corresponding to a large value of $\hat{\beta}$.

In any case, inserting the estimated parameter values above into the individual demand Function (7.10), one can then estimate the numerical value of the expected individual demands for any traveler, given his or her travel times and costs to the alternative destinations. By doing this for the ten travelers in the sample, and comparing the predicted average demands at each of the destinations with the observed values, one may evaluate the performance of the calibrated model with respect to the observed demands. (See Exercise 7.10.) More formal methods for model performance evaluation will be discussed in section 7.8.

If observations of the individual values for travel costs t_{nj}, and so on, are not available, but rather, their observed values, the above procedure may be applied on an aggregate basis. The resulting calibrated parameter values will then be different, and reflect an "aggregation error." (See Exercise 7.5.)

7.3 MAXIMUM LIKELIHOOD METHOD WITH AVERAGE UTILITY SPECIFICATION

7.3.1 Single Level of Choice

In the previous section, each individual traveler was assumed to have a different systematic utility function of his or her personal characteristics. If the systematic utilities of the choices facing individual travelers in a given demand zone are, or are assumed to be, the same for each individual in the zone, then in the example above,

$$\begin{aligned} c_{nj} &= c_{ij}; \qquad \forall\, n \in i,\, i,\, j \\ t_{nj} &= t_{ij}; \qquad \forall\, n \in i,\, i,\, j \end{aligned} \tag{7.11}$$

This is in fact the assumption which underlies all of the travel demand models developed in this text. Consequently, the probability of choosing destination j may be specified as

$$P_{j/i} = \frac{e^{(\alpha_j - \beta c_{ij} - \tau t_{ij})}}{\sum_j e^{(\alpha_j - \beta c_{ij} - \tau t_{ij})}}; \qquad \forall\, i, j \tag{7.12a}$$

that is, Formula (7.10), in which we assume that the destination costs are included in g_{ij}. Therefore, the predicted aggregate demands are equal to

$$T_{ij} = T_i \frac{e^{(\alpha_j - \beta c_{ij} - \tau t_{ij})}}{\sum_j e^{(\alpha_j - \beta c_{ij} - \tau t_{ij})}}; \qquad \forall\, i, j \tag{7.12b}$$

where the T_i's, c_{ij} and t_{ij}'s are given.

7.3.1.1 Maximum Likelihood Equations Given a given sample of observed origin-destination demands $\hat{T}_{ij}$, the corresponding M.L. equations may be determined using the same maximum likelihood principle as above, by maximizing Λ, the logarithm of the sample's probability of occurrence. If travelers are assumed to behave independently, from the multinomial probability distribution function (see section 3.5.3 in Appendix A), the probability that a *given* number of travelers T_i from zone i are distributed into the observed $\hat{T}_{ij}$ is equal to

$$L_i = \frac{T_i!}{\hat{T}_{i1}!\ \hat{T}_{i2}!, \ldots, \hat{T}_{in}!} P_{1/i}^{\hat{T}_{i1}} P_{2/i}^{\hat{T}_{i2}}, \ldots, P_{n/i}^{\hat{T}_{in}} = \frac{T_i! \prod_j P_{j/i}^{\hat{T}_{ij}}}{\prod_j \hat{T}_{ij}!}; \qquad \forall\, i \tag{7.13a}$$

in which the probabilities $P_{j/i}$ are as given by Formula (7.12a). Since the T_i's are given and thus the subsamples of travelers from each demand (origin) zone are independent, the probability of observing the entire sample of travelers from all zones is the product of the above probabilities:

$$L = \prod_i \mathrm{L}_i = \prod_i \frac{T_i! \prod_j P_{j/i}^{\hat{T}_{ij}}}{\prod_j \hat{T}_{ij}!} \tag{7.13b}$$

The unknown parameter values are then the solution to problem

$$\operatorname*{Max}_{\boldsymbol{\beta}} \Lambda = \ln \prod_i \frac{T_i! \prod_j P_{j/i}^{\hat{T}_{ij}}}{\prod_j \hat{T}_{ij}!} = \sum_i \left[\ln \hat{T}_i! + \sum_j (\ln P_{j/i}^{\hat{T}_{ij}} - \ln \hat{T}_{ij}!) \right]$$

in which $\boldsymbol{\beta}$ refers generically to the set of parameters for the probabilities $P_{j/i}$, β, α, τ. Since the terms $\ln (\hat{T}_i!)$ and $\ln (\hat{T}_{ij}!)$ are constants (i.e., each is a function of the fixed, observed values, and not of the parameter values), the maximization is then equivalent to

$$\text{Max } \Lambda_{(\boldsymbol{\beta})} \left[\sum_i \sum_j \ln P_{j/i}^{\hat{T}_{ij}} \right] = \text{Max } \Lambda_{(\boldsymbol{\beta})} \left[\sum_{ij} \hat{T}_{ij} \ln P_{j/i}(\boldsymbol{\beta}) \right] \tag{7.14}$$

The "first-order" conditions may be formulated as

$$\frac{\partial \Lambda}{\partial \boldsymbol{\beta}} = \frac{\partial}{\partial \boldsymbol{\beta}} \sum_{ij} \hat{T}_{ij} \ln P_{j/i} = \sum_{ij} \frac{\hat{T}_{ij}}{P_{j/i}} \frac{\partial P_{j/i}}{\partial \boldsymbol{\beta}} \tag{7.15}$$

since the $\hat{T}_{ij}$'s are constant with respect to the parameters. Carrying out the derivations, one obtains the following result, in terms of predicted demands rather than probabilities (see Exercise 7.2).

FACT 7.2

The ML equations with common utility specification are

$$\sum_{ij} T_{ij} t_{ij} = \sum_{ij} \hat{T}_{ij} t_{ij}; \tag{7.16a}$$

$$\sum_{ij} T_{ij} c_{ij} = \sum_{ij} \hat{T}_{ij} c_{ij}; \tag{7.16b}$$

$$\sum_i T_{ij} = \sum_i \hat{T}_{ij}; \qquad \forall j \tag{7.16c}$$

where, given the T_i's

$$T_{ij} = T_i P_{j/i}(\boldsymbol{\beta}); \qquad \forall\, i, j$$

It should be noticed that the form of these equations is entirely similar to those obtained with disaggregate data. Specifically, in both cases the M.L. equations state that the respective observed (r.h.s.) and predicted (l.h.s.) total monetary and time costs (or, more generally, component utilities) for all travelers are equal. Accordingly, if h_{ij} is a function of traveler and zonal characteristics, as specified in Formula (4.20c)

$$h_{ij} = \sum_k \eta_{ki} R_{ki} + \sum_k \gamma_{kj} A_{kj}; \qquad \forall\ i, j \tag{7.17}$$

in which the η_{ki}'s and γ_{kj}'s must be calibrated, the M.L. equations for these parameters would similarly be

$$\sum_{ij} \hat{T}_{ij} \eta_{ki} = \sum_{ij} T_{ij} \eta_{ki}; \qquad \forall\ k \tag{7.18a}$$

$$\sum_{ij} \hat{T}_{ij} \gamma_{kj} = \sum_{ij} T_{ij} \gamma_{kj}; \qquad \forall\ k \tag{7.18b}$$

Thus, in general, there will be one (nonlinear) equation to solve for each of the unknown parameter values. However, note that if, for instance,

$$h_{ij} = h_i; \qquad \forall\ i, j \tag{7.19}$$

that is, a given factor of choice is common to all demand zones, Equations (7.16) become identities, and the values of the corresponding parameters cannot be determined. In all these equations, the unknowns, in this case β and τ, appear through the expression of the predicted values T_{ij}'s in Formula (7.12b).

7.3.1.2 Illustrative Example To illustrate, let us now go back to the illustrative example above, and assume that all individual travelers have a common utility specification, so that the individual demand model is as given by Formula (7.12a). We assume that the only available information about origin-destination travel times and costs comes from the sample of ten travelers as specified in Tables 7.1 to 7.4, and not from general information about the network, such as represented in Table 1.8.[1] The values of the origin-destination travel times and travel costs to the respective destinations are then taken to be the average over all travelers in a given demand zone, and are consequently as given in Tables 7.6 and 7.7.

[1]The procedure would be the same with any given t_{ij} and c_{ij} values.

TABLE 7.6 Average Travel Times t_{ij}

$j =$	1	2	3	4
$i = 1$	2	6.5	7.25	11.5
2	8	1	12	7
3	8	7	1.5	7
4	9.3	8.3	8.6	5.33

TABLE 7.7 Average Travel Costs c_{ij}

$j =$	1	2	3	4
$i = 1$	0	1	0	25
2	0	0	0	0
3	0.75	0	0	0
4	0	2	0	0

The observed travel demands are in Tables 7.8 and 7.9.

The corresponding M.L. Equations (7.16) are, then, again setting $\alpha_1 = 0$

$$4\left[2\,\frac{e^{(0-0\beta-2\tau)}}{e^{(0-0\beta-2\tau)}+e^{(\alpha_2-1\beta-6.5\tau)}+e^{(\alpha_3-0\beta-7.25\tau)}+e^{(\alpha_4-0\beta-11.5\tau)}}\right.$$

$$+\,6.5\,\frac{e^{(\alpha_2-1\beta-6.5\tau)}}{e^{(0-0\beta-2\tau)}+e^{(\alpha_2-1\beta-6.5\tau)}+e^{(\alpha_3-0\beta-7.25\tau)}+e^{(\alpha_4-0\beta-11.5\tau)}}$$

$$+\,7.25\,\frac{e^{(\alpha_3-0\beta-7.25\tau)}}{e^{(0-0\beta-2\tau)}+e^{(\alpha_2-1\beta-6.5\tau)}+e^{(\alpha_3-0\beta-7.25\tau)}+e^{(\alpha_4-0\beta-11.5\tau)}}$$

$$\left.+\,11.5\,\frac{e^{(\alpha_4-0\beta-11.5\tau)}}{e^{(0-0\beta-2\tau)}+e^{(\alpha_2-1\beta-6.5\tau)}+e^{(\alpha_3-0\beta-7.25\tau)}+e^{(\alpha_4-0\beta-11.5\tau)}}\right]$$

$$+\,1[\ldots$$

$$= (4(2) + 1(1) + 2(1.5) + 3(5.33)) = 28$$

TABLE 7.8 Observed value of T_i

$i =$	1	2	3	4
T_i	4	1	2	3

TABLE 7.9 Observed T_{ij}

$j =$	1	2	3	4
$i = 1$	4	0	1	0
2	0	1	0	0
3	0	0	2	0
4	0	0	0	3

for τ, and

$$4\left[0\frac{e^{(0-0\beta-2\tau)}}{e^{(0-0\beta-2\tau)}+e^{(\alpha_2-1\beta-6.5\tau)}+e^{(\alpha_3-0\beta-7.25\tau)}+e^{(\alpha_4-0\beta-11.5\tau)}}\right.$$

$$+1\frac{e^{(\alpha_2-1\beta-6.5\tau)}}{e^{(0-0\beta-2\tau)}+e^{(\alpha_2-1\beta-6.5\tau)}+e^{(\alpha_3-0\beta-7.25\tau)}+e^{(\alpha_4-0\beta-11.5\tau)}}$$

$$+0\frac{e^{(\alpha_3-0\beta-7.25\tau)}}{e^{(0-0\beta-2\tau)}+e^{(\alpha_2-1\beta-6.5\tau)}+e^{(\alpha_3-0\beta-7.25\tau)}+e^{(\alpha_4-0\beta-11.5\tau)}}$$

$$\left.+0\frac{e^{(\alpha_4-0\beta-11.5\tau)}}{e^{(0-0\beta-2\tau)}+e^{(\alpha_2-1\beta-6.5\tau)}+e^{(\alpha_3-0\beta-7.25\tau)}+e^{(\alpha_4-0\beta-11.5\tau)}}\right]$$

$$+1[\ldots$$

$$=(4(0)+1(0)+2(0)+3(0))=0$$

for β.[2]

The equations for α_2, α_3 and α_4, in that order are

$$4\left[\frac{e^{(\alpha_2-1\beta-6.5\tau)}}{e^{(\alpha_2-1\beta-6.5\tau)}+e^{(\alpha_2-1\beta-6.5\tau)}+e^{(\alpha_3-0\beta-7.25\tau)}+e^{(\alpha_4-2\beta-8.3\tau)}}\right.$$

$$\left.+\frac{e^{(0-0\beta-2\tau)}}{e^{(\alpha_2-1\beta-6.5\tau)}+e^{(\alpha_2-1\beta-6.5\tau)}+e^{(\alpha_3-0\beta-7.25\tau)}+e^{(\alpha_4-2\beta-8.3\tau)}}\right]$$

$$+1[\ldots=1$$

$$4\left[\frac{e^{(\alpha_3-0\beta-7.25\tau)}}{e^{(\alpha_2-1\beta-6.5\tau)}+e^{(\alpha_2-1\beta-6.5\tau)}+e^{(\alpha_3-0\beta-7.25\tau)}+e^{(\alpha_4-2\beta-8.3\tau)}}\right]$$

$$+1[\ldots=2$$

$$4\left[\frac{e^{(\alpha_4-0\beta-5.33\tau)}}{e^{(\alpha_2-1\beta-6.5\tau)}+e^{(\alpha_2-1\beta-6.5\tau)}+e^{(\alpha_3-0\beta-7.25\tau)}+e^{(\alpha_4-2\beta-8.3\tau)}}\right]$$

$$+1[\ldots=3$$

The solution to these equations is

$$\hat{\beta}=2.029 \qquad \hat{\alpha}_3=-1.406$$

$$\hat{\tau}=2.008 \qquad \hat{\alpha}_4=1.036$$

$$\hat{\alpha}_2=-6.146$$

[2]Note that mathematically, this equation would appear impossible to solve, since the sum of positive values cannot equal zero. However, calibration means finding parameter values which bring the r.h.s. as close as possible to zero.

As expected, these values are different from the disaggregate values. The calibrated destination-route choice model may then be specified as shown on page 271.

With these values, the predicted aggregate demands are as represented in Table 7.10. These may be compared with the observed values in Table 7.9.

As expected, the parameter estimates are different from those obtained in the "full information" procedure above. The difference is sometimes referred to as the "aggregation bias." In this case, due to the small sample size, the difference is substantial. It should be intuitively clear that since averages summarize individual values, calibration based on the basis of an individual specification incorporates more information that calibration based on an average specification, and is therefore, ceteris paribus, "better." The effect of individual traveler aggregation on the predicted aggregate destination demands may be evaluated from comparing them with the "disaggregate" predicted demands. (See Exercises 7.5 and 7.6.)

7.3.2 Multiple Levels of Choice

7.3.2.1 Sequential Calibration The ML calibration method was described in the case of a single level of choice. When calibrating models of combined choices/travel demand, the joint probability of a combined choice is given by a nested logit model, as opposed to a multinomial logit model, as above. For instance, when route utilities are assumed to be random, and consequently route choice is assumed to be probabilistic, the model of joint destination/route choice is formulated as

$$P_{jr/i} = \frac{e^{\beta_d(\alpha_j + \tilde{W}_{j/i})}}{\sum_j e^{\beta_d(\alpha_j + \tilde{W}_{j/i})}} \frac{e^{-\beta_r(c_{ijr} + \tau t_{ijr})}}{\sum_r e^{-\beta_r(c_{ijr} + \tau t_{ijr})}}; \qquad \forall\ i, j, r \tag{7.20}$$

in which

$$\tilde{W}_{j/i} = \frac{1}{\beta_r} \ln \sum_r e^{-\beta_r(c_{ijr} + \tau t_{ijr})} \qquad \forall\ i, j \tag{7.21}$$

TABLE 7.10 Predicted Aggregate Origin-Destination Demands

	T_{ij}			
$j =$	1	2	3	4
$i = 1$	4			
2		1		
3			2	
4				3

$$P_{1/i} = \frac{e^{(-2.029c_{ij} - 2.008t_{ij})}}{e^{(-2.029c_{ij} - 2.008t_{ij})} + e^{(-6.146 - 2.029c_{ij} - 2.008t_{ij})} + e^{(-1.406 - 2.029c_{ij} - 2.008t_{ij})} + e^{(1.036 - 2.029c_{ij} - 2.008t_{ij})}}; \qquad \forall\ i$$

$$P_{2/i} = \frac{e^{(-6.146 - 2.029c_{ij} - 2.008t_{ij})}}{e^{(-2.029c_{ij} - 2.008t_{ij})} + e^{(-6.146 - 2.029c_{ij} - 2.008t_{ij})} + e^{(-1.406 - 2.029c_{ij} - 2.008t_{ij})} + e^{(1.036 - 2.029c_{ij} - 2.008t_{ij})}}; \qquad \forall\ i$$

$$P_{3/i} = \frac{e^{(-1.406 - 2.029c_{ij} - 2.008t_{ij})}}{e^{(-2.029c_{ij} - 2.008t_{ij})} + e^{(-6.146 - 2.029c_{ij} - 2.008t_{ij})} + e^{(-1.406 - 2.029c_{ij} - 2.008t_{ij})} + e^{(1.036 - 2.029c_{ij} - 2.008t_{ij})}}; \qquad \forall\ i$$

$$P_{4/i} = \frac{e^{(1.036 - 2.029c_{ij} - 2.008t_{ij})}}{e^{(-2.029c_{ij} - 2.008t_{ij})} + e^{(-6.146 - 2.029c_{ij} - 2.008t_{ij})} + e^{(-1.406 - 2.029c_{ij} - 2.008t_{ij})} + e^{(1.036 - 2.029c_{ij} - 2.008t_{ij})}}; \qquad \forall\ i$$

The same general M.L. principle as used above may again be applied, but this time *sequentially*, one level at a time. The estimation at one level is conditioned on those at the subsequent, "downstream" levels. In this multilevel case, the value of "second-level" parameter β_r would first be estimated, on the basis of the single-level, conditional probability

$$P_{r/ij} = \frac{e^{-\beta_r(c_{ijr} + \tau t_{ijr})}}{\sum_r e^{-\beta_r(c_{ijr} + \tau t_{ijr})}}; \qquad \forall\ i, j, r \tag{7.22}$$

using the same techniques as above, on the basis of route choice observations of individual travelers. The resulting equations for the two parameters β_r and $\tau' = \beta_r \tau$ would then be

$$\sum_i \sum_j \sum_r c_{ijr} T_{ijr} = \sum_i \sum_j \sum_r c_{ijr} \hat{T}_{ijr} = \hat{c} \tag{7.23a}$$

$$\sum_i \sum_j \sum_r t_{ijr} T_{ijr} = \sum_i \sum_j \sum_r t_{ijr} \hat{T}_{ijr} = \hat{t} \tag{7.23b}$$

Next, knowing the value of β_r, the values $\hat{\tilde{W}}_{j/i}$ may then be estimated from Formula (7.21). Thus, all utility components at the next level, intervening in the probability of destination choice

$$P_{j/i} = \frac{e^{\beta_d(\alpha_j + \tilde{W}_{j/i})}}{\sum_j e^{\beta_d(\alpha_j + \tilde{W}_{j/i})}}; \qquad \forall\ i, j \tag{7.24}$$

(in this simple case only the $\tilde{W}_{j/i}$'s), would have a known, numerical value. Finally, the "first-level" parameters would be estimated, again using the same technique. The resulting equations for β_d and the α's would then be

$$\sum_i \sum_j \tilde{W}_{j/i} T_{ij} = \sum_i \sum_j \tilde{W}_{j/i} \hat{T}_{ij} \tag{7.25a}$$

$$\sum_i T_j = \sum_i \hat{T}_j; \qquad \forall\ j \tag{7.25b}$$

The same sequential technique would be applicable to any number of choice levels.

7.3.2.2 Simultaneous Calibration An alternative approach, which may be more efficient in some cases, is to calibrate all parameters at the various choice levels simultaneously. This approach is sometimes referred to as "full information calibration." For instance, in the formulation above, the M.L. equations would be derived with respect to the log-likelihood function in terms of the T_{ijr} and T_{ij}'s *together*, and not separately, as in the sequential method. Consequently, the M.L. condition for those parameters (such as β_r in the present case), which intervene in the probabilities at several levels, will be different, as well as the resulting values. (See Exercises 7.14 and 7.15.)

7.4 MAXIMUM LIKELIHOOD METHOD FOR VARIABLE UTILITY

The application of the M.L. method described in the previous section is, as we have seen, predicated on the ability to obtain the partial derivatives of the log-likelihood function $\Lambda(\beta)$ with respect to its parameters. However, when there is congestion at the destinations, the values of the destination costs s_j are a function of the destination volumes, and therefore of the probabilities $P_{j/i}$, which are in turn functions of the destination costs.

$$s_j = s_j\left(\sum_i T_i P_{j/i}(\beta, \tau, \alpha)\right); \qquad \forall j \tag{7.26}$$

Thus, variable destination costs are a function of themselves. The same problem arises when there is congestion on the network. For instance, under the deterministic route utility assumption, it is not even possible to express the values of the minimum travel times t_{ijr} as a differentiable function of T_{ijr}. Nevertheless, the former are a function of the latter, and consequently the t_{ijr}'s are functions of themselves. Also, for models of stochastic route choice, the variable route costs are the (implicit) solution of the system of Equations (5.37), and they are again functions of themselves. To date, M.L. equations have not formally been developed for models of congested route choice, let alone combined travel demands.[3]

However, the M.L. calibration problem in the presence of demand externalities may be conceptually stated as

$$\underset{\boldsymbol{\beta}}{\text{Max}} \ln(\Lambda) = \sum_{ijmr} \hat{T}_{ijmr} \ln P_{ijmr}(U(\boldsymbol{\beta})) \tag{7.27}$$

such that the probabilities have numerical values corresponding to the user-equilibrium solution of the R.T.'s U.M. problem in which the parameters β intervene

$$\underset{T_{ijmr}}{\text{Max}}\ U(T_{ijmr}, \boldsymbol{\beta}(T_{ijmr})) \tag{7.28}$$

subject to the standard constraints. The first, ''top'' problem (TP) of likelihood maximization is constrained by the optimal solution of the second, ''bottom''

[3]Attempts in the deterministic case have been unfruitful, possibly because of the discontinuous nature of the route demands. Such derivation in the ''smooth'' but analytically much more complex stochastic case might be possible, using computerized symbolic computation.

problem (BP) of the R.T.'s utility maximization. Such an optimization is called a "bilevel" program.

Standard multivariate constrained optimization techniques, such as those we have been using so far, are not applicable to multilevel optimization problems, as they are predicated on the ability to express all constraints explicitly, for example, in equality or inequality form, and not as in the present case, in the form of another optimization problem. Since, as discussed earlier, the functional form of the relationship of P_{ijmr} to $\boldsymbol{\beta}$ is not known explicitly, one cannot estimate the gradient of the UP's objective function and apply techniques which require "first-order" information. To date, solving bilevel programs has proved difficult, and to sometimes provide unreliable solutions.

One possible remedy is to solve the TP with an algorithm which does not require the computation of the gradient of its objective function. An algorithm which only requires the numerical value of the objective function is the Hooke and Jeeves algorithm (Hooke and Jeeves, 1962). The algorithm performs exploratory searches and pattern moves, during which the values of the lower problem variables are estimated. For each value $\boldsymbol{\beta}$ at which the function U is being evaluated, the value of $P(\boldsymbol{\beta})$ is then obtained by solving the lower problem Max $B(P, \boldsymbol{\beta})$. This approach has been applied to calibration problems when the number of parameters is small. (See, for instance, Lee, 1987.) These attempts have apparently met with mixed success so far.

An alternative method is to solve formally the system of Equations (7.10) and (7.11) at each iteration, instead of merely adjusting consecutively the parameter values. The system to solve is

$$\sum_{ij} t_{ij}(\boldsymbol{\beta})\hat{T}_{ij} - \sum_{ij} t_{ij}(\boldsymbol{\beta})T_{ij}(\boldsymbol{\beta}) = 0 \tag{7.29}$$

and so on for each unknown parameter. A suitable method in this connection is Powell's method (Powell, 1970), which is a Newton-type method for solving square systems of equations (i.e., systems with the same number of equations as variables), but does not require the derivatives of the functions t_{ij} and T_{ij} with respect to the unknowns $\boldsymbol{\beta}$. Instead, it estimates their numerical value, based on the change in the l.h.s. of Equations (7.20) and (7.21). Thus, at each iteration of Powell's method, the current value of the parameters are used to solve the model, and the numerical value of the l.h.s. of the above equations is estimated, leading to the next iteration of Powell's algorithm, and so on, until convergence to a solution.

7.5 MAXIMUM ENTROPY METHOD (R.T.'S PARTIAL UTILITY MAXIMIZATION)

As has been repeatedly demonstrated throughout this text, when individual travelers are assumed to have a common systematic utility, aggregate demands are produced by the representative traveler (R.T.) who behaves similarly (i.e., makes utility maximizing choices) to the individual travelers. This alternative approach is particularly valuable in connection with model calibration, in the presence of demand externalities.

7.5.1 Fixed Utilities

7.5.1.1 The R.T's Partial Utility Maximization In this section, an alternative, and, when there are no demand externalities, *equivalent* model estimation method is described. Since bilevel problems are typically difficult to solve, this alternative formulation to the calibration problem may be attractive on practical grounds as well. We again refer to the model of combined probabilistic destination and deterministic route choice above, and begin with the case of fixed travel costs and destination costs. The case of variable utilities will be considered in the next section.

Consider the following problem in which it is assumed that parameters β, τ and α_j are known, that is, are given constants,

$$\operatorname*{Min}_{(T_{ij})} = \tau \sum_i \sum_j T_{ij} t_{ij} + \beta \sum_i \sum_j T_{ij} c_{ij} + \sum_i \sum_j T_{ij} \ln T_{ij} + \sum_j \alpha_j \sum T_{ij} \tag{7.30a}$$

subject to the usual consistency Constraints:

$$\sum_j T_{ij} = T_i; \qquad \forall\, i \tag{7.30b}$$

If $\hat{t}$, $\hat{c}$ and $\hat{k}_j$ are constants, this problem is equivalent to

$$\begin{aligned} \operatorname*{Min}_{(T_{ij})} = \tau \left[\sum_i \sum_j T_{ij} t_{ij} - \hat{t} \right] + \beta \left[\sum_i \sum_j T_{ij} c_{ij} - \hat{c} \right] \\ + \sum_i \sum_j T_{ij} \ln T_{ij} + \sum_j \alpha_j \left[\sum T_{ij} - \hat{k}_j \right] \end{aligned} \tag{7.30c}$$

subject to Constraints (7.30b). This is because multiplying or dividing the objective function by a constant C_1, and subtracting another constant C_2 does not affect the solution of a mathematical optimization problem such as the one above. That is, the solution to

$$\operatorname*{Min}_{x} \{ f(x)/C_1 - C_2 \}$$

is the same as the solution to

$$\operatorname*{Min}_{x} \{ f(x) \}$$

when the C's are independent of x.

Now consider the following "partial" R.T.'s utility maximization problem

$$\text{Max } U'_{(T_{ij})}\left[-\sum_i \sum_j T_{ij} \ln T_{ij}\right] = \text{Min} \sum_{ij} T_{ij} \ln T_{ij} \tag{7.31}$$

subject to

$$\sum_i \sum_j T_{ij} c_{ij} \leq \hat{c} \qquad (\beta) \tag{7.32a}$$

$$\sum_i \sum_j T_{ij} t_{ij} \leq \hat{t} \qquad (\tau) \tag{7.32b}$$

$$\sum_i T_{ij} = \hat{k}_j; \qquad \forall j \quad (\alpha_j) \tag{7.32c}$$

as well as to the usual requirement

$$\sum_j T_{ij} = T_i; \qquad \forall i; \tag{7.32d}$$

in which the given constants $\hat{c}$, $\hat{t}$ are the observed total monetary and time cost expended by the travelers,

$$\hat{c} = \sum_i \sum_j \hat{T}_{ij} c_{ij} \tag{7.33a}$$

$$\hat{t} = \sum_i \sum_j \hat{T}_{ij} t_{ij} \tag{7.33b}$$

and $\hat{k}_j$ are the observed destination volumes,

$$\hat{k}_j = \hat{T}_j; \qquad \forall j \tag{7.33c}$$

Objective Function (7.31) may be considered the R.T.'s "partial" utility. As discussed in section 8 of Chapter 2, the single term in it corresponds to the utility derived from the diversity, or dispersion of choices. The first set of Constraints (7.32a) require the R.T.'s demands to be such that the predicted total monetary expenditure is no more than that observed. The second set of Constraints (7.32b) require similarly that the predicted total travel be no more than that observed. The third set of Constraints (7.32c) require the R.T.'s destination demands to be respectively equal to the observed destination demands.[4] Finally, the R.T.'s predicted choices must conform to the given travel volumes T_i, as always, as stated by Constraint (7.32d).

[4]One of these J constraints is redundant, since the sums of the observed and predicted demands, respectively, are equal. However, a redundant constraint has no effect on an optimization problem, whereas a redundant equation in a square system makes it indeterminate.

Now, if we form the (partial) Lagrangian for Problem (7.31)–(7.32) (in the familiar minimization form), with Lagrange coefficients as listed next to the constraints, we obtain the Problem

$$\underset{(T_{ij},\beta,\tau)}{\text{Min}} = \tau\left[\sum_i \sum_j T_{ij}t_{ij} - \hat{t}\right] + \beta\left[\sum_i \sum_j T_{ij}c_{ij} - \hat{c}\right]$$

$$+ \sum_i \sum_j T_{ij} \ln T_{ij} + \sum_j \alpha_j \left[\sum_i T_{ij} - \hat{k}_j\right] \qquad (7.35)$$

again, subject to Constraints (7.32d).

In this form, this is exactly Problem (7.30c), (7.30b), which is in turn precisely our original Problem (7.30a)–(7.30b). Thus, the aggregate travel demands T_{ij} and T_{ijr} produced by the solution of Problem (7.31)–(7.32) conform to Formula (7.12b). However, the key point is that now the calibrated parameter values are also produced by the problem's solution, as the *unknown* values of the Lagrangian coefficients (or dual variables) for Constraints (7.32a), (7.32b), and (7.32c), respectively.

Let us now examine in greater detail Constraints (7.32a) and (7.32b). First, they are written as "less than" inequalities. From KKT theory, this will insure that the values of the corresponding dual variables τ and β will be nonnegative, as required. (See section 2.2 in Appendix A.) However, these constraints may be expected to be active, binding at the solution of Problem (7.31). Otherwise, again from the "complementary conditions" Formula (A.25a) in Appendix A of KKT theory, the values of τ and β would be zero, contradicting our assumptions. In contrast, Constraints (7.32c) are written as equalities, so that the α's have an unrestricted sign.

Objective Function (7.31) is convex, and all Constraints (7.32) are linear. Consequently, Problem (7.31)–(7.32) is convex, and its solution, including the set of calibrated parameter values, will be unique. Therefore, the *unique* solution to the constrained R.T.'s problem are the required aggregate demands specified by Formula (7.10), in which the parameters values α_j, β and τ are equal to the dual variables for Constraints (7.32c), (7.32a) and (7.32b), respectively.

Thus, a *calibrated* travel demand model, together with corresponding estimates of travel demand, may be obtained from solving the "partial" R.T. utility maximization problem, given the observed values of the total (or equivalently average) values of the utility factors, here travel time, money, and the "unit" factor corresponding to parameters α_j. Since the "partial" R.T. utility is in fact the entropy of the distribution of origin-destination demands, this method is functionally equivalent to the "maximum entropy" approach (Wilson, 1970).

These various results are summarized as follows.

PROPOSITION 7.1

Parameter estimates may be estimated as the dual variables of the R.T. "partial utility" maximization problem

$$\text{Min } V_{(T_{ij})} = \left[\sum_i \sum_j T_{ij} \ln T_{ij} \right] \tag{7.35a}$$

subject to

$$\sum_i \sum_j T_{ij} c_{ij} = \hat{c} \tag{7.35b}$$

$$\sum_i \sum_j T_{ij} t_{ij} = \hat{t} \tag{7.35c}$$

$$\sum_i T_{ij} = \hat{T}_j; \qquad \forall j \tag{7.35d}$$

$$\sum_j T_{ij} = T_i; \qquad \forall i \tag{7.35e}$$

Several points should be noted. First, the constrained R.T. utility maximization problem is a nonlinear minimization with a convex objective function and linear constraints, and may therefore be solved with the partial linearization algorithm.

Also, it is important to note that the M.L. Equations (7.16) are in fact the same as the constraints in the "partial" R.T. problem above. Thus the two methods are equivalent.[5] This is significant, since, as will be seen in section 7.7.1, the M.L. estimated values possess useful statistical properties, which will then also be shared by the "partial U.M." values. However, as will be seen in the next section, this is no longer true when demand externalities are present.

From a practical point of view, calibration and demand estimation are thus performed *simultaneously*, by solving the R.T.'s problem, and are therefore totally compatible. Perhaps more importantly, calibration is now theoretically consistent with utility maximization. From a practical standpoint, the R.T. approach requires solving only a single problem, instead of two, that is, demand estimation and calibration as in the M.L. method. This particular feature will become rather useful when dealing with the variable utility case, as we shall see in the next section.

Also, it should be clear from the above discussion that when the right-hand sides of all Constraints (7.32) are given values which reflect anticipated, or

[5]Anas (1983) was the first to demonstrate this *numerical* equivalence between the conceptually different M.L. and utility maximization approaches.

even for that matter, desired, conditions rather than presently observed ones, as specified by Equalities (7.33), then the model's predictions will be maximally compatible with these expectations.

7.5.1.2 Illustrative Example To illustrate the R.T.'s approach to model calibration, let us apply it to the same situation as addressed in the illustrative example of section 7.3.1. The "partial" utility maximization problem to solve is

$$\underset{(T_{ij})}{\text{Min}} \left[\sum_i \sum_j T_{ij} \ln T_{ij} \right]$$

subject to

$$\sum_j T_{1j} = 4; \quad \sum_j T_{2j} = 1; \quad \sum_j T_{3j} = 2; \quad \sum_j T_{4j} = 3$$

$$\sum_i \sum_j T_{ij} c_{ij} = \sum_i \sum_j \hat{T}_{ij} c_{ij} = 0$$

$$\sum_i \sum_j T_{ij} t_{ij} = \sum_i \sum_j \hat{T}_{ij} t_{ij} = 28$$

$$\sum_i T_{i2} = 1; \quad \sum_i T_{i3} = 2; \quad \sum_i T_{i4} = 3$$

$$T_{ij} \geq 0; \quad \forall\, i, j$$

The solution to this problem, in terms of its dual variables, is

$$\hat{\tau} = 2.008$$

$$\hat{\beta} = 2.029$$

$$\hat{\alpha}_2 = -6.146, \ \hat{\alpha}_3 = -1.406, \ \hat{\alpha}_4 = 1.036$$

As expected, we retrieve the values obtained with the aggregate M.L. method. The model specification, and the resulting predicted aggregate demands, are consequently the same.

An important implication of these findings is that since the values of $\hat{\tau}$ and $\hat{\beta}$ are respectively equal to the dual variables of Constraints (7.32b) and (7.32d), according to duality theory, they represent the effect on the value of the objective function, the total utility received by the R.T., of a change of one unit in the value of the right-hand side of the corresponding constraint. For instance, in the present case, if the total travel time spent were to be 27 minutes, the R.T.'s received utility, or total traveler welfare, would go up by 1.77 units for ten travelers, or 0.177 unit per traveler. This value then represents the marginal utility of travel time, which in the present case is constant, since there is no congestion. Similarly, the marginal utility of money, or income, is equal to 2.304. Consequently, in *monetary* units, the value of travel time is equal to 0.177/0.2304 per minute, or $46/hr.

In closing this section, it might be pointed out that the R.T.'s "partial" utility maximization problem in effect determines aggregate travel demands under the assumption that travelers maximize their individual received utilities while constrained not only by a monetary budget, as usual, but also by a time budget, Constraint (7.32b). This formulation may be compared with various travel demand models which incorporate time constraints, including the UMOT model (Zahavi, 1979), as well as "entropy-maximizing" (Wilson, 1967), "information minimizing," and "minimum dispersion" (Erlander, 1977) formulations. (See Exercises 7.7 and 7.8.)

7.5.1.3 Combined Demands The same "partial" R.T. U.M. problem may be used to calibrate a hierarchical model of combined travel demands. For instance, it was shown in Exercise 4.8 that the R.T.'s utility maximization problem which produces the aggregate destination-mode demands when route choices are deterministic,[6] is

$$\operatorname*{Min}_{(T_{ij},\, T_{ijm})} \left[\sum_{ijm} T_{ijm} t_{ijm} + \sum_{ijm} T_{ijm} \ln T_{ijm} + \frac{1}{\beta'_d} \sum_{ij} T_{ij} \ln T_{ij} + \sum_{ijm} T_{ijm} \alpha_{ij} + \sum_j \sum_i T_{ij} \alpha_j \right] \tag{7.36a}$$

such that

$$\sum_m T_{ijm} = T_{ij}; \qquad \forall\, i, j \tag{7.36b}$$

$$\sum_j T_{ij} = T_i; \qquad \forall\, i \tag{7.36c}$$

$$T_i \geq 0; \qquad T_{ij} \geq 0,\ T_{ijm} \geq 0; \qquad \forall\, i, j, m \tag{7.36d}$$

Following the same approach as demonstrated in the single-level case above, we have the following result.

PROPOSITION 7.2

The "partial" R.T. U.M. problem in the case of combined destination, mode and deterministic route demands, in that order, is

$$\operatorname*{Min}_{(T_{ij},\, T_{ijm})} \left[\sum_{ijm} T_{ijm} \ln T_{ijm} \right] \tag{7.37}$$

[6] Lower-level parameter β_m has been set to 1, without loss of generality.

subject to

$$\sum_{ij} T_{ij} \ln T_{ij} = \sum_{ij} \hat{T}_{ij} \ln \hat{T}_{ij} \tag{7.38a}$$

$$\sum_{ijm} T_{ijm} t_{ijm} = \sum_{ijm} \hat{T}_{ijm} t_{ijm} \tag{7.38b}$$

$$\sum_{m} T_{ijm} = \hat{T}_{ij}; \qquad \forall\, i, j \tag{7.38c}$$

$$\sum_{i} T_{ij} = \hat{T}_{j}; \qquad \forall\, j \tag{7.38d}$$

$$T_i \geq 0; \qquad T_{ij} \geq 0,\ T_{ijm} \geq 0; \qquad \forall\, i, j, m \tag{7.38e}$$

as well as to definitional constraints (7.36b,c). The dual variable corresponding to the first constraint is equal to $(1/\beta'_d)$. The dual variables for the subsequent constraints are, in order, τ, α_{ij}, α_j. In contrast with the single-level case, specifically due to the presence of Constraint (7.38a), the constraints in this hierarchical R.T. problem are *not* the same as the M.L. equations. Consequently, the calibrated aggregate demands produced by the reduced R.T. problem do not possess the statistical properties of M.L. estimates.

7.5.2 Variable Utilities

7.5.2.1 The R.T.'s Partial Utility Maximization Problem The same approach is applicable in the case of variable utilities, that is, in the presence of demand externalities. In this case, we have the following.[7]

PROPOSITION 7.3

The R.T.'s "partial" utility maximization problem is

$$\text{Min } V_{(T_{ij})} = \sum_{i} \sum_{j} T_{ij} \ln T_{ij} \tag{7.39}$$

subject to

$$\sum_{i} \sum_{j} T_{ij} c_{ij} \geq \sum_{i} \sum_{j} \hat{T}_{ij} c_{ij} \tag{7.40a}$$

$$\sum_{a} \int_{0}^{v_a} t_a(x)\, dx \geq \sum_{a} \int_{0}^{\hat{v}_a} t_a(x)\, dx \tag{7.40b}$$

$$\sum_{j} \int_{0}^{\Sigma_i T_{ij}} s_j(x)\, dx = \sum_{j} \int_{0}^{\Sigma_i \hat{T}_{ij}} s_j(x)\, dx \tag{7.40c}$$

(Continued)

[7]Note that, Constraints (7.40c) are written as equalities, so that π may have an unrestricted sign, e.g., to allow for the possibility of positive destination externalities.

where

$$v_a = \sum_i \sum_j \sum_r (T_{ijr} + S_{ijr})\delta^a_{ijr}; \qquad \forall\, a$$

as well as to Constraints (7.38c to e).

Objective Function (7.39) is convex, and Constraints (7.38a–e) are linear. Furthermore, since the l.h.s. of Constraints (7.40a) and (7.40b) are both concave functions, and since the direction of the Constraints is $\geq$, the region they define is convex. (See section 1.8 of Appendix A.) Consequently, the Problem is convex, and its solution, including the set of calibrated parameter values, will be unique. Because of the presence of nonlinear Constraints (7.40b,c), the linearization method may not be applied to this problem. The problem may be solved with "reduced gradient" techniques (Murtaugh and Saunders, 1987; Lasdon and Warren, 1978).

It should be pointed out that, as mentioned earlier, M.L. equations in the present case of variable utilities have not been derived. Thus, it may no longer be the case that constraints (7.40) on the partial R.T. problem are the same as the M.L. equations. Since, in addition, there are no exact methods to obtain the numerical estimates of calibrated parameter values in this case, one no longer has the assurance that the estimates under the two methods are the same. Consequently, there is no guarantee that the estimates obtained from the R.T.'s problem possess the statistical properties discussed in section 7.7.

In closing this section, it may be mentioned that because it may be difficult in practice to obtain observed values for the r.h.s. of Constraints (7.40b) and (7.40c), an alternative formulation of the "partial" R.T. problem may be formulated by interchanging the roles of the cumulative travel cost and the entropy of origin-destination flows,

$$\text{Min } V_{(T_{ij})} = \sum_a \int_0^{v_a} t_a(x)\, dx \tag{7.41}$$

subject to

$$-\sum_i \sum_j T_{ij} \ln T_{ij} \geq -\sum_i \sum_j \hat{T}_{ij} \ln \hat{T}_{ij} \tag{7.42}$$

as well as the other Constraints (7.38c)–(7.38e), (7.40a)–(7.40c).

It may be shown that the solution to this problem will also take the functional form of the logit combined demands (7.12d), (see Exercise 7.16), in which, again, travel and destination costs have their equilibrium values. However, the calibrated parameter values and the associated aggregate demands will be numerically different from those obtained under the previous formulation, since they are respectively based on different information. Furthermore, these estimates will have different variances. See Anas (1988) for an investigation in the case of stochastic route choice.

7.6 MODEL VALIDATION AND APPLICATION

After the model is specified, it may then be applied to produce predictions of the levels of its endogenous, or internal variables (i.e., the outputs). However, the model must be *validated* before it can be applied for forecasting purposes. Validating a model essentially means comparing its predictions with observations, and determining to what extent the two agree. It is important to note that validation must be performed on another set of observations than that which was used to calibrate the model.

Indeed, successful calibration means, by definition, that the model reproduces well the observed values used in the calibration. This does not imply that it will also perform well with respect to another data set, for two main reasons. The first is that even though it may be assumed that underlying conditions remain the same, inherent statistical fluctuations in the values of the model's variables (e.g., travel demands) will result in variability in calibration performance. How to estimate the potential level of such statistical fluctuations as a byproduct of the calibration process is examined in the next section.

The second reason is that prevailing conditions (e.g., socioeconomic) may change. For instance, the preferences individual travelers have with respect to various factors of choice (i.e., utility parameters), such as π for destination costs, τ for travel costs, and so on, may be a function of economic conditions. Consequently, changes in prevailing conditions will be reflected in changes in parameter values. The magnitude of such changes, and their effects on model predictions must then be assessed experimentally. However, some formal techniques have been developed to adjust parameter values calibrated on the basis of a given data set to reflect other conditions represented by another data set. These "parameter transfer" techniques are, however, somewhat beyond the scope of the book. The interested reader may refer, for instance, to Koppelman and Wilmot (1982).

In any case, an indication of the magnitude of the prediction error which may be expected, both from inherent statistical fluctuations, as well as from changes in the underlying conditions, may be very useful in connection with the application of the model to future conditions, or to locales other than the one in which the model was calibrated. In either type of application, a decision will have to be made as to whether the calibrated parameter values are stable enough to be used, and/or whether they should be adjusted in some fashion.

Once the model is validated, it may then be used for the purpose of forecasting, or predicting travel demand under future conditions.

7.7 SAMPLING EFFECTS

As discussed above, the main purpose of calibrating a travel demand model is to fit it to given conditions, present or future, so that it may then be applied to analysis and prediction under the same conditions. An issue of great practical interest in this connection is how "good" the model is, how well it may be expected to replicate observed conditions. In fact, it may be remembered that the basic rationale underlying the calibration process was to make the fit between the model's predictions and corresponding observations as close as possible, in some sense yet to be defined specifically.

In general, the fit between the model's predictions and the observations will not be perfect. That is, the model may or may not be able to replicate exactly each and every observation of traveler demands, whether at the individual or at the aggregate level. This may be seen in Tables 7.5 and 7.9. The reason is simply that a model is a theoretical and simplified mathematical formulation of traveler behavior, which clearly is not easily described in mathematical terms.

As a result, if such a theoretical model is used in a predictive mode, these predictions will then be subject to some error. These errors may be called modeling errors. For instance, we have assumed that the determinants of a traveler's choice of destination are limited to the destination and travel costs. The absence of other destination characteristics, notably price of the activity to be conducted at the destination, may introduce a modeling error.

Specification errors may also be related to other causes than the nature of the model's variables, including the nature of the assumptions underlying the formulation. For instance, in the examples above, we have assumed that every traveler considers every destination, that the "destination set" is the same from all demand areas. Should this not be so, another modeling error would result. In general, modeling errors result from the need to simplify reality so that it may be represented mathematically.

There is, however, another source of prediction error. Specifically, even if the calibrated model were perfect, and managed to replicate a given sample of observations obtained under given conditions, it may not be able to replicate well another sample of observations obtained under the same conditions.

Indeed, since individual travelers are observed to randomly choose a given destination, the δ_{nj}'s as defined in Formula (7.2) vary with each sample of observations. As a result, the calibrated parameter values using the M.L. method to which these δ_{nj}'s are input would then be different. For the same reason, the observed trip ends $\hat{T}_j$, as well as the other observed entropy and cumulative travel and destination costs intervening in the maximum likelihood or R.T. partial utility maximization methods, would also fluctuate from sample to sample.

Calibrated parameter values, being functions of random variables, are themselves random variables, whose value may be expected to fluctuate from model calibration to calibration. Consequently, the model's output itself, the predicted travel demands, which are functions of the random parameter values, are also random variables.[8] This creates the second type of error in model prediction, which may be called estimation error. This type of error is related to the characteristics of the data on which the model is specified, rather than those of the model itself, as for the specification error above.

It is impossible to separate the effects of one type of error from those of the other, as the difference between predicted and observed values is attributable to both. However, for a *given* type of model, it is possible to analyze the statistical effects of specification errors. In this section, we examine the effects of this type of error on the calibration results, and by extension, on the performance, or efficiency, of the calibrated model. Specifically, the issue is the magnitude of the statistical, or sampling error in the calibrated parameter values.

In nontechnical terms, the question is: How reliable (stable) are the calibrated parameter values $\hat{\beta}$, $\hat{\tau}$, and so on, or generically, $\hat{\boldsymbol{\beta}}$? Specifically, these values constitute *estimates* of the exact parameter values, β, τ, and so on. In difference, the exact parameter values would be obtained on the basis of an infinite sample of observations, which, although theoretically desirable, is obviously infeasible in practice.

7.7.1 Effects on Maximum Likelihood Estimates

Using probability theory, it may be shown that the parameter estimates $\hat{\boldsymbol{\beta}}$, *when obtained with the M.L. method*, possess several remarkable and very useful features. Thus, the M.L. calibration method offers in this regard a significant advantage over the other method.

First, the M.L. estimates will be "unbiased." This means that the average of n estimated values of $\hat{\boldsymbol{\beta}}$, $\hat{\boldsymbol{\beta}}_1, \hat{\boldsymbol{\beta}}_2, \ldots \hat{\boldsymbol{\beta}}_k, \ldots, \hat{\boldsymbol{\beta}}_n$ will tend to the exact value $\boldsymbol{\beta}$ as n tends to ∞. Also, they will be "consistent," meaning that the variance of the resulting estimator will become smaller as n becomes larger. Moreover, they will be "efficient," meaning that when the size N of the sample of observations δ_{jn} $(n = 1, 2, \ldots N)$ from which the parameters $\boldsymbol{\beta}$ are estimated is infinite (or "large" in practice), the variances for the estimates $\hat{\boldsymbol{\beta}}$ will be the smallest of estimates obtained with *any* method.

Finally, again as the size of the sample of observations becomes infinite, the joint probability distribution function for the various parameters will become increasingly well approximated by a multivariate normal probability distribution function. Consequently, the (marginal) distributions of the individual parameters are also normal. (See section 3.5.4 in Appendix A.) For instance, for say, a given individual parameter τ,

$$\text{Prob. } (\hat{\tau} \leq \hat{\tau}_0) \approx \Phi\left(\frac{\tau_0 - E(\tau)}{\sigma_\tau}\right) \tag{7.43}$$

[8]This has nothing to do with the fact that the model's output is in fact cast in a probabilistic form itself, i.e., as probabilities of travel choices, e.g., $P_{j/i}$.

where $\Phi(x)$ is the value at x of the cumulative Normal distribution function. $E(\tau)$ is the mean of the distribution, the exact, but unknown value of the parameter, and σ_τ is its standard deviation. The approximation becomes (asymptotically) an equality when $n = \infty$.

In practical terms, these features may be used to conduct several types of statistical analysis about the exact value of the parameters, including the estimation of "confidence intervals" for them, and the testing of hypotheses about their value. The knowledge of the standard deviation σ_τ of the estimate is required, however. The corresponding variances may be obtained from the parameter estimates' "variance-covariance" matrix, which is (asymptotically) equal to the negative of the inverse of the Hessian of the log-likelihood function. (See Formula (A.17) in Appendix A.)

FACT 7.3

$$\begin{bmatrix} \sigma_\beta^2 & \text{Cov}(\beta, \alpha_1) & \ldots & \text{Cov}(\beta, \tau) \\ \text{Cov}(\alpha_1, \beta) & \sigma_{\alpha_1}^2 & \ldots & \text{Cov}(\alpha_1, \tau) \\ \ldots & & & \ldots \\ \text{Cov}(\tau, \beta) & \text{Cov}(\tau, \alpha_1) & \ldots & \sigma_\tau^2 \end{bmatrix}$$

$$\simeq - \begin{bmatrix} \dfrac{\partial^2 \Lambda(\hat{\boldsymbol{\beta}})}{\partial \beta^2} & \dfrac{\partial^2 \Lambda(\hat{\boldsymbol{\beta}})}{\partial \beta \, \partial \alpha_1} & \ldots & \dfrac{\partial^2 \Lambda(\hat{\boldsymbol{\beta}})}{\partial \beta \, \partial \tau} \\ \dfrac{\partial^2 \Lambda(\hat{\boldsymbol{\beta}})}{\partial \alpha_1 \, \partial \beta} & \dfrac{\partial^2 \Lambda(\hat{\boldsymbol{\beta}})}{\partial \alpha'} & \ldots & \dfrac{\partial^2 \Lambda(\hat{\boldsymbol{\beta}})}{\partial \alpha_1 \, \partial \tau} \\ \ldots & & & \ldots \\ \dfrac{\partial^2 \Lambda(\hat{\boldsymbol{\beta}})}{\partial \tau \, \partial \beta} & \dfrac{\partial^2 \Lambda(\hat{\boldsymbol{\beta}})}{\partial \tau \, \partial \alpha_1} & \ldots & \dfrac{\partial^2 \Lambda(\hat{\boldsymbol{\beta}})}{\partial \tau^2} \end{bmatrix}^{-1} \tag{7.44}$$

In words, the procedure for the estimation of the variances of the estimated values for the parameters, $\sigma_{\boldsymbol{\beta}}^2$, is to compute the Hessian of the log-likelihood function Λ (the matrix of the second derivatives with respect to the parameters), compute its numerical value for the estimated values of the parameters, compute its inverse, and take the negative of the result. The variances and covariances of the calibrated parameter values are then arrayed in the corresponding locations of the resulting matrix, as represented in Formula (7.44). In particular, the terms on the diagonal of the variance-covariance matrix provide the values for the estimated variances $\sigma_{\boldsymbol{\beta}}^2$ of the respective parameters.[9]

[9] It may be noted that when there are many parameters to estimate, the large size of the matrix may pose computational problems.

Using the above results, standard statistical tests of hypotheses may be conducted, and confidence intervals for the "true" values of the parameters may be constructed. This will provide answers to questions such as the following:

- If we are willing to run a risk of being wrong of, say, 5% or 10%, etc., do we accept the proposition that the true value of parameter μ is equal to zero?
- What are the lowest (highest) values which the true value of μ could take, again at a risk of being wrong, of say, 5% (i.e., with 95% confidence)?
- What is the likelihood that the value of μ could be as low as, or lower than a given level, μ_0?

The respective answers to these various questions may be used to select variables for inclusion in the demand model, as well as to provide some bounds on the predicted travel demands. For instance, if the numerical value of the parameter for a given factor of choice is very small, it implies that factor is irrelevant, and should be dismissed from the model's specification. Similarly, confidence intervals for individual parameter values,

$$\tau_m \leq \tau \leq \tau_M$$

or confidence regions for sets of parameter values

$$\boldsymbol{\beta} \in \boldsymbol{\beta}_0$$

may be (numerically) translated into their counterparts in terms of predicted travel demands:

$$T_{ijm} \leq T_{ij} \leq T_{ijM}$$

7.7.2 Effects on Maximum Entropy Estimates

As mentioned earlier, the estimated parameter values possess the various statistical properties above only if they have been obtained with the M.L. method, when there are no externalities on the factors of traveler choice. In particular, if they have been obtained with the R.T.'s partial utility maximization method, the probability distribution function (p.d.f.) of the calibrated parameter values, and their variances will be unknown, so that Formulas (7.44) and (7.43) would not be applicable. This drawback of the R.T. method with respect to the M.L. method is the price one must pay for the ability to rigorously estimate the parameter values when externalities are present.

The recourse in this case is to estimate empirically the probability distribution function of the calibrated parameter values, through computer simulations. The method consists essentially of generating a very large series of hypothetical "observations," for instance, normally distributed around the actual observations. From repeated individual applications of the method on each set of observations, a large number of independently calibrated parameter values is generated, from which a p.d.f. may then be constructed using, for instance, a "Monte Carlo" method. This method exploits the fact that transforming any random variable into its cumulative p.d.f. function results in uniformly distributed random variable on the interval (0,1). Transforming the values of the latter through the inverse of observed cumulative p.d.f. thus results in the (unknown) original distribution. An example of application of this method to a similar problem is provided in Knudsen and Fotheringham (1985) and Openshaw (1979). This is known as the "bootstrap" method, for somewhat obvious reasons. In turn, having the knowledge of the p.d.f. allows one to construct confidence intervals and test hypotheses regarding calibrated values, as above.

7.8 MEASURES OF CALIBRATION EFFECTIVENESS

However useful they may be, the various analyses described above do not provide a specific answer to the basic question: "How well does the calibrated model perform in reproducing the observed travel demands?" In this section, we present methods which are designed to provide an answer to it.

It is logical to compare the calibrated travel demand model to the best model, as well as to the worst model, one might possibly obtain. We therefore define what these two extreme cases are. Remembering that the parameters we are calibrating multiply the values of the basic factors of choice (i.e., travel time and costs) in a utility function, it is clear that the worst model in this respect is one in which all parameter values are equal to zero. In this case, no factors of travel choice whatsoever are taken into account by the predictive model. Consequently, using, for instance, Logit Function (7.1b), the probability that a given traveler chooses a given destination would be estimated as

$$P_{j/n} = \frac{e^{-\beta(\alpha_j - \tau t_{ij} - c_{nj})}}{\sum_j e^{-\beta(\alpha_j - \tau t_{ij} - c_{nj})}} = \frac{e^{-0}}{\sum_j e^{-0}} = 1/J; \qquad \forall\, j, n \tag{7.45}$$

That is as expected; since no factors of choice differentiate destinations, they all have the same probability of being chosen by any traveler.

On the other hand, the "least bad" prediction of destination demands one can make *without* the given model, or more generally without any information, is to predict that travelers will choose destinations totally at random, and will choose all destinations in equal proportions. Consequently $P_{j/i}$ as predicted by the worst model, will be equal to $1/J$. Thus, a calibrated travel demand model where all parameters are set equal to zero is worthless, in that the model's predictions are no better than (reasonable) predictions without it.[10] Such a

[10] "Unreasonable" predictions might, for instance, be negative valued, which would make the model arbitrarily worse.

model must then be considered the worst, the least informational. Conversely, a given model will obviously be the best if its predictions of travel demands, the T_{ij} values, are all precisely equal to the corresponding observed values $\hat{T}_{ij}$.

The efficiency of a given calibrated travel demand model may then be characterized by its performance, relative to these two extreme cases. It is natural to use the same criterion as was used for calibration to assess calibration performance. Since the respective objective functions used in the M.L. and the M.E. calibration procedures are different (i.e., log-likelihood and entropy, respectively), evaluation of model performance will take two different forms. We begin with the M.L. method.

7.8.1 Maximum Likelihood Method

For the M.L. method, the criterion is the value of the log-likelihood function Λ, as specified by Formula (7.3b). We will illustrate the approach in the most general case, using disaggregate data. As seen above, the case of aggregate data follows directly.

At calibration, Λ has the value corresponding to the calibrated parameter values, $\Lambda(\hat{\boldsymbol{\beta}})$. It is useful to note that given its definition in Formula (7.3b), $\Lambda(\boldsymbol{\beta})$ will be nonpositive for any value of $\boldsymbol{\beta}$, since Log $x \leq 0$ for $x \leq 1$. The log-likelihood value for the worst model defined above is

$$\Lambda(0) = \sum_{n=1}^{N} \sum_{j=1}^{J} \delta_{nj} \ln P_{j/n}(0, 0, \ldots 0) = \sum_{n=1}^{N} \sum_{j=1}^{J} \delta_{nj} \ln \left(\frac{1}{J}\right)$$

$$= \ln \left(\frac{1}{J}\right) \sum_{n=1}^{N} \sum_{j=1}^{J} \delta_{nj} = -\ln (J) \sum_{j=1}^{J} \hat{Y}_j$$

since, as we have seen above, $P_{j/i}(\mathbf{0}) = 1/J$, so that finally

$$\Lambda(\mathbf{0}) = -\hat{Y} \ln (J) \tag{7.46}$$

Conversely, the value of the log-likelihood function for the best model, corresponding to the optimal parameter values $\boldsymbol{\beta}^*$, is obtained by equating each $P_{j/n}$ to the observed value $\hat{P}_{j/n}$. The observed value of the probability with which traveler n chooses destination j is obviously

$$\hat{P}_{j/n} = \begin{cases} 1; & \text{if traveler } n \text{ chooses destination } j \\ 0; & \text{if not} \end{cases}; \qquad \forall\, j, n$$

In this form, $\hat{P}_{j/n} = \delta_{nj}$ as defined above. Consequently,

$$\Lambda(\boldsymbol{\beta}^*) = \sum_{n=1}^{N} \sum_{j=1}^{J} \delta_{nj} \ln P_{j/n}(\boldsymbol{\beta}^*) = \sum_{n=1}^{N} \sum_{j=1}^{J} \delta_{nj} \ln \delta_{nj} = 0 \tag{7.47}$$

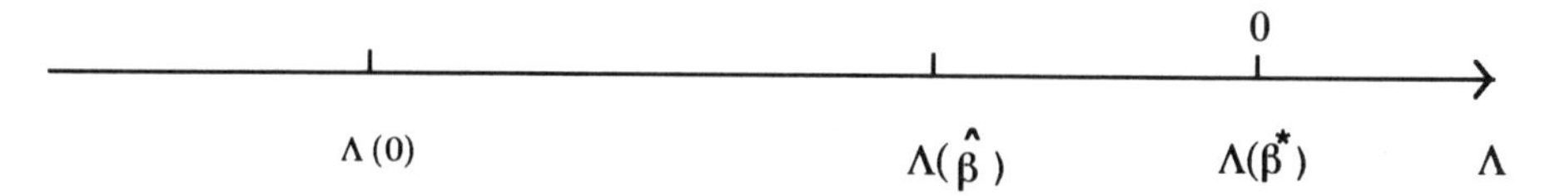

Figure 7.1 Relative positions of the log-likelihood function values.

since $0 \ln 0 = 0$. Thus, the maximum value for Λ, corresponding to a perfectly calibrated model, is zero.

The respective values of the log-likelihood function for the worst, best, and calibrated models, respectively, may then be graphically represented as in Figure 7.1.

The model's performance may then be measured numerically by the "log-likelihood ratio":

$$\rho^2 = \frac{\Lambda(\hat{\boldsymbol{\beta}}) - \Lambda(\mathbf{0})}{\Lambda(\boldsymbol{\beta}^*) - \Lambda(\mathbf{0})} = 1 - \frac{\Lambda(\hat{\boldsymbol{\beta}})}{\Lambda(\mathbf{0})} \tag{7.48}$$

The value of ρ^2 will always be between zero and one. When the calibrated model is as good as possible, then $\Lambda(\hat{\boldsymbol{\beta}}) = \Lambda(\boldsymbol{\beta}^*)$, and ρ^2 is equal to 1. Conversely, when the calibrated model is as bad as (reasonably) possible, $\Lambda(\hat{\boldsymbol{\beta}}) = \Lambda(\mathbf{0})$ and ρ^2 is equal to 0. The value of ρ^2 may then be used as an indicator of model performance, in the same manner as the square of the coefficient of correlation R^2 is a measure of performance for linear regression. It should be noted that, as for R^2, the performance of the model does not improve proportionately to the value of ρ^2. In other words, this performance scale is not linear.

It is important to keep in mind that the value of ρ^2, as that of all other calibration statistics above, fluctuates from calibration to calibration. However, its p.d.f. is not analytically known, so that an observed value would have to be tested using the kind of "bootstrap" method described above.

Another way to assess a calibrated model's performance is based on the following feature of the log-likelihood value, which is a random variable, being a function of the random variables $\hat{\boldsymbol{\beta}}$.

FACT 7.4

The statistic

$$\chi_* = 2[\Lambda(\boldsymbol{\beta}^*) - \Lambda(\hat{\boldsymbol{\beta}})] \tag{7.49}$$

has a Chi-square p.d.f., with a number of degrees of freedom equal to the number of observations.

This statistic may be interpreted as representing the "distance," measured in log-likelihood values, between the best model and the calibrated model. The hypothesis that the calibrated model performs as well as the best model is equivalent to the hypothesis that the true value of this statistic is equal to zero, and may therefore be tested on the observed value for χ_*.

FACT 7.5

The statistic

$$\chi_0 = 2[\Lambda(\hat{\boldsymbol{\beta}}) - \Lambda(\mathbf{0})] \tag{7.50}$$

is distributed with a Chi-square p.d.f. with a number of degrees of freedom equal to the number of parameters.

It may then be used in the same manner as above to test the hypothesis that the calibrated model performs as badly as the worst model, that is, the hypothesis that the true value of this statistic may be assumed to be equal to zero. (See Exercise 7.18.)

As mentioned earlier, an important question is which explanatory variables should be included in a given model (e.g. which factors of supply site choice should intervene in the specification of the traveler utility function). The same type of statistic as above may also be used to investigate the contribution to the performance of the model of a given, single parameter. If, for example, the true value of parameter π may be assumed to be equal to zero, this would imply that destination costs are irrelevant to the travel demand being modeled, and may therefore be deleted from the model. This particular assumption may be tested using the following result.

FACT 7.6

The statistic

$$\chi_1 = 2[\Lambda(\hat{\boldsymbol{\beta}}, \hat{\pi}) - \Lambda(\hat{\boldsymbol{\beta}}, \mathbf{0})] \tag{7.51}$$

is distributed with a Chi-square p.d.f., with one degree of freedom.

More generally, the contribution of a set of k parameters $\boldsymbol{\beta}'$ may be tested in the following way.

FACT 7.7

The statistic

$$\chi_k = 2[\Lambda(\hat{\boldsymbol{\beta}}, \hat{\boldsymbol{\beta}}') - \Lambda(\hat{\boldsymbol{\beta}}, \mathbf{0})] \tag{7.52}$$

is distributed with a Chi-square p.d.f., with k degrees of freedom.

7.8.2 Maximum Entropy Method

In this case, the model's performance must be assessed on the basis of the comparison of the entropy of the predicted values to that of the observed values. The entropy function

$$E = -\sum_i \sum_j T_{ij} \ln T_{ij}$$

attains its maximum value when the distribution of travel demands are as dispersed as can be, that is, when they are uniform,

$$T_{ij} = T_i/J; \qquad \forall\, i, j \tag{7.53}$$

for which it takes on a value of

$$E_u = \ln J - \sum_j T_i(\ln T_i) \tag{7.54}$$

Conversely, the entropy function attains its minimum value when the distribution of demands are as concentrated as can be, as when

$$T_{ij} = \begin{cases} T_i; & \text{for } j = j_i \\ 0; & \text{for all other } j \end{cases}; \qquad \forall\, i \tag{7.55}$$

for which it takes on a value of

$$E_l = -\sum_i T_i \ln T_i \tag{7.56}$$

The entropy of the predicted demands, under a set of calibrated parameter values $\hat{\boldsymbol{\beta}}$ is

$$E_{\hat{\beta}} = -\sum_i \sum_j T_{ij}(\hat{\boldsymbol{\beta}}) \ln T_{ij}(\hat{\boldsymbol{\beta}}) \tag{7.57}$$

Similarly, the entropy of the observed demands is

$$E_{\hat{T}} = -\sum_i \sum_j \hat{T}_{ij} \ln \hat{T}_{ij}$$

Then, in analogy with the ρ^2 statistic defined in Formula (7.45), a similar indicator may be defined in the following way:[11]

$$\eta^2 = 1 - \frac{|E_{\hat{\beta}} - E_{\hat{T}}|}{E_u - E_l} \tag{7.58}$$

[11]The symbol $|x|$ means "absolute value" of x, i.e., $|x|$ is x if $x > 0$ and $-x$ if x is negative.

The value of η^2 will be between zero, when the calibration performance is worst, and 1.0, when the observed and predicted entropies are equal.

Also, the variances of the respective entropies $E_{\hat{\beta}}$ and $E_{\hat{T}}$ may be estimated as follows (Hutcheson, 1970).

THEOREM 7.1

$$\sigma^2_{E\hat{\beta}} = \sum_i T_i^3 \left[\sum_j P_{j/i}(\beta) \ln^2 P_{j/i}(\beta) - \left(\sum_j (\beta) \ln (\beta) \right)^2 \right] + \frac{J-1}{2} \sum_i T_i^2 \tag{7.59a}$$

$$\sigma^2_{E\hat{T}} = \sum_i T_i^3 \left[\sum_j \hat{P}_{j/i}(\beta) \ln^2 \hat{P}_{j/i}(\beta) - \left(\sum_j \hat{P}_{j/i}(\beta) \ln \hat{P}_{j/i}(\beta) \right)^2 \right] + \frac{J-1}{2} \sum_i T_i^2 \tag{7.59b}$$

where $P_{j/i}$ is given by Formula (7.2) with all parameters at their calibrated values, and

$$\hat{P}_{j/i} = \frac{\hat{T}_{ij}}{T_i}; \qquad \forall\ i, j \tag{7.56c}$$

If the number IJ of terms in the respective entropies is large, it may be assumed that their observed values will be normally distributed with the above variances. A T-test may then be conducted on the statistic

$$t = \frac{E_{\hat{\beta}} - E_{\hat{T}}}{[\sigma^2_{E\hat{\beta}} - \sigma^2_{E\hat{T}}]^{1/2}} \tag{7.60}$$

to determine whether the assumption that the predicted and observed distributions of traveler demands are similar, in terms of entropy, must be rejected, or may be retained, at a chosen level of risk. (See section 6 of Appendix A.) Alternatively, another statistic, the "minimum discriminant information" (Bishop, Feinberg, and Holland, 1975), may also be used.

THEOREM 7.2

The statistic

$$M = 2\hat{T} \sum_i \sum_i \hat{T}_{ij} \ln (\hat{T}_{ij}/T_{ij}) \tag{7.61}$$

(*Continued*)

where

$$\hat{T} = \sum_i \sum_j \hat{T}_{ij}$$

has an asymptotic Chi-square distribution with IJ–k degrees of freedom, where k is the number of constraints on the R.T.'s partial problem.

M takes on a value of zero when the model is perfect, when all predictions are equal to the corresponding observation, and a positive value for less than perfect model fit.

The statistical significance of the assessed model performance may then be tested by comparing the observed value of M as estimated from Formula (7.61) with the threshold value from the Chi-square table with the appropriate number of degrees of freedom at the chosen level of confidence. (See section 6 of Appendix A.) If the former is greater than the latter, the hypothesis that the distribution of predicted values is not significantly different from that of the observed values must then be rejected.

7.9 SUMMARY

In this chapter, various methods have been represented for calibrating a travel demand model to specific conditions of application. Two main approaches may be used in this connection. The first, the maximum likelihood method (M.L.), offers the advantage of allowing formal statistical analysis of the calibration results, both with respect to parameter values and the model's performance, using several statistical measures, which can be tested for their significance. It is thus the preferable method, in particular if disaggregate (individual) information is available about travelers and their choice of destination. It is difficult to apply, however, when the factors of choice are variable, that is, subject to congestion externalities, or to suppliers actions, as with price setting.

The other method, solving the "partial" R.T. U.M. problem, allows simultaneous model calibration and prediction, with the advantage that it may be applied to situations where factors of travel choice are variable. In addition, aggregate demands will naturally reflect observed, as well as expected (e.g., future) conditions, and are totally compatible with utility maximization. This method is easier to apply in the case of variable factors of choice. When these factors are fixed, the two methods are strictly equivalent.

Using either method, the calibrated model then reproduces, to varying degrees of efficiency, given observations of travel demands. Several indicators of model (calibration) performance were developed for each of the two methods above. Statistical tests may then be conducted to assess the significance of individual, or multiple parameters, and by implication, that of the factors of travel choice they correspond to.

7.10 EXERCISES

7.1 Derive Equations (7.7) and (7.8a).

7.2 Derive Equations (7.16).

7.3 Modal choice is assumed to be specified by a logit model:

$$P_m = \exp\,(\beta\tilde{U}_m) \Big/ \sum_m \exp\,(\beta\tilde{U}_m)$$

The fixed utilities of two given modes in a certain travel corridor are

$$m = 1 \quad 2$$
$$\tilde{U}_m = 3 \quad 5$$

The observed numbers N_m of users of the two modes are

$$\hat{T}_{im} = 40 \qquad 60$$

a. Based on these data, use the maximum likelihood method to determine the estimated value $\hat{\beta}$ of parameter β. (Hint: You may want to use a graphical method to solve for $\hat{\beta}$.)

b. Determine the standard deviation of $\hat{\beta}$.

c. Determine a 95% confidence interval for the true value of β.

d. Test, at the 95% confidence level, the hypothesis that the true value of β is equal to 0.

e. Describe qualitatively (i.e., in words) the implication of the findings in (d).

f. Assess the performance of the calibrated logit model, using the statistic ρ^2.

g. Describe qualitatively the implication of the findings in (f).

h. Test, at the 95% confidence level, and using the statistic χ_0 specified in Formula (7.51), the hypothesis that the true value of β is equal to 0.

7.4 Consider the following model of destination choice

$$P_{j/n} = \frac{e^{(\alpha_j + \tau t_{nj})}}{\sum_j e^{(\alpha_j + \tau t_{nj})}}; \qquad \forall\, n, j$$

The following individual observations were made for three individuals

δ_{nj}	$j = 1$	2	t_{nj}	$j = 1$	2
$n = 1$	1	0	$n = 1$	2	1
2	0	1	2	1.5	2
3	1	0	3	0.5	2.5

a. Write the two equations in the two unknown parameter values.
b. Solve for τ assuming $\alpha_1 = 0$.
c. Estimate the value of the log-likelihood function Λ for the best and worst models in this situation.
d. Estimate the value of the likelihood ratio ρ^2.
e. Estimate the value of the standard deviations of the two parameters from Formula (7.44).
f. Build a 95% confidence interval for the true value of parameter τ.
g. Build a 95% confidence interval for the true value of parameter α_2.
h. Test at a level of confidence of 98% the hypothesis that the true value of parameter τ is equal to zero. What are the implications of such a hypothesis in terms of the effect of travel time on destination choice?
i. Test at a level of confidence of 98% the hypothesis that the true value of parameter α_2 is equal to zero. What are the implications of such a hypothesis in terms of individual utility specification?
j. Test at a level of confidence of 98% the hypothesis that the true values of both parameters τ and α_2 are equal to zero. What are the implications of such a hypothesis in terms of the model specification?

7.5 On the basis of the average values of travel time

$$t_1 = (2 + 0.5)/2 = 1.25$$

$$t_2 = 2/1 = 2$$

calibrate the model in the previous exercise in the "aggregate form":

$$P_j = \frac{e^{(\alpha_j + \tau t_j)}}{\sum_j e^{(\alpha_j + \tau t_j)}}; \qquad \forall j$$

Again, solve for τ assuming $\alpha_1 = 0$. Compare with the previous, "disaggregate" calibration results.

7.6 If now, in the same situation as the two previous exercises, the individual choices are now known, but only the numbers of choosers of each alternative, i.e., $T_1 = 2$, $T_2 = 1$, and the observed average values

$$t_1 = (2 + 1.5 + 0.5)/3 = 1.33$$

$$t_2 = (1 + 2 + 2.5)/3 = 1.83$$

calibrate the model:

$$P_j = \frac{e^{(\alpha_j + \tau t_j)}}{\sum_j e^{(\alpha_j + \tau t_j)}}; \qquad \forall\, j$$

Compare with the above results.

7.7 Provide an alternative interpretation of the R.T.'s constrained utility maximization problem, based on the concept of "entropy" (see section 3.6 of Appendix A).

7.8 Provide an alternative interpretation of the R.T.'s constrained utility maximization problem, based on the concept of "information" (see section 3.6 of Appendix A).

7.9 Interpret the estimated values of β and τ in the illustrative example of section 7.2.2 to estimate the "value of time" to the travelers in the sample.

7.10 Using the predicted individual probabilities of choice represented in Table 7.5, evaluate the expected destination demands and compare with the given demands. Characterize as formally as possible the calibrated model's performance. Also, compare with the predicted demands in Table 7.10, and draw conclusions about the performance of "full information" calibration, versus "average" calibration.

7.11 With reference to the illustrative example of section 3.1.2, assume that transit has recently been introduced in the area, with the origin-destination travel times as in Table 7.11.

Transit fares are the same for all trips. Origin-destination car travel times are assumed equal to the minimum route times in Table 7.6, and the car costs equal to those in Table 7.7.

TABLE 7.11 Average Transit Travel Times t_{ij}

$j =$	1	2	3	4
$i = 1$	3	8	9	14
2	10	3	18	10
3	13	11	3	11
4	13	12	14	8

TABLE 7.12 Observed Modal Demands for Exercise 7.11

$i =$	1	2	3	4
T_{ic}	115	225	65	95
T_{it}	35	50	5	10

A sample of 600 travelers has been observed. The modal origin demands T_{ic} and T_{it} are given in Table 7.12.

The observed T_{ijm}'s are given in Table 7.13. Given these data, use the sequential calibration procedure described in section 7.3.2.1 to calibrate a model of joint modal, destination, and route choice of the form

$$P_{imj} = \frac{e^{(\gamma_m + \beta_m \tilde{W}_{m/i})}}{\sum_m e^{(\gamma_m + \beta_m \tilde{W}_{m/i})}} \frac{e^{(\alpha_j - \beta_d c_{ijm} - \tau t_{ijm})}}{\sum_j e^{(\alpha_j - \beta_d c_{ijm} - \tau t_{ijm})}}; \qquad \forall\ i, j, m$$

where, as usual

$$\tilde{W}_{m/i} = \frac{1}{\beta_d} \ln \sum_j e^{(\alpha_j - \beta_d c_{ijm} - \tau t_{ijm})}; \qquad \forall\ i, m$$

7.12 Show that the solution to the "partial" R.T.'s utility maximization problem in the destination-route case is of the form

$$T_{ij} = T_i \hat{T}_j \mu_i \lambda_j e^{-\beta(c_{ij} + \tau t_{ij})}; \qquad \forall\ i, j$$

7.13 Continuing the preceding exercise, devise a solution procedure to solve the problem in this case, i.e., identify the values of the μ_i and λ_j's.

TABE 7.13 Observed Modal Flows for Exercise 7.11

T_{icj} $j =$	1	2	3	4
$i = 1$	200	25	0	0
2	0	20	0	5
3	10	10	55	0
4	10	15	40	110
T_{imj}				
$i = 1$	35	2	2	0
2	0	5	0	0
3	2	4	20	0
4	0	0	20	30

7.14 Derive the ''full information'' M.L. equations for the respective parameters in the hierarchical destination-route choice model Formula (7.20), i.e., when the destination and route demands are treated simultaneously, as opposed to sequentially.

7.15 Apply the result of Exercise 7.14 to Exercise 7.11.

7.16 Show that the solution to Problem (7.41), (7.42), (7.38c–e), (7.40a–c) is the calibrated equilibrium demand model, Formula (7.12b).

7.17 Show that objective function (7.4) is everywhere convex. (Hint: Apply the results of section 1.7 in Appendix A.)

7.18 Two travel modes are available between two given locations. The fixed travel times on them are respectively

$$t_1 = 10;\ t_2 = 5$$

The observed ridership on the two modes are

$$\hat{T}_1 = 400;\ \hat{T}_2 = 600$$

a. Derive the M.L. equation for the calibration of a model of mode choice of the form

$$P_m = \frac{e^{-\beta t_m}}{\sum_m e^{-\beta t_m}}; \qquad \forall\ m$$

b. Construct a 95% confidence interval for the true value of β.

c. Test whether the calibrated model may be assumed to be the ''best'' model at the same level of confidence.

d. Test whether the calibrated model may be assumed to be the ''worst'' model.

7.19 In the previous exercise, the travel times are now assumed to be variable, and specified as

$$t_1 = 10 + 0.015v$$

$$t_2 = 15 + 0.01v$$

Given the same observed values for the modal demands, formulate the R.T.'s ''partial'' U.M. problem which solves the combined prediction-calibration problem. Solve the problem, i.e., obtain the numerical values of the parameters.

CHAPTER 8

JOINT EQUILIBRIUM MODELING OF ACTIVITY AND TRAVEL SYSTEMS

8.1 INTRODUCTION: GENERALIZED ACTIVITY-TRAVEL EQUILIBRIUM

In this chapter, we consider the connection between urban activities and urban travel. Indeed, the underlying purpose of any travel is always the conduct of some activity, be it work, shopping, or recreation. Consequently, urban travel demand is ultimately a reflection of the demand for "urban activities." Thus, formulating in greater detail the relationships between urban activities and travel will reinforce our travel demand models. While the full investigation of these relationships is beyond the scope of this book, as it would take us into the analysis of the economic activity system which transportation supports, including industrial organization, economic competition, housing, and labor markets, it is important to place travel into a somewhat larger perspective than that in which we have kept it so far for purposes of expository simplicity.

In such a perspective, both travel *and* the given activity are demanded by individual travelers/consumers and supplied by transportation and activity providers, respectively. In the same manner that travel choices are influenced by travel costs, activity-related choices will be influenced by activity costs. Accordingly, equilibrium must now be formulated not only with respect to travel in the network sense used so far, but also, in the economic sense, between supply and demand for the activity. Furthermore, these equilibria must coexist *jointly*. This implies that activity and travel costs will interact, since they both mediate the decisions of the various actors in the activity/travel system.

The conceptualization of this enlarged system may be articulated in terms of its function, the actors in it, and the facilities they use, or operate. The function of the system is to allow the conduct of a given activity by residents of an urban area, such as shopping, recreation, education, and generally, any activity for which they must travel and pay a price.[1] The conduct of the activity may also require the consumption of some generic commodity (e.g., food), in which case the commodity's price will be included in the activity price.[2] A prime example of such an activity would be shopping, for which the associated commodity is retail goods.

The activity may be conducted at a number of facilities located at various locations in the urban area. These locations are nodes on the transportation system serving the area, so that a choice of facility may be equated with a choice of destination. These facilities have different levels of service (e.g., physical or staff size), and different activity prices. These facilities are in competition with one another for travelers' patronage, who choose an activity site on the basis of its level of service, activity price, and access costs.

The activity suppliers (A.S.) have a dual role. First, they provide for profit the activity to travelers, including the accompanying commodity, if applicable. Also, they acquire at least cost the supporting commodity from their own suppliers, which we may call commodity suppliers (C.S.). The commodity is shipped from warehouses to activity facilities. Commodity suppliers also have a dual role. First, they sell, also for profit, the commodity to the activity suppliers. For simplification, they are assumed to also produce the commodity. This last assumption will be relaxed in section 8.5.

[1]Initially, for simplification purposes, *single* activity systems will be addressed. Multiple activity systems may be generalized from the single activity case.

[2]Note that work (i.e., the conduct of a job) is not considered such an activity for our purposes.

The transportation system (e.g., street system) is used both for person and freight transportation, and is subject to congestion. As usual, the network is assumed to be used both for activity-related and general travel, that is, travel in connection with other urban activities (e.g., travel to work).

8.2 A COMBINED MODEL OF ACTIVITY ALLOCATION, PERSONAL TRAVEL, AND GOODS MOVEMENTS

In this section, we develop a model of joint equilibrium activity/travel demand modeling, as a specific illustration of the above concepts. This is but one example of a possible approach. Others may be found in the transportation modeling and/or regional science literature. (See, for instance, Kim, 1989, and Anas, 1984.) We first describe the assumptions underlying the model.

8.2.1 Model Assumptions

The general assumptions underlying the model are as follows. (See comments in section 8.4 for possible ways to extend the model by relaxing some of these assumptions).

1. The demand for activity/commodity originating in demand areas is assumed fixed and known. This means that the model does not address the first level of traveler/consumer choice, whether to consume/travel or not. This may be reasonable for such commodities as food, where demand is not discretionary. Thus, the number of trips per unit time is fixed.
2. Traveler choices concern only *where* to conduct the activity (the choice of destination), and which route to follow. The amount of activity and/or commodity to be consumed per trip is assumed fixed.
3. A single mode is assumed available, both for personal and freight transportation. This implies that the model does not address modal choice. This assumption may be reasonable for such activities as shopping, which are typically conducted by car.
4. Activity-related travel demands are assumed to depend on non-activity-related travel demands, but not reciprocally. Specifically, non-activity-related car and truck volumes are assumed known, and constitute the fixed backdrop upon which activity-related car and truck volumes are distributed. As mentioned earlier, ''work'' is not considered an activity in the sense of this model.
5. All actors in the transportation/economic system are assumed to maximize the utility of their actions/choices, subject to budgetary constraints.

6. Route choice is assumed deterministic, and location choice is assumed to be probabilistic for all actors.
7. Car occupancy is assumed equal to one, for simplicity and without loss of generality. Also, the commodity is assumed shipped in full truckloads.
8. Logistical aspects of urban goods movements are simplified by assuming that commodities are delivered in single-stop trips, and are routed along the minimum cost path from a given warehouse to a given activity location.
9. Activity suppliers are assumed to play the role of freight carriers. Also, commodity suppliers are assumed to play the role of producers, that is, the commodity is assumed to initially originate with commodity suppliers.
10. Activity/commodity supply is assumed to be oligopolistic, that is, there are only a few retailers (suppliers) and wholesalers (producers), specifically only one at any given location.
11. The car and truck route flows for general traffic are assumed given. If route utilities are the same for all travelers, irrespective of purpose, given origin-destination flows for general traffic may be used alternatively, and simply added to the unknown origin-destination demands for the given purpose in the developments below.

In this section, following the same approach as in the development of the previous models, we first define the various utilities facing the network users, which now include the travelers as well as the A.S.'s and the C.S.'s.

8.2.2 Model Variables

We now specify the model's variables, and their symbols, beginning with those related to the given activity. There are I demand zones and J activity sites (i.e., commercial areas, recreation areas, etc.). The demand for activity by residents in zone i during the time period under consideration (e.g., peak, rush, or a given clock hour) is T_i, measured in car trips, and is given. It is assumed that a fixed, unit amount of activity is conducted per trip (e.g., buying one unit of the commodity, eating one meal, etc.).

The corresponding travel demand from the centroid zone i to destination j is T_{ij}, and the resulting demand for activity at location j is T_j. The activity-related car traffic on route r between zones i and j is T_{ijr}. These respective three variables are endogenous. Non-activity-related origin-destination car demands during the same time period are S_{ij}, and are given.

There are K warehouse locations for the commodity. The amount of goods shipped out of warehouse k is G_k. The corresponding flow of goods between given warehouse k and a given facility j is G_{kj}. The flow of goods on truck route r is G_{kjr}. These respective three variables are endogenous.

It is assumed that both passenger cars and freight vehicles use the same network.[3] However, some links may be reserved for cars only. Consequently, car and truck routes from a given origin to a given destination may be different, necessitating the use of two link-path incidence variables. δ_{ijr}^{a} is equal to 1 if link a is part of car route r between i and j, and zero otherwise. ρ_{ijr}^{a} is equal to 1 if link a is part of truck route r between i and j, and zero otherwise. Car volume on link a is v_{a}^{c}, and truck volume is v_{a}^{t}, in car equivalents. General origin-destination truck volumes are L_{ij}, measured in car equivalents, and are given.

8.2.3 Individual Utilities and Aggregate Demands

We begin with the A.S.'s utilities. First, the A.S.'s decisions concern only the location of his or her C.S., and setting the activity's price to travelers. An A.S.'s demand for the commodity is equal to the amount he or she sells to travelers. This is assumed unknown from the viewpoint of the modeler, but known from that of the A.S. Activity price is equal to the wholesale commodity price plus a *given* mark up, whose level is set by the A.S., externally to the model, so as to maximize profit.[4]

Accordingly, we assume that the indirect utility to the A.S. at site j of purchasing one unit of the commodity from the C.S. at site k is equal to

$$\tilde{V}_{kj}^{s} = b_{j}^{s} - c_{kj} - R_{k}^{s} - \tau t_{kj}^{*} + \epsilon_{kj} = b_{j}^{s} - g_{kj}^{*} - R_{k}^{s} + \epsilon_{kj}; \qquad \forall\, j, k \quad (8.1)$$

The term b_{j}^{s} is the A.C.'s budget. c_{kj} is the given network user charge from k to j. R_{k}^{s} is the wholesale price for the commodity at k, and is given. t_{kj}^{*} is the equilibrium travel time from k to j, the minimum of the variable route travel times between these two points.[5] Thus, the systematic part of the indirect utility, net of budget, is the delivered unit cost from j to k. The random term ϵ_{kj} is assumed to have a Gumbel probability distribution, with mean zero and standard deviation σ_1.

Following a now familiar approach, the probability $P_{k/j}$ that the A.S. at j

[3]We assume for simplification that the clock time for the conduct of the activity corresponds to the clock time for commodity deliveries to support the given activity. If not, truck volumes may be prorated by the ratio of the percentage of a longer-term distribution (e.g., daily) of commodity delivery trips during the time period, to the corresponding percentage of the daily (or longer time period, e.g., weekly, or monthly) distribution of activity-related car trips during the same period, depending on data availability and the temporal pattern of transport specific to the given commodity.

[4]The assumption of a *given* profit greatly simplifies the model, as in effect, A.S.'s may be considered consumers of the (wholesale) commodity.

[5]For simplifying the notation, in the remainder of the chapter the asterisk will be deleted from travel times and link volumes, which are understood to have their equilibrium values, as in Chapters 5 and 6.

will choose the C.S. at k is equal to

$$= P(-R_k^s - g_{kj} + \epsilon_{kj} \geq -R_{k'}^s - g_{k'j} + \epsilon_{k'j};\ \forall_j\, k'); \qquad \forall\, j$$

The above assumptions about the random utility terms ϵ_{kj} imply that this probability is equal to

$$P_{k/j} = \frac{e^{-\gamma(g_{kj} + R_k^s)}}{\sum_k e^{-\gamma(g_{kj} + R_k^s)}}; \qquad \forall\, k, j \tag{8.2}$$

The positive value of parameter γ is related to the standard deviation σ_1 of the random term ϵ_{kj} in Formula (8.1):

$$\gamma = \pi(\sigma_1 \sqrt{6})^{-1} \tag{8.3}$$

Consequently, given the aggregate demand T_j for the commodity at activity site j, the expected demand by the A.S. at j from the wholesaler at site k is then equal to

$$G_{kj} = T_j \frac{e^{-\gamma(g_{kj} + R_k^s)}}{\sum_k e^{-\gamma(g_{kj} + R_k^s)}}; \qquad \forall\, k, j \tag{8.4}$$

Finally, the demand G_k for the commodity at wholesale site k to equal to

$$G_k = \sum_j G_{kj} = \sum_j T_j \frac{e^{-\gamma(g_{kj} + R_k^s)}}{\sum_k e^{-\gamma(g_{kj} + R_k^s)}}; \qquad \forall\, k \tag{8.5}$$

We now specify the utility and demand functions for travelers. The utility received by an individual traveler/consumer in demand zone i from making one trip to conduct the given activity at site j is specified, using standard notation, as

$$\tilde{V}_{ij}^c = b_i^c + h_{ij} - R_j^c - \tau t_{ij} - c_{ij} + \epsilon_{ij}; \qquad \forall\, i, j \tag{8.6}$$

c_{ij} is the transportation cost (network user charge) from i to j. The term R_j^c is the commodity's *expected* price to consumers at j. h_{ij} is a site attractiveness term. Note that the utility specification above assumes that the value of travel time is the same for consumers as it is for A.S.'s. (See Exercise 8.1.)

Since from the modeler's viewpoint the individual C.S. supplying a given A.S. is not known, R_j^c is equal to the expected value $\tilde{W}_j^s$ of the delivered price paid by the A.S., which is passed on to the traveler/consumer, plus a mark up, r_j, whose value is set by the A.S. such that it maximizes his or her profit:

$$R_j^c = \tilde{W}_j^s + r_j; \qquad \forall j \tag{8.7}$$

and where

$$\tilde{W}_j^s = E_{\epsilon_{kj}}\{\underset{k}{\text{Min}}\,[R_k^s + g_{kj} + \epsilon_{kj}]\}; \qquad \forall j \tag{8.8}$$

Given the above assumptions about the random utility terms ϵ_{kj}, $\tilde{W}_j^s$ is equal to

$$\tilde{W}_j^s = -\frac{1}{\gamma} \ln \left[\sum_k e^{-\gamma(R_k^s + g_{kj})} \right]; \qquad \forall j \tag{8.9}$$

As noted above, the markup r_j is given, but the expected activity price R_j^c at j is determined endogenously. The random term ϵ_{ij} in Formula (8.6) is assumed to have a Gumbel probability distribution, with mean zero and standard deviation σ_2. Consequently, the probability $P_{j/i}$ that a traveler/consumer from demand zone i will choose activity site j conforms to a logit function, and is equal to

$$P_{j/i} = \frac{e^{\beta(h_{ij} - R_j^c - g_{ij})}}{\sum_j e^{\beta(h_{ij} - R_j^c - g_{ij})}} = \frac{e^{\beta(h_{ij} - \tilde{W}_j^s - r_j - g_{ij})}}{\sum_j e^{\beta(h_{ij} - \tilde{W}_j^s - r_j - g_{ij})}}; \qquad \forall i, j \tag{8.10}$$

The positive value of parameter β is related to the standard deviation σ_2 of the random term ϵ_{ij} in Formula (8.6):

$$\beta = \pi(\sigma_2\sqrt{6})^{-1} \tag{8.11}$$

Consequently, the aggregate demand from demand area i to activity site j is

$$T_{ij} = T_i P_{j/i} = T_i \frac{e^{\beta(h_{ij} - g_{ij} - R_j^c)}}{\sum_j e^{\beta(h_{ij} - g_{ij} - R_j^c)}}; \qquad \forall i, j \tag{8.12}$$

and the total demand at site j is consequently equal to

$$T_j = \sum_i T_{ij} = \sum_i T_i \frac{e^{\beta(h_{ij} - g_{ij} - R_j^c)}}{\sum_j e^{\beta(h_{ij} - g_{ij} - R_j^c)}}; \qquad \forall j \tag{8.13}$$

Replacing this value in the expression of the demand G_k at wholesale site k, Formula (8.5), we have

$$G_k = \sum_i \sum_j T_i \frac{e^{\beta(h_{ij} - g_{ij} - R_j^c)}}{\sum_j e^{\beta(h_{ij} - g_{ij} - R_j^c)}} \frac{e^{-\gamma(g_{kj} - R_k^s)}}{\sum_k e^{\gamma(g_{kj} - R_k^s)}}; \qquad \forall\, k \tag{8.14}$$

8.2.4 The R.S. and R.T. Utility Maximization Problems

As described above, two kinds of actors are maximizing their utility in the present situation, A.S.'s and travelers, respectively. The aggregate received utility by all A.S.'s is, according to Formula (6.11a), equal to

$$\tilde{W}_S = B_s + \frac{1}{\gamma} \sum_j T_j \ln \sum_k e^{-\gamma(g_{kj} + R_k^s)} - \sum_a \int_0^{v_a^*} g_a(v)\, dv + \sum_i \sum_j \sum_r G_{ijr}^* g_{ijr}^* \tag{8.15a}$$

in which B_s is the total budget for all A.S.'s

$$B_s = \sum_j b_j^s \tag{8.15b}$$

In light of the definition of the expected $\tilde{W}_j^s$ in Formula (8.9), this may be restated as

$$\tilde{W}_S = B_S - \sum_j T_j^* \tilde{W}_j^s - \sum_a \int_0^{v_a^*} g_a(v)\, dv + \sum_i \sum_j \sum_r G_{ijr} g_{ijr}^* \tag{8.15c}$$

It is straightforward to verify, using the approach in Appendix 6.5, that, as required:

$$-\frac{\partial \tilde{W}_S}{\partial R_k^s} \Big/ \frac{\partial \tilde{W}_S}{\partial b_s} = \sum_j T_j^* P_{k/j} = G_k^*; \qquad \forall\, k \tag{8.16}$$

In the same fashion, the R.T.'s direct utility is equal to

$$\tilde{W}_C = B_c + \frac{1}{\beta} \sum_i T_i \ln \sum_j e^{\beta(h_{ij} - g_{ij}^* - R_j^c)} - \sum_a \int_0^{v_a^*} g_a(v)\, dv + \sum_i \sum_j \sum_r T_{ijr}^* g_{ijr}^* \tag{8.17}$$

where B_c is the total budget for all activity-performing travelers:

$$B_c = \sum_i b_i^c \tag{8.18}$$

Similarly, one has, as required:

$$-\frac{\partial \tilde{W}_C}{\partial R_j^c} \Big/ \frac{\partial \tilde{W}_C}{\partial B_c} = \sum_i T_i P_{j/i} = T_j^*; \qquad \forall j \tag{8.19}$$

PROPOSITION 8.1

For *given* T_i's and v_j^c's, the direct utility of the R.S. representing all suppliers in their role of consumers of the commodity is equal to

$$U_S = -\tau \sum_a \int_0^{v_a^t + v_a^{c*} + v_a^0} t_a(v)\, dv - \frac{1}{\gamma} \sum_k \sum_j G_{kj} \ln \frac{G_{kj}}{T_j^*} + G_0 \tag{8.20a}$$

where v_a^0 is the given "background" link volume.

Consequently, the aggregate commodity demands, the amount of commodity G_{kj} purchased by the A.S. at site j from wholesalers at sites k, as specified in Formula (8.4), are the solution to the R.S. utility maximization problem:

$$\underset{(G_{kj}, G_{kjr}, G_0)}{\text{Max}} \quad U_S = -\tau \sum_a \int_0^{v_a^t + v_a^{c*} + v_a^0} t_a(v)\, dv - \frac{1}{\gamma} \sum_k \sum_j G_{kj} \ln \frac{G_{kj}}{T_j^*} + G_0 \tag{8.20b}$$

such that commodity volumes correspond to the amounts purchased by travelers

$$\sum_k G_{kj} = \sum_i T_{ij}^*; \qquad \forall j \tag{8.21}$$

subject to the standard balance of flow constraints

$$\sum_r G_{kjr} = G_{kj}; \qquad \forall k, j \tag{8.22}$$

and subject to the suppliers' budgetary constraint

$$\sum_j \sum_k G_{kj}(R_k^s + c_{kj}) + G_0 = B_s \tag{8.23}$$

as well as the definitional relationship between link volumes, in car equivalents, and commodity volumes

$$v_a^t = \sum_k \sum_j \sum_r (\phi G_{kjr} + L_{kjr}) \rho_{ijr}^a; \qquad \forall\, a \tag{8.24}$$

where ϕ is the conversion factor. In this problem, the *given* aggregate consumer demands T_j and car link flows v_a^c are the solution of the R.T.'s direct utility maximization problem. Consequently, Problem (8.20b)–(8.23) is constrained by the representative traveler/consumer (R.T.)'s utility maximization problem, whose solution produces the T_j's and v_a^c's.

PROPOSITION 8.2

The R.T.'s utility is equal to

$$U_C = \sum_i \sum_j T_{ij} h_{ij} - \tau \sum_a \int_0^{v_a^t + v_a^c + v_a^0} t_a(v)\, dv - \frac{1}{\beta} \sum_i \sum_j T_{ij} \ln T_{ij} + T_0 \tag{8.25a}$$

Consequently, the aggregate demands T_{ij} are produced by the solution of the R.T.'s problem

$$\max_{(T_{ij}, T_{ijr}, T_0)} \left[\sum_i \sum_j T_{ij} h_{ij} - \tau \sum_a \int_0^{v_a^t + v_a^c + v_a^0} t_a(v)\, dv - \frac{1}{\beta} \sum_i \sum_j T_{ij} \ln T_{ij} + T_0 \right] \tag{8.25b}$$

subject to the balance of flow constraints

$$\sum_j T_{ij} = T_i; \qquad \forall\, i \tag{8.26a}$$

$$\sum_r T_{ijr} = T_{ij}; \qquad \forall\, i, j \tag{8.26b}$$

and to the budgetary constraint

$$\sum_i \sum_j T_{ij}(R_j^c + c_{ij}) + T_0 = \sum_j T_j r_j + \sum_{ij} T_{ij} c_{ij} + \sum_j T_j \tilde{W}_{ij}^s + T_0 = B_c \tag{8.26c}$$

as well as the definitional Relationships

$$v_a^c = \sum_i \sum_j \sum_r (T_{ijr} + H_{ijr}) \delta_{ijr}^a; \qquad \forall\, a \tag{8.26d}$$

$$\sum_i T_{ij} = T_j; \qquad \forall\, j \tag{8.26e}$$

Finally, all variables must be nonnegative, or positive, as indicated.

$$T_{ij} > 0,\ T_{ijr} \geq 0,\ v_a^c \geq 0,\ G_{kj} > 0,\ G_{kjr} \geq 0;$$
$$H_{ijr} \geq 0,\ L_{kjr} \geq 0,\ v_a^l \geq 0; \qquad \forall\, i, j, a_l, a_c, r \tag{8.27}$$

Parameters β, γ, ω and τ have positive values.

$$\omega > 0; \quad \beta > 0; \quad \gamma > 0; \quad \tau > 0 \tag{8.28}$$

If we assume that A.S.'s react to traveler/consumer decisions, the joint equilibrium leading to the determination of suppliers' and consumers' demands may be identified by solving a "bilevel" problem, in which the top problem is the R.T.'s utility maximization, Problem (8.25b)–(8.26), and the bottom problem is the R.S.'s utility maximization, Problem (8.20b)–(8.23).

However, this bilevel problem may be transformed into a joint, single-level problem in the following manner. Constraint (8.26c) may be rewritten as

$$T_0 = B_c - \sum_j T_j r_j - \sum_{ij} T_{ij} c_{ij} - \sum_j T_j \tilde{W}_j^s$$

and the term T_0 replaced by its expression above in Formula (8.26c), in the expression of the Objective Function (8.25b), which then becomes

$$\begin{aligned} \underset{(T_{ij}, T_{ijr}, T_0)}{\text{Max}} \quad U_C = {} & \sum_i \sum_j T_{ij} h_{ij} - \tau \sum_a \int_0^{v_a^l + v_a^c + v_a^0} t_a(v)\, dv \\ & - \frac{1}{\beta} \sum_i \sum_j T_{ij} \ln T_{ij} + B_c - \sum_j T_j r_j - \sum_{ij} T_{ij} c_{ij} - \sum_j T_j \tilde{W}_j^s \end{aligned} \tag{8.29}$$

subject to Constraints (8.26a, b, d, e).

Because this optimization problem is constrained by Problem (8.20b) to (8.23), the value of the last term in Objective Function (8.29) is equal to its value as given by Constraint (8.15c) up to terms constant for the RT:

$$-\sum_j T_j \tilde{W}_j = \tilde{W}_s$$

where $\tilde{W}_S$ is the suppliers' total received utility, that is, the maximum value of the R.S.' direct utility, the solution to Problem

$$\operatorname*{Max}_{(G_{kj}, G_{kjr}, G_0)} \quad U_S = -\tau \sum_a \int_0^{v_a^{*t} + v_a^c + v_a^0} t_a(v)\, dv$$
$$- \frac{1}{\gamma} \sum_k \sum_j G_{kj} \ln \frac{G_{kj}}{T_j} + G_0 \tag{8.30}$$

subject to Constraints (8.21)–(8.23). Replacing the expression of U_S^* in Formula (8.30) in the Objective Function (8.29), the R.T.'s U.M. problem may then be rewritten as

$$\operatorname*{Max}_{(T_{ij}, T_{ijr}, T_0)} \quad U_{SC} = \Bigg\{ \sum_i \sum_j T_{ij}(h_{ij} - c_{ij}) - \tau \sum_a \int_0^{v_a^t + v_a^c + v_a^0} t_a(v)\, dv$$
$$- \frac{1}{\beta} \sum_i \sum_j T_{ij} \ln T_{ij} + B_c - \sum_j T_j r_j \tag{8.31}$$
$$+ \operatorname*{Max}_{(G_{kj}, G_{kjr}, G_0)} \Bigg(- \sum_a \int_0^{v_a^{*t} + v_a^c + v_a^0} t_a(v)\, dv$$
$$- \frac{1}{\gamma} \sum_k \sum_j G_{kj} \ln \frac{G_{kj}}{T_j} + G_0 \Bigg) \Bigg\}$$

subject to its own constraints, as well as to the constraints on the R.S. problem. This discussion may finally be summarized as

PROPOSITION 8.3

The joint utility of the R.T. and A.S. is equal to

$$U_{SC} = \sum_i \sum_j T_{ij}(h_{ij} - c_{ij}) - \tau \sum_a \int_0^{v_a} t_a(v)\, dv$$
$$- \frac{1}{\beta} \sum_i \sum_j T_{ij} \ln T_{ij} - \sum_j T_j r_j$$
$$- \frac{1}{\gamma} \sum_k \sum_j G_{kj} \ln \frac{G_{kj}}{T_j} + G_0 \tag{8.32}$$

It is shown in Appendices 8.1 and 8.2 at the end of this chapter that maximization of this function subject to the RS budget and all consistency constraints above always has a unique solution, the aggregate demands as specified in Formulas (8.4) and (8.12).

8.2.5 Spatial Activity Price/Travel Cost Equilibrium

In summary of the above analysis, the solution to Problem (8.32) and (8.21)–(8.24) produces the aggregate, equilibrium demands for the activity and the associated commodity, as well as for personal travel and freight transport, consistent with the principles of utility maximization on the part of all actors in the general activity/transport system. Specifically, activity travelers/travelers maximize the utility of their joint choice of an activity site and travel route to it. Activity suppliers maximize the utility of their joint choice of C.S. and freight shipping routes. In both cases, activity utilities are random, and transport utilities are deterministic, from the modeler's standpoint. All transport costs and destination costs are determined endogenously. Activity and commodity supply meets demand at activity sites. The transportation network, in which a given link may or may not be shared by private cars and trucks, is in user equilibrium for all its users.

When both traveler/consumer and A.C.'s utilities are random, the activity and travel systems are in joint stochastic equilibrium. More precisely, the expected activity price R_j^c which equilibrates supply and demand for the activity is related to the expected commodity's cost

$$R_j^c = \tilde{W}_j^s + r_j; \qquad \forall\, j \tag{8.33}$$

In reference to Conditions (8.56d) in Appendix 8.2 in this chapter, when $\beta \neq \infty$ and $\gamma = \infty$, that is, when the A.S.'s utilities are random, the optimality conditions for the G_{kj}'s are now

$$G_{kj}(R_k^s + g_{kj} - \varphi_j) = 0; \qquad \forall\, k, j \tag{8.34a}$$

$$R_k^s + g_{kj} \geq \varphi_j; \qquad \forall\, k, j \tag{8.34b}$$

Equations (8.34b) show that in this case φ_j is the minimum delivered price to site j from any wholesaler k. In turn, Equations (8.34a) show that if wholesaler k supplies activity site j, that is, G_{kj} is different from zero, then the generalized delivered price from wholesaler k to site j is equal to the minimum delivered price for the commodity at facility j. Conversely, if the delivered price of the commodity from warehouse k to facility j is greater than the minimum delivered price at j, wholesaler k does not supply retailer j, and $G_{kj} = 0$.

This constitutes the classical statement of ''spatial price equilibrium'' (Takayama and Judge, 1971). It is worth emphasizing that wholesale commodity demands are in this case no longer stochastic, represented by the Probabilities (8.2) that a given A.S. will be supplied from a given wholesaler. Instead, the probability that A.S. j will acquire the commodity from that wholesaler offering the minimum delivered price is 100%. Consequently, there is no possibility that he might acquire it from any other wholesaler.

Thus, in the general case when γ is not equal to infinity, the two activity systems, retail and wholesale, may be said to be stochastically coupled. When γ is equal to infinity, there is a one-to-one, deterministic coupling between them. In this case, the expression for travel demands (8.12) becomes

$$T_{ij} = T_i \frac{e^{\beta(h_{ij} - W_j^s - r_j - \overset{*}{g}_{ij})}}{\sum_j e^{\beta(h_{ij} - W_j^s - r_j - \overset{*}{g}_{ij})}}; \qquad \forall\, i, j \tag{8.35}$$

where W_j is the minimum delivered price of the commodity at site j;

$$W_j^s = \operatorname*{Min}_k \,(R_k^s + g_{kj}); \qquad \forall\, j \tag{8.36}$$

8.3 MODEL APPLICATION

In this section, issues pertaining to the implementation of the model are examined, beginning with an algorithm for obtaining solutions to Problem (8.32), (8.21), (8.23), (8.26a, b, d to f).

8.3.1 Solution Algorithm

First, the problem may be solved with an algorithm based on the same partial linearization method (Evans, 1976) which we have systematically used for the other combined models. At iteration l, the auxiliary problem to solve for obtaining a descent direction for the main problem is, in minimization form,

$$\begin{aligned}
\text{Min } U_{SC}^l = {} & \tau \left[\sum_{a_c} y_{a_c}^l t_a(y_a^{l-1}) + \sum_{a_t} y_{a_t}^l t_a(y_a^{l-1}) \right] \\
& + \frac{1}{\beta} \sum_{ij} Z_{ij}^l \ln Z_{ij}^l + \sum_i \sum_j Z_{ij}^l (r_j + c_{ij} - h_{ij}) \\
& + \frac{1}{\gamma} \sum_k \sum_j Q_{kj}^l \ln \frac{Q_{kj}^l}{Z_{ij}^l} + \sum_k Q_{kj}^l \sum_j (R_k^s + c_{kj})
\end{aligned} \tag{8.37}$$

subject to Constraints (8.21), (8.23), (8.26a, b, d to f) in the auxiliary variables Z and Q. It may also be shown that this auxiliary problem is a convex problem. Therefore, it may be solved directly from its first-order conditions

$$Z_{ij}^l = \frac{T_i e^{\beta(h_{ij} - r_j - g_{ij}^{l-1} - \tilde{W}_j^{s,l-1})}}{\sum_j e^{\beta(h_{ij} - r_j - g_{ij}^{l-1} - \tilde{W}_j^{s,l-1})}}; \qquad \forall\, i, j, l \tag{8.38}$$

where

$$\tilde{W}_j^{sl} = -\frac{1}{\gamma} \ln \left[\sum_k e^{-\gamma(g_{kj}^l + R_k^s)} \right]; \qquad \forall\, j, l \tag{8.39}$$

$$Q_{kj}^l = \frac{Z_j^l e^{-\gamma(g_{kj}^{l-1} + R_k^s)}}{\sum_k e^{-\gamma(g_{kj}^{l-1} + R_k^s)}}; \qquad \forall\, k, j, l \tag{8.40}$$

and

$$\begin{aligned} g_{ij}^l &= \sum_a g_a(y_a^l)\delta_{ijr}^a \quad \text{if } Z_{ijr}^l \geq 0; \\ & \qquad\qquad\qquad\qquad\qquad \forall\, i, j, l \\ g_{ij}^l &\geq \sum_a g_a(y_a^l)\delta_{ijr}^a \quad \text{if } Z_{ijr}^l = 0 \end{aligned} \tag{8.41}$$

where Z_{ijr}^l is the auxiliary flow on car path r between i and j at iteration l, and

$$\begin{aligned} g_{kj}^l &= \sum_a g_a(y_a^l)\rho_{ijr}^a \quad \text{if } Q_{kjr}^l \geq 0 \\ & \qquad\qquad\qquad\qquad\qquad \forall\, k, j, l \\ g_{kj}^l &\geq \sum_a g_a(y_a^l)\rho_{ijr}^a \quad \text{if } Q_{kjr}^l = 0; \end{aligned} \tag{8.42}$$

where Q_{kjr}^l is the auxiliary flow on truck path r between k and j at iteration l.

From Relationships (8.41/42), it is apparent that both car and truck flows between two locations i and j at iteration l are assigned to the minimum generalized cost car and truck routes respectively between these points. Thus, solving subproblem (8.37) subject to its constraints requires distributing the given T_i's from a given origin i to destinations j according to Formula (8.38), in which the travel and activity costs are set at their current values, distributing the resulting G_j's from a given j to supply points k, according to Formula (8.40), in which the travel and activity costs are also set at their current values, and assigning the resulting origin-destination flows to the corresponding minimum cost paths.

The optimal size of the move along the descent direction is obtained in the same manner as in the Frank-Wolfe algorithm by minimizing, with respect to θ, the function

$$U'_{SC}(v_{a_m}^l + \theta(y_{a_m}^l - v_{a_m}^l),\ T_{ij}^l + \theta(Z_{ij}^l - T_{ij}^l),\ G_{kj}^l + \theta(Q_{kj}^l - G_{kj}^l)) \tag{8.43}$$

Once the optimal value of θ is found, the new, improved solution is

$$v_{a_m}^{l+1} = v_{a_m}^l + \theta(y_{a_m}^l - v_{a_m}^l); \qquad \forall\, a_m \tag{8.44a}$$

$$T_{ij}^{l+1} = T_{ij}^l + \theta(Z_{ij}^l - T_{ij}^l); \qquad \forall\, i, j \tag{8.44b}$$

$$G_{kj}^{l+1} = G_{kj}^l + \theta(Q_{kj}^l - G_{kj}^l); \qquad \forall\, k, j \tag{8.44c}$$

The iterative step of the solution algorithm for Problem (8.32), (8.21)–(8.23), (8.26a, b, d to f) is then as follows.

ALGORITHM 8.1 PARTIAL LINEARIZATION FOR JOINT ACTIVITY-TRAVEL DEMANDS

Given a feasible solution (v_{a_m}, T_{ij}, G_{kj}), the corresponding link travel costs and activity costs, and the given T_i's:

A. Determine the minimum cost paths between origins i and destinations j, and the corresponding g_{ij}'s and g_{kj}'s.

B. Estimate the auxiliary Z_{ij}'s, and Q_{kj}'s, using Formulas (8.38) and (8.40).

C. Assign them respectively to the minimum cost paths from i to j.

D. Determine the corresponding auxiliary link volumes y_a^c and y_a^t for cars and trucks, respectively.

E. Solve for θ which minimizes $U'_{sc}(\theta)$ as defined in Formula (8.43).

F. Update the solution as defined by Relationships (8.44).

G. Check convergence, whether

$$|(\boldsymbol{y}^k - \boldsymbol{v}^k) \cdot \nabla_v U'_{SC}(\boldsymbol{T}^k, \boldsymbol{v}^k) + (\boldsymbol{Z}^k - \boldsymbol{T}^k) \cdot \nabla_T U'_{SC}(\boldsymbol{T}^k, \boldsymbol{v}^k)| \leq \epsilon \tag{8.45}$$

where ϵ is a given tolerance. If so stop; if not iterate.

We now turn to data requirements for practical application of the model. This necessitates the input of a number of pieces of data, in addition to the standard network description (i.e., link capacities, "free flow" travel times, car-truck equivalency, etc.), including:

1. Activity-related trip origins in each of the demand zones. These data may be obtained from suppliers' records, for instance, in the case of shopping, credit card, or point-of-sale bills.
2. Origin-destination car and truck flows for general travel. This "all other purposes" category might be approximated by travel to work, and may be obtained from surveys conducted by local transportation agencies, U.S. Census, or the output of external travel forecasts.
3. Fixed and variable A.S. and C.S. costs, by facility. This information may be supplied by facility operators.
4. The hourly cost of trucking in the area.

In addition, the values of all model parameters must be input to the model. These values may have to be obtained through a prior calibration of the model.

However, the calibration of the model requires the availability of observed values for the model's output variables. This includes activity-related car volumes between demand areas (e.g., Census tracks, Zip Code areas), and facilities, and activity-related truck volumes between warehouses and facilities. These data may be obtained from suppliers and C.S.'s records.

8.3.2 Illustrative Example

To illustrate the application of the model, we again utilize the same spatial structure and network as we have consistently since the beginning. For simplification, but without loss of generality, we assume that 200 consumers of a given product are only located in zone 3, that there are only two retail suppliers respectively located in zones 2 and 4, only one wholesale supplier located in area 1, that there is no general traffic, and that consumer and truck traffic takes place simultaneously. We assume that trucks transporting the commodity are only allowed to use the network in Figure 8.1, and that ϕ in Formula (8.24) is equal to 10, that is, the car equivalent of one truck transports 10 units of the commodity.

The travel costs are equal to

$$c_{32} = 1$$

$$c_{34} = 3$$

The unit wholesale cost of the commodity is equal to

$$R_1^2 = 1$$

The unit costs of transporting the commodity are

$$c_{12} = 3$$

$$c_{14} = 5$$

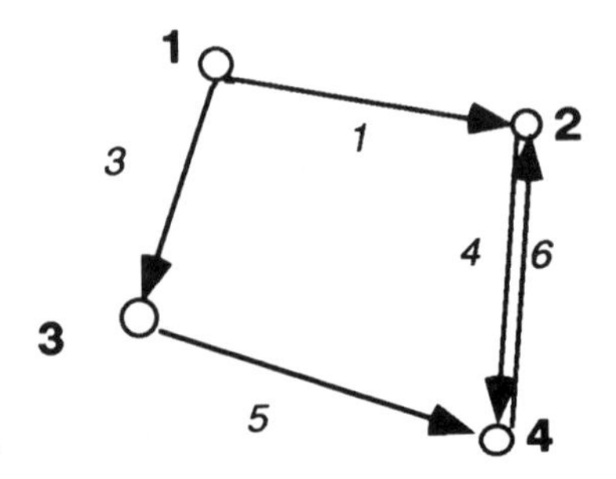

Figure 8.1 Truck network.

TABLE 8.1 Origin-Destination Consumer and Truck (Car Equivalents) Flows

$T_{32} = 87$; $T_{34} = 113$
$G_{12} = 9$; $G_{14} = 11$

The unit retail markups are equal to

$$r_2 = 2.5$$

$$r_4 = 1.5$$

All parameters h_{ij} are assumed equal to zero.

The value of travel time is assumed to be the same for individuals and shippers. Finally, the parameter values are equal to

$$\beta = 0.1$$

$$\tau = 1$$

$$\gamma = 0.08$$

Under these conditions, the equilibrium consumer and freight flows, in numbers of cars and trucks respectively, are represented in Table 8.1 and Figure 8.2.

The equilibrium modal link travel times are in Table 8.2.

The equilibrium route travel times are in Table 8.3.

In the present case, because there is only one wholesaler supplying the two retailers, the expected delivered prices to the respective retail sites are simply

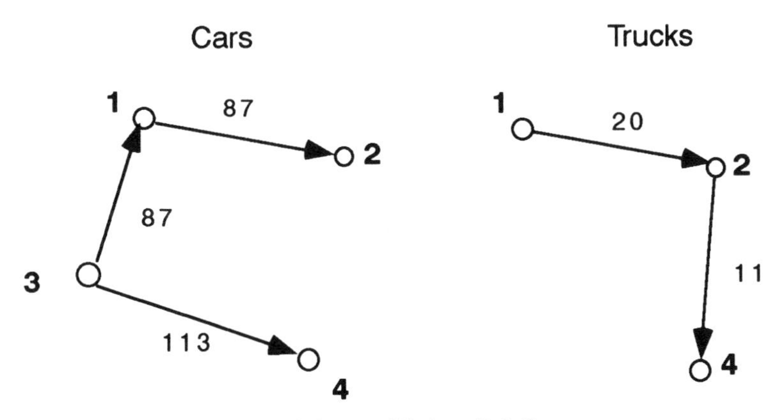

Figure 8.2 Equilibrium link flows.

TABLE 8.2 Link Travel Times

$$t_1 = 16,$$
$$t_2 = 15; \quad t_3 = 6.0; \quad t_4 = 8; \quad t_5 = 25.4; \quad t_{10} = 11.8$$

TABLE 8.3 Equilibrium Route Travel Times

$$t_{32} = 27.7$$
$$t_{34} = 25.3$$
$$t_{12} = 16.0$$
$$t_{14} = 24.0$$

equal to the sum of the wholesale price plus the equilibrium generalized transport cost;

$$\tilde{W}_2 = 1 + 1(16) + 3 + 2 = 22$$

$$\tilde{W}_4 = 1 + (24) + 5 + 0 = 29$$

This solution, in fact, corresponds to a deterministic spatial price equilibrium.

8.4 GENERAL VERTICAL (MULTISUPPLIER) DEMAND STRUCTURES

The activity allocation and travel demand model described above may be extended along several directions. First, as already mentioned, a full description of C.S. behavior should include their interactions with commodity producers, and in turn, the producers' interaction with producers of other goods required in the commodity's production, and so forth. The same approach as was used here could be ''cascaded'' at these subsequent levels, within an ''input-output'' framework. In addition, workers at the respective facilities will in the long term generate demand for residential activity, as well as induce additional personal travel on the network, and ''second-level'' demand for the activity itself. Thus the model's formulation fits naturally into a spatial economic base, or ''Lowry-type'' framework (Oppenheim, 1980). More generally, the model may be extended from its present version for a single, generic activity, to the case of multiple urban activities, leading to a general model of a comprehensive activity and travel system.

In this section, we generalize the two-tiered activity/commodity demand system described earlier to the case of general multiple-commodity demands, in which other, lower levels of supply of the *same commodity* are also present. This may be called a ''vertical'' hierarchical demand structure. Specifically, wholesalers generally obtain the commodity from producers, or manufacturers.

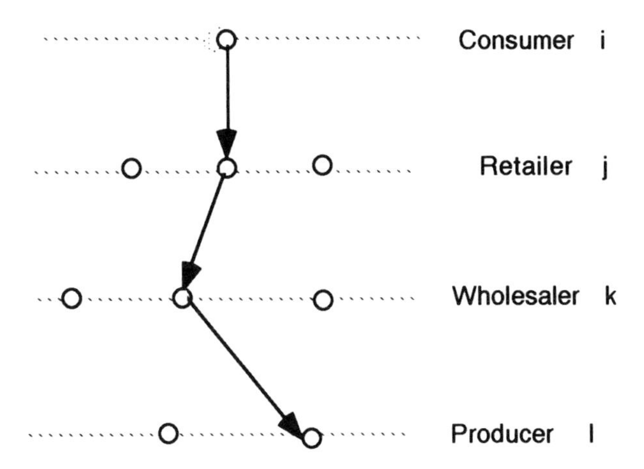

Figure 8.3 Vertical hierarchical demand structure.

Producers may then be placed at a tertiary level, and so on. This type of demand structure may be graphically represented in general as in Figure 8.3.

We assume in this case, without loss of generality, that origin-destination freight costs are given, independent of network travel times as above, and are equal to f_{kj}. Thus, truck link volumes are not needed, nor are their relationship to the origin-destination commodity demands. Accordingly, the aggregate demands for the multiple commodities may be estimated, in the presence of demand externalities, as the solution to the problem

$$\begin{aligned}\max_{(T_{ij}, G^s_{kj}, G^p_{lk}, v_a)} \quad U_{SC} = & \sum_i \sum_j T_{ij}(h_{ij} - C_{ij} - r^s_j) - \sum_a \int_0^{v_a} g_a(v)\, dv \\ & - \frac{1}{\beta} \sum_i \sum_j T_{ij} \ln T_{ij} \\ & - \frac{1}{\gamma} \sum_k \sum_j G^s_{kj} \ln \frac{G^s_{kj}}{T_j} - \frac{1}{\nu} \sum_l \sum_k G^p_{lk} \ln \frac{G^p_{lk}}{G^s_k} \\ & - \sum_k \sum_j (r^w_k + f_{kj}) G^s_{kj} - \sum_l \sum_k (R^p_l + f_{lk}) G^p_{lk} \end{aligned} \tag{8.46}$$

subject to

$$\sum_j T_{ij} = T_i; \qquad \forall\, i \tag{8.47a}$$

$$\sum_k G^s_{kj} = \sum_i T_{ij}; \qquad \forall\, j \tag{8.47b}$$

$$\sum_l G^p_{lk} = \sum_j G^s_{kj}; \qquad \forall\, k \tag{8.47c}$$

$$\sum_r T_{ijr} = T_{ij}; \qquad \forall\, i, j \tag{8.47d}$$

$$v_a = \sum_i \sum_j \sum_r (T_{ijr} + S_{ijr} + L_{ijr} + \phi(G^p_{ijr} + G^s_{ijr}))\delta^a_{ijr}; \qquad \forall\, a \tag{8.47e}$$

G^p_{lk} is the commodity flow from producer l to wholesaler k. G^s_{kj}, as before, is the commodity flow from wholesaler k to retailer j. ν is a parameter which specifies the magnitude of the random error in producer utilities, and is the counterpart to parameter γ. R^p_l is the unit producer price at l. r^s_j is the unit mark up on retail price at j. r^w_j is the unit markup on wholesale price at k. It is shown in Appendix 8.3 that the solution to this problem is

$$T_{ij} = T_i \frac{e^{\beta(h_{ij} - \overset{*}{g}_{ij} - \tilde{W}^s_j - r^s_j)}}{\sum_j e^{\beta(h_{ij} - \overset{*}{g}_{ij} - \tilde{W}^s_j - r^s_j)}}; \qquad \forall\, i, j \tag{8.48}$$

where now

$$\tilde{W}^s_j = -\frac{1}{\gamma} \ln \sum_k \exp\left(-\gamma\left(r^w_k + \frac{1}{\nu} \ln \sum_l e^{-\nu(R^p_l + f_{lk})}\right)\right); \qquad \forall\, j \tag{8.49}$$

and

$$G^s_{kj} = T_j \frac{e^{-\gamma(\tilde{W}^p_k + r^w_k)}}{\sum_k e^{-\gamma(\tilde{W}^p_k + r^w_k)}}; \qquad \forall\, k, j \tag{8.50}$$

where

$$\tilde{W}^p_k = -\frac{1}{\nu} \ln \left[\sum_l e^{-\nu(R^p_l + f_{lk})}\right]; \qquad \forall\, k \tag{8.51}$$

and

$$G^p_{lk} = G^s_k \frac{e^{-\nu(R^p_l + f_{lk})}}{\sum_l e^{-\nu(R^p_l + f_{lk})}}; \qquad \forall\, l, k \tag{8.52}$$

8.5 GENERAL HORIZONTAL (MULTICOMMODITY) DEMAND STRUCTURES

Another possible type of multiple commodity demand structure is when the A.S. requires inputs from several other suppliers of *different* commodities. For instance, in the case of restaurant operators, when the commodity is meals,

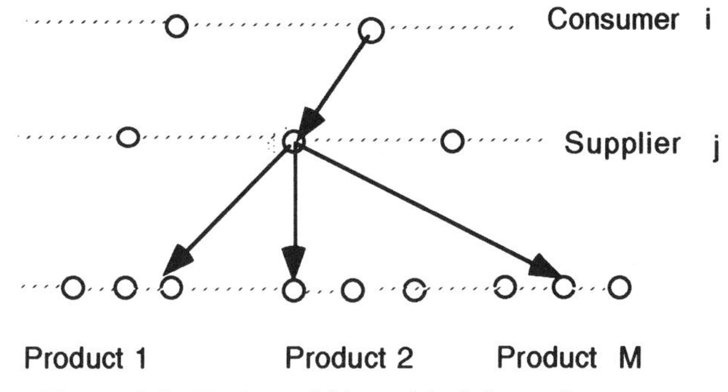

Figure 8.4 Horizontal hierarchical demand structure.

several types of wholesale foods will be required, such as, bread, wine, and fruit. This would constitute an example of a "horizontal" enlargement of the demand structure. Such a demand structure may be graphically represented in general as in Figure 8.4. The aggregate demands for the multiple commodities may be estimated, again in the presence of demand externalities, as the solution to the following Problem:

$$\begin{aligned}\max_{(T_{ij}, G^n_{kj}, v_a)} \; U_{RC} = & \sum_i \sum_j T_{ij}(h_{ij} - c_{ij} - r_j) - \sum_a \int_0^{v_a} g_a(v)\, dv \\ & - \sum_n \frac{1}{\gamma_n} \sum_k \sum_j G^n_{kj} \ln \frac{G^n_{kj}}{a_n T_j} - \frac{1}{\beta} \sum_i \sum_j T_{ij} \ln T_{ij} \\ & - \sum_n \sum_k \sum_j (R^n_j + f_{kj}) G^n_{kj} \end{aligned} \tag{8.53}$$

subject to

$$\sum_j T_{ij} = T_i; \qquad \forall\, i \tag{8.54a}$$

$$\sum_k G^n_{kj} = a_n T_j; \qquad \forall\, j, n \tag{8.54b}$$

$$\sum_r T_{ijr} = T_{ij}; \qquad \forall\, i, j \tag{8.54c}$$

$$v_a = \sum_i \sum_j \sum_r \left(T_{ijr} + S_{ijr} + L_{ijr} + \phi \sum_n G^n_{ijr} \right) \delta^a_{ijr}; \qquad \forall\, a \tag{8.54d}$$

in which the index n refers to the commodity type. Accordingly, variables G^n_{kj} represent the demand from the supplier at j at warehouse k for commodity type n, and similarly, the other symbols retain the same meaning they had previously. The given coefficients a_n represent the respective amounts of commodities n required for one unit of primary commodity.

It is shown in Appendix 8.4 that the demands for the respective products, as provided by the solution to this problem, are

$$G_{kj}^n = a_n T_j \frac{e^{-\gamma_n(f_{kj} + R_k^n)}}{\sum_k e^{-\gamma_n(f_{kj} + R_k^n)}}; \qquad \forall\, n, k, j \tag{8.55a}$$

and

$$T_{ij} = T_i \frac{e^{\beta(h_{ij} - g_{ij} - r_j - \Sigma_n a_n \tilde{W}_j^n)}}{\sum_j e^{\beta(h_{ij} - g_{ij} - r_j - \Sigma_n a_n \tilde{W}_j^n)}}; \qquad \forall\, i, j \tag{8.55b}$$

where the t_{ij}^*'s are the endogenously determined, equilibrium values of travel times, reflecting network congestion, and

$$\tilde{W}_j^n = -\frac{1}{\gamma_n} \ln \left[\sum_k e^{-\gamma_n(R_k^n + f_{kj})} \right]; \qquad \forall\, n, j \tag{8.55c}$$

In conclusion of this section, it may be pointed out that the above respective models may be combined, to interrelate the multiple horizontal, as well as vertical, relationships between the various commodities and services intervening in a comprehensive spatial activity system. In particular, a secondary "commodity" which is always required is employees. The resulting formulation would then lead to equilibrium models of general spatial activity systems, including commercial, industrial, and the resultant employment and transportation activity.

8.6 TOWARD GENERAL ACTIVITY AND TRAVEL SYSTEMS MODELING

The models developed in the previous sections treat personal travel and freight flows as the respective components of a general economic, or activity system, interacting with a congestible transportation network. Activity and travel distributions are assumed to result from utility maximization. Other models have been developed which use the same paradigm. For instance, a model which combines the assumptions of minimum (deterministic) aggregate cost and maximum entropy, within a spatial economic base model for commodity production and consumption served by a congestible network, has been developed by Kim (1983). Another general activity and travel spatial equilibrium including commodity consumption, employment, shopping and transportation, also on a congestible network, is due to Anas (1984). This modeling approach, which may be termed joint equilibrium, represents the descendent of Lowry's (1964) activity allocation modeling.

The model developed in the previous sections may be refined in several directions to extend its range of application. First, it was assumed that there is only one mode available for both personal and freight transportation. Urban goods deliveries are overwhelmingly conducted by truck. Also, urban residents will primarily use the car to conduct many activities, particularly shopping. Nevertheless, alternative personal travel modes (e.g., transit) may be introduced into the model in a relatively standard manner, as second-level alternatives given the choice of an activity location.

Furthermore, shipping costs were assumed to be either proportional to the truck's travel time or given. These assumptions obviously simplify reality, for analytical convenience. The case when shipping costs are represented by a general function of freight volumes may also be modeled, through an appropriate reformulation of Objective Function (8.10). The solution algorithm, however, becomes significantly more burdensome, as it now requires diagonalization. (See, for instance, Florian and Spiess, 1982).

In addition, the demand for activity originating in demand areas was assumed constant. While this assumption may be appropriate for some urban activities, including shopping for nondiscretionary items such as food, price-sensitive demands would be more realistic for others. Also, as mentioned earlier, the assumption that suppliers maximize utilities for given profits r_j may be replaced by the more realistic assumption that they maximize profits, where r_j would be endogenously determined. This, however, leads to a multilevel optimization problem, in which each supplier's profit objective is maximized simultaneously, subject to the reaction of the consumers/travelers, which is represented by Problem (8.25b)–(8.26) (Oppenheim, 1995). Such problems are difficult to solve, as we'll see in the next chapter.

8.7 SUMMARY

In this chapter, a combined equilibrium model of urban personal travel and goods movements was developed, in which commodity flows are generated by the need to support a given generic urban activity undertaken by individual travelers, which involves consumption of a given commodity. A fundamental feature of the model is explicit, full representation of the interacting behaviors of travelers and commodity suppliers/shippers within the framework of a spatial competitive economy. Concurrently, passenger and freight flows take place on a common congestible network, which is also used for general travel.

Travelers are assumed to maximize their utilities, through their joint choice of an activity site and travel route to it. Activity suppliers also maximize their utilities, through their joint choice of C.S. and freight shipping routes. In both cases, site-related utilities are random, and travel-related utilities are deterministic. Commodity supply meets demand at each activity site at which it is traded, primary or secondary. The transportation network, in which a given link may or may not be shared by private cars and freight trucks, is in user equilibrium for all users. As a special case, a spatial price equilibrium for the commodity is obtained. It was shown that, under general conditions, the model always possesses a unique solution. An algorithm for obtaining that solution is described.

The model was then extended to the case of multiple trading levels, and of multiple commodities. In conclusion, several areas for further extensions toward comprehensive models of urban activity and travel systems were discussed.

8.8 EXERCISES

8.1 Generalize the formulation of R.S.'s utility to allow for a different value of travel time, say ω for shippers, than τ for consumers.

8.2* Generalize the formulation of Problem (8.46), (8.47) to include congestion externalities.

8.3* Generalize the formulation of Problem (8.53), (8.54) to include congestion externalities.

8.4* Derive the expression of the combined received utility for the R.T.'s Problem (8.32) and the corresponding traveler/consumer/consumer surplus.

8.5 Derive Formula (8.58)

8.6 Derive Formula (8.59)

APPENDIX 8.1 EXISTENCE AND UNIQUENESS OF MODEL SOLUTION

Referring to Problem (8.32), (8.21)–(8.23), (8.26a, b, d to f), it is clear that the feasible region R in the multidimensional space (F, H, T, G), where F is the vector of path flows for activity-related trips, H is the vector of path flows for fixed trips, T is the vector of origin-destination flows, and G is the vector of origin-destination freight flows, is convex, closed, and bounded.

A feasible solution may be constructed as follows by distributing given flows evenly:

$$T_{ij} = T_i/J; \qquad \forall\, i, j$$

$$T_{ijr} = T_{ij}/R_{ij}; \qquad \forall\, i, j, r$$

$$G_{kj} = T_j/K; \qquad \forall\, j, k$$

$$G_{kjr} = G_{kj}/R_{ij}; \qquad \forall\, k, j, r$$

$$H_{ijq} = W_{ij}/Q_{ij}; \qquad \forall\, i, j, q$$

$$L_{kjs} = L_{kj}/S_{ij}; \qquad \forall\, k, j, s$$

Since, as shown above, the feasible region R for the problem is compact, convex, and nonempty, if the direct utility U_{rc}, Function (8.32) is everywhere strictly concave, then the solution will be unique (see section 2 of Appendix A). Without loss of generality, U_{rc} may be written as:

$$U_{rc}(v_a, T_{ij}, G_{kj}) = F_1(v_a) + F_2(T_{ij}) + F_3(G_{kj})$$

Referring to Appendix 2 in chapter 6, it was shown that $F_1(v_a) + F_2(T_{ij})$ is strictly concave. The expression of $F_3(G_{kj})$ is similar to that of $F_2(Y_{ij})$, so that by the same type of arguments which were used to establish the strict concavity of $F_2(Y_{ij})$, $F_3(G_{kj})$ may be shown to be strictly concave. $U_{rc}(v_a, T_{ij}, G_{kj})$ is then strictly concave, guaranteeing the uniqueness of the model's solution.

APPENDIX 8.2 RETRIEVING EQUILIBRIUM JOINT AGGREGATE ACTIVITY AND COMMODITY DEMANDS

Following the standard approach, we examine the implications of the Karush-Kuhn-Tucker (KKT) conditions at the solution. Since the constraints are either linear equalities or nonnegativity restrictions on the variables, the KKT conditions with respect to flow variables T_{ijr}, G_{kjr}, T_{ij}, G_{kj}, T_0, G_0 are, with respect to the Lagrangean L:[6]

$$T_{ijr}\frac{\partial L}{\partial T_{ijr}} = 0; \frac{\partial L}{\partial T_{ijr}} \geq 0; \qquad \forall\, i, j, r \tag{8.56a}$$

$$T_{ij}\frac{\partial L}{\partial T_{ij}} = 0; \frac{\partial L}{\partial T_{ij}} \geq 0; \qquad \forall\, i, j \tag{8.56b}$$

$$G_{kjr}\frac{\partial L}{\partial G_{kjr}} = 0; \frac{\partial L}{\partial G_{kjr}} \geq 0; \qquad \forall\, k, j, r \tag{8.56c}$$

$$G_{kj}\frac{\partial L}{\partial G_{kj}} = 0; \frac{\partial L}{\partial G_{kj}} \geq 0; \qquad \forall\, k, j \tag{8.56d}$$

$$G_0\frac{\partial L}{\partial G_0} = 0; \frac{\partial L}{\partial G_0} \geq 0 \tag{8.56e}$$

$$T_0\frac{\partial L}{\partial T_0} = 0; \frac{\partial L}{\partial T_0} \geq 0 \tag{8.56f}$$

We now express these conditions analytically. We first determine the derivatives of L with respect to activity-related car route flows T_{ijr}. To that effect,

[6]The KKT conditions with respect to variables H_{ijr}, and L_{ijr}, are not stated, since it is not necessary to develop them.

we use the fact that the derivative of the disutility function $F(T_{ij}, G_{kj}, v_a^c, v_a^t)$ with respect to a given link volume v_a is equal to the travel cost on the link:

$$\frac{\partial F}{\partial v_a} = g_a(v_a); \qquad \forall\ a$$

Therefore,

$$\frac{\partial F}{\partial T_{ijr}} = \sum_a \frac{\partial F}{\partial v_a} \cdot \frac{\partial v_a}{\partial T_{ijr}} = \sum_a g_a(v_a) \frac{\partial (v_a^c + v_a^t)}{\partial T_{ijr}} = \sum_a g_a(v_a)\delta_{ijr}^a; \qquad \forall\ i, j, r$$

so that the derivative of the Lagrangean for the minimization problem is

$$\frac{\partial L}{\partial T_{ijr}} = \sum_a g_a(v_a)\delta_{ijr}^a - \mu_{ij}; \qquad \forall\ i, j, r$$

The summation in this equation represents the travel cost g_{ijr} on route r between origin i and destination j. The KKT conditions then become

$$(t_{ijr} - \mu_{ij})T_{ijr} = 0; \qquad \forall\ i, j, r$$

$$t_{ijr} \geq \mu_{ij}; \qquad \forall\ i, j, r$$

These equations are the familiar user equilibrium conditions for activity-related car trips. Thus, activity travelers travel to their activity locations so as to minimize their travel costs. In the same manner, conditions with respect to the G_{kjr} may be rewritten as

$$(t_{kjr} - \theta_{kj})G_{kjr} = 0; \qquad \forall\ k, j, r$$

$$t_{kjr} \geq \theta_{kj}; \qquad \forall\ k, j, r$$

These equations are similar to the equations above with respect to the T_{ijr}, and show that goods are shipped from a given warehouse to a given activity supply location according to a user equilibrium, minimum cost principle.

Finally, although not presented here, by expressing the KKT conditions with respect to H_{ijr} and L_{ijr}, the user equilibrium conditions for non-activity-related car and truck flows would similarly be established. Thus, all trips (i.e., activity-related and non-activity-related car trips and truck trips) are routed through the network in accordance with the user equilibrium principle, as required.

We now turn to the derivatives of L with respect to the origin-destination activity flows.

$$\frac{\partial L}{\partial T_{ij}} = \frac{1}{\beta}(\ln T_{ij} + 1) + \frac{1}{\gamma} - h_{ij} + c_{ij} + r_j - \lambda_i + \mu_{ij} + \varphi_j; \qquad \forall\ i, j$$

After some derivations similar to many we have performed throughout the text, the KKT conditions with respect to the T_{ij} may be shown to imply

$$T_{ij} = T_i \frac{e^{\beta(h_{ij} - r_j - c_{ij} - \tau t_{ij} - \varphi_j)}}{\sum_j e^{\beta(h_{ij} - r_j - c_{ij} - \tau t_{ij} - \varphi_j)}}; \qquad \forall\, i, j \tag{8.57}$$

Thus, as expected from Formula (8.12), activity-related car trips originating in zone i are distributed to activity locations j according to a logit function. The exact meaning of the last term in the argument, and the fact that Formula (8.57) is equivalent to Formula (8.12), will become apparent after the examination of the next and last conditions, with respect to the origin-destination activity-related truck flows G_{kj}.

Next, we have

$$\frac{\partial L}{\partial G_{kj}} = \frac{1}{\gamma}\left(\ln \frac{G_{kj}}{T_j} + 1\right) + R_k^s + c_{kj} + \theta_{kj} - \varphi_j; \qquad \forall\, k, j$$

It is straightforward to show (see Exercise 8.5) that these conditions combined with Constraints (8.21) imply

$$\varphi_j = -\frac{1}{\gamma} \ln \left[\sum_k e^{-\gamma(c_{kj} + R_k^s + \tau t_{kj})}\right]; \qquad \forall\, j \tag{8.58}$$

and (see Exercise 8.6), that consequently

$$G_{kj} = \frac{T_j e^{-\gamma(c_{kj} + R_k^s + \tau t_{kj})}}{\sum_k e^{-\gamma(c_{kj} + R_k^s + \tau t_{kj})}}; \qquad \forall\, k, j \tag{8.59}$$

Therefore, Formula (8.59) implies that the flow of goods from warehouse k to facility j conforms to Formula (8.4), as required. Also, we now see that Formula (8.58) implies that the value of dual variable φ_j is equal to the expected cost to supplier j of buying and transporting one unit of the commodity, $\tilde{W}_j^s$ in Formula (8.9). Replacing φ_j by $\tilde{W}_j^s$ in Formula (8.57), we obtain

$$T_{ij} = \frac{T_i\, e^{\beta(h_{ij} - r_j - c_{ij} - \tau t_{ij} - \tilde{W}_j^s)}}{\sum_j e^{\beta(h_{ij} - r_j - c_{ij} - \tau t_{ij} - \tilde{W}_j^s)}}; \qquad \forall\, i, j \tag{8.60}$$

in which the travel times t_{ij} have their equilibrium, endogenously determined values. t_{ij} is the dual variable for Constraints (8.26b) and (8.22).

Letting

$$R_j^c = \tilde{W}_j^s + r_j; \qquad \forall\, j$$

Formula (8.60) may then be written

$$T_{ij} = \frac{T_i \, e^{\beta(h_{ij} - R_j^c - c_{ij} - \tau t_{ij})}}{\sum_j e^{\beta(h_{ij} - R_j^c - c_{ij} - \tau t_{ij})}}; \qquad \forall\, i, j.$$

It is apparent in this form that the expression for the origin-destination activity-related car trips conforms exactly to the required form of Formula (8.12). In summary, then, Objective Function (8.32) may be considered a "joint" direct utility function underlying consumers' and suppliers' aggregate commodity demands.

APPENDIX 8.3 OBTAINING EQUILIBRIUM DEMANDS IN THE MULTIPLE VERTICAL DEMANDS CASE

The objective function for Problem (8.46) has the same terms as that for Problem (8.32), plus additional terms in the new variables G_{lk}^p. These new terms are, in minimization form:

$$\frac{1}{\nu} \sum_l \sum_k G_{lk}^p \ln \frac{G_{lk}^p}{G_k^s} + \sum_l \sum_k (R_l^p + f_{lk}) G_{lk}^p$$

Similarly, the constraints for the problem include Constraints (8.21)–(8.24), plus the additional Constraint (8.47c) in the new variables G_{lk}^2. Thus, the Lagrangean for the current problem is the same as for Problem (8.32), with the addition of the above term, and of the term[7]

$$\sum_k \sigma_k \left(\sum_l G_{kj}^s - \sum_l G_{lk}^p \right)$$

in which the σ_k's are the KKT coefficients for Constraints (8.47c). Following once more the same derivations as for the original Problem (8.32), we have successively

$$\frac{\partial L}{\partial G_{kj}^s} = \frac{1}{\gamma} \left(\ln \frac{G_{kj}^s}{T_j} + 1 \right) + \frac{1}{\gamma} + r_k^w + f_{kj} - \varphi_j + \sigma_k; \text{ for} \qquad \forall\, k, j$$

which leads to

$$\varphi_j = -\frac{1}{\gamma} \ln \left[\sum_k e^{-\gamma(r_k^w + f_{kj} - \sigma_k)} \right]; \qquad \forall\, j \qquad (8.61)$$

[7]Constraints (8.47e) are definitions, and therefore do not intervene in the specification of the Lagrangian.

and

$$G_{kj}^{s} = \frac{T_j \, e^{-\gamma(r_k^{w} + f_{kj} - \sigma_k)}}{\sum_k e^{-\gamma(r_k^{w} + f_{kj} - \sigma_k)}}; \qquad \forall\, k, j$$

Next, we have

$$\frac{\partial L}{\partial G_{lk}^{p}} = \frac{1}{\nu} \ln \frac{G_{lk}^{p}}{G_k^{1}} - R_l^{p} - f_{lk} - \sigma_k; \qquad \forall\, k, l$$

which leads to

$$\sigma_k = -\frac{1}{\nu} \ln \left[\sum_l e^{-\nu(R_l^{p} + f_{lk})} \right]; \qquad \forall\, k \tag{8.62}$$

and shows that $\sigma_k = \tilde{W}_k^{p}$ as defined in Formula (8.51). Replacing σ_k in Formula (8.61) by its expression in Formula (8.62), we get Formula (8.49). Finally, both problems are formulated in the same manner with respect to the T_{ij}, both in terms of constraints and objective function. Thus, T_{ij} has the expression in Formula (8.48), but where now $\tilde{W}_j^{s}$ is specified, as we have seen, by Formula (8.49).

APPENDIX 8.4 RETRIEVING EQUILIBRIUM DEMANDS IN THE MULTIPLE HORIZONTAL DEMANDS CASE

The Lagrangian for the problem is again the same, except for the third summation, which reflects the new term in the objective function, and is now

$$\sum_n \frac{1}{\gamma_n} \sum_k \sum_j G_{kj}^{n} \ln \frac{G_{kj}^{n}}{a_n T_j}$$

and for the penultimate term, which reflects the new Constraints (8.54b) and is now

$$\sum_n \sum_j \varphi_{jn} \left(a_n \sum_i T_{ij} - \sum_k G_{kj}^{n} \right)$$

The φ_{jn}'s are the KKT coefficients corresponding to Constraints (8.54b). Repeating the derivations, $\partial L / \partial T_{ij}$ is again the same as above, except that φ_j is now replaced by $\Sigma_n a_n \varphi_j$. The single change carries through to Formula (8.57) for T_{ij}:

$$T_{ij} = T_i \frac{\exp\left(\beta\left(h_{ij} - r_j - g_{ij} - \sum_n a_n \varphi_{jn}\right)\right)}{\sum_j \exp\left(\beta\left(h_{ij} - r_j - g_{ij} - \sum_n a_n \varphi_{jn}\right)\right)}; \qquad \forall\, i, j \tag{8.63}$$

We have

$$\frac{\partial L}{\partial G_{kj}^n} = \frac{1}{\gamma_n} \ln \frac{G_{kj}^n}{a_n T_j} + R_k^n + f_{kj} - \varphi_{jn}; \qquad \forall\, k, j, n$$

Following the same exact derivations as above, we obtain

$$\varphi_{jn} = -\frac{1}{\gamma_n} \ln \left[\sum_k e^{-\gamma_n (R_k^n + f_{kj})} \right]; \qquad \forall\, j, n \tag{8.64}$$

and

$$G_{kj}^n = \frac{a_n T_j e^{-\gamma_n (R_k^n + f_{kj})}}{\sum_k e^{-\gamma_n (R_k^n + f_{kj})}}; \qquad \forall\, k, j, n$$

which is Formula (8.55a). Finally, replacing φ_{jn} in Formula (8.63) by its expression in Formula (8.64), we get Formula (8.55b).

CHAPTER 9

OPTIMAL TRANSPORTATION SUPPLY

9.1 INTRODUCTION: A GENERAL FRAMEWORK FOR INTEGRATED SUPPLY/DEMAND ANALYSIS

We have now completed the description of the demand side, the behavior of individual travelers in connection with the various travel choices facing them.

This resulted in the development of models for travel demand forecasting and their estimation. While this book is focused on the demand side, in this last chapter we examine briefly some of the decisions facing the suppliers of urban transportation, including the operators of the street and transit systems.

The formulation of both the demand and the supply side of the urban transportation system, is necessary for the determination of the "satisfied," or "realized" travel demands. Nevertheless, this chapter constitutes only a brief introduction to the topic of transportation supply analysis. There are many aspects we will not have the opportunity to examine in any depth, or at all. To do this topic justice, another book would be required.

In any case, for our purposes the supply side is defined as those decisions of the operators of the transportation system which affect the decisions of individual travelers. These operators include the public authorities responsible for the transportation network's design and/or operation, typically the municipal government and the transit properties. There are many different such decisions, and we will not be able to examine each and every example. For instance, we will not address traffic control or management, nor land use planning, which may be considered supply decisions in the general sense above.

Supply problems may generally be classified in terms of a simple typology, organized in terms of the following dimensions and their corresponding references:

- *Supply item (I)*
  - network layout (d)
  - level of service (l)
  - pricing (p)
- *Supplier's objectives (O)*
  - minimization of societal travel costs (t)
  - minimization of supplier costs (s)
  - maximization of ridership (r)
- *Travelers' response (R)*
  - deterministic (d)
  - probabilistic (p)
- *Externalities (E)*
  - absent (n)
  - present (y)
- *Time frame (T)*
  - short term (s)
  - medium term (m)
  - long term (l)

In addition to these categories, combinations of categories in the first two dimensions may also be formed, such as combined network design and level of service, and minimization of combined travel costs and supplier cost.

It may also be useful to contrast the nature of the models pertaining to the supply and demand sides, respectively. The demand models we have formulated in the previous chapters may be said to be *descriptive*, in the sense that they formulate what travelers *would* do if they pursue a rational objective, e.g., maximizing the utility of their choices. On the other hand, the supply models we will develop in this chapter may be said to be *normative*, or prescriptive, in the sense that they formulate what suppliers *should* do to best achieve whatever particular objective(s) have been selected. For instance, in the case of private transportation suppliers, this will be profit maximization. In the case of a transit property, this might be maximization of ridership, and so on.

Whatever the particular nature of the supply problem it is clear that suppliers' decisions must be taken in consideration of travelers' behavior. For instance, a contemplated change in capacity between location i and location j, or in the frequency of service, will affect the travel time between these two points, and thus the demand for travel. This in turn will affect the profit, or ridership, and so on. Consequently, suppliers' *actions* must take into consideration travelers' *reaction*, specifically, that of the R.T. which represents them as a group.

This type of situation may also be formulated in terms of "game theory" (Fisk, 1986; Owen, 1982). Specifically, it corresponds to a "Stackelberg game" (Stackelberg, 1952), in which the supplier is the leader, and the R.T. is the follower. The first player in this game is assumed to have the knowledge of the reaction of the second player to his actions.[1]

In addition, we will assume that each player makes only one move. This means that the integrated supply-demand model is a static, not a dynamic one. Each action may then take place in the short, medium, or long term, as noted above. The solution to this "one-shot" game characterizes a *generalized* supply-demand equilibrium, in which travelers' choices reflect suppliers' decisions, and reciprocally.

It should be pointed out in this connection that there has been an ongoing debate about the appropriateness, or even the validity, of the equilibrium concept, in general economics as well as in urban travel analysis. By definition, this concept implies that a state has been reached in the distribution of travel activity (demand and supply) in which no further changes, either on the part of individual travelers, or individual travel suppliers, may be expected, if prevailing conditions do not change. It is clear that such situations are difficult, if not impossible, to observe in reality.

To remedy this, alternative, "disequilibrium" approaches to the analysis of supply and demand interactions have been proposed. Without entering the argument, it may be pointed out that the equilibrium concept is the dominant one in theoretical as well as practical work, and so is the one we shall use. This is in great measure due to the fact that it leads to tractable formulations, while translating a rationale which is acceptable, at least in the short term. In particular, dynamic ("real-time") formulations of travel demand are also based

[1]This means that the supplier has formulated and calibrated travel demand models and uses them.

on the equilibrium principle (see, for instance, (Ran and Boyce, 1994), (Janson, 1991), and (Mahmassani and Herman, 1984)).

In any event, the qualitative framework above must be represented analytically, as we will have to formally incorporate, in some fashion, the demand models previously developed into the analysis of optimal supplier actions. Since the demand side was formulated as the solution to the R.T.'s utility maximization problem, it is apparent that determining optimal supply will require solving a constrained optimization problem, in which some of the constraints take the form of another optimization problem. This type of optimization problem, which we already encountered in connection with model estimation in Chapter 7, is known as a bilevel problem (Boyce, 1987).

Formally, the structure of "bilevel" problems may be represented as

$$\operatorname*{Max}_{s} O(s, \boldsymbol{T})$$

such that (9.1a)

$$g(s, \boldsymbol{T}) \leq \boldsymbol{a}$$

such that

$$\operatorname*{Max}_{\boldsymbol{T}} U(s, \boldsymbol{T})$$

such that (9.1b)

$$f(s, \boldsymbol{T}) \leq \boldsymbol{b}$$

That is, the supply problem (SP) with objective function O is constrained by the demand problem (DP) with aggregate utility function U. This particular framework has been used to formulate various versions of the optimal transportation supply problem, which are briefly reviewed in the next section.

In conclusion of this introductory section, two characteristics of bilevel programs should be noted. First, at the theoretical level, it is in general difficult to establish rigorously the existence and uniqueness of optimal, global solutions, to bilevel programs of the form in Formula (9.1) in the same manner, for instance, as we were able to for most of the demand side problems, in particular in the congested case. Consequently, for instance, welfare analysis, in which the effect on travel demand of various supply actions is analyzed, may not be reliable.

Second, standard optimization algorithms are in principle not applicable for obtaining solutions to programs of the form in Formula (9.1), since such techniques require that all constraints be explicitly stated (e.g., in (in)equality form).

9.2 AN OVERVIEW OF SOLUTION PROCEDURES FOR MULTILEVEL PROGRAMS

Special techniques for multilevel optimization programs have been devised over the past few years. The variety of methods is rather large. In this section, we briefly review some of these methods, in terms of several representative types of approach.

9.2.1 Transformation into an Unconstrained Problem

The first possible approach to solving bilevel problems is to try and eliminate the constraint represented by the Demand Problem, by incorporating it into the objective function of the supply problem (SP). Specifically, one may try and formulate the SP as the unconstrained problem

$$\operatorname*{Max}_{s}\ O(s,\ \boldsymbol{T}(s))$$

where $\boldsymbol{T}(s)$ represents the demand side, the travelers' reaction to the provision of transportation supply level s. The difficulty with this approach is that, for the type of situation described above, the functional form of $T(s)$ is not known explicitly. That is, it is not possible to derive the analytical expression of user equilibrium travel times between given origins and destinations. Instead, they must be obtained numerically, as part of the solution of the user equilibrium (lower) problem. Consequently, the numerical values of $\boldsymbol{T}(s)$ must be systematically evaluated while optimizing O, through repeated solutions of the DP. This particular difficulty will disappear, as we shall see in section 9.5, in the case of stochastic route choice.

The "Hooke and Jeeves" (1962) method is particularly appropriate to the optimization of a function $O(s,\ \boldsymbol{T}(s))$, where $\boldsymbol{T}(s)$ is the (vector) argument of the optimum value of another function $U(s,\ \boldsymbol{T})$, and where, for that reason, the partial derivatives of $O(s,\ \boldsymbol{T}(s))$ with respect to s are difficult, or impossible, to obtain. For each value s at which the function O is being evaluated, the value of $\boldsymbol{T}(s)$ is must then be obtained by solving the lower problem Max/Min $U(\boldsymbol{T},\ s)$. The method thus does not require the computation of the gradient of the objective function O, but only its values, as the algorithm is based on exploratory searches and pattern moves.

The main drawback to the method is the resultant computational burden. However, its major advantage in connection with the optimal supply problem is that it is easily adapted to multilevel programs, since this structure may be extended to more than two levels, in a "cascading" fashion. In turn, the DP may be solved with the appropriate algorithms described in the preceding chapters. Several researchers have used this method, and have provided practical guidelines for its application (Abdulaal and LeBlanc, 1979; Friesz and Harker, 1983). It has been found to perform well when the number of supply variables is small. However, because of its heavy computational burden, this particular approach is not very efficient for realistic large-size networks, where the number of such variables is large. Accordingly, an attempt to circumvent the problem of the unknown functional form of the $T(s)$ relationship, through application of a multivariate "interval reduction" method, was made by Suwansirikul et al. (1987).

9.2.2 Diagonalization Methods

Perhaps the most intuitive approach to solving bilevel programs is to solve the SP with values for the variables of the DP fixed at their previous (or initial values), solving the DP with the values of the SP variables fixed at their last solution, and so on iteratively, until convergence. This is effect amounts to ''block diagonalizing'' the variables in the respective SP and DP. This method is similar to the block Gauss-Seidel and block Jacobi methods for solving equation systems (Ortega and Rheinboldt, 1970).

However, the solution obtained in this fashion will be one in which the decisions of one player are made without the knowledge of, or taking into account, the reaction of the other player, only its previous decision. In effect then, the solution, if it exists, will not correspond to that of the Stackelberg game described above, but rather to the solution of a ''Cournot-Nash'' game (Nash, 1951). It is intuitively clear that the Nash solution constitutes an upper bound to the former (in the case of a minimization SP). Lower bounds for the Stackelberg solution may also be identified, for example, by replacing the ''user optimum'' DP by a ''system optimum'' problem (Fisk, 1986). It may also be expected intuitively that if the interaction between the SP and DP is ''weak,'' the solution of the Nash game will be a good approximation to the solution of the Stackelberg game. However, for strong interactions between the two levels, the two solutions may be completely different (Fisk, 1984).

In spite of these potential pitfalls, a major advantage of the diagonalization method is that it can readily be extended to multilevel problems.

9.2.3 Incorporation of DP's Optimality Conditions into SP's Constraints

Another possible approach is to replace the DP by its optimality conditions and then incorporate them into the SP, as constraints added to whatever constraints the SP may already have. The difficulty, however, is that the resulting feasible region usually becomes nonconvex. Consequently, the application of standard constrained optimization techniques (e.g., Lagrangian relaxation, or feasible descent techniques) is impractical. Thus, in general, the resulting single level problem is no easier to solve than the original, two-level problem, particularly for problems of realistically large size.

Nevertheless, an early application of this approach was made by (Tan et al., 1979). Bard and Falk (1987), Bard and Moore (1990) developed Branch and Bound-type algorithms for the DP's optimality conditions. Fortuny-Amat and McCarl (1981) alternatively used mixed integer programming. Aiyoshi and Shizimu (1984) apply a penalty function method to the penalized DP.

9.2.4 Sensitivity Analysis Techniques

This approach is based on the work of Fiacco (1976) on the sensitivity of the solutions of nonlinear programs to changes in parameters of its objective function and/or constraints. The essence of these results is that, under certain mildly restrictive conditions, the first-order derivatives of the solution values $\boldsymbol{T}$ and the objective function's optimal value for a constrained nonlinear program parameterized by a vector s, as well as the derivatives of its associated Lagrangian variables $\boldsymbol{\mu}$, may be expressed in terms of the first and second derivatives, with respect to both $\boldsymbol{T}$ and $\boldsymbol{\mu}$, of the objective function and of the constraints. The relationship may be efficiently expressed in matrix form.[2]

The required conditions are that all functions defining the objective and the constraints be twice continuously differentiable near the (locally unique) optima, that the second-order conditions for a local optimum hold (see section 2.1 of Appendix A), that strict complementary slackness conditions hold at the optimum (these are part of the KKT conditions described in section 2.2 of Appendix A), and finally, that the gradients of the constraints be linearly independent, meaning that none of the respective directions defined in the multidimensional space by the first derivatives of the constraints is a linear function of any other.

The resultant relationships may then be used to formulate the effect of changes in the solution of the SP on the optimal values for the DP, by letting the former play the role of the parameter s. This provides the analytical expression for the local, approximate effect on the optimal values of the SP's variables of changes in the DPs variables. This "reaction function" takes the form of the product of two matrices. One is the negative of the inverse of the Jacobian of the LP's KKT conditions with respect to its variables. The other is the Jacobian of the same system with respect to the "perturbing," SP variables.

This reaction function may then be included as a constraint in the SP, to replace the DP. Thus effectively, the bilevel program is again approximated by a single-level problem, to which standard algorithms may be applied. In addition, the derivative information obtained in this fashion may be used in gradient-based searches for the SP (Kim, 1989).

There are, however, several caveats to this procedure. First, because the objective function of the DP is not quadratic, its gradient will not be linear, so that nonlinear constraints will be added to the definition of the feasible region of the SP. This may make it nonconvex, and therefore difficult to solve. Another potential problem is that due to the large number of variables and constraints in the DP, the resulting size of the matrices described above may in practice create computational problems of storage and inversion. Examples of how to deal with these issues may be found in Miller et al. (1992) and Kim (1989).

Chao and Friesz (1984), and Tobin and Friesz, (1988), were the first to apply this approach to network flow problems formulated as single-level optimization problems. Subsequently, Kim (1989) applied it to bilevel problems,

[2]For the specific formulas, see, for instance, Fiacco (1976), or Tobin and Friesz (1988) in the case of network-related problems.

and developed a related iterative technique, the "bilevel descent algorithm." Another approach, the "simulated annealing" method, has also been applied by (Friesz et al., 1990, 1991) to network design problems.

9.2.5 Simulated Annealing Method

This method, which has its origins in statistical mechanics, is based on systematic random perturbations of the current solution (starting from a feasible configuration of values), while remaining in a suitably defined feasible region. If the new solution is improved, it is of course accepted. Otherwise, this (worse) solution is accepted with a probability provided by a Boltzman probability distribution function. This key feature of the technique allows moving away from local solutions, so that global solutions have a high probability of being identified.

The search for the optimum is shaped by a key parameter, whose value determines the position of the method between a totally random search and a deterministic, steepest descent search. Its value is adjusted during the procedure. Other parameters can also be used to speed up the search. Practical guidelines for their specification are given in Vanderbilt and Louie (1984). The basic algorithm is described in Friesz et al. (1991), and the underlying rationale is described in detail in Friesz et al. (1990).

The procedure does not require systematic computations of the solution of the DP, but only the examination of the feasibility, with respect to optimality conditions of the DP, of successive candidate solutions. As mentioned, global optima have a significant probability of being identified, a major advantage when applied to nonconvex problems. The drawback of the method is that the number of iterations can potentially be several thousand times larger than in the methods above (Friesz et al., 1990, 1991). However, each of these iterations does not entail the solution of an optimization problem, as in the other methods, but rather the direct examination of feasibility requirements, which is much faster.

9.2.6 Enumeration Methods

These methods are based on the direct examination of selected solution points. For instance, Bialas and Karwan (1984) have shown that when both the SP and DP are linear, a global solution will lie at an extreme point on the constraint set. They also devised an algorithm based on explicit partial enumeration of these points. For linear problems, Candler and Townsley (1982) have proposed procedures based on exploration of feasible dual SP solutions. This may only identify a local optimum. Bard (1982) used a grid search, which, however, was shown to fail for certain problems (Ben Ayed, Boyce, and Blair, 1988).

9.2.7 Miscellaneous Methods

Finally, a large variety of standard numerical techniques has been applied recently to solve multilevel programs, including subgradient projection methods (Constantin, 1990), parametric methods (Judice and Faustino, 1988), approximation methods (Loridan and Morgan, 1988a), and quadratic programming (Gauvin and Savard, 1990). (See for instance Bard and Falk, 1982). A method which has seen a resurgence of applications recently is TABU search (Glover, 1989, 1990).

In conclusion of this brief review, it may be emphasized that multilevel programming, for all its conceptual appeal in connection with the formulation of generalized equilibria, presents many theoretical as well as computational difficulties. Many of these difficulties offer daunting challenges. For instance, attempts to derive formal optimality conditions comparable to the KKT conditions for single-level problems have so far apparently been unfruitful. Problems which were thought resolved later turned out to be still open. Such difficulties must then of necessity be circumvented with various heuristics, approximations, or simplifications whose results may be flawed.

Nevertheless, multilevel programming remains an active research area as of this writing (see, for instance, Anandalingham and Friesz, 1992), and it may be hoped that improved techniques will be available in the near future.

9.3 A BRIEF HISTORICAL REVIEW OF TRANSPORTATION SYSTEMS DESIGN

In this section, we briefly review some of the formulations of the optimal transportation supply problems which have been presented in the past. This review is not intended to be exhaustive, but merely to place the formulations in the subsequent sections in perspective. Also, in order not to burden the exposition, the various analytical formulations will only be described qualitatively.

In the early formulations of the optimal transportation supply problem, supply was equated with physical capacity. Thus, in this case the suppliers' decision variables are the capacities of the network's arterials. The supplier's objective is the minimization of a generalized cost, combining aggregate traveler cost and construction cost. Travel demands, i.e., origin-destination flows, are assumed given. The demand side is represented by travelers' choice of routes through the network, which are assumed to minimize travel costs, resulting in deterministic UE route flows. In terms of the typology above, this type of problem may be characterized as (Id, Os, Rd, Ey, Ti).

In its earliest statement (Dafermos, 1968), the demand side was represented by the inclusion of the optimality KKT conditions, in the form of constraints added to the network flow conservation constraints. A few years later, the same problem was investigated by Abdulaal and Leblanc (1979). In that case, the demand side was represented as the implicit solution of the network equilibrium problem itself. However, the problem was not yet explicitly stated as a multilevel program.

Subsequently, drawing on contemporary formulations of bilevel programs in economics (Fortuny-Amat and McCarl, 1981), and of multilevel programs (Bard and Falk, 1982; Candler and Townsley, 1982), Kim (1988) formally stated the problem as a bilevel problem, in which the SP is the minimization of the generalized cost above, and the DP is the user equilibrium Problem (5.9). LeBlanc and Boyce (1986) proposed a linear bilevel formulation based on piecewise linearization, which allows the use of an exact algorithm, as opposed to a heuristic solution.

Later, other formulations were developed, in which the definition of transportation supply became more general. In several examples, supply was equated with level of transportation service, a problem of the type (Il, Os, Rd, Ey, Ti). In the case of urban road networks, the decision variables were traffic signal settings (i.e., the red and green times) at the network's nodes representing street intersections. Marcotte (1983) formulated the demand side as a variational inequality (VI) equivalent to the KKT conditions for network user equilibrium (Harker and Pang, 1990). The same problem was later investigated by Fisk (1984), and formulated as a ''maximin'' problem, using other equivalence results (Zukovitsky et al., 1973).

In the case of transit networks, the corresponding decision variables are fares and frequencies. In the version of Fisk (1986), which extended previous work by Fisk and Boyce (1983), the objective function is the transit operator's profit, a problem of the type (Il, Op,Rd,Ey,Ti). In an important contrast with the versions above, the origin-destination demands for transit travel are variable, and function of fares and level of service. It was shown that the problem could be stated under several analytical forms, which are equivalent under certain general conditions. The first form was as a standard nonlinear optimization problem, subject to the demand side represented by nonlinear ''complementarity conditions''. Another form was that of an ''infinitely constrained'' problem (John, 1948). Another form was as a ''maximin'' problem (Glowinsky et al. 1981). Still another form was as a bilevel program, where the SP is the objective function above, and the DP is the nonlinear optimization problem in the link flows and variable origin-destination flows.

The same case of transit networks was addressed later by Constantin (1990). The SP's objective is the total travel time for all travelers, including waiting time. The decision variables are the frequencies of service. The DP is a transit user equilibrium network assignment problem.

In the case of freight transportation networks, the level of service is defined by the private sector, under conditions of competition between multiple carriers. This implies that the transportation supply side is now represented by the solution to several optimization problems, corresponding to individual carriers' profits. Freight shippers play the role of the travelers, or network users, and are assumed to choose shipment routes in a user equilibrium manner, where link costs are a function of the rates charged by carriers, as well as the level of carrier service at the link level (e.g., frequency and reliability). In the short term, the demand for shipments may be assumed given. In the long term, it may be specified from a "spatial price" equilibrium which may be formulated as an optimization problem, in which the decision variables are the link activity flows and supply and demand volumes at the network's nodes (Harker and Friesz, 1986).

Several formulations of this situation were developed, corresponding to various assumptions about the nature of the competition between carriers. In the model proposed by Harker and Friesz (1983), a single monopolistic carrier is assumed. In the model of Friesz and Harker (1983), perfect competition is assumed instead. In both cases, the interaction between carriers and shippers was represented as a Nash game (Nash, 1951). Fisk (1986), building again on previous work (Fisk and Boyce, 1983), developed a more general model, under the assumption of oligopolistic competition between carriers, with carriers being leaders, and the generic shipper being the follower in a multiperson Stackelberg game.

In the next section, we formulate a few selected examples of optimal supply problems. Again, this is by no means a comprehensive, or detailed exposition, but rather, an introduction to this vast area. It might be emphasized in this regard that progress is difficult, due to the fact that while it is relatively easy to formalize the optimal supply problem as various forms of multilevel optimization programs, it is significantly more difficult to examine the theoretical properties of their solution, as well as to obtain reliable numerical estimates for them.

9.4 DESIGN OF UNCONGESTED NETWORKS

9.4.1 Road Networks

In this section, we consider the specific problem of designing a future street network, under uncongested conditions. Specifically, the supplier's problem is to decide whether or not a given potential link a on a nonexistent network should be built. Link capacities are assumed to be given, so that the decision variables may conveniently be defined as

$$\delta_a = \begin{cases} 1; & \text{if link } a \text{ is to be constructed} \\ 0; & \text{if not} \end{cases} \quad ; \qquad \forall\, a \tag{9.2}$$

Several alternative supplier objectives are possible. The first type of supplier objective may be the minimization of supplier cost. In this case, the problem's type is (Id, Oc, Rd, En, Tl). The optimal supply problem may then be formulated as

$$\underset{\delta_a}{\text{Min}} \left[\sum_a \delta_a s_a \right] \tag{9.3a}$$

subject to

$$\delta_a = 0, 1; \qquad \forall\, a \tag{9.3b}$$

where s_a is the supplier's cost of providing link a, and to the R.T.'s utility maximization problem which produces the aggregate travel demands

$$\operatorname*{Min}_{v_a(\boldsymbol{\delta})} \left[\sum_a g_a v_a \right] \tag{9.4}$$

$$\sum_{a \in i^-} v_{aj} = \sum_{a \in i^+} v_{aj} + T_{ij}; \qquad \forall\, i, j \tag{9.5a}$$

$$v_a = \sum_j v_{aj}; \qquad \forall\, a, j \tag{9.5b}$$

$$v_{aj} \leq \delta_a M; \qquad \forall\, j, a \tag{9.5c}$$

$$v_{aj} \geq 0; \qquad \forall\, a, j \tag{9.5d}$$

where M is a given number at least as large as the total given origin-destination demands

$$M \geq \Sigma_{ij} T_{ij} \tag{9.6}$$

so that Constraint (9.5c) prevents nonexisting links to carry any flow.

Both the SP and DP problems have a linear objective function and constraints. It may be shown (Bialas and Karwan, 1984) that a global solution for this type of problem will lie at an extreme point on the constraint set. A special algorithm based on explicit, but partial, enumeration of these points may then be used. The algorithm, however, may only identify a local optimum. Another approach is to use a grid search (Bard, 1982). This approach has its own pitfalls, since it was shown to fail for certain problems (Ben Ayed, Boyce, and Blair, 1988; Ben Ayed, et al. 1992). These respective problems illustrate the particular difficulties posed by bilevel problems mentioned in section 9.1.

Alternatively, the diagonalization method, as described in section 9.3.2, may be used. This allows the SP, which is a combinatorial "binary program" in which the unknowns may only take the values 0 or 1, to be solved with several general algorithms for integer programs. One such method is the "Greedy" heuristic, which has proved surprisingly efficient in a wide variety of situations. The essence of method is to (conveniently) ignore the combinatorial nature of the problem. In qualitative terms, the procedure consists of starting with all δ's equal to zero, and at each iteration give a value of 1 to the *single* δ_a which maximizes a given criterion, while of course observing the constraints. Iterations stop when the criterion's value is positive for all unexamined δ_a's (for a minimization problem as in the present case), or when all constraints are met.

In the present case, since the SP's objective is to minimize the total cost of link construction, the criterion function may, for instance, simply be defined as the change in supplier cost per number of travelers transported through the single added link.

It is important to note that other definitions of the criterion function may lead to other solutions. Thus, the Greedy algorithm is a heuristic procedure, and as such will only provide an upper bound (for a minimization problem), on the exact optimal value of the objective function.

Alternatively, an exact algorithm for identifying the optimal solution to the SP at a given iteration of the diagonalization method is the "Branch and Bound" algorithm. The essence of the Branch and Bound algorithm is intelligent enumeration of solutions, so as to divide the original solution space into several smaller ones. These smaller regions may in turn themselves be divided similarly, in a recursive fashion. At each such "branching" an upper bound is placed on the value of the objective function for all the feasible solutions which branch out of the node. Branches whose upper bounds are higher than the current optimum are deleted. Various rules may be used to obtain the bound. This algorithm is described in detail in most introductory texts to mathematical programming and operations research, Hillier and Lieberman (1986), in introductory texts to integer programming (see, for instance, Garfinkel and Nemhauser, 1972), as well as in some more specialized references (see, for instance, Papadimitriou and Steiglitz, 1982; Nemhauser and Wolsey, 1988; Parker and Rardin, 1988).

The above optimal supply problem may be generalized to combine both the determination of optimal link capacities (which were assumed given) with the layout problem. The supply problem type then becomes (Id/1, Ot/s, Rd, En, Tl). The objective function is

$$\underset{(\delta, \Delta K)}{\text{Min}} \left[\sum_a s_a \delta_a + \sum_a k_a(\Delta K_a) \right] \tag{9.7}$$

The constraints are

$$\delta_a = 0, 1; \qquad \forall\, a \tag{9.8a}$$

$$\Delta K_a \geq 0; \qquad \forall\, a \tag{9.8b}$$

where $k_a()$ is the cost of increasing the existing or minimum capacity K_a on link a by an increment of ΔK_a. The other constraint on the supplier's problem is, as always, the R.T.'s problem, which is represented by the same DP (9.4)–(9.5).

In this "mixed integer" program (i.e., with integer and continuous variables) the SP and the DP are again linear. Thus, the same algorithm for linear bilevel problems as above may also be applied. If the diagonalization method is used, adaptations of the Branch and Bound approach are available for mixed integer linear programs. (See, for instance, Martin and Schrage, 1985.)

9.4.2 Transit Networks

In this section, the above formulation is adapted to the case of a public transportation network. We retain the basic assumptions that the network is uncongested and that expected travel times are deterministic. The supplier's problem is the joint determination of the optimal route layout, together with that of the optimal frequencies f_a on the network's links.

Several alternative supplier objectives are again possible. The first type of supplier objective may be the minimization of total expected travel time for all riders. In this case, the problem's type is (Id/s, Ot, Rd, En, Tm). Accordingly, the SP's objective function is

$$\underset{(\delta,f)}{\text{Min}} \left[\sum_a v_a(\delta, f)\, g_a(\delta) + \overline{\omega} \sum_i \sum_j w_{ij}(f) \right] \tag{9.9}$$

in which $\overline{\omega}$ is a given parameter (i.e., from calibration), translating the importance of waiting time relative to travel time, as well as the effects of random arrivals of passengers and vehicles at the transit stops, as we have seen in section 4 of Chapter 3. The constraints are

$$\sum_a f_a s_a \leq S \tag{9.10a}$$

$$\delta_a = 0, 1; \qquad \forall\, a \tag{9.10b}$$

$$f_a \geq 0; \qquad \forall\, a \tag{9.10c}$$

where s_a is the operational cost of providing frequency of service f_a on link a. Alternatively, budgetary Constraint (9.10a) may be replaced by a constraint on the size of the available fleet of vehicles N:

$$\sum_a f_a t_a \leq N \tag{9.10d}$$

The other constraint on the supplier's problem is, as always, the R.T.'s U.M. problem, which is represented by the DP

$$\underset{(v_{aj}(\delta,f),\, w_{ij}(\delta,f))}{\text{Min}} \left[\sum_a g_a \sum_j v_{aj} + \overline{\omega} \sum_i \sum_j w_{ij} \right] \tag{9.11}$$

subject to

$$v_{aj} \leq \delta_a f_a w_{ij}; \qquad \forall\, i, j, a \in i^- \tag{9.12a}$$

$$\sum_{a \in i^-} v_{aj} - \sum_{a \in i^+} v_{aj} = T_{ij}; \qquad \forall\, i, j \tag{9.12b}$$

$$T_{jj} = -\sum_{i \neq j} T_{ij}; \qquad \forall\, j \tag{9.12c}$$

$$v_{aj} \geq 0; \qquad \forall\, a, j \tag{9.12d}$$

It is worth noting that, in this particular case, the objective functions for the supply side and for the demand side are the same, namely aggregate travel time. This is because there is no congestion, and consequently, user optimum coincides with system optimum; if aggregate travel costs are minimized, then each individual traveler's costs are also minimized. Nevertheless, because supply actions precede traveler's reactions (i.e., the respective supply and demand problems are not simultaneous), and because the respective decision variables are not the same, the joint problem must still be treated as a bilevel problem, of the general form in Formula (9.1).

9.5 DESIGN OF CONGESTED NETWORKS

9.5.1 Road Layout

It was assumed in the previous sections that congestion is not present on the transportation network. Such an assumption may be adequate when designing new networks. However, when redesigning existing networks, as when changing existing capacities to reflect changed travel demands, congestion is usually, if not by definition, present, and must therefore be taken into account. In this case, neither the SP nor the DP are linear problems. Under the assumption of deterministic utilities, obtaining the optimal solution to the supply problem will be correspondingly more difficult. (See the historical review in section 9.3.)

However, under the random utility assumption, when users' behavior is limited to route choice, the problem solution is simpler. Specifically, the route demands are in this case continuous. The integrated supply/demand problem is of type (Id, Ot/c, Rs, Ey, Tl), and may be stated as

$$\underset{\delta}{\text{Min}} \left[\sum_a v_a g_a(\boldsymbol{\delta}) + \sum_a s_a(\delta_a) \right] \tag{9.13}$$

subject to

$$\delta_a = 0, 1; \qquad \forall\, a \tag{9.14a}$$

$$\text{Min } F_{(T,\delta)} = \sum_a \int_0^{v_a} g_a(x)\, dx + \frac{1}{\beta} \sum_i \sum_j \sum_r T_{ijr} \ln T_{ijr} \tag{9.15}$$

$$\sum_r T_{ijr} = T_{ij}; \qquad \forall\, i, j \tag{9.16a}$$

$$v_a = \delta_a \sum_i \sum_j \sum_r T_{ijr} \delta_{ijr}^a; \qquad \forall\, a \tag{9.16b}$$

$$T_{ijr} > 0; \qquad \forall\, i, j, r \tag{9.16c}$$

In this case, as we have seen in section 3.3 of Chapter 3, the first-order conditions for the DP take on the form of a standard square system of continuous Equations (5.37), which is reproduced here for convenience:

$$T_{ijr} = \frac{T_{ij} \exp\left\{-\beta \sum_a \delta^a_{ijr} g_a\left(\sum_{ijr} T_{ijr}\delta^a_{ijr}\right)\right\}}{\sum_r \exp\left\{-\beta \sum_a \delta^a_{ijr} g_a\left(\sum_{ijr} T_{ijr}\delta^a_{ijr}\right)\right\}}; \qquad \forall\, i, j, r \qquad (9.17)$$

Thus, in the stochastic case, the DP may equivalently be replaced by this system of "smooth" demand functions, which are inherently easier to deal with than the deterministic discontinuous demands. This equation system is added to the SP. It then constitutes a system of definitional relationships between route volumes and route costs, in the same manner as Relationships (9.16b) between link and route volumes. The SP is then a nonlinear binary problem with nonlinear constraints. Such problems may be solved with a Branch and Bound algorithm. Chen and Alfa (1991) developed an algorithm which combines a Branch and Bound algorithm with an incremental stochastic assignment, using the STOCH algorithm.

9.5.2 Setting Link Capacities

In this section, we consider the problem of optimal determination of link capacities. This may in a sense be considered the continuous version of the preceding problem. Specifically, the supplier's decision variables are the changes ΔK_a in link capacities from their present levels K_a. The problem type is thus (Il, Ot, Rs, Ey, Tl). The UP's objective function would be formulated as

$$\underset{\Delta K}{\text{Min}} \left(\sum_a v_a t_a(v_a, \Delta K_a)\right) \qquad (9.18)$$

in which

$$t_a(v_a, \Delta K_a) = t^0_a\left(1 + 0.15\left(\frac{v_a}{K_a + \Delta K_a}\right)^4\right) + c_a; \qquad \forall\, a \qquad (9.19)$$

if, for instance, the B.P.R. link cost function is used. The SP's constraints are

$$\sum_a s_a(\Delta K_a) \leq S \qquad (9.20)$$

$$\Delta K_a \geq 0; \qquad \forall\, a \qquad (9.21)$$

in which $s_a(\Delta K_a)$ is the given cost function for capacity increases on link a, and to the R.T.'s demands formulated as Conditions (9.17).

Davis (1991) developed a procedure for the computation of the derivatives of Conditions (9.17), using the STOCH algorithm. This procedure is combined with a standard "reduced gradient" search for solving the SP. Alternatively, a sequential quadratic programming approach may also be used.

9.6 TRANSPORTATION PRICING

Urban roads "congestion pricing" has recently emerged as a potentially efficient means of controlling the negative externalities of car travel, increasing municipal revenues, and bringing some equity in the consumption of a public good, that is, "urban travel," or trips. This is due in part to recent technological improvements such as electronic toll collecting, which greatly increase the feasibility of such schemes. In this section we consider the problem of optimal pricing for transportation service. We assume that transportation system user charges take two main forms, respectively road tolls and transit fares. The interesting case of parking charges, in particular, will not be examined here, as it involves both the public and private sector, with a multiplicity of different objectives.

Setting charges for intraurban travel, either in the form of transit fares and/or road fees (e.g., bridge or tunnel tolls) may be considered a form of transportation supply, in the same manner that setting capacities or levels of service. Indeed, in the latter case, the effects of supply actions are translated into travel and/or waiting time, while in the former, they result in monetary costs. In either case, these respective supply actions have an impact on individual traveler utilities, and consequently, on their travel choices. Thus, the optimal transportation pricing problem is conceptually similar to the problems considered above. However, the supplier's control variables are now economic, as opposed to physical (e.g., capacities or frequencies).

Transportation providers may pursue various objectives through their pricing actions. We will successively examine two main types of situations, specifically when the objective is revenue enhancement, and congestion mitigation. Other, more complex, objectives are also possible, for instance, profit maximization and improvement of community welfare (traveler "surplus").

9.6.1 Congestion Pricing for Maximizing Revenues/Ridership

In the first type of situation, we assume that the objective is the maximization of revenue for a given supplier in a bimodal situation, either the road *or* transit operator. We assume that *only* the given supplier is making pricing decisions; the other supplier does not react, and keeps its own prices fixed. This particular assumption will be relaxed next. The supplier's decision variables representing the fare structure to be designed, are the charges c_a for each trip made through link a (e.g., segment fare). The problem type is then (Ip, Or, Rd, En, Ts).

There may be constraints on the suppliers' actions. For instance, there may be a limit M_m on the maximum total charge which may be accrued for any modal trip, for instance, because of political considerations.

The SP may then be stated as

$$\text{Max } O(c_a^m, v_{aj}^m, w_{ij}, T_{ijm}) \left[\sum_{a_m} v_a^m c_a^m \right]; \qquad m = \text{car } or \text{ transit} \tag{9.22}$$

such that

$$\sum_a c_a^m \delta_{i,j,r}^{a_m} \leq M_m; \qquad \forall\, i, j, r; \qquad m = \text{car } or \text{ transit} \tag{9.23}$$

The summation on the left-hand side of Constraint (9.23) represents the sum of the link charges which are incurred on route r from i to j.

The other constraint to the supplier problem is the R.T.'s reaction to the charges. In the present case, this is a model of mode and route choice, with given origin-destination flows T_{ij}, and may be derived from the model of destination, mode and route choice for scheduled modes (see Exercise 6.25). In terms of the demands for strategies T_{is}, the DP is thus

$$\text{Min } U_{(v_{aj}^c, v_{as}^t, w_{is}, T_{ijm}, c_a^m)} = \sum_m \sum_a \int^{v_a^m} g_a^m(c_a^m)\, dx + \bar{\omega} \sum_i \sum_s w_{is} + \frac{1}{\beta_m} \sum_{ijm} T_{ijm} \ln T_{ijm} + \sum_{ijm} T_{ijm} h_{ijm} \tag{9.24}$$

and subject to

$$v_{as}^t \leq \delta_s^a f_a w_{is}; \qquad \forall\, a \in i^-, s \in (i, j) \tag{9.25a}$$

$$\sum_{a \in i^-} v_{aj}^c - \sum_{a \in i^+} v_{aj}^c = T_{ij}; \qquad \forall\, i, j \tag{9.25b}$$

$$\sum_{a \in i^-} v_{as}^t - \sum_{a \in i^+} v_{as}^t = T_{is}; \qquad \forall\, i, s \in (i, j) \tag{9.25c}$$

$$\sum_m T_{ijm} = T_{ij}; \qquad \forall\, i, j \tag{9.25d}$$

$$\sum_{s \in (i,j)} T_{is} = T_{ijt}; \qquad \forall\, i, j \tag{9.25e}$$

$$T_{iim} = -\sum_{k \neq i} T_{kim}; \qquad \forall\, i, m \tag{9.25f}$$

$$v_a^t = \sum_j \sum_{s \in (i,j)} v_{as}^t; \qquad \forall\, a \tag{9.25g}$$

$$v_a^c = \sum_j v_{aj}^c; \qquad \forall\, a \tag{9.25h}$$

$$v_{as}^t,\ v_{aj}^c,\ w_{is},\ T_{is} \geq 0,\ T_{ijc} > 0; \qquad \forall\, i, j, a, s \tag{9.25i}$$

This problem may be solved with the same techniques as for the previous problems.

The same formulation may be adapted to represent the problem of maximizing usage of a given mode, say transit, for instance, instead of revenues. In such a case, the SP's objective function would be

$$\text{Max } O(c_a^m,\ v_{aj}^c,\ w_{is},\ T_{is},\ T_{ijc}) = \sum_{is} T_{is} \tag{9.26}$$

subject to Constraints (9.25).

If *both* modal suppliers set user charges at the same time, and each with the objective of maximizing their revenues, the DP remains the same, but now the SP is composed of *two* individual problems

$$\text{Max } U_c(c_a^c,\ v_{aj}^c,\ T_{ijc}) \left[\sum_{a_c} v_a^c c_a^c \right] \tag{9.27a}$$

$$\text{Max } U_t(c_a^t,\ v_{as}^t,\ T_{is}) \left[\sum_{a_t} v_a^t c_a^t \right] \tag{9.27b}$$

such that

$$\sum_a c_a^m \delta_{i,j,r}^{a_m} \leq M_m; \qquad \forall\, i, j, r, m \tag{9.28}$$

Various assumptions are possible about the nature of this "multiple-person" game, including the order, if any, in which the respective suppliers make their price moves, leading to different optimal solutions. In practice this multilevel model may be solved using, for instance, a hierarchical "Hooke and Jeeves" search, or alternatively, a diagonalization method with another "outer loop" added. The same caveats about the practical difficulties of such an approach and the reliability of the numerical solution apply. Finally, the same approach may be extended to the case of unknown destination volumes, i.e., with variable T_{ijm} as shown in the next section.

9.6.2 Congestion Abatement

In the next possible type of pricing situation, we assume that there are two types of car traffic, single occupant (mode s), and multiple occupant (mode c), using the same network. Tolls in the unknown amounts c_a^s (the decision variables) are to be charged to single-occupant cars only, with the objective of minimizing total travel time. In addition, we assume in this particular case that given the unit costs s_a of collecting road tolls on given links, the agency must stay within a given budget B.

Mathematically, the SP may then be formulated as

$$\text{Min } O(c_a^s) = \sum_a v_a t_a(v_a) \tag{9.29}$$

subject to

$$\sum_a c_a v_a \leq S \tag{9.30}$$

The DP is the model of destination, mode, and route choice in Formulas (6.22)–(6.24).

$$\text{Min } F_{(v_a^m, T_{ij}, T_{ijm})} = \tau \sum_a \left[\int_0^{v_a} g_a(c_a)\, dx \right] + \frac{1}{\beta_m} \sum_{ijm} T_{ijm} \ln T_{ijm}$$
$$+ \frac{1}{\beta'_d} \sum_{ij} T_{ij} \ln T_{ij} - \sum_{ijm} T_{ijm} h_{ijm} - \sum_{ij} h_{ij} T_{ij} + \sum_j R_j \sum_i T_{ij} \tag{9.31}$$

where R_j is the *given* average price of the activity to be conducted at destination j, and which is the result of profit-maximizing decisions of *activity* suppliers.

The variables, or unknowns, are the link flows, by travel mode, v_a^m (or alternatively, path flows T_{ijmr} and S_{ijmr}), the origin-destination flows T_{ij} and their modal components T_{ijm}. The constraints on the problem are

$$\sum_r T_{ijmr} = T_{ijm}; \qquad \forall\, i, j, m \tag{9.32a}$$

$$\sum_m T_{ijm} = T_{ij}; \qquad \forall\, i, j \tag{9.32b}$$

$$\sum_j T_{ij} = T_i; \qquad \forall\, i \tag{9.32c}$$

$$v_a^m = \sum_i \sum_j \sum_r (T_{ijmr} + S_{ijmr})\, \delta_{ijr}^{am}; \qquad \forall\, a, m \tag{9.32d}$$

$$v_a = v_a^s + \frac{1}{e} v_a^c; \qquad \forall\, a \tag{9.32e}$$

$$T_{ij} > 0,\ T_{ijm} > 0,\ T_{ijmr} \geq 0,\ v_a^m \geq 0; \qquad \forall\, i, j, r, a \tag{9.32f}$$

where e is the average car occupancy for H.O.V.'s.

As in the previous case, if origin-destination volumes T_{ij} have given values, variables T_{ij} are replaced by these constants, resulting in the deletion of the last two terms in the objective function, since they are now a constant, and of Constraint (9.32c), which has become an identity.

Also, it should be noted that the respective objectives above may also be combined. For instance, it might be to minimize aggregate travel, while at the same time minimizing implementation costs

$$\text{Min } O = \nu \sum_a v_a s_a + \sum_a v_a t_a \tag{9.33}$$

The value of coefficient ν translates the relative importance of these respective objectives in this multiobjective program (Cohon, 1978). It may also be interpreted as a coefficient in the Lagrangian of Problem (9.29)–(9.30). Another type of combined objective might be to minimize aggregate travel, while at the same time maximizing net revenues. In this case, the SP's objective function would be

$$\text{Max } O = \nu \sum_a v_a s_a - \sum_a v_a t_a(v_a) \tag{9.34}$$

9.7 SUMMARY

In this chapter, several examples of models for identifying optimal transportation supply decisions so as to meet various objectives regarding ridership, revenues, and congestion, were developed. In all cases, these models were integrated with demand models formulating the reaction of travelers as a group to the level of service offered, whether represented by links, capacities, or user charges. Consequently, the integrated supply/demand equilibrium model takes the general form of a bilevel optimization program. These programs are significantly more difficult to analyze, as well as to solve, than the single-level problems we have formulated for the demand side only.

Application of the models developed in this chapter may provide some insight into the multiple, complex impacts of taxing mobility on individual and public welfare, as well as on the private (e.g., commercial) sector. In particular, the reciprocal effects of congestion on the transportation network and of prices for the activities whose conduct underlies travel (e.g., shopping) may be explored. Basic theoretical issues raised by road user fees, including the importance of accessibility to the economy, fair pricing of intraurban mobility, and the role of spatial congestion as an economic mediator, may thus be addressed.

The supply models developed in this chapter may also be extended to the more general case of *competitive* network-based supply problems. (See, for instance, De Palma and Thisse, 1993.)

APPENDIX A

MATHEMATICAL BACKGROUND

In the course of this text, we will apply various mathematical concepts to the analysis of urban travel demand and supply. To make the overall exposition as reasonably self-sufficient as possible, in this appendix we present some background material which might be useful to those readers who may not be familiar with these concepts, or who need to re-acquaint themselves with them. We begin with the basic elements of calculus, namely functions, derivatives, and integrals.

A.1 ELEMENTS OF CALCULUS

A.1.1 Functions

A function may be thought of as a *continuous* correspondence between the values of two variable quantities, the function y and the variable x. Symbolically,

$$y = y(x) \tag{A.1}$$

For instance, if the travel time, say t, on a link depends on the volume, say x, on the link, the cost will be specified as the function $t(x)$. A function may have various mathematical expressions. For instance, if the link cost increases proportionately with the volume, starting from a base cost, the appropriate function would be:

$$t(x) = t_0 + ax \tag{A.2}$$

This is a so-called *linear* function, because its graph has the shape of a straight line, as represented in Figure A.1.

A.1.2 Rates of Change and Derivatives

A fundamental characteristic of a function is its *rate of change*. Formally, the rate of change of function y, for a change in the value of the variable from x_1 to x_2, is equal to

$$\frac{\Delta y}{\Delta x} = \frac{y(x_2) - y(x_1)}{x_2 - x_1} \tag{A.3a}$$

Graphically, the rate of change may be represented as the slope of the straight line connecting the two points, the rise, or fall, in the value of the function y, per unit change in the value of the variable x, as represented in Figure A.2.

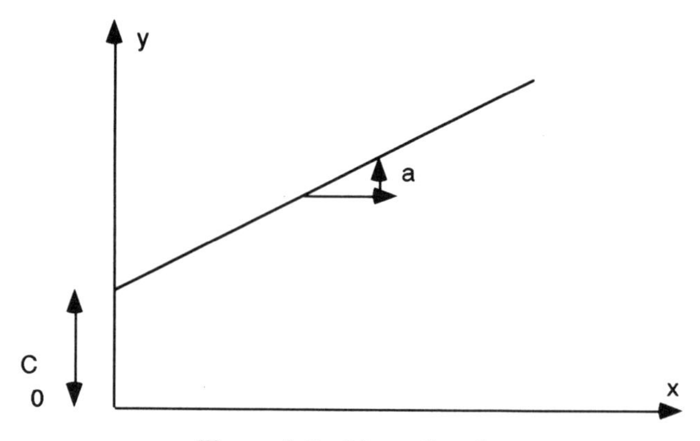

Figure A.1 Linear function.

The *instantaneous* rate of change may be defined as the value of the above expression, when $\Delta x = (x_2 - x_1)$ becomes infinitesimally small, when x_2 is infinitely close to x_1, in which case Δx will be represented as dx. The resulting infinitesimally small change in the function y is similarly noted dy. Thus, the instantaneous rate of change in the function $y(x)$ is defined as the limit, when dx goes to zero, of the quantity[1]

$$\frac{dy}{dx} = \frac{y(x + dx) - y(x)}{dx} \tag{A.3b}$$

[1]In some cases, which will not arise in this book, the above limit may not exist or may have two different values, depending on whether dx is positive or negative, i.e., on whether the point of coordinates x and $y(x)$ is approached from above or from below.

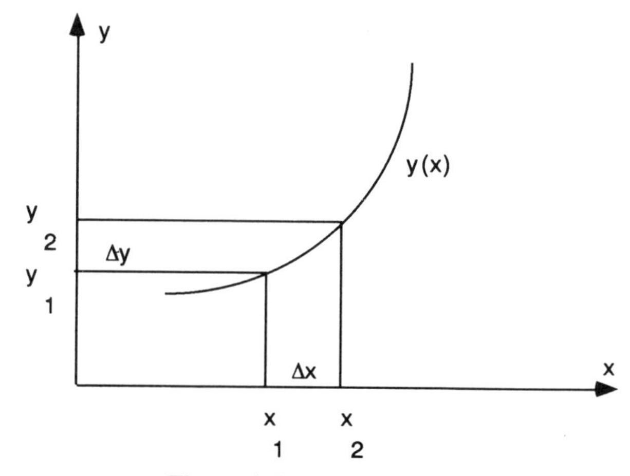

Figure A.2 Rate of change.

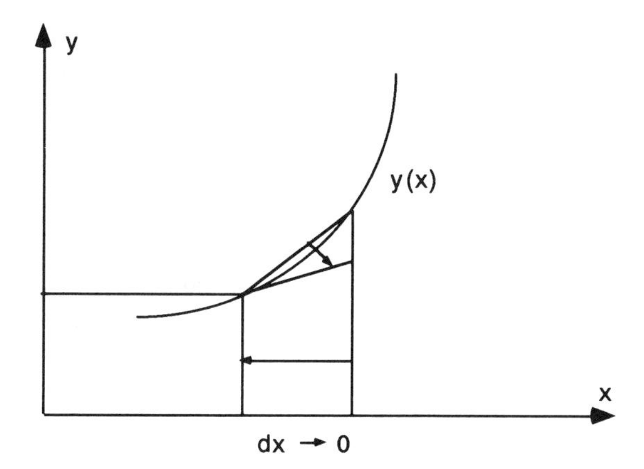

Figure A.3 Geometric interpretation of derivative.

The instantaneous rate of change depends in general (except for linear functions) on the value of x at which it is evaluated, and is thus itself a function of x. This function is called the "derivative" function of $y(x)$, and is noted $y'(x)$. Its value represents the slope of the tangent to the plot of the curve $y(x)$ at point x, as represented in Figure A.3.

It may be helpful to visualize these definitions by thinking of function $y(x)$ as the distance traveled as a function of time x. The derivative then represents the instantaneous speed, measured at a given time x.

Given the specification of the function $y(x)$, the derivative function may be determined, using some simple formulas derived from the application of the definition represented by Formula (A.3b). Some of the relevant rules for our purposes are summarized in Table A.1.

The *second* derivative of the function $y(x)$, which is noted $y''(x)$ or $\partial^2 y/\partial x^2$, is simply the derivative of the derivative.

TABLE A.1 Derivatives of Common Functions

Function $y(x)$	Derivative $\frac{dy}{dx} \equiv y'(x)$
a (constant)	0
x^n	nx^{n-1}
$\ln x$ (natural logarithm)	$1/x$
e^{ax}	ae^{ax}
$\Sigma_i\, a_i y_i(x)$	$\Sigma_i\, a_i y_i'(x)$
$y_1(x)y_2(x)$	$y_1'(x)y_2(x) + y_1(x)y_2'(x)$
$y_1(x)/y_2(x)$	$[y_1'(x)y_2(x) - y_2'(x)y_1(x)]/[y_2(x)]^2$
$y_1(y_2(x))$	$y_1'(y_2)\, y_2'(x)$

Given the value of the derivative, or instantaneous rate of change, the above definition may be used to estimate the variation in the function $y(x)$ which corresponds to an infinitesimally small variation dx in the variable x, as

$$dy = \frac{dy}{dx} dx = y' dx \tag{A.4a}$$

For small variations in x, the *linear* approximation to the corresponding variation in y is

$$y_2 - y_1 = \Delta y \approx y' \Delta x \tag{A.4b}$$

Graphically, this is represented in Figure A.3. Using the analogy of the derivative with speed above, the distance traveled in a small period of time starting at time x may be approximated as the product of the instantaneous speed at the time, and of the small time period. It is a linear approximation because the function is replaced, over the small interval dx, by a straight line with slope equal to the derivative, that is, the rate of change. The larger the interval Δx, the more erroneous the approximation becomes. In the limit, the approximation is only good for $dx \to 0$. Graphically, the approximation may be represented as in Figure A.4.

A.1.3 Partial Derivatives and the Chain Rule

The functions above are *univariate* functions, because they depend on a single variable x. When the function depends on several variables $x_1, x_2, \cdots, x_n$, it is called a *multivariate* function. For instance, the probability that a traveler n chooses a given travel alternative j may be specified as a multivariate func-

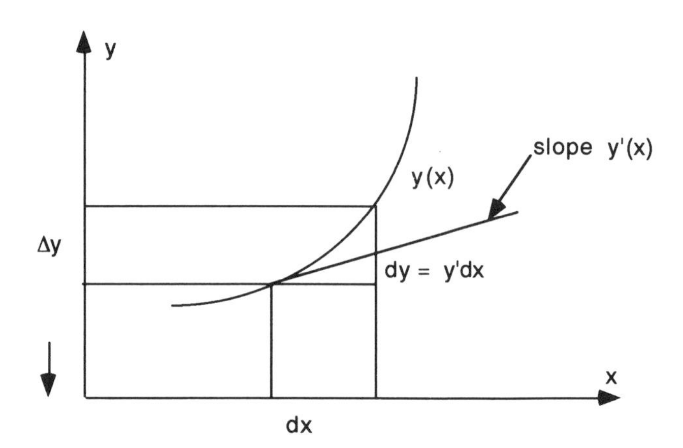

Figure A.4 Linear approximation to a function's variation.

tion of the characteristics c_j and t_j of the respective alternatives

$$P_j = \frac{e^{-(\alpha_j + \beta c_j + \tau t_j)}}{\sum_j e^{-(\alpha_j + \beta c_j + \tau t_j)}}; \qquad \forall j \tag{A.5}$$

In order to represent graphically the variations of a multivariate function as above, more than one dimension is necessary for the several variables x. In the case of only two x's, this might look something like Figure A.5. (Obviously, the case of more than two x's cannot be represented in the conventional three-dimensional space.)

In the case of a multivariate function,, the *partial* derivatives of the function, each with respect to a given variable x_j, are computed in the same manner as above, as the instantaneous rates of change of the function when x_j varies by an infinitesimal amount, and the other variables do not vary (are fixed). Formally,

$$\frac{\partial y(x_1, \cdots, x_n)}{\partial x_j}$$

$$= \underset{(dx_j) \to 0}{\text{limit}} \left\{ \frac{y(x_1, \cdots, x_j + dx_j, \cdots, x_n) - y(x_1, \cdots, x_j, \cdots, x_n)}{dx_j} \right\}; \qquad \forall j \tag{A.6}$$

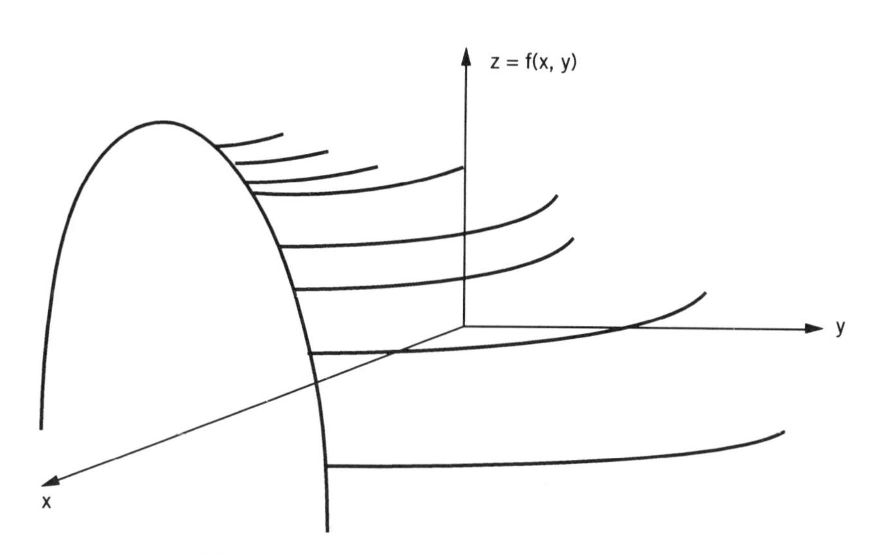

Figure A.5 Variations of a bivariate function.

This is a function of x_j, as well as of the other x_k's, and is thus a multivariate function itself, which may be noted $y_j'(\mathbf{x})$ for short, with $\mathbf{x}$ representing the vector of variables x, $[x_1, x_2, \cdots, x_j, \cdots, x_n]$.[2]

The same rules as above for the computation of the derivatives of univariate functions may then be used also for the computation of the derivatives of multivariate functions. Indeed, in effect, the latter becomes a univariate function for purposes of partial derivation, the values of all variables, expect the one under consideration, being kept constant.

When there exists relationships between the individual variables, a small variation in x_j will create a variation in all other variables x_k. In such a case, the resultant variation in the function $y(\mathbf{x})$ would then cumulate the direct variation due to x_j, and the indirect variations due to the x_k's. Using the symbolic notation introduced above, the total variation in the function is then equal to

$$dy = y_j'\, dx_j + \sum_{k \neq j} y_k'\, dx_k \tag{A.7}$$

Dividing both sides of the equality by dx_j, and replacing the y's by their full expression, we get

$$\frac{dy}{dx_j} = \frac{\partial y}{\partial x_j} + \sum_{k \neq j} \frac{\partial y}{\partial x_k} \cdot \frac{\partial x_k}{\partial x_j}; \qquad \forall\, j \tag{A.8}$$

This expression for the partial derivative when the variables are not independent is sometimes called the "chain rule."

A.1.4 Integrals

Given the rate of change, or derivative $y'(x)$ of a function of the variable x, the total variation in y for any interval x_1 to x_2 may then be estimated as the sum of variations in y for each of an infinite number of infinitesimally small intervals dx. Graphically, this operation may be represented as in Figure A.6. It is clear visually that the value of the variation in y is equal to the area under the plot of $y'(x)$ over the relevant interval of values for x, since it is equal to the sum of all the infinitesimally small rectangles whose bases cover the interval of values for x, and heights approximate the value of the function $y(x)$ at x.

In general, the function whose derivative is equal to a given function $y(x)$ is called the "integral" of the function, and is noted

$$\int y(x)\, dx \tag{A.9a}$$

[2]In general, bold symbols represent vectors, or matrices.

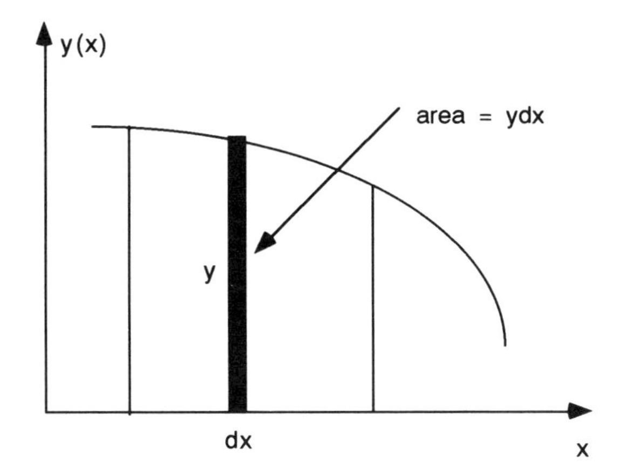

Figure A.6 Geometric interpretation of an integral.

The problem of finding the integral may be considered the reverse of the problem of determining the derivative given the function, and is called "integrating" the function. In the analogy above, this would correspond to finding the distance traveled over a given period of time, given the specification of the speed, as a function of time. Reading Table A.1 in reverse, from right to left, provides some simple rules for integrating basic functions. These are summarized in Table A.2.

It may be noted that the value of the *indefinite* integrals above is defined up to an arbitrary constant k. This is because no matter when the value of k is, it will vanish in the computation of the derivative, since the derivative of a constant is zero, as noted above. The variation of a given function $y(x)$, over a range of values x_1 to x_2, may then be computed as

$$\Delta y = y_1 - y_2 = \int_{x1}^{x2} dy = \int_{x1}^{x2} y'(x)\, dx \qquad \text{(A.9b)}$$

TABLE A.2 Integrals of Common Functions

Function $y(x)$	Integral $Y(x) \equiv \int y(x)\, dx$
a (constant)	$ax + \mathrm{k}$ (k is an arbitrary constant)
x^n $(n \neq -1)$	$x^{n+1}/(n+1) + \mathrm{k}$
$1/x$	$\mathrm{Ln}\, x + \mathrm{k}$ (natural logarithm, i.e., of base e).
$\ln x$	$x \ln x - x + \mathrm{k}$
$\Sigma_i\, a_i y_i(x)$	$\Sigma_i\, a_i \int y(x)\, dx + \mathrm{k}$
e^{ax}	$e^{ax}/a + \mathrm{k}$

The symbol

$$\int_{x_1}^{x_2} y(x)\, dx \tag{A.9c}$$

is called the *definite* integral, and represents a numerical value, as opposed to the indefinite integral defined above, which is a function of its bounds.

The concept of integral may be extended to the case of multivariate function, leading to an *indefinite multiple* integral. This is defined as

$$\int\int y(x_1, x_2) d(x_1) d(x_2) \tag{A.10}$$

The value of the multiple *definite* integral $\int_{a_1}^{a_2} \int_{b_1}^{b_2} y(x_1, x_2) d(x_1) d(x_2)$ represents the volume under the surface represented by the values of the function, over the rectangle defined by the values a_1, a_2, b_1, and b_2. This is represented in Figure A.7.

A.1.5 Logarithmic and Exponential Functions

The "logarithmic" function $\ln x$ is defined as the function whose derivative is $1/x = x^{-1}$. Indeed, for the value of $n = -1$ in the second line of Table A.2, the integral would be undefined (i.e., $x^0/0$). The exponential function is defined as the mathematical "inverse" (i.e., the specification of the expression of $x(y)$ as a function of y), of the logarithmic function

$$y = \ln x \leftrightarrow x = e^y \tag{A.11}$$

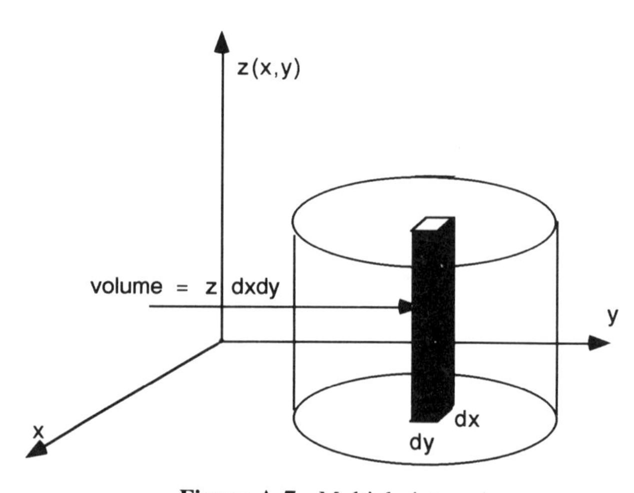

Figure A.7 Multiple integral.

in which e is the "base" of "natural logarithms", and is equal to 2.7218. Due to the additive property of powers, or exponents, we may write

$$e^{y_1 + y_1} = e^{y_1} e^{y_2} \tag{A.12}$$

so that the above Relationship (A.11) between the exponential function and the logarithmic function implies

$$\ln (x_1 x_2) = \ln x_1 + \ln x_2 \tag{A.13}$$

A.1.6 Inequalities and Feasible Regions

The above relationships between a variable y and other variables $\mathbf{x}$ were specified as an *equality*, of the form $y = y(\mathbf{x})$. It is sometimes useful to specify the relationship as an inequality. In this case, a given function of x, $y(\mathbf{x})$, is required to be strictly less than, or less than or equal to, a given quantity, say b (conversely, greater than, or greater than or equal to b). In this case, the relationships between given functions and given variables take one of the alternative forms:

$$\begin{aligned} \mathbf{y}(\mathbf{x}) &< \mathbf{b} \\ \mathbf{y}(\mathbf{x}) &\leq \mathbf{b} \\ \mathbf{y}(\mathbf{x}) &> \mathbf{b} \\ \mathbf{y}(\mathbf{x}) &\geq \mathbf{b} \end{aligned} \tag{A.14}$$

We'll be dealing essentially with the case when the functions $\mathbf{y}(\mathbf{x})$ are linear. Graphically, the set of points which satisfy the inequality does not lie on the graph $\mathbf{y}(\mathbf{x})$, as above, but rather in one or the other of the two regions defined in the two-dimensional plane by the straight line (or by the hyperplane in multidimensional space). This is illustrated in Figure A.8.

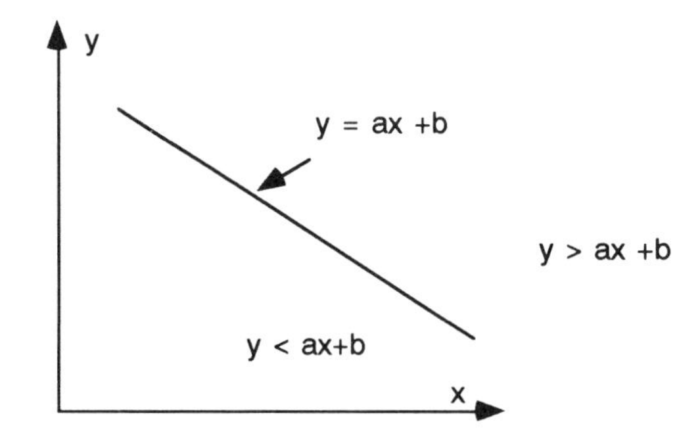

Figure A.8 Regions defined by a single inequality.

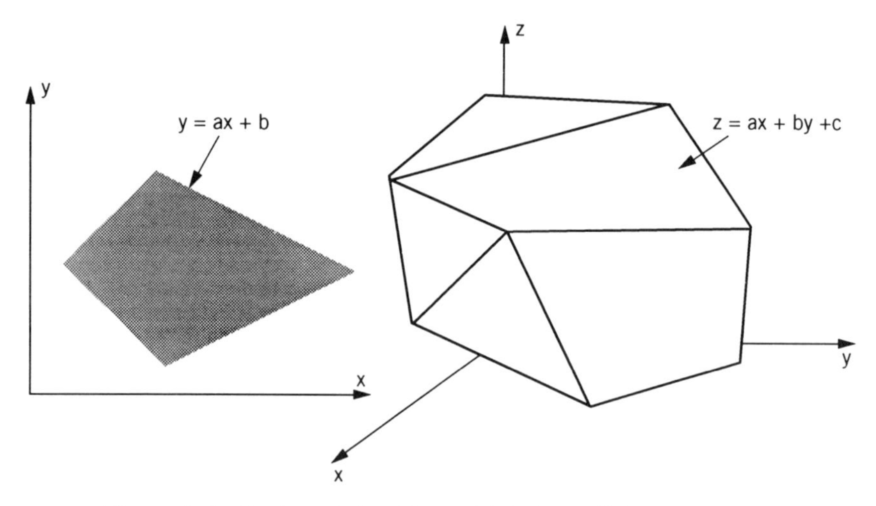

Figure A.9 Regions defined by multiple inequalities and equalities.

Which of these two regions is the acceptable, or *feasible* one may be determined simply by checking whether a point with given coordinates (for instance, the origin of the coordinates, i.e., the point for which $y = 0$ and $x = 0$) satisfies the relationships or not. That is, if in the case of the first inequality above, $\mathbf{y}(\mathbf{0}) < \mathbf{b}$, then the origin, together with all the other points on the same side of the curve, belongs to the feasible region.

In general, a region may be defined by several such inequalities as represented in Figure A.9.

A.1.7 Convexity

A function $y(x)$ is said to be everywhere *strictly convex* if all points on the straight line joining any two points on the graph of the function are above the graph. Graphically, this is represented as in Figure A.10 for the case of a univariate function. The same definition applies to multivariate functions, the straight line being replaced by a plane.

If some of the points on such a straight line are *on* the curve, the function is convex (i.e., not strictly). An example of such a function is also provided in Figure A.11. More formally, a strictly convex function is such that

$$y(\lambda x_1 + (1 - \lambda)x_2) < \lambda y(x_1) + (1 - \lambda)y(x_2);$$

$$\text{for any } \lambda \text{ such that } 0 \leq \lambda \leq 1 \tag{A.15}$$

For a *convex* function, the "<" sign is replaced by a "≤" sign. The converse of a convex function is a *concave* function, for which the "≤" sign is replaced by a "≥" sign in Formula (A.15).

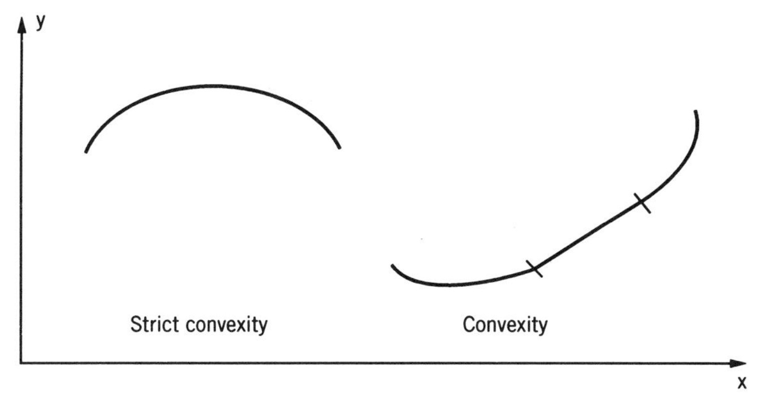

Figure A.10 Convexity and strict convexity.

Alternatively, strictly convex functions may also be defined as being such that

$$y(x + \Delta x) < y(x) + y'(x)\Delta x; \qquad \text{for any } x, \Delta x \tag{A.16}$$

In words, this property states that the value of a strictly convex function is always underestimated by a linear approximation. That is, the graph of a strictly convex function always lies above its tangent line (in the case of a univariate function), or tangent plane (in the case of a multivariate function). This is graphically represented for the former case in Figure A.11.

Alternatively, in the univariate case, strictly convex functions may be characterized as having a positive second derivative:

$$\frac{\partial^2 y}{\partial x^2} > 0$$

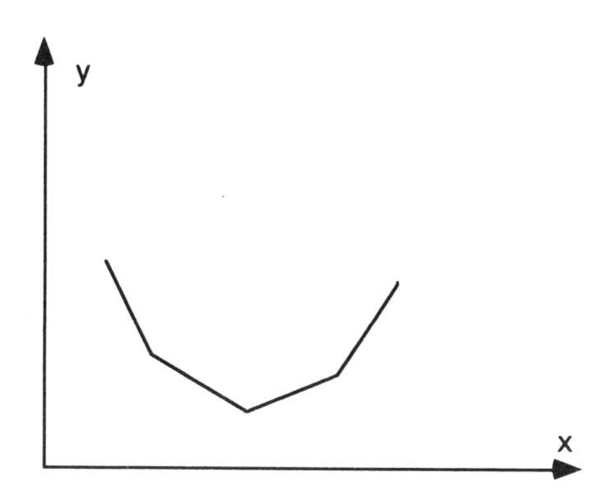

Figure A.11 Linear approximation for a convex function.

In the multivariate case, convexity may alternatively be determined from a *matrix* in which the respective second derivatives of the function with respect to its variables are arrayed, and which is called the Hessian.

$$H = \begin{bmatrix} \frac{\partial^2 y}{\partial x_1^2}, \frac{\partial^2 y}{\partial x_1 \partial x_2}, & \cdots & , \frac{\partial^2 y}{\partial x_1 \partial x_n} \\ & \vdots & \\ \cdots, & \frac{\partial^2 y}{\partial x_i \partial x_j}, & \cdots \\ & \vdots & \\ \frac{\partial^2 y}{\partial x_n \partial x_1}, & \cdots\cdots\cdots & , \frac{\partial^2 y}{\partial x_n^2} \end{bmatrix} \quad \text{(A.17)}$$

This matrix must be "positive definite" for a function to be strictly convex. This means that the product $\mathbf{xHx^t}$, which is a number, must be positive for any nonzero vector $\mathbf{x}$. For our purposes, there are several criteria for testing positive-definiteness. One is that if the matrix is *diagonal*, as below (in the case of a 3×3 matrix),

$$\begin{bmatrix} a_{11} & 0 & 0 \\ 0 & a_{22} & 0 \\ 0 & 0 & a_{33} \end{bmatrix}$$

all elements a_{ii} must be > 0. Another is that if the matrix is symmetric, of the form

$$\begin{bmatrix} a_{11} & \cdots & a_{1j} & \cdots & a_{1n} \\ \cdots & \cdots & \cdots & \cdots & \cdots \\ a_{i1} & \cdots & a_{ii} & \cdots & a_{in} \\ \cdots & \cdots & \cdots & \cdots & \cdots \\ a_{n1} & \cdots & a_{n2} & \cdots & a_{nn} \end{bmatrix}$$

where $a_{ij} = a_{ji}$ for all i and j, a *sufficient* condition for positive-definiteness is

$$a_{ii} > \sum_{j \neq i} |a_{ij}|; \qquad \forall\, i \quad \text{(A.18)}$$

Finally, another condition is that all "principal minor determinants" be positive. Convexity in multivariate functions correspond to semipositive definiteness of their Hessian, for which the $>$ sign is replaced by $\geq$ in the above conditions.

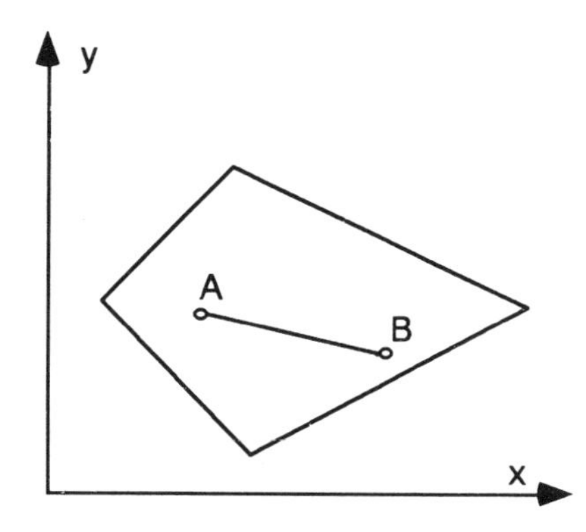

Figure A.12 Convex region.

The concept of convexity also applies to regions in an n-dimensional space. A region will be convex if all the points on a segment joining any two points belonging to the region are also in the region. Graphically, this is represented in Figure A.12.

Mathematically, this condition states that if points $\mathbf{x}_1$ and $\mathbf{x}_2$ are in the region, then the line segment $\lambda \mathbf{x}_1 + (1 - \lambda)\, \mathbf{x}_2$ will entirely lie in the region, for any λ such that $0 \leq \lambda \leq 1$.

If a function $y(\mathbf{x})$ is convex, then the region defined by the inequality $y(\mathbf{x}) \leq \geq b$ is convex. Consequently, feasible regions defined by linear equalities or inequalities are convex regions.

A.2 OPTIMALITY CONDITIONS FOR NONLINEAR MATHEMATICAL PROGRAMS

Many of the models developed in this text are stated as constrained, nonlinear minimization problem, that is, as the solution of the problem of finding the minimum of a nonlinear, multivariate function.

$$\min_{\mathbf{x}} F(x_1, \cdots x_i, \cdots, x_n) \tag{A.19}$$

subject to m constraints of the form

$$\begin{array}{l} C_1(x_1, x_2, \cdots x_i, \cdots, x_n) = C_1(x) \geq L_1 \\ \cdots\cdots\cdots\cdots\cdots\cdots\cdots\cdots\cdots\cdots \\ C_j(x_1, x_2, \cdots x_i, \cdots, x_n) = C_j(x) \geq L_j \\ \cdots\cdots\cdots\cdots\cdots\cdots\cdots\cdots\cdots\cdots \\ C_m(x_1, x_2, \cdots x_i, \cdots, x_n) = C_m(\mathbf{x}) \geq L_m \end{array} \tag{A.20}$$

The expressions $C_i(\mathbf{x})$ may or may not be linear. Also, the n variables x_i are assumed to be nonnegative:

$$x_i \geq 0; \qquad \forall\ i \tag{A.21}$$

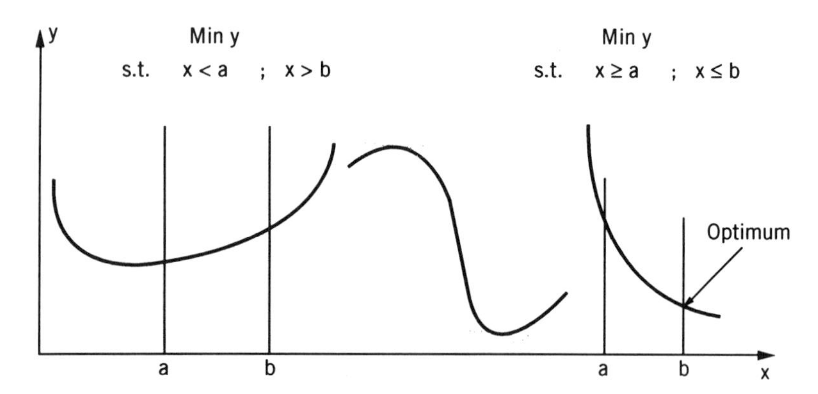

Figure A.13 Nonexisting, multiple, and boundary solutions.

The L_i's are constants, which may be positive, negative, or equal to zero. The $(m + n)$ Constraints (A.201) define the "feasible region" in the n-dimensional space of the variables.

This particular approach raises two basic questions. The first one is whether a given optimization program will have a solution. The existence of a solution is by no means always guaranteed for any program, as may be seen in Figure A.13, in which the feasible region is empty, that is, it does not have any points in it. However theoretically correct, a model without any solutions would obviously be of little use.

The other question is whether there is a unique solution, or whether there are multiple solutions to the optimization problem. This is illustrated in Figure A.13. While there is inherently nothing wrong about the possibility of multiple solutions, this nevertheless brings the problem of determining which of these is the "right," or the most likely one. Thus, if possible, it is in general preferable to use models which have a unique solution.

A.2.1 Unconstrained Programs

An important special case of mathematical programs is when they are unconstrained, of the form

$$\text{Min } F(x_1, \cdots x_i, \cdots, x_n) \tag{A.22}$$

In this case, for the Objective Function (A.22) to be at an optimum (i.e., maximum or minimum), its first-order derivatives with respect to all of its variables must be equal to zero:

$$\frac{\partial F}{\partial x_i} = 0; \qquad \forall\ i \tag{A.23}$$

The reason for these *necessary* first-order[3] conditions is graphically represented, in the case of an univariate function, in Figure A.13. This is a system of n equations in n unknowns.

However, it is also clear that this condition would also be met for a maximum. We thus need another criterion for detecting whether the solution is a minimum or a maximum. As can be seen in the figure, if the function is strictly convex at the solution, the optimum will be a minimum, and conversely, if it is strictly concave, it will be a maximum. Therefore, if the Hessian of $F(x)$ is positive definite at the solution, it will be a minimum. In the case of a univariate function, this second-order condition is that the function's second derivative be positive.

Regarding the uniqueness of the minimum, if $F(x)$ is convex (or *a fortiori* strictly convex) everywhere else than at the solution, the minimum will be unique, global. This may again be tested from the value of the Hessian, applying any of the tests described above.

A.2.2 Constrained Programs

In the case of constrained optimization problems, it is apparent that the first-order Conditions (A.23) are no longer valid, as a minimum may take place at a point where the gradient of the function is not zero, for instance, a boundary solution, on one of the constraints. Such a situation is represented in Figure A.13.

For constrained programs, the first-order optimality conditions are the so-called Karush-Kuhn-Tucker (KKT) conditions, from the names of the mathematicians who first derived them. These are *necessary* conditions, which must be observed by the solution of any constrained multivariate optimization program. It should be noted that there may be rare cases (i.e., when the gradients of the constraint functions $C_j(x^*)$ are not independent) when the KKT conditions do not hold at optimality. We will state the KKT conditions with respect to Program (A.19)–(A.20), which is the standard form for a minimization problem. Conditions for maximization problems, which we are not using, take a symmetrical form.

The KKT conditions state that if a point $x^* = (x_1^*, \cdots x_i^*, \cdots, x_n^*)$ is an optimal solution to this problem, then there exist a set of quantities λ_i, called the dual variables of the program, such that, at the solution[4]

$$\frac{\partial F(x^*)}{\partial x_i} - \sum_j \lambda_j \frac{\partial C_j}{\partial x_i}(x^*) \geq 0; \qquad \forall\, i \tag{A.24a}$$

$$x_i^* \left[\frac{\partial F(x^*)}{\partial x_i} - \sum_j \lambda_j \frac{\partial C_j}{\partial x_i}(x^*) \right] = 0; \qquad \forall\, i \tag{A.24b}$$

[3]So named because it concerns the first derivatives.

[4]The symbol ∀ means "for all values of."

$$\lambda_j[C_j(x^*) - L_j] = 0; \qquad \forall j \qquad \text{(A.25a)}$$

$$C_j(x^*) \geq L_j; \qquad \forall j \qquad \text{(A.25b)}$$

$$x_i \geq 0; \qquad \forall i \qquad \text{(A.26a)}$$

$$\lambda_j \geq 0; \qquad \forall j \qquad \text{(A.26b)}$$

For a maximization problem, the direction of Inequations (A.24a) and (A.25b) is reversed, to $\leq$.

The first set of Conditions (A.24a) and (A.24b), taken together, state that at the minimum, either the gradient of the objective function is a linear combination with positive weights of the gradients of the constraints, or the minimum solution is on one of the coordinate axes (i.e., some x_i^*'s = 0). The next two sets of conditions are called the "complementary slackness conditions." Taken together again, they state that if constraint j is effective, or binding (i.e., $C_j(x^*) = L_j$), the value of dual variable λ_j is strictly positive, and conversely, if it is not binding, ($C_j(x^*) > L_j$), then λ_j is equal to zero. Finally, the last two sets of conditions state that all variables, primal and dual, are positive-valued.

The KKT conditions may be considered as the generalization of the conditions for optimality of unconstrained programs, Formula (A.23). In fact, this formula is a special case of Formulas (A.24) and (A.25) when there are no constraints, when the λ_j's are all equal to zero. Also, when the Objective Function $F(x)$ (A.19) in the above minimization problem is convex for all $\mathbf{x}$, and if in addition the feasible region is also convex, the KKT conditions are necessary and sufficient, that is, they may be used to solve the problem directly. In practice, however, this approach is not very efficient, given the nature of (In) Equations (A.24) and (A.25).

When all constraints in the optimization problem (other than nonnegativity requirements for the variables) are equalities, the KKT conditions may be expressed equivalently in a somewhat simpler form. In this case, one may form the Lagrangian of the optimization problem:

$$G(x, \lambda) = F(x) - \sum_j \lambda_j[C_j(x^*) - L_j] \qquad \text{(A.27)}$$

The KKT conditions may then be stated, with respect to the Lagrangian, as:

$$\frac{\partial G(x, \lambda)}{\partial \lambda_j} = 0; \qquad \forall j \qquad \text{(A.28a)}$$

$$x_i \frac{\partial G(x, \lambda)}{\partial x_i} = 0; \qquad \forall i \qquad \text{(A.28b)}$$

$$\frac{\partial G(x, \lambda)}{\partial x_i} \geq 0; \qquad \forall\, i \tag{A.28c}$$

Regarding the existence of solutions to constrained optimization programs, if the feasible region is defined by linear (in)equalities, which will be our case, and is not empty, (i.e., there exists at least one set of values for the program's variables which observe all constraints), then the program will have at least one (optimal) solution. More generally, a sufficient condition is that the feasible region be closed and bounded (i.e., compact).

Regarding uniqueness, if the Objective Function $F(\mathbf{x})$ (A.19) is the above minimization problem is strictly convex for all $\mathbf{x}$, and if, in addition, the feasible region is also convex, then the solution will be unique.

A.3 ELEMENTS OF PROBABILITY THEORY

A.3.1 Random Events

Many of the models of travel demand we have developed in this text are probabilistic in the sense that the predicted traveler demands are not known with certainty. That is, in contrast with the variables discussed above, we do not know with certainty what the demands will be, only their average Y_j in the notation of the exposition. Their actual value is random. Standard, classical calculus of the kind reviewed above is therefore inappropriate to deal with such random variables. Special concepts are thus required, which we now review in this section.

We begin with the basic concept of probability. This applies to situations in which some potential events may or may not occur. These respective events may be referred to, *exhaustively*, as $E_1, E_2, \cdots, E_j, \cdots, E_n$. For instance, Event 1 may correspond to the fact that a given traveler does not choose a given destination, while Event 2 will be that he or she does. Graphically, this may be represented in the form of the "sample space" for the events, as in Figure A.14.

The probability of a given event is defined as a function, in the sense of section A1.1 above, of the event, and will be noted $P(E_j)$. This function must observe several basic requirements:

$$0 \leq P(E_j) \leq 1.0; \qquad \forall\, j \tag{A.29a}$$

$$P(E) = 1.0 \text{ where } E \text{ is the entire sample space} \tag{A.29b}$$

$$P(\varnothing) = 0 \tag{A.29c}$$

where $\varnothing$ is the complement of the sample space (the empty space).

The first axiom states that the probability of any event has a value always between zero and one, inclusive. The second states that the probability of the

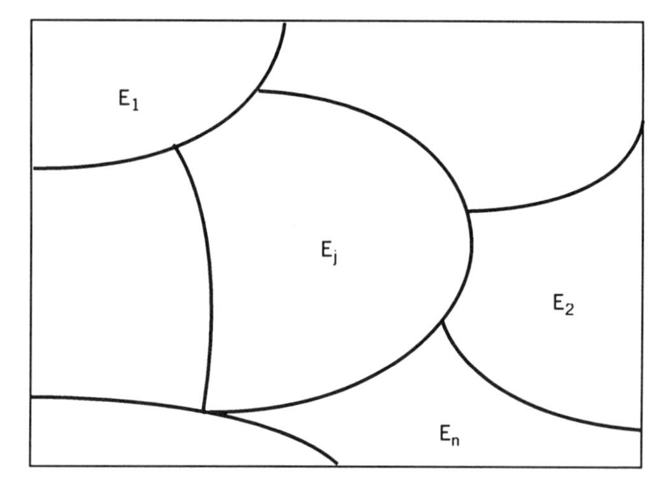

Figure A.14 Sample space for events.

"certain" event is always one. The third states that the probability of the "impossible" event is always zero. These last two statements may also be taken as a definition of the "certain" and "impossible" choices, respectively.

Having defined probabilities of basic events, we may now define various compound events, by combining, in the form of unions ("either or"), and intersections ("and"), individual basic events as defined above. The events thus constructed may be called "outcomes." This is graphically illustrated in Figure A.15.

The fundamental rule of probability calculus is that the probability $P(E_i \cup E_j)$ of the outcome defined as "*either* one *or* the other" of two events E_i and E_j is equal to the sum of the individual events' probabilities, minus the probability $P(E_i \cap E_j)$ that the events occur simultaneously. Formally,

$$P(E_i \cup E_j) = P(E_i) + P(E_j) - P(E_i \cap E_j); \qquad \forall\, i, j \qquad \text{(A.30a)}$$

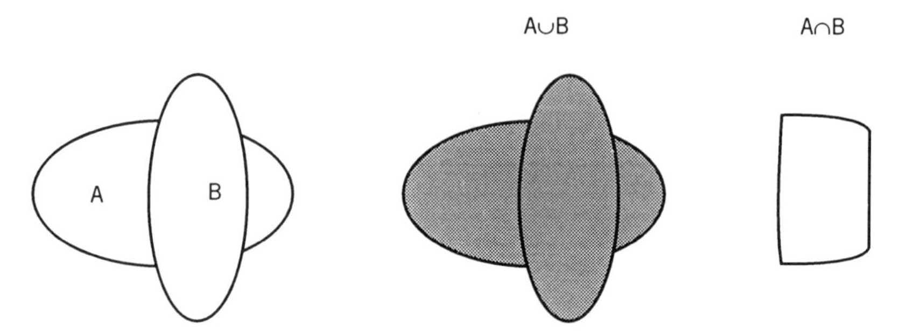

Figure A.15 Union and intersection of events.

$P(E_i \cap E_j)$, or $P(E_iE_j)$ for short, is the *joint* probability of the two events, the probability that they occur simultaneously. Note that conversely, this may be restated as the probability that two events occur simultaneously is equal to the sum of the probabilities that they occur individually, minus the probability that "*either* one *or* the other occur." This illustrated graphically in Figure A.15.

If two events are mutually exclusive (cannot occur simultaneously), then the probability of their joint occurrence is of course zero, and thus

$$P(E_i \cup E_j) = P(E_i) + P(E_j) \tag{A.30b}$$

The probabilities of events may be estimated either empirically, or, in some cases, theoretically. The *observed* probability that event E_j occurs may be defined as

$$P_j = N_j/N \tag{A.30c}$$

where N_j is the number of times event E_j has been observed to occur, out of a total number of observations (i.e., events) N. Thus, P_j as defined by Formula (A.30) represents the *frequency* with which event j has been observed to occur, and constitutes an approximation to the exact, or "true," theoretical probability. When the number of occurrences/observations becomes infinite, the observed probability as defined by Formula (A.30c) tends to the theoretical probability.

In some cases, it is possible to estimate probabilities theoretically, without recourse to empirical observations, but solely based on logical derivations. This is the case, incidentally, for a very important probabilistic model for our purposes, the logit model (see section 2.3 in Chapter 2).

Given the above definitions, the calculus of random variables, or "probability theory," may then be developed. We now present its most useful results.

A.3.2 Conditional Probabilities

The *conditional* probability of event E_i *given* event E_j is defined as

$$P(E_i/E_j) = P(E_iE_j)/P(E_j); \qquad \forall\, i, j \tag{A.31}$$

Note the special symbol $P(E_i/E_j)$ for conditional probabilities. The conditional probability may be interpreted as the original, unconditional, probability, rescaled to a sample space limited to event E_j. This concept is graphically represented in Figure A.16.

In practical terms, conditional probabilities are estimated, or derived, in the same manner as unconditional probabilities, but with a sample space limited to the conditioning event. For instance, the conditional probability that a given traveler makes a choice of destination j, given that he or she has already chosen access mode m, would be the frequency with which destination j is chosen by

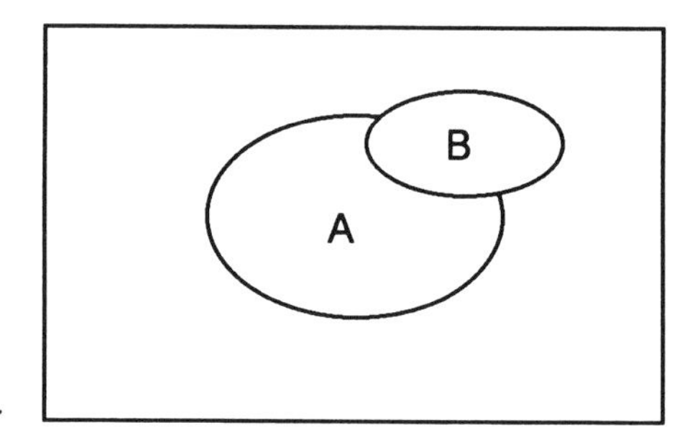

Figure A.16 Conditional probability.

choosers of mode m, and would thus be estimated within mode m's choosers only.

From Formula (A.31) we have

$$P(E_i E_j) = P(E_i/E_j)P(E_j); \quad \forall\, i, j \tag{A.32a}$$

or symmetrically, since changing the order of the two events obviously does not affect the probability of their joint occurrence:

$$P(E_i E_j) = P(E_j/E_i)P(E_i); \quad \forall\, i, j \tag{A.32b}$$

Repeating this process with a third event E_k, the joint probability of the three events would similarly be

$$P(E_i E_j E_k) = P(E_k/E_i E_j)P(E_i E_j/E_i)P(E_i); \quad \forall\, i, j, k \tag{A.32c}$$

and so on.

Two events are statistically independent if

$$P(E_j/E_i) = P(E_j) \tag{A.33a}$$

or, equivalently, if

$$P(E_i E_j) = P(E_j)P(E_i) \tag{A.33b}$$

A.3.3 Random Variables

In many situations, the probabilistic events discussed above may be associated to, or represented by, a numerical value. In this case, the occurrence of the event may conveniently be thought of as the event that a given numerical value is taken. The one-to-one correspondence (or function, in the sense of section A.1.1), which associates events with (real-valued) numbers is called "random variable" Y. That is,

$$Y(E_i) = y_i; \quad \forall\, i.$$

where Y is the random variable and y_i the value corresponding to event i.

For instance, the value of a random variable corresponding to the event that a destination's volume on a given day equals 300 might most conveniently, but not necessarily, be 300. In any event, the sample space in such a case corresponds to the set of values for the random variable. The random variable is called "discrete" or "continuous," depending on whether these values are discrete (and thus in finite number) or continuous (and in infinite number). This is illustrated in Figure A.17.

In the same manner as for events, to each value of the random variable Y is attached a probability of occurrence. When the y_i's are discrete, these probabilities are specified in the form of a *probability distribution function*, a set of discrete probabilities $P(y_i)$. These probabilities must observe the axioms of probability above, that is, be such that

$$0 \leq P(y_i) \leq 1.0; \qquad \forall\, i \tag{A.34a}$$

$$\sum_i P(y_i) = 1.0 \tag{A.34b}$$

$$P(a \leq y \leq b) = \sum_{a \leq y_i \leq b} P(y_i) \tag{A.34c}$$

When the values of the random variable are continuous, the probability of a single value is always zero. This might be expected intuitively, since in this case there is an infinite number of values, among which the total probability of 1.0 is distributed. However, the probability that the value of the random variable is in an infinitesimally small interval starting at a given value, say y_0 is *defined* as

$$P(y_0 \leq y \leq y_0 + dy) = f_Y(y_0)\, dy \tag{A.35}$$

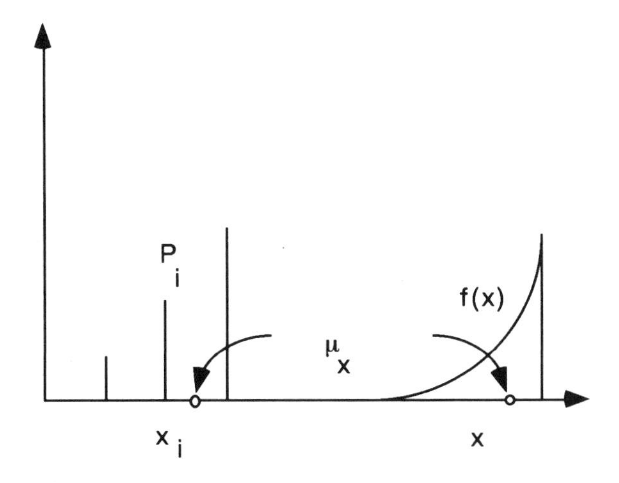

Figure A.17 Values for discrete and continuous random variables.

The function $f_Y(y)$ is called the "probability density function," or p.d.f. for short. It may be helpful to think of probabilities as weights. In this analogy, the p.d.f. plays the role of the weight density function in physics. Remembering the discussion about functions and their integrals above, it may be seen that definition (A.35) above implies the equivalent definition

$$P(a \leq y \leq b) = \int_a^b f_Y(y)\, dy \tag{A.36}$$

That is, the probability that the value of the continuous random variable falls in an interval from a and b is equal to the integral of the p.d.f. over the given interval. Accordingly, if the "*cumulative* probability function" (c.p.f.) $F_Y(y_0)$ is defined as the probability that the random variable's value is *less than or equal* to a given value, say y, then

$$F_Y(y_0) = \int_{-\infty}^{y_0} f_Y(y)\, dy \tag{A.37}$$

Remembering the earlier discussion, we see that the c.p.f. is mathematically the integral of the p.d.f. This implies that the p.d.f. is the derivative of the c.p.f.:

$$f_Y(y) = \frac{dF_Y(y)}{dy}; \qquad \forall\, y \tag{A.38}$$

In the case of continuous random variables, the basic probability axioms (A.28) imply that the p.d.f. must be such that

$$0 \leq f_Y(y) \leq 0; \qquad \forall\, y \tag{A.39a}$$

$$\int_{-\infty}^{+\infty} f_Y(y)\, dy = 1.0 \tag{A.39b}$$

This, in terms of the c.p.f., implies that

$$F_Y(-\infty) = 0 \tag{A.40a}$$

$$F_Y(+\infty) = 1 \tag{A.40b}$$

$$\frac{dF_Y(y)}{dy} \geq 0: \qquad \forall\, y \tag{A.40c}$$

The concept of a univariate random variable as discussed above may be extended to the multivariate case. A multivariate random variable is defined as a *vector* function

$$Y = \begin{bmatrix} Y_1 \\ Y_2 \\ \vdots \\ Y_i \\ Y_n \end{bmatrix}$$

in which each of the Y_i's is a random variable. The probability that *each* of the component random variables' value falls into specified, infinitesimally small intervals is defined as

$$\begin{aligned} P(y_{10} &\leq Y_1 \leq y_{10} + dy_1; \cdots, y_{i0} \leq Y_i \leq y_{i0} + dy_i \cdots; \\ &y_{n0} \leq Y_n \leq y_{n0} + dy_n) \\ &= f_Y(y_{10}; \cdots, y_{i0}, \cdots; y_{n0})\, dy_1 \cdots dy_i \cdots dy_n \qquad \text{(A.41)} \end{aligned}$$

The function $f_Y(y_1; \cdots, y_i, \cdots; y_n)$ is the multivariate p.d.f. Equivalently, the joint cumulative probability $F_Y(y_o)$ may be computed, in similarity with the univariate case, as

$$\begin{aligned} F_Y(y_o) &= F_Y(y_{10}, \cdots, y_{i0}, \cdots; y_{n0}) \\ &= P(y_1 \leq y_{10}; \cdots, y_i \leq y_{i0}, \cdots; y_n \leq y_{n0}) \\ &= \int_{-\infty}^{y_{10}} \cdots \int_{-\infty}^{y_{i0}} \cdots \int_{-\infty}^{y_{n0}} \\ &\quad \cdot f_Y(y_1, \cdots, y_i, \cdots; y_n)\, dy_1 \cdots dy_i \cdots dy_n \qquad \text{(A.42)} \end{aligned}$$

In other words, the multivariate p.d.f. is the nth-order partial derivative of the c.p.f., or, the c.p.f. is the n-th dimensional multiple integral of the p.d.f. Given their definition, cumulative probability distribution functions must be such that

$$F(\infty) = 1$$

$$F(-\infty) = 0$$

Finally, n random variables Y_i are said to be statistically independent if

$$f(\mathbf{y}) = f_{Y1(y1)} f_{Y2(y2)} \cdots f_{Yn(yn)}; \qquad \forall\, \mathbf{y} \qquad \text{(A.43)}$$

where $fy_i(y_i)$ is the *marginal*, or individual p.d.f. of y_i.

A.3.4 Means, Variances, and Covariances of Random Variables

We now define several summary (i.e., single-valued) descriptors of probability distributions, or equivalently of the corresponding random variables. Given the values of a discrete random variable, the corresponding probabilities, the *mean*, or expected value, or average, μ_Y, is defined as

$$\mu_Y = \sum_i y_i P_i \tag{A.44a}$$

This quantity may be interpreted as the weighted average of the random variable's values, where the weights are equal to the values' probabilities of occurrence. In the analogy above, the value of the mean along the axis of y values represents the position of the center of gravity of the probabilities. This is graphically represented in Figure A.17.

In the case of continuous random variables, the mean is estimated in the same manner, replacing the P_i's at location i by element $f_Y(y)\,dy$ for infinitesimal interval $(y, y + dy)$.

$$\mu_Y = \int_{-\infty}^{+\infty} y f_Y(y\,dy) \tag{A.44b}$$

The mean is similarly located at the center of gravity of the mass corresponding to the p.d.f. (as also represented in Figure A.20). The distribution of probabilities, either discrete or continuous, is thus in equilibrium around the position of the mean μ_Y.

The variance σ_Y^2 of a random variable is defined as the average value, in the sense above, of the squared deviations of the random variable's values from their mean. The main reason for using the square of the deviations is that, as is easy to show, the average of the deviations themselves is always equal to zero. Thus, for discrete and continuous random variables, respectively:

$$\sigma_Y^2 = \sum_i (y_i - \mu_Y)^2 P_i \tag{A.45a}$$

$$\sigma_Y^2 = \int_{-\infty}^{+\infty} (y - \mu_Y)^2 f_Y(y)\,dy \tag{A.45b}$$

The standard deviation σ is defined as the square root of the variance, and is measured in the same units as for the random variable's values, which is not the case for the variance. The value of the standard deviation may be loosely interpreted as a proxy for the average deviation of the random variable's value from its mean.

More rigorously, the value of the standard deviation σ may be used to *underestimate* the probability that the random variable's value will lie within an interval of half-width $k\sigma$ centered on the mean, as

$$P(\mu - k\sigma \leq Y \leq \mu + k\sigma) \geq 1 - \frac{1}{k^2}; \qquad \forall\, k \tag{A.46}$$

For instance, if the standard deviation of daily demand is equal to 40, the probability that demand on a given day will be somewhere between $300 - 2(40) = 220$ and $300 + 2(40) = 380$ is at least equal to $1 - 1/(2^2) = 0.75$, or 75%. Of course, if the probability distribution function itself is used, instead of its two summaries μ and σ, then the probability of the range of values above can be estimated exactly, using Formulas (A.43a) or (A.43b). Thus, indicators such as μ and σ provide compact, but incomplete, information about the random variable.

The *covariance* of two random variables is defined as the mean, or expected value, of the product of their individual deviations from their respective means.

$$\text{Cov}\,(Y_1, Y_2) = \int_{-\infty}^{+\infty} \int_{-\infty}^{+\infty} (y_1 - \mu_{Y_1})(y_2 - \mu_{Y_2}) f_{Y_1Y_2}(y_1; y_2)\, dy_1\, dy_2 \tag{A.47}$$

where $f(y_1; y_2)$ is the joint p.d.f. for the two variables, as defined above. It may be noted that the variance of a single random variable is in fact the covariance with itself.

Two random variables y_1 and y_2 are statistically independent if

$$P(Y_1 = y_{1i}; Y_2 = y_{2j}) = P(Y_1 = y_{1i})P(Y_2 = y_{2j}); \qquad \forall\, y_{1i}, y_{2j}$$

That is, any combination of events corresponding to the occurrences of values for the respective variables are independent, in the sense of Definition (A.33a).

It is easy to see that if two random variables are statistically independent, their covariance will be equal to zero. This is not true reciprocally. That is, if two variables have zero covariance, it only means that they are *linearly* independent, there is no linear relationship between the values they respectively and randomly take. There might, however, be another, nonlinear correspondence. If the random variables are statistically independent, however, there is no relationship, linear or otherwise, between their realizations.

The coefficient of correlation ρ is defined as

$$\rho(Y_1, Y_2) = \text{Cov}\,(Y_1, Y_2)/\sigma_{Y_1}\sigma_{Y_2} \tag{A.48}$$

and is thus a rescaled covariance, so that its values are between -1 and $+1$. The interpretation of the values of ρ is represented in Figure A.18.

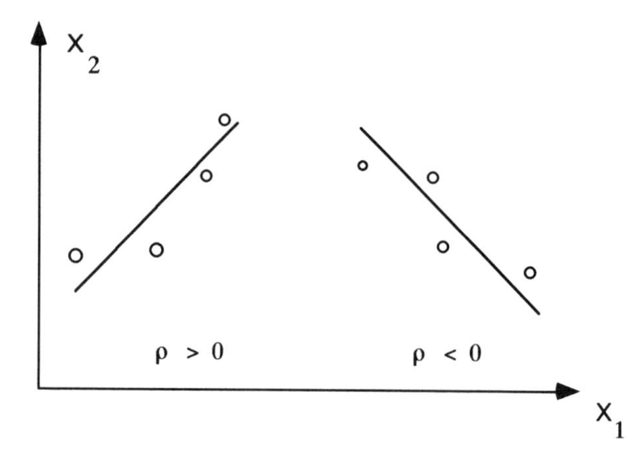

Figure A.18 Coefficient of linear correlation between two random variables.

A.3.5 Several Useful Probability Distribution Functions

A.3.5.1 The Bernoulli Distribution This is perhaps the simplest, most basic distribution, the "coin toss" distribution. It applies to the outcome of a "binary" experiment. There are only two discrete values for the variable, 1 (or "yes," "open," "tails," etc.), and 0 (or "no," "closed," "heads," etc.). $P_1 = p$ and $P_2 = q = 1 - p$. The distribution is thus characterized by the single parameter p. Such a distribution applies to the variables δ_{nj} used in section 2.4 of Chapter 2 and 7.2 of Chapter 7, and which described whether traveler n chooses destination j or not. The distribution is represented in Figure A.19.

It is easy to show that the mean and variance of the Bernoulli distribution/

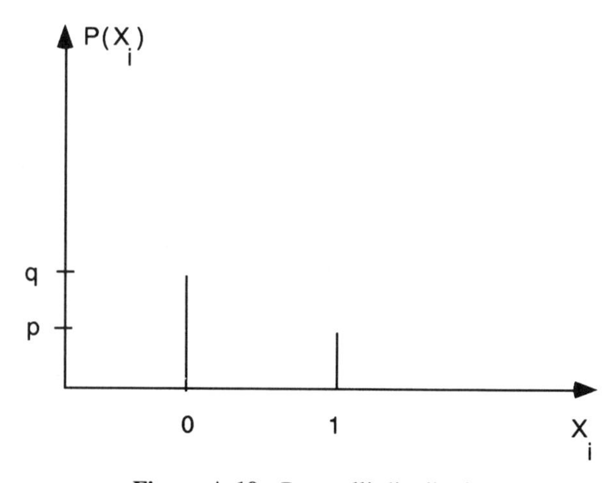

Figure A.19 Bernoulli distribution.

random variable are respectively

$$\mu = p \quad \text{(A.49a)}$$

$$\sigma^2 = p(1 - p) \quad \text{(A.49b)}$$

A.3.5.2 The Binomial Distribution This distribution applies to the random variable representing the total number of ''1'' 's (''yesses,'' or ''heads,'' etc.), in a series of n similar Bernoulli experiments. It applies, for instance, to the total number of travelers observed as patronizing a given destination in a sample of size n, assuming that all travelers have the same probability of choosing destination j, and the experiment is, again, a ''yes/no'' one. Consequently, the range of values for the variable is from 0 to n, inclusive. The probability that the total number of 1's, in effect the sum of the outcomes, if the other value is ''zero,'' is equal to, say, x is

$$P(Y = x) = \frac{n!}{x!(n - x)!} p^x(1 - p)^{n-x}; \; x = 0, 1, \cdots, n \quad \text{(A.50)}$$

where $x! = x(x - 1)(x - 2) \cdots 2.1$, the product of all integers from 1 up to and including x.[5] The distribution is thus characterized by two parameters, p and n. It is graphically represented in Figure A.20.

The mean and variance of the binomial distribution are

$$\mu = np \quad \text{(A.51a)}$$

$$\sigma^2 = np(1 - p) \quad \text{(A.51b)}$$

[5] $x!$ is referred to as ''factorial'' x.

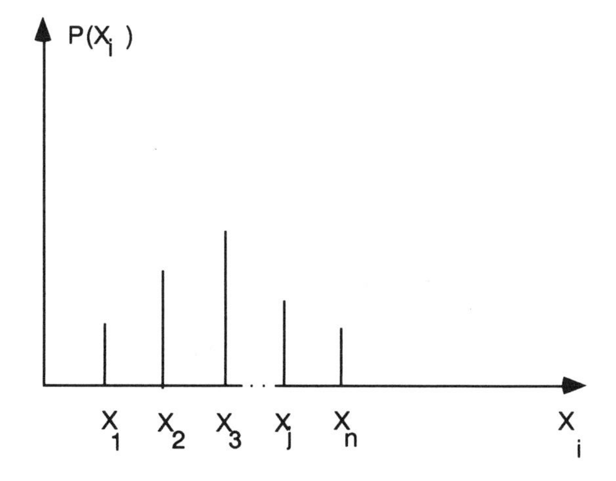

Figure A.20 Binomial distribution.

A.3.5.3 The Multinomial Distribution The multinomial distribution function may be thought of as a generalization of the binomial, in which the underlying Bernoulli variable is replaced with a variable which takes one of m values, y_1 through y_m. The probability of value y_i is p_i. The multivariate multinomial distribution function describes the *joint* probability that, in n realizations of the variable, the total number of outcomes y_1 of kind 1 is x_1, *and* the total number of outcomes y_2 of kind 2 is x_2, · · · , and so on. This probability is equal to[6]

$$P(Y_1 = x_1, \cdots, Y_i = x_i \cdots, Y_m = x_m)$$

$$= \frac{n!}{x_1!x_2! \cdots x_m!} p_1^{x_1} p_2^{x_2} \cdots p_m^{x_m} = \frac{n!}{\prod_{i=1}^{m} x_i!} \prod_{i=1}^{m} p_i^{x_i} \tag{A.52}$$

Of course

$$0 \leq x_i \leq n; \qquad \forall\, i$$

and

$$n = \sum_{i=1}^{m} x_i$$

It is easy to see that the binomial distribution above corresponds to the special case $m = 2$. In any case, this distribution would apply to the probability that, in a sample of n travelers, given numbers n_j of travelers choose a given destination j out of J possible alternatives.

The mean and variance of the multinomial variables y_i are

$$\mu_{Y_i} = np_i \tag{A.53}$$

$$\forall\, i$$

$$\sigma_{Y_i}^2 = np_i(1 - p_i); \tag{A.54}$$

The covariance between any two variables Y_i and Y_j is

$$\text{Cov}\,(Y_i; Y_j) = -np_i p_j; \qquad \forall\, i, j \tag{A.55}$$

A.3.5.4 The Normal Distribution The normal distribution function may be derived in its own right. However, it may also be thought of as an approximation to a special case of the binomial, specifically when the value of n is very large (or theoretically, infinite). The normal random variable is a contin-

[6]The symbol Π means "product of" (i.e., in the same manner as Σ means "sum of").

uous variable, which can take on any real value, between $-\infty$ and $+\infty$. The probability that the variable y takes a value less than y_0 is given by:

$$P(Y \leq y_0) = \frac{1}{\sqrt{2\pi}\,\sigma} \int_{-\infty}^{y_0} \exp\left(-\frac{(y-\mu)^2}{2\sigma^2}\right) dy \tag{A.56}$$

The p.d.f. and c.p.f. for the normal distribution are represented in Figure A.21.

The value of $P(Y \leq y_0)$ above for a given y_0, μ and σ, does not have to be estimated from the computation of the definite integral, but may be read from a ''standard normal'' table, after rescaling the value of y_0 into the corresponding ''Z value.'' (Such tables, as well as further details on their use, may be found in any standard text on probability theory and/or statistics.) In any case, the mean and variance of the normal variable are:

$$\mu = \mu \tag{A.57a}$$

$$\sigma^2 = \sigma^2 \tag{A.57b}$$

Thus, the two parameters of the distribution function are its actual mean and standard deviation. The univariate normal distribution function may be generalized to a multivariate version.

The importance of the normal distribution function comes from the fact that the distribution of the sum, or average, of the values taken by *any* random variable will become more and more closely approximated by a normal distribution, as the size of the sample of values becomes increasingly larger. Therefore, the normal distribution function may be applied to entities which result from the addition, or compounding, of several identical random factors, such as measurement, or perception errors. For this reason, the normal distribution is in principle the most appropriate in connection with random formulations of utility, as discussed in section 2.3 of Chapter 2.

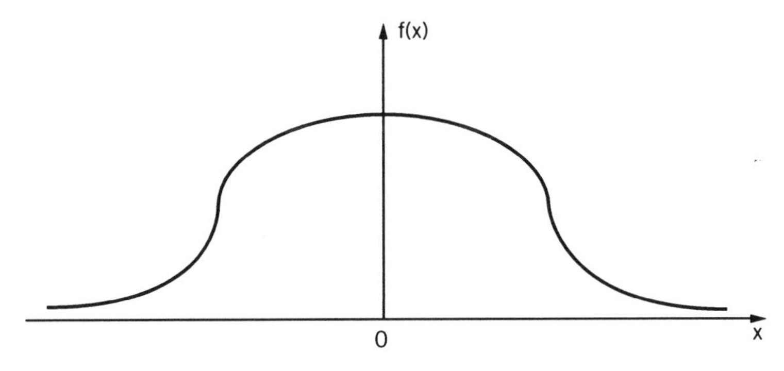

Figure A.21 The normal distribution function.

A.3.5.5 The Gumbel Distribution For reasons also discussed in section 2.3 of Chapter 2, this presents analytical difficulties. It turns out that another probability distribution function, whose general shape closely resembles that of the normal distribution function, is much more convenient analytically. This, however, entails some conceptual drawbacks, which are discussed in the same section.

This distribution is the Gumbel or ''double exponential'' distribution, which is a member of the family of ''extreme value'' distributions, as it describes the probability density function of the maximum of an infinite number of i.i.d., continuous random variables.

The distribution is, as the normal, continuous, and covers the same range, all real numbers. The c.p.f. of a Gumbel random variable with mean zero is defined as

$$P(X \leq x) = F_X(x) = e^{-e^{-\beta x - E}}; \quad -\infty < x < \infty \qquad \text{(A.58a)}$$

where e is the base of natural logs ($e = 2.718 \cdots$) and E is Euler's constant ($E = 0.5772 \cdots$). The mean and variance of the distribution are[7]

$$\mu = 0 \qquad \text{(A.59a)}$$

$$\sigma^2 = \pi^2/6\beta^2 \qquad \text{(A.59b)}$$

The c.p.f. of the general Gumbel random variable with location parameter K and standard deviation σ is

$$P(X \leq x) = F_X(x) = e^{-e^{-\beta(x - K)}}; \quad -\infty < x < \infty \qquad \text{(A.58b)}$$

The expected value of this variable is equal to

$$\mu = E/\beta + K \qquad \text{(A.59c)}$$

The variance is the same.

A.3.6 Entropy of a Distribution

We now proceed to a concept which will intervene in the derivation and calibration of travel demand models. This is the concept of the entropy of a probability distribution function. Consider the set of values T_j representing, for instance, the numbers of travelers who patronize given destinations j. The total number of travelers is given, and is equal to T. Clearly, it is possible to assign, or allocate in several different ways, the same total number of travelers T to the various destinations j, so as to obtain the given distribution $\{T_j\}$ (the sym-

[7]See, for instance, Anderson et al. (1992), p. 57 for derivations.

bol $\{x_i\}$ means " the set of values x_i). For instance, the following distribution of travelers to two destinations

$$T_1 = 1;\ T_2 = 2$$

may be obtained from the given total of three travelers *A*, *B*, and *C* through the following three different assignments:

Assignment	$j = 1$	$j = 2$
1	A	B, C
2	B	A, C
3	C	A, B

Thus, in general, the number of different assignments N of T travelers which leads to a given distribution $\{T_j\}$ is in general a function of the values T_j. It is well known that the number of ways in which, out of a total of T objects, T_1 are assigned to class 1, T_2 to class 2, T_j to class j, and so on, is equal to

$$N = \frac{T!}{\prod_j T_j!} \tag{A.60a}$$

$x!$ (factorial x), is equal to $x.(x - 1) \cdots 1$. In the example above, N is equal to:

$$N = \frac{3!}{2!1!} = 3$$

Since N is typically very large, its natural logarithm, which is called the "entropy," is used instead to characterize the distribution. In general, therefore, the entropy E of a distribution of values is thus equal to

$$E = \text{Log}\, \frac{T!}{\prod_j T_j!} \tag{A.60b}$$

The importance of the entropy stems from the observation that if individual travelers are assumed to be similar, or interchangeable, then the most "probable" distribution of destination demands is that corresponding to the highest number of assignments N, the maximum entropy, since the logarithmic function increases monotonically with its argument.

The entropy of a given distribution $\{T_j\}$ is maximum when the distribution is uniform (when all T_j's have the same value). Conversely, it achieves its minimum value when the distribution is concentrated in a single value, $T_j = 0$ for all j 's except one. Thus, this value also measures the amount of "information'' the distribution of values contains. That is, when the distribution is uniform, it contains no information about the values whatsoever. At the other end of the scale the information is maximum, since one value has been singled out. Alternatively still, the entropy of a distribution may also be interpreted as a measure of the dispersion of the T_j values across the facilities. Maximum entropy/minimum information corresponds to maximum dispersion, and vice versa. The most probable distribution is also the least informative one.

In practical situations, the magnitude of the T_{ij}, measured in numbers of travelers, for instance, is typically in the hundreds. We may then use ''Sterling's approximation.'' This states that when x is large,

$$\ln (x!) \simeq x \ln x - x = x(\ln x - 1) \tag{A.61a}$$

With this approximation, the entropy is equal to

$$T(\ln T - 1) - \sum_j T_j(\ln T_j - 1) \tag{A.61b}$$

The same concept may, of course, be applied to a probability distribution function, instead of a distribution of values. In this case, the sum of the values is simply equal to 1.0, and the entropy is equal to

$$\begin{aligned} 1(\ln 1 - 1) - \sum_j P_j(\ln P_j - 1) &= -1 - \sum_j P_j(\ln P_j) + \sum_j P_j \\ &= -1 - \sum_j P_j \ln P_j + 1 = -\sum_j P_j \ln P_j \end{aligned}$$

A.4 ELEMENTS OF CONSUMER DEMAND THEORY

In this section, we briefly review basic elements of consumer demand theory, since we will conceive of individual travelers as ''consumers'' of trips. This review is not intended as a self-contained or detailed exposition, but to present some background on the mathematical results, theorems, and formulas which are invoked in the course of the development of our travel demand models.[8]

A.4.1 Direct Utility Function

As discussed in section 2.1 of Chapter 2, in our framework, an individual traveler is assumed to be faced with several alternatives j, such as modes, routes, destinations. The demands for each of these individual alternatives may be interpreted as the demands for various different versions of a same generic

[8]For further explanation, the reader is referred, for instance, to Varian (1992), the reference of choice in this area, from which this review is drawn.

"product," that is, one trip on a given mode, or route, or to a given destination, or using combinations thereof.

Under certain general conditions, it is possible to define a function which represents the preferences of an individual traveler in origin zone i for making specified numbers of trips T_{jmr} on given modes, routes, and so on, as well as the amount T_0 of "all other consumption" as a whole. These preferences are assumed to depend on the traveler's characteristics, X, and on the attributes of the trips. The function which represents these preferences is the *direct utility* function:

$$U = U(\mathbf{X}, \mathbf{T}) \tag{A.62}$$

Given an individual traveler's direct utility function U, the numbers of trips taken according to the various travel alternatives maximize the resultant utility, subject to a budgetary constraint. Therefore, the individual traveler's demands for the various types of trips may be obtained through solving the optimization problem

$$\begin{aligned} &\operatorname*{Max}_{\mathbf{T}} \ U(\mathbf{T}, \mathbf{X}) \\ &\text{s.t. } \mathbf{tc} + T_0 = b \end{aligned} \tag{A.63}$$

where $\mathbf{X}$ is the vector of traveler characteristics and of trip attributes facing him or her, $\mathbf{c}$ is the vector of trip costs, and b is the individual's budget. This implies that the price of "all other consumption" is arbitrarily set at 1, for convenience. Consequently, all other prices are scaled to that price, which, in practice, may be the consumer price index.

A.4.2 Demand Function

The demand functions T_j for trips of type j are defined as the optimal trip making levels, the multidimensional (vector) solution to Problem (A.63):

$$\mathbf{T}^*(\mathbf{X}, \mathbf{c}, b) \tag{A.64}$$

Demand functions possess certain analytical properties. These properties must be verified when demand functions are specified *a priori*, for analytical convenience, rather than derived in the above fashion. A basic requirement is that the function be continuous in its variables. Another important feature is that the matrix with general term

$$a_{ij} = \left[\frac{\partial T_j}{\partial c_i} + T_i \frac{\partial T_j}{\partial b}\right]; \qquad \forall\ i, j$$

be symmetric, negative, semidefinite.

At the optimal solution of Problem (A.63), the KKT conditions (see section A.2.2) imply that

$$\frac{\partial U(\mathbf{T})/\partial c_i}{\partial U(\mathbf{T})/\partial c_j} = \frac{c_i}{c_j}; \qquad \forall\, i, j \tag{A.65}$$

The fraction on the left represents the marginal rate of substitution between trips of type i and trips of type j.

Inverse demand functions are defined as specifying the levels of price, given the demands. These functions are equal to

$$c_j = \frac{\partial U}{\partial T_j} \Big/ \sum_k \frac{\partial U}{\partial T_k}; \qquad \forall\, j \tag{A.66}$$

A.4.3 Indirect Utility Function

The maximum value of the direct utility corresponding to the solution of Problem (A.63), which is the utility corresponding to the optimal trip making (i.e., the travel demands), is called the indirect utility, and is noted $\tilde{U}(\mathbf{X}, \mathbf{c}, b)$. Thus

$$\tilde{U}(\mathbf{X}, \mathbf{c}, b) = U^*(\mathbf{X}, \mathbf{c}, b) \tag{A.67}$$

A major property of the indirect utility function is "Roy's identity," which connects it with the demand functions, and which is stated as

$$T_j = -\frac{\partial \tilde{U}}{\partial c_j} \Big/ \frac{\partial \tilde{U}}{\partial b}; \qquad \forall\, j \tag{A.68}$$

This identity provides a way to obtain the demands, given indirect, rather than direct utility functions. Consequently, as far as demand specification is concerned, these two types of utility functions are equivalent. It should therefore not be surprising that given either one, the other may be retrieved. Specifically, the direct utility function is the solution of the following optimization problem:

$$U = \min_{c_j} \tilde{U}(\mathbf{c})$$

$$\text{s.t. } \mathbf{cT} = b \tag{A.69}$$

The relationships which close the circle among these three types of entities are

$$T_j = \frac{\partial \tilde{U}}{\partial c_j} \Big/ \sum_k \frac{\partial \tilde{U}}{\partial c_k}; \qquad \forall\, j \tag{A.70}$$

These formulas provide an alternative way to estimate demand functions, from indirect, rather than direct utility functions.

Finally, the *conditional* indirect utility function is defined as the indirect utility, *given that* trip j has been made.

A.4.4 The Representative Traveler Problem

Given several individual travelers, each with their own direct or indirect utility functions, one may ask whether it is possible to construct a direct or indirect utility function from which the aggregate travel demands of these travelers may be retrieved, using the relationships or definitions above. If so, the *single* traveler characterized by these utilities is termed the "representative traveler" (R.T.) for the individual travelers.

In some cases, the answer to this question is affirmative. In particular, it is necessary, as well as sufficient, that the indirect utility function for individual traveler n be of the so-called "Gorman" form:

$$\tilde{U}_n(\mathbf{c}, \mathbf{X}_n, b_n) = f_n(\mathbf{c}, \mathbf{X}_n) + b_n g(\mathbf{c}); \qquad \forall\, n \tag{A.71}$$

where $\mathbf{c}$ is the vector of prices, and $\mathbf{X}_n$ that of the traveler's characteristics.

The existence of the R.T. is of theoretical interest, as it endows the aggregate travel demands and aggregate utilities with microeconomic justification. In particular, its indirect utility may be used as a measure of social or community welfare for the given population of travelers. It is also of practical interest, as the aggregate demands may then be obtained as the solution of a single maximization problem subject to the single constraint reflecting the total budget.

In particular, this aggregate-level approach is the only possible one when "externalities" between individual travelers demands must be taken into account. Externalities are said to be present when an individual traveler's trip making affects the utility, and therefore the travel demand, of all others. An important example of negative externalities is congestion effects on the network's links, or at the destinations, whereby increasing numbers of users of such facilities increase the time it takes to use them.

APPENDIX B

SOLUTIONS TO SELECTED EXERCISES

1.2

Link	1	2	3	4	5	6	7	8	9	10
Route i, j, r										
1.2.1	1	0	0	0	0	0	0	0	0	0
1.2.2	0	0	1	0	1	1	0	0	0	0
1.2.3	0	1	0	0	0	1	0	0	0	0
1.3.4	0	0	1	0	0	0	0	0	0	0
1.3.5	0	1	0	0	0	0	0	1	0	0
1.3.6	1	0	0	1	0	0	0	1	0	0
1.4.7	0	1	0	0	0	0	0	0	0	0
1.4.8	1	0	0	1	0	0	0	0	0	0
1.4.9	0	0	1	0	1	0	0	0	0	0
2.1.10	0	0	0	0	0	0	0	0	0	1
2.1.11	0	0	0	1	0	0	1	0	0	0
2.1.12	0	0	0	1	0	0	0	1	1	0
2.3.13	0	0	1	0	0	0	0	0	0	1
2.3.14	0	0	0	1	0	0	0	1	0	0
2.3.15	0	1	0	0	0	0	0	1	0	1
2.3.16	0	0	0	1	0	0	1	0	1	0
2.4.17	0	0	0	1	0	0	0	0	0	0
2.4.18	0	1	0	0	0	0	0	0	0	1
2.4.19	0	0	1	0	1	0	0	0	0	1
3.1.20	0	0	0	0	0	0	0	0	1	0
3.1.21	0	0	0	0	1	0	1	0	0	0
3.1.22	0	0	0	0	1	1	0	0	0	1
3.2.23	1	0	0	0	0	0	0	0	1	0

Link	1	2	3	4	5	6	7	8	9	10
Route i, j, r										
3.2.24	0	0	0	0	1	1	0	0	0	0
3.2.25	0	1	0	0	0	1	0	0	1	0
3.2.26	1	0	0	0	1	0	1	0	0	0
3.4.27	0	0	0	0	1	0	0	0	0	0
3.4.28	0	1	0	0	0	0	0	0	1	0
3.4.29	1	0	0	1	0	0	0	0	1	0
4.1.30	0	0	0	0	0	0	1	0	0	0
4.1.31	0	0	0	0	0	1	0	0	0	1
4.1.32	0	0	0	0	0	0	0	1	1	0
4.2.33	0	0	0	0	0	1	0	0	0	0
4.2.34	1	0	0	0	0	0	1	0	0	0
4.2.35	1	0	0	0	0	0	0	1	1	0
4.3.36	0	0	0	0	0	0	0	1	0	0
4.3.37	0	0	1	0	0	0	1	0	0	0
4.3.38	0	0	1	0	0	1	0	0	0	1

1.3 Taking the logarithms of both sides of Equation (1.12), we get

$$\ln T_{ijmr} = \ln \alpha_{ijmr} + \sum_k \beta^k_{ijmr} \ln X^k_{ijmr}; \qquad \forall\, i, j, m, r \tag{B.1}$$

With the change of variable $T' = \text{Log}\ T$ and $X' = \text{Log}\ X$, this is now a standard multivariate regression of T' in the X's. The calibrated values of parameters β_k in Formula (1.12) are obtained directly as the regression coefficients (slopes), while that of parameter Log α_{ijmr} is the constant (intercept), say a_{ijmr}, so that

$$\alpha_{ijmr} = e^{a_{ijmr}} \tag{B.2}$$

1.4 The "Furness" procedure (Furness, 1965), consists of giving initial arbitrary values to the a_i's, inserting these values into Formulas (1.8b), and comparing them with the current values for the b_j's. If (with great luck) they are equal, then the set of values for the a_i's, and b_j's constitute a solution. If not, insert the last values for the b_j's into Formulas (1.8a), estimate the a_i's and compare their values with the last values for the a_i's. If these are equal to the previous values, the solution has been found, and so on. The process stops when consecutive values are within a given tolerance. In practice, the process will converge relatively quickly in most cases, provided that

$$\sum_i T_i = \sum_j T_j = T \tag{B.3}$$

i.e., the given trip origins and ends are consistent with one another in terms of total number of trips. If this condition is not met, an artificial, fictitious node should be introduced to take up the slack.

1.5

a. $T_1 = 3.37;\ T_2 = 2.74;\ T_3 = 2.89$

b. $T_{ij} =$

$j =$	1	2	3
$i = 1$	1.86	0.51	1.08
2	0.04	1.59	1.08
3	0.08	0.90	1.80

1.6 T_{ij}

$j =$	1	2	3
$i = 1$	1.56	3.19	0.26
2	3.21	1.31	0.47
3	15.24	25.49	4.26

1.8

a. $c_t = \$1.21$

b. 47 minutes

1.9 $\beta \approx 1.$

1.12 According to Formula (2.43), the elasticity may be estimated as

$$e_{T_{ijmr}/X^k_{ijmr}} = \frac{\partial \ln T_{ijmr}}{\partial \ln X^k_{ijmr}}; \qquad \forall\ i, j, m, r$$

In the present case, given the expression of T_{ijmr} in Formula (1.12), we see immediately that

$$\frac{\partial \ln T_{ijmr}}{\partial \ln X^k_{ijmr}} = \beta^k_{ijmr}; \qquad \forall\ i, j, m, r \tag{B.4}$$

2.1 The region of variation of the two random utility terms may be represented as the interior of the rectangle in Figure B.1. The probability that the first alternative is selected is equal to

$$P_1 = \Pr(\tilde{V}_1 > \tilde{V}_2) = \Pr(\tilde{U}_1 + \epsilon_1 > \tilde{U}_2 + \epsilon_2)$$
$$= \Pr(-3 + \epsilon_1 > -2.5 + \epsilon_2) = \Pr(\epsilon_1 - \epsilon_2 > 0.5)$$

The region defined by the condition $(\epsilon_1 - \epsilon_2 > 0.5)$ lies on one side of the straight line of equation $\epsilon_1 - \epsilon_2 = 0.5$, i.e., $\epsilon_2 = \epsilon_1 - 0.5$. It is easy to see that the origin of the coordinates, with $\epsilon_1 = \epsilon_2 = 0$ does not

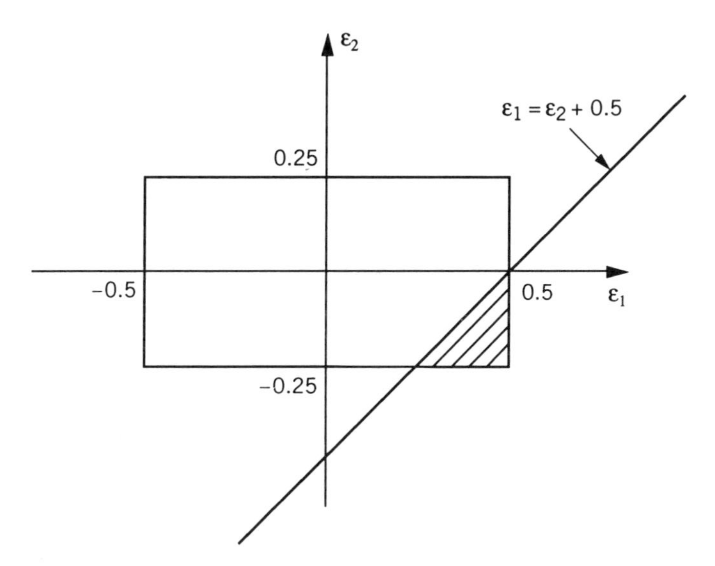

Figure B.1 Solution for Exercise 2.1.

meet the condition. The appropriate region is thus the shaded triangle. Since the ϵ's are independent, their joint p.d.f. is equal to the product of their individual p.d.f.'s, i.e., $f(\epsilon_1, \epsilon_2) = (1)(0.5) = 0.5$ and is constant at any point within the region, including the rectangle in Figure B.1. The probability Pr $(\epsilon_1 - \epsilon_2 > 0.5)$ represents the volume under the uniformly flat surface at height 0.5, and with base equal to the triangle (see section 1.4 of Appendix A). It may then be estimated simply as the ratio of the triangle's area to the total rectangle area. This is equal to 0.0625. $P_2 = 1 - P_1 = 0.9375$.

2.2 When there are only two alternatives of choice, the probability of choosing, say alternative 1 may be estimated as

$$P_1 = \Pr(\tilde{U}_1 + \epsilon_1 > \tilde{U}_2 + \epsilon_2) = \Pr(\epsilon_1 - \epsilon_2 > \tilde{U}_2 - \tilde{U}_1)$$

According to a well-known property of the normal distribution, if ϵ_1 and ϵ_2 are both normally distributed random variables, their difference is also normally distributed, with a mean equal to the difference in the means, in this case zero, and a variance equal to

$$\sigma^2 = \sigma_1^2 + \sigma_2^2 - 2\rho\sigma_1\sigma_2 \tag{B.5}$$

Thus,

$$P_2 = P\{Z > (\tilde{U}_2 - \tilde{U}_1)/(\sigma_1^2 + \sigma_2^2 - 2\rho\sigma_1\sigma_2)^{1/2}\}$$

Given the numerical values of the fixed utilities $\tilde{U}$ and that of the σ's and ρ, this probability may then be estimated from tables of the standard normal distribution.

2.3 We first estimate the systematic utilities, i.e., the means of the V's

$$\tilde{U}_1 = -2(2) - 1.5(2) = -7.0$$

$$\tilde{U}_2 = -2(3) - 1.5(1) = -7.5$$

$$\tilde{U}_3 = -2(4) - 1.5(1) = -9.5$$

Next, from the variance-covariance matrix $[\sigma_{ij}]$, the correlations between the random utilities are equal to

$$\rho_{13} = \rho_{13}/\sigma_1\sigma_3 = 3/\sqrt{4}\sqrt{4} = 0.75$$

$$\rho_{23} = 0.50$$

$$\rho_{12} = 0$$

Next, the variances of the various differences between two of the random utilities, i.e., the denominators of the expressions for the α's in Formula (2.58), are equal to

$$\text{Var}\,(\tilde{V}_1 - \tilde{V}_2) = 8$$

$$\text{Var}\,(\tilde{V}_1 - \tilde{V}_3) = 2$$

$$\text{Var}\,(\tilde{V}_2 - \tilde{V}_2) = 3$$

Thus, using Formula (2.58), we have

$$\alpha_{12} = 0.5/\sqrt{8} = 0.19$$

$$\alpha_{13} = 2.5/\sqrt{2} = 1.77$$

$$\alpha_{23} = 2/\sqrt{4} = 1.0$$

From a table of the normal probability function, we obtain the values of the $\Phi(\alpha)$

$$\Phi(0.19) = 0.575; \quad \Phi(-0.19) = 0.427; \quad \phi(0.19) = 0.392$$

$$\Phi(1.77) = 0.962; \quad \Phi(-1.77) = 0.038; \quad \phi(1.77) = 0.083$$

$$\Phi(1.00) = 0.841; \quad \Phi(-1.00) = 0.159; \quad \phi(1.00) = 0.242$$

and consequently, using Formula (2.59), we have

$$\omega_{12} = (49 + 4)\Phi(0.19) + (56 + 4)\Phi(-0.19) - 14.5\sqrt{8}\phi(0.19) = 41.15$$

$$\omega_{13} = (49 + 4)\Phi(1.77) + (90 + 4)\Phi(-1.77) - 16.5\sqrt{2}\phi(1.77) = 52.62$$

$$\omega_{23} = (56 + 4)\Phi(1.0) + (590 + 4)\Phi(-1.0) - 17\sqrt{4}\phi(1.0) = 57.24$$

Next, using Formula (2.57), we have

$$\mu_{12} = -7(0.575) - 7.5(0.427) + \sqrt{8}(0.392) = -6.12$$

$$\mu_{23} = -7.5(0.841) - 9.5(0.159) + \sqrt{4}(0.242) = -7.33$$

$$\mu_{13} = -7(0.962) - 7.5(0.038) + \sqrt{2}(0.083) = -6.91$$

Next, the variances of the maximums of two given random utilities are estimated using Formula (2.60) as

$$\text{Var}(\text{Max}(\tilde{V}_1, \tilde{V}_2)) = \omega_{12} - \mu_{12}^2 = 41.15 - (6.12)^2 = 3.70$$

$$\text{Var}(\text{Max}(\tilde{V}_2, \tilde{V}_3)) = \omega_{23} - \mu_{23}^2 = 52.05 - (7.36)^2 = 3.75$$

$$\text{Var}(\text{Max}(\tilde{V}_1, \tilde{V}_3)) = \omega_{13} - \mu_{13}^2 = 52.62 - (6.91)^2 = 4.87$$

Next, the correlations of these maximums with the third utilities are estimated using Formula (2.61), as

$$\rho_3' = [2(0.75)0.575 + 2(0.5)0.427]/\sqrt{3.7} = 0.67$$

$$\rho_1' = [2(0)0.841 + 2(0.75)0.159]/\sqrt{3.51} = 0.13$$

$$\rho_2' = [2(0)0.962 + 2(0.5)0.038]/\sqrt{4.87} = 0.02$$

We are finally in a position to estimate the probabilities of choice using Formula (2.62). Specifically, we have

$$P_1 = \Pr(\tilde{V}_1 > \text{Max}(\tilde{V}_2, \tilde{V}_3)) = \Pr(\tilde{V}_1 - \text{Max}(\tilde{V}_2, \tilde{V}_3) > 0) \tag{B.6a}$$

where ($\tilde{V}_1$ – Max ($\tilde{V}_2$, $\tilde{V}_3$)) is a random variable with a normal distribution with mean

$$\mu = \tilde{U}_1 - \mu_{23} = -7.00 + 7.33 = 0.33$$

and standard deviation

$$\sigma^2 = [\sigma_1^2 + \sigma_{23}^2 - 2\rho_1' \sigma_1 \sqrt{\sigma_{23}^2}]^{1/2}$$
$$= (4 + 3.51 - 2(2)1.87(0.13))^{1/2} = 2.56$$

Consequently, Pr ($\tilde{V}_1$ – Max ($\tilde{V}_2$, $\tilde{V}_3$) > 0) = $P(Z > (0 - 0.33)/2.56)$ = $P(Z > -0.129) = 1 - \Phi(-0.129) = 0.552$.

Similarly,

$$\mu(\tilde{V}_2 - \text{Max }(\tilde{V}_1, \tilde{V}_3)) = -0.59$$
$$\sigma(\tilde{V}_2 - \text{Max }(\tilde{V}_1, \tilde{V}_3)) = 2.95$$

and

$$P_2 = \text{Pr }(\tilde{V}_2 - \text{Max }(\tilde{V}_1, \tilde{V}_3) > 0) = 0.421 \tag{B.6b}$$

Finally,

$$\mu(\tilde{V}_3 - \text{Max }(\tilde{V}_1, \tilde{V}_2)) = -3.38$$
$$\sigma(\tilde{V}_3 - \text{Max }(\tilde{V}_1, \tilde{V}_2)) = 1.60$$

and

$$P_3 = \text{Pr }(\tilde{V}_3 - \text{Max }(\tilde{V}_1, \tilde{V}_2) > 0) = 0.017. \tag{B.6c}$$

As expected, the three probabilities add up to 1.0. It may easily be seen that when there is a large number of alternatives, "cascading" this process in a hierarchical manner may quickly become cumbersome, even when implemented with a computer for simulation purposes.

2.8 First we must determine the cumulative probability distribution function of $\tilde{V}$, the maximum of the two random utilities. The range of variation of $\tilde{V}$ is from -3.5 to -2.25. We have: $F_{\tilde{V}}(x) = P(\tilde{V} < x) = P(3 + \epsilon_1 < x;\ 2 + \epsilon_2 < x) = P(3 + \epsilon_1 < x)P(2 + \epsilon_2 < x) = F\epsilon_1(x)F\epsilon_2(x)$, the product of the individual cumulative p.d.f.'s of the ϵ's, since they are independent. $F_1(x)$ and $F_2(x)$ may be derived from the specification of the p.d.f. of the ϵ's. From this, the probability density function $f_{\tilde{V}}(x)$ of $\tilde{V}$ may be obtained as the derivative of $F_{\tilde{V}}(x)$ with respect to x. The expected value $\tilde{W}$ of $\tilde{V}$ may then be obtained as the integral

$$\int_{-3.5}^{-2.25} \tilde{V} f_{\tilde{V}}(x)\, dx$$

2.9 This property was in fact utilized in the derivation of Formula (2.12a), when simplifying it by deleting the term b representing the individual budget, which is common to all utilities. Formally,

$$\frac{e^{\beta(\tilde{U}_j + K)}}{\sum_j e^{\beta(\tilde{U}_j + K)}} = \frac{e^{\beta K} e^{\beta \tilde{U}_j}}{\sum_j e^{\beta K} e^{\beta \tilde{U}_j}} = \frac{e^{\beta K} e^{\beta \tilde{U}_j}}{e^{\beta K} \sum_j e^{\beta \tilde{U}_j}} = \frac{e^{\beta \tilde{U}_j}}{\sum_j e^{\beta \tilde{U}_j}}$$

Consequently, for convenience, one of the utilities may always be set to zero, by subtracting its value from all the utilities, becoming the "base" utility. This makes intuitive sense, since alternatives are chosen relative to one another, not in the absolute. This property is to be expected from the logit model, or for that matter any model of utility maximizing choice, since probabilities of choice are determined by the probability that a given alternative offer the maximum utility. One implication of this fact is that the expected received utility $\tilde{W}$ and the traveler surplus TS are respectively only measured up to such an arbitrary constant. (See Exercise 2.13 below.)

2.13 It was shown in Exercise 2.9 that the logit model is unchanged if a constant K is added or subtracted to all systematic utilities $\tilde{U}_j$. The addition of the constant K affects the indirect utility $\tilde{W}$ as follows:

$$\tilde{W}' = \frac{1}{\beta} \ln \sum_j e^{\beta(\tilde{U}_j + K)} = \frac{1}{\beta} \ln \left[e^{\beta K} \sum_j e^{\beta \tilde{U}_j} \right]$$

$$= \frac{1}{\beta} \left[\ln \sum_j e^{\beta \tilde{U}_j} + \ln e^{\beta K} \right] = \tilde{W} + \frac{1}{\beta} (\beta K) = \tilde{W} + K$$

Consequently, it is always possible to insure that $\tilde{W}$, as well as the "traveler surplus" TS will be positive. Therefore, these respective indicators of transportation system performance should be used in a *comparative* rather than absolute fashion.

2.14 In this case the problem is

$$\max_{(P_j)} \left[3P_1 + 2.75P_2 - \frac{1}{\beta} (P_1 \ln P_1 + P_2 \ln P_2) \right] \qquad \text{(B.7a)}$$

subject to the constraint that

$$P_1 + P_2 = 1. \tag{B.7b}$$

Replacing P_2 by $(1 - P_1)$ in the objective function, we obtain an unconstrained problem:

$$\underset{(P_1)}{\text{Max}} \left[3P_1 + 2.75(1 - P_1) - \frac{1}{\beta}(P_1 \ln P_1 + (1 - P_1)\ln(1 - P_1)) \right]$$

Since the objective function is everywhere strictly concave (see section 2.2 of Appendix A), there is a unique solution. Setting the derivative with respect to P_1 equal to 0, we obtain the equation

$$0.25 - \frac{1}{\beta}(\ln P_1 - \ln(1 - P_1)) = 0$$

With an assumed value of 0.64 for β, as in the illustrative example of section 2.3.2, the equation becomes

$$\ln \frac{P_1}{(1 - P_1)} = 0.16$$

or $P_1/(1 - P_1) = 1.17$ from which $P_1 = 0.54$ and $P_2 = 1 - P_1 = 0.46$.

When there are more than two alternatives, the R.T.'s U.M. problem cannot be solved in this simple fashion. Instead, numerical algorithms for nonlinear problems in !inear constraints must be used, as we shall see, beginning in Chapter 5.

2.16 We may write successively

$$Y_j^a = N \frac{e^{\beta(\tilde{U}_j + \Delta\tilde{U}_j)}}{\sum_{k=1}^{J} e^{\beta(\tilde{U}_k + \Delta\tilde{U}_k)}} = N \frac{e^{\beta\tilde{U}_j} e^{\beta\Delta\tilde{U}_j}}{\sum_{k=1}^{J} e^{\beta\tilde{U}_k} e^{\beta\Delta\tilde{U}_k}}$$

$$= N \frac{e^{\beta\tilde{U}_j} e^{\beta\Delta\tilde{U}_j} \sum_{l=1}^{J} e^{\beta\tilde{U}_l}}{\sum_{k=1}^{J} e^{\beta\tilde{U}_k} \sum_{k=1}^{J} e^{\beta\tilde{U}_k} e^{\beta\Delta\tilde{U}_k}} = \frac{Y_j^b e^{\beta\Delta\tilde{U}_j}}{\sum_{k=1}^{J} \frac{e^{\beta\tilde{U}_k}}{\sum_{l=1}^{J} e^{\beta\tilde{U}_l}} e^{\beta\Delta\tilde{U}_k}} = \frac{Y_j^b e^{\beta\Delta\tilde{U}_j}}{\sum_{k=1}^{J} P_k^b e^{\beta\Delta\tilde{U}_k}}; \quad \forall j$$

This particular expression is sometimes referred to as the "incremental logit." Its use obviates the need to recalculate the "after" probabilities.

3.1 Shortest route: 1, 2, 5, 7. Length: 6.

3.2 Shortest route: A, B, C, G. Length: 4.

3.3 The problem may be interpreted, or translated, as the problem of routing in a cost-minimizing manner, a unit volume of flow from origin i to destination j. Thus, the problem may be formulated as

$$\operatorname*{Min}_{(v_a)} \left[\sum_a v_a t_a \right] \tag{B.8}$$

$$\sum_{a \in k^-} v_a = \sum_{a \in k^+} v_a; \qquad \forall\, k \neq i, j \tag{B.9a}$$

$$\sum_{a \in i^-} v_a = 1 \tag{B.9b}$$

$$\sum_{a \in j^+} v_a = 1 \tag{B.9c}$$

Constraints (B.9) translate conservation of flow at the general mode, different from either the origin or the destination, and set the amount leaving the origin and entering the destination to one. This is a linear program in the variables v_a. This formulation may be generalized to allow for the simultaneous identification of the minimum cost routes for several origin destination combinations.

3.4 For simplification, let us formulate the problem in the case of a single origin and destination, i.e., $T_{ij} = 0$ for all $i \neq 1$, and $T_{1J} = 1$, so that we'll be able to compare to the formulation in the preceding exercise. When all f_a's are infinite, the second term in the objective function becomes equal to zero:

$$\operatorname*{Min}_{(v_a)} \left[\sum_a t_a v_a \right] \tag{B.10}$$

Similarly, the constraints become

$$v_a^j \leq f_a w_{ij} = \infty; \qquad \forall\, a, i, j \tag{B.11a}$$

which is always verified, and may thus be dismissed, and

$$\sum_{a \in i^-} v_a - \sum_{a \in i^+} v_a = 0; \qquad \forall\, i \neq 1, \mathrm{J}; \qquad \text{for the general node} \tag{B.11b}$$

$$\sum_{a \in 1^-} v_a = 1; \qquad \text{for the first node} \tag{B.11c}$$

$$-\sum_{a \in \mathrm{J}^+} v_a = T_{\mathrm{J}} = -\sum_i T_i = -1; \qquad \text{for the last node}$$

according to Formula (3.27), or

$$\sum_{a \in J^+} v_a = 1 \tag{B.11d}$$

In this form, the problem is indeed an MCR problem. This fact would be demonstrated in the same manner for any number of simultaneous demands.

3.7

a. The simplified network is represented in Figure B.2. (Spiess and Florian, 1989). Squares represents stations/stops, while circles represent platforms. Note that this network allows for travelers from X and Y to B. If only travelers from A to B are considered, the network may be simplified even further as in Figure B.3.

b. The various iterations are summarized in Table B.1, using the format in the exposition, and further detailed below.

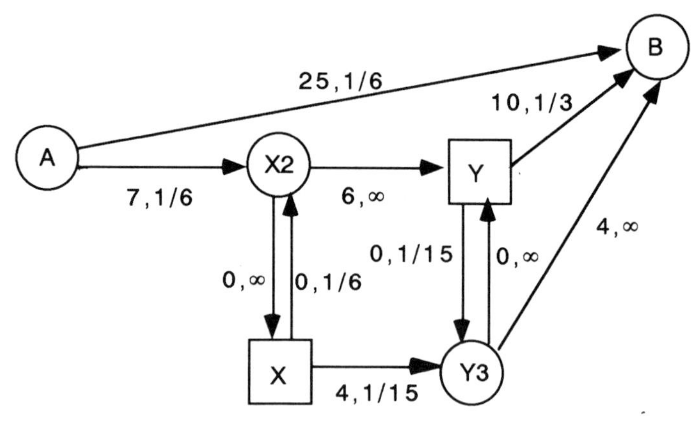

Figure B.2 Simplified transit network for Exercise 3.7.

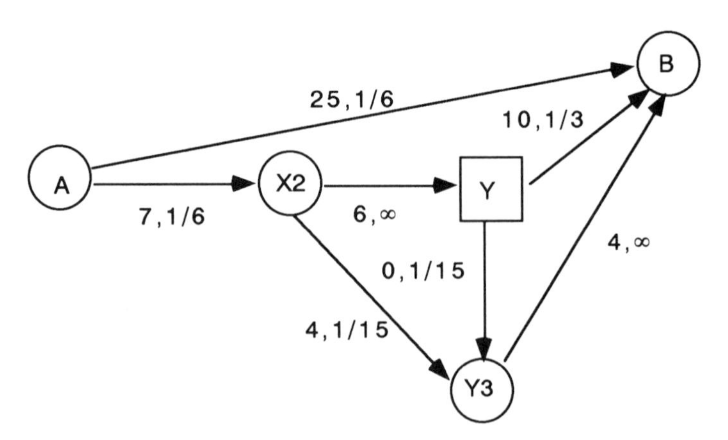

Figure B.3 Further simplified transit network for Exercise 3.7.

TABLE B.1 Iterations for Transit Network SPA

It.#	A	X_2	X	Y_3	Y	B	Link	f_a	$U_j + c_a$	a in A^{+}?
0	∞,0	∞,0	∞,0	∞,0	∞,0	0,0	—	—	—	—
1	∞,0	∞,0	∞,0	4,∞	∞,0	0,0	Y_3B	∞	4	Yes
2	∞,0	∞,0	∞,0	4,∞	11.5,1/15	0,0	YY_3	1/15	4	Yes
3	∞,0	∞,0	15.5,1/15	4,∞	11.5,1/15	0,0	XY_3	1/15	8	Yes
4	∞,0	∞,0	15.5,1/15	4,∞	10.25,1/2.5	0,0	YB	1/2.5	10	No
5	∞,0	∞,0	15.5,1/15	4,∞	10.25,1/2.5	0,0	Y_3Y	∞	10.25	No
6	∞,0	15.5,∞	15.5,1/15	4,∞	10.25,1/2.5	0,0	X_2X	∞	15.5	Yes
7	∞,0	15.5,∞	15.5,1/15	4,∞	10.25,1/2.5	0,0	XX_2	1/6	15.5	No
8	∞,0	15.5,∞	15.5,1/15	4,∞	10.25,1/2.5	0,0	X_2Y	∞	16.25	No
9	25.5,1/6	15.5,∞	15.5,1/15	4,∞	10.25,1/2.5	0,0	AX_2	1/6	22.5	Yes
10	25.25,1/5	15.5,∞	15.5,1/15	4,∞	10.25,1/2.5	0,0	AB	1/6	25.25	Yes

Iteration 1 (Special initial case)

1. Select link a to examine, such that $(g_a + u_{ik}) < (g_{a'} + u_{ik'}) \forall k'$

(i, j)	c_a	U_j	$U_j + c_a$	
Y_3B	4	0	4	
YB	10	0	10	SELECT Y_3 B
AB	25	0	25	

2. Test selected link Y_3 B, i.e., $(g_a + u_{ik}) <? u_{ij}$
   $U_i > U_j + c_a$
   $\infty > 4$ Yes. Include Y_3 B in A^+. Take out of A.
3. Update
   $f_i = f_i + f_a = 0 + \infty = \infty$
   $u_{ij} = [u_{ij} f_i + (t_a + u_{ik}) f_a]/(f_i + f_a) = (0 \cdot \infty + 4 \cdot \infty)/(0 + \infty)$
   $= (0.5 + 4 \cdot \infty)/\infty = 4$
   New label for Y_3: $(4, \infty)$.

Iteration 2

1.

(i, j)	c_a	U_j	$U_j + c_a$	
AB	25	0	25	
YB	10	0	10	
YY_3	0	4	4	SELECT
XY_3	4	4	8	
X_2Y	6	∞	∞	
X_2X	0	∞	∞	
XX_2	0	∞	∞	
AX_2	7	∞	∞	

2. Test
   $U_i > U_j + c_a$
   $\infty > 4$ Yes. Include YY_3 in A^+. Take out of A.
3. Update
   $f_i = 0 + 1/15 = 1/15$
   $u_{ij} = (0 \cdot \infty + 4 \cdot 1/15)/(0 + 1/15) = (0.5 + 4/15)/(1/15) =$
   11.5
   New label for Y: $(11.5, 1/15)$

Iteration 3

1.

(i, j)	c_a	U_j	$U_j + c_a$	
AB	25	0	25	
YB	10	0	10	
Y_3Y	0	11.5	11.5	
XY_3	4	4	8	SELECT
X_2Y	6	11.5	17.5	

X_2X	0	∞	∞
XX_2	0	∞	∞
AX_2	7	∞	∞

2. Test
   $U_i > U_j + c_a$
   $\infty > 8$ Yes. Include XY_3 in A^+. Take out of A.
3. Update
   $f_i = 0 + 1/15 = 1/15$
   $u_{ij} = (0 \cdot \infty + 8 \cdot 1/15)/(0 + 1/15) = (0.5 + 8/15)/(1/15) = 15.5$
   New label for X: (15.5, 1/15)

Iteration 4

1.

(i, j)	c_a	U_j	$U_j + c_a$	
AB	25	0	25	
YB	10	0	10	SELECT
Y_3Y	0	11.5	11.5	
X_2Y	6	11.5	17.5	
X_2X	0	15.5	15.5	
XX_2	0	∞	∞	
AX_2	7	∞	∞	

2. Test
   $U_i > U_j + c_a$
   $11.5 > 10$ Yes. Include XY_3 in A^+. Take out of A.
3. Update
   $f_i = 1/15 + 1/3 = 1/2.5$
   $u_{ij} = (11.5 \cdot 1/15 + 10 \cdot 1/3)/(1/15 + 1/3) = 10.25$
   New label for X: (10.25, 1/2.5)

Iteration 5

1.

(i, j)	c_a	U_j	$U_j + c_a$	
AB	25	0	25	
Y_3Y	0	10.25	10.25	SELECT
X_2Y	6	10.25	16.25	
X_2X	0	15.5	15.5	
XX_2	0	∞	∞	
AX_2	7	∞	∞	

2. Test
   $U_i > U_j + c_a$
   $4 > 10.25$ No. Go on to next iteration.

Iteration 7

1.

(i, j)	c_a	U_j	$U_j + c_a$	
AB	25	0	25	
X_2Y	6	10.25	16.25	
XX_2	0	∞	∞	SELECT
AX_2	7	∞	∞	

2. Test
   $U_i > U_j + c_a$

15.5 = 15.5. Yes/no. In general, in such a situation, one may, or may not, include XX_2 in A^+, thus providing alternate solutions. However, in this particular case, this link is not useful for travelers from A to B, and will not be included.

Iteration 8

1.

(i, j)	c_a	U_j	$U_j + c_a$	
AB	25	0	25	
X_2Y	6	10.25	16.25	SELECT
AX_2	7	15.5	15.5	

2. Test
   $U_i > U_j + c_a$
   $15.7 > 16.25$ NO

Iteration 9

1.

(i, j)	c_a	U_j	$U_j + c_a$	
AB	25	0	25	
AX_2	7	15.5	22.5	SELECT

2. Test
   $U_i > U_j + c_a$
   $\infty > 22.5$ Yes. Include AX_2 in A^+. Take out of A.
3. Update
   $f_i = 0 + 1/6 = 1/6$
   $u_{ij} = (0 \cdot \infty + 22.5 \cdot 1/6)/(0 + 1/6) = 25.5$
   New label for A: (25.5, 1/6)

Iteration 10

1.

(i, j)	c_a	U_j	$U_j + c_a$	
AB	25	0	25	SELECT

2. Test
   $U_i > U_j + c_a$
   25.5 > 25 Yes. Include AB in A^+. Take out of A.
3. Update
   $f_i = 1/6 + 1/6 = 1/5$
   $u_{ij} = (25.5 \cdot 1/6 + 25 \cdot 1/6)/(1/6 + 1/6) = 25.25$
   New label for A: (25.25, 1/3)

At this point there are no more links to examine, since set A is empty. The optimal total expected travel time is the first entry in node A's label, i.e., 25.25 minutes. The volumes on the links are then obtained by performing the forward pass. Equivalently, the optimal routes from A to B, in probability terms, are represented in Figure B.4. The optimal strategy involves taking line 1 directly to B 50% of the time, and taking line 2 to X2 and changing to line 3 to go to B the other 50% of the time. The average waiting time to leave A (i.e., $1/2(1/f_A)$ at the solution), is 1.5 minutes. The average waiting time to leave X2 after changing lines is $1/2(1/15) = 7.5$ minutes. Consequently, the expected total travel time is

$$1.5 + \{0.5(25) + 0.5(7 + 7.5 + 8)\} = 25.25 \text{ minutes}$$

which, as expected, is equal to the first entry in the label u_A at the solution.

This solution may be compared to the value of 27.75 in Spiess and Florian (1989), where parameter $\overline{\omega}$ is given a value of 1, corresponding to Poisson-distributed vehicle arrivals (resulting in exponentially distributed headways), instead of uniform arrivals in the present case, and a value for $\overline{\omega}$ of 0.5. Average link frequencies, or

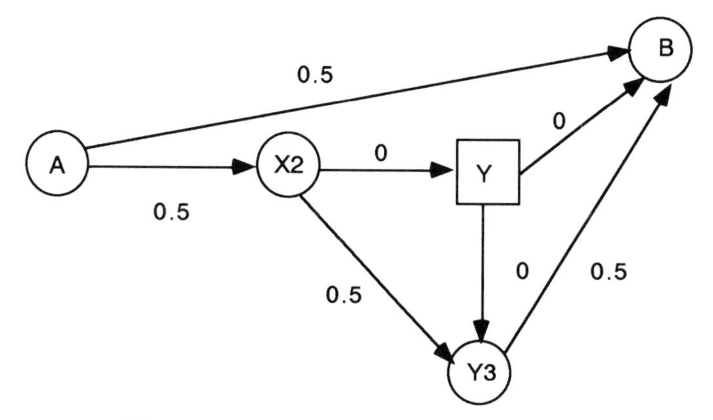

Figure B.4 Optimal routes for Exercise 3.7.

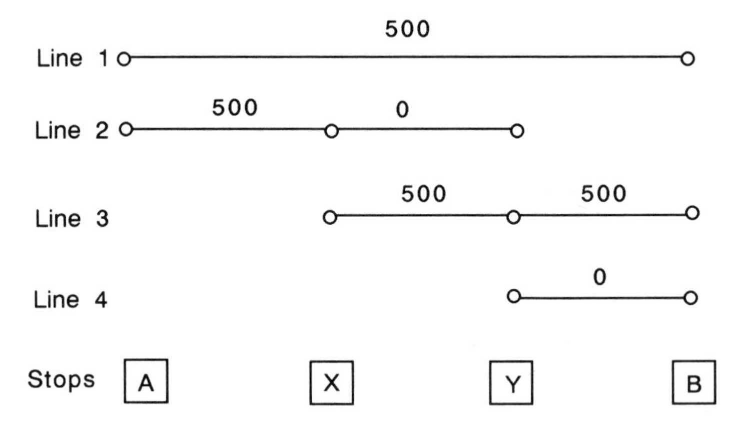

Figure B.5 Transit assignment for Exercise 3.7.

headways are the same in both cases, but the variance of the headways is smaller in the uniform than in the Poisson case.

c. For 1,000 travelers, the volumes on the links, in passengers per hour, will then be as represented in Figure B.5.

3.13 We have

$$h(1) = e^{0.1(5-0-5)} = 1$$
$$h(2) = e^{0.1(13-0-15)} = 0.82$$
$$h(3) = e^{0.1(6-0-6)} = 1$$
$$h(4) = 0$$
$$h(5) = e^{0.1(13-6-7)} = 1$$
$$h(6) = 0$$

Consequently

$$p(1) = 1$$
$$p(2) = 0.82$$
$$p(3) = 1$$
$$p(4) = 0$$
$$p(5) = 1(1) = 1$$
$$p(6) = 0$$

Finally, we perform the backward pass, and estimate the link volumes, according to Formula (3.24c). We have

$$v_1 = (250 + 0)\left(\frac{1}{1 + 0}\right) = 250$$

$$v_6 = (250 + 0)\frac{0}{1 + 0} = 0$$

$$v_2 = (0 + 0)\frac{0.82}{1 + 0.82} = 0$$

$$v_3 = 0$$

$$v_5 = (0 + 66)\frac{1}{1 + 1 + 0.82} = 36$$

3.14

$$\underset{(v_a, T_0)}{\text{Max}}\ U_m = \left[-\tau \sum_a t_a v_a + T_0\right] \tag{B.12a}$$

such that

$$\sum_a v_a c_a + T_0 = B \tag{B.13a}$$

$$\sum_{a \in i^-} v_{aj} = \sum_{a \in i^+} v_{aj} + T_{ij}; \qquad \forall\ i, j \tag{B.13b}$$

$$v_a = \sum_j v_{aj}; \qquad \forall\ a \tag{B.13c}$$

$$T_{ii} = -\sum_{j \neq i} T_{ji}; \qquad \forall\ i \tag{B.13d}$$

$$v_{aj} \geq 0; \qquad \forall\ a, j \tag{B.13e}$$

In the same manner as above, variable T_0 and the first budgetary constraint may be deleted from the problem's formulation. The new objective function would then be

$$\underset{(v_a)}{\text{Max}}\left[-\tau \sum_a t_a v_a + B - \sum_a v_a c_a\right] = \underset{(v_a)}{\text{Min}}\left[\sum_a (\tau t_a + c_a) v_a\right] \tag{B.12b}$$

This formulation, however, while functionally equivalent, has somewhat less behavioral content, since it is more logical to think of the R.T. as choosing routes, rather than links.

3.15 From the expression of $\tilde{W}_R$, Formula (3.16b), we have

$$\frac{\partial \tilde{W}_R}{\partial c_{ijr}} = T_{ij} \frac{\partial}{\partial c_{ijr}} \ln \sum_r e^{-\beta_r(\tau t_{ijr} + c_{ijr})}$$

$$= T_{ij} \frac{1}{\beta_r \sum_r e^{-\beta_r(\tau t_{ijr} + c_{ijr})}} \frac{\partial}{\partial c_{ijr}} \sum_r e^{-\beta_r(\tau t_{ijr} + c_{ijr})}$$

$$= T_{ij} \frac{1}{\beta_r \sum_r e^{-\beta_r(\tau t_{ijr} + c_{ijr})}} \frac{\partial e^{-\beta_r(\tau t_{ijr} + c_{ijr})}}{\partial c_{ijr}} = -T_{ij} \frac{\beta_r e^{-\beta_r(\tau t_{ijr} + c_{ijr})}}{\beta_r \sum_r e^{-\beta_r(\tau t_{ijr} + c_{ijr})}}$$

$$= -T_{ij} P_{r/ij} = -T_{ijr}; \qquad \forall\, i, j, r$$

It is furthermore obvious that $(\partial \tilde{W}_R / \partial B) = 1$. Hence, Formula (3.18b).

3.16 First, write the utility function in terms of the budget B and the travel costs c_{ijr}, by substituting T_0, as done previously. Then, simply replace in that expression the T_{ijr}'s by their logit expression, Formula (3.15b). After some simplifications, Formula (3.16b) emerges.

3.18 In the probabilistic case, individual travelers are assigned to routes in such a manner that they minimize the *random* expected cost of travel. The randomness may be due to uncertainty about link travel times, *not* vehicle or passenger arrival times, which is already incorporated into the f_a's and $\overline{\omega}$ values. In order to estimate expected costs as a model unknown, and not externally as in section 3.4.2, a *strategy*-based formulation, as opposed to a route-based one, is necessary. This is described in detail in the general, congested case in section 5.5.1 of Chapter 5. The generalization to the stochastic case is then effected with respect to *strategies*, not routes, as in the car case, in section 3.3. See the solution to Exercises 5.28 and 5.29 in Appendix B, for a specific illustration.

3.23 A *strategy* from i to j may be defined as a sequence of links. In difference with a *route*, however, at each of the nodes, if there are several optimal links, the *first* which becomes available, (i.e. is served by a transit vehicle), is taken. For instance, in the case of the hypothetical network, the strategies from 1 to 4 may be described as follows.
From node 1, take:
- link 1 to node 2
- link 2 to node 4
- link 3 to node 3
- link 1 to node 2, *or* link 2 to node 4, whichever comes first

- link 1 to node 2, *or* link 3 to node 3, whichever comes first
- link 2 to node 4, *or* link 3 to node 3, whichever comes first
- link 1 to node 2, *or* link 2 to node 4, *or* link 3 to node 3, whichever comes first

From node 2, take:
- link 4 to node 4

From node 3, take:
- link 5 to node 4

Combining these "local" alternatives, the seven strategies from 1 to 4 are defined as the respective sets of links

	Links
Strategy # 1	2
2	1, 4
3	3, 5
4	1, 2, 4
5	2, 3, 5
6	1, 3, 4, 5
7	1, 2, 3, 4, 5

(Note that, in difference, there are only 3 routes from 1 and 4). The composition of individual strategies, and their relationship to individual links, may then be represented in the form of values of the indicator δ_s^a

$$\delta_s^a = \begin{cases} 1, \text{ if } a \in s \\ 0, \text{ otherwise} \end{cases}$$

For instance, the values for the illustrative network are described in the solution to Exercise 5.29 later in this Appendix.

The R.T.'s U.M. problem in the strategy-based formulation is

$$\text{Min } U_{MR}(T_0, T_{is}, w_{is}, v_{as}) = +\sum_a v_a t_a + \overline{\omega} \sum_i \sum_{s \in S_j} w_{is}$$

subject to

$$v_{as} = f_a w_{is} \delta_s^a; \qquad \forall\, i, a \in i^-, s$$

$$\sum_{a \in i^-} v_{as} - \sum_{a \in i^+} v_{as} = T_{is}; \qquad \forall\, i, s \in S_j$$

$$\sum_s v_{as} = v_a; \qquad \forall\, a$$

$$\sum_{s \in S_j} T_{is} = T_{ij}; \qquad \forall\, i, j$$

$$v_{as}, w_{is}, T_{is}, T_0 \geq 0, \qquad \forall\, i, j, a, s$$

In the case of the illustrative example in section 3.4, the solution is

$$T_{1,14} = 200$$

$$v_{1,14} = 54;\ v_{2,14} = 81;\ v_{3,14} = 65;\ v_{3,14} = 54;\ v_{5,14} = 65;$$

$$w_{1,14} = 1.35$$

4.1

a.

$$T_{13} = 100 \frac{e^{-1(0.5 \cdot 2 + 1)}}{e^{-1(0.5 \cdot 2 + 1)} + e^{-1(4 \cdot 0.5 + 0.5)}} = 62;$$

$$T_{14} = 100 - T_{13} = 38$$

$$T_{23} = 200 \frac{e^{-1(4 \cdot 0.5 + 1)}}{e^{-1(4 \cdot 0.5 + 1)} + e^{-1(0.5 \cdot 2 + 0.5)}} = 37;$$

$$T_{24} = 200 - T_{23} = 163$$

c. The rate of change of T_{13} with respect to t_{13} is, according to Formula (2.49), equal to

$$\frac{\partial T_{13}}{\partial t_{13}} = \frac{\partial T_{13}}{\partial \tilde{U}_{13}} \frac{\partial \tilde{U}_{13}}{\partial t_{13}} = \beta N_1 P_{3/1}(1 - P_{3/1})(-0.5)$$

$$= 1 \cdot 100 \cdot 0.62 \cdot 0.38 \cdot 0.5 = -12$$

Consequently, the direct elasticity is equal to

$$\frac{\partial T_{13}}{\partial t_{13}} \left(\frac{t_{13}}{T_{13}} \right) = -12 \frac{2}{62} = -0.39$$

Thus, for each percent increase in t_{13}, there is a 0.39 percent decrease in T_{13}. Similarly, the rate of change of T_{13} with respect to t_{14} is, according to Formula (2.50) equal to

$$\frac{\partial T_{13}}{\partial t_{14}} = \frac{\partial T_{13}}{\partial \tilde{U}_{14}} \frac{\partial \tilde{U}_{14}}{\partial t_{14}} = -\beta N_1 P_{3/1} P_{4/1}(-0.5)$$

$$= -1 \cdot 100 \cdot 0.62 \cdot 0.38(-0.5) = 12$$

Consequently, the cross elasticity is equal to

$$\frac{\partial T_{13}}{\partial t_{14}} \left(\frac{t_{14}}{T_{13}} \right) = 12 \frac{4}{62} = 0.78$$

Thus, for each percent increase in t_{14}, there is a 0.78 percent increase in T_{13}.

d. The probability that an individual traveler from zone 1 goes to zone 3 is equal to $p = 0.62$, as seen above. Consequently, the probability that 5 out of 10 travelers chosen at random go to zone 3 is given by Formula (2.19):

$$P = \frac{10!}{5!5!} 0.62^5 \cdot 0.38^5 = 0.18 \text{ or } 18\%$$

The probability that 50 out of 100 travelers chosen at random go to zone 3 may be estimated from the normal distribution with a mean of $100(0.62) = 62$ and a standard deviation of $\sqrt{100(0.62)(0.38)} = 4.85$, as is thus equal to

$$P\{(49.5 - 62)/4.85 \leq Z \leq (51.5 - 62)/4.85\}$$
$$= P\{-2.58 \leq Z \leq -2.16\} \approx 0.01$$

where Z is the standard normal variable.

4.4

a. The various routes and their costs may be enumerated as follows

(i, j)	links	route #	mode	Cost
1,1	1	1	1	10
1,1	1	1	2	8
1,1	2,5	2	1	22
1,1	2,5	2	2	17
2,1	3	1	1	8
2,1	3	1	2	9
2,1	4,5	2	1	21
2,1	4,5	2	2	18
1,2	2	1	1	13
1,2	2	1	2	10
1,2	1,6	2	1	25
1,2	1,6	2	2	14
2,2	4	1	1	12
2,2	4	1	2	11
2,2	3,6	2	1	23
2,2	3,6	2	2	15

b. Using Formula (3.15a), the conditional probabilities $P_{r/ijm}$ are equal to

$$P_{1/111} = \frac{e^{-\beta_r C_{1111}}}{e^{-\beta_r C_{1111}} + e^{-\beta_r C_{1112}}} = \frac{e^{0.5(-10)}}{e^{0.5(-10)} + e^{0.5(-22)}} = 0.9975$$

$$P_{2/111} = \frac{e^{-\beta_r C_{1112}}}{e^{-\beta_r C_{1111}} + e^{-\beta_r C_{1112}}} = \frac{e^{0.5(-22)}}{e^{0.5(-10)} + e^{0.5(-22)}}$$

$$= 1 - 0.9975 = 0.0025$$

$$P_{1/112} = \frac{e^{-\beta_r C_{1121}}}{e^{-\beta_r C_{1121}} + e^{-\beta_r C_{1122}}} = \frac{e^{0.5(-8)}}{e^{0.5(-8)} + e^{0.5(-17)}} = 0.989;$$

$$P_{2/112} = 0.011$$

$$P_{1/211} = \frac{e^{-\beta_r C_{2111}}}{e^{-\beta_r C_{2111}} + e^{-\beta_r C_{2112}}} = \frac{e^{0.5(-8)}}{e^{0.5(-8)} + e^{0.5(-21)}} = 0.998;$$

$$P_{2/211} = 0.002$$

$$P_{1/212} = \frac{e^{-\beta_r C_{2121}}}{e^{-\beta_r C_{2121}} + e^{-\beta_r C_{2122}}} = \frac{e^{0.5(-9)}}{e^{0.5(-9)} + e^{0.5(-18)}} = 0.989;$$

$$P_{2/212} = 0.011$$

$$P_{1/121} = 0.997;\ P_{2/121} = 0.03;\ P_{1/122} = 0.880;\ P_{2/122} = 0.12$$

$$P_{1/221} = 0.996;\ P_{2/221} = 0.004;\ P_{1/222} = 0.880;\ P_{2/222} = 0.12$$

c. In order to estimate the probabilities $P_{m/ij}$, we first estimate the expected utilities received from route choices:

$$\tilde{W}_{1/11} = \frac{1}{\beta_r} \ln (e^{-\beta_r C_{1111}} + e^{-\beta_r C_{1112}}) = \frac{1}{0.5} \ln (e^{0.5(-10)} + e^{0.5(-22)})$$

$$= -9.995$$

$$\tilde{W}_{2/11} = \frac{1}{\beta_r} \ln (e^{-\beta_r C_{2111}} + e^{-\beta_r C_{2112}}) = \frac{1}{0.5} \ln (e^{0.5(-8)} + e^{0.5(-17)})$$

$$= -7.978$$

so that

$$P_{1/11} = \frac{e^{\beta_m(h_{111} + \tilde{W}_{1/11})}}{e^{\beta_m(h_{111} + \tilde{W}_{1/11})} + e^{\beta_m(h_{112} + \tilde{W}_{2/11})}}$$

$$= \frac{e^{1.5(-3-9.995)}}{e^{1.5(-3-9.995)} + e^{1.5(-5-7.978)}} = 0.494$$

$$P_{2/11} = 0.506$$

Similarly,

$$\tilde{W}_{1/21} = \frac{1}{0.5} \ln (e^{0.5(-8)} + e^{0.5(-21)}) = -7.997$$

$$\tilde{W}_{2/21} = \frac{1}{0.5} \ln (e^{0.5(-9)} + e^{0.5(-18)}) = -8.978$$

$$P_{1/21} = \frac{e^{1.5(-3-7.997)}}{e^{1.5(-3-7.997)} + e^{1.5(-5-8.978)}} = 0.989; \ P_{2/21} = 0.011$$

$$\tilde{W}_{1/12} = -13; \ \tilde{W}_{2/12} = -9.75; \ P_{1/12} = 0.13; \ P_{2/12} = 0.87$$

$$\tilde{W}_{1/22} = -12; \ \tilde{W}_{2/22} = -10.75; \ P_{1/22} = 0.75; \ P_{2/22} = 0.25$$

e. The probabilities $P_{j/i}$ are similarly estimated by first evaluating the expected utilities received from mode choices.

$$\tilde{W}_{1/1} = \frac{1}{\beta_m} \ln (e^{\beta_m(h_{111} + \tilde{W}_{1/11})} + e^{\beta_m(h_{112} + \tilde{W}_{2/11})})$$

$$= \frac{1}{1.5} \ln (e^{1.5(-3-9.95)} + e^{1.5(-5-7.978)}) = -12.52$$

$$\tilde{W}_{1/2} = \frac{1}{\beta_m} \ln (e^{\beta_m(h_{211} + \tilde{W}_{1/21})} + e^{\beta_m(h_{212} + \tilde{W}_{2/21})})$$

$$= \frac{1}{1.5} \ln (e^{1.5(-3-7.997)} + e^{1.5(-5-8.978)}) = -10.99$$

$$\tilde{W}_{2/1} = \frac{1}{1.5} \ln (e^{1.5(-3-13)} + e^{1.5(-5-9.75)}) = -14.65$$

$$\tilde{W}_{1/2} = \frac{1}{1.5} \ln (e^{1.5(-3-12)} + e^{1.5(-5-10.75)}) = -14.81$$

$$P_{1/1} = \frac{e^{\beta_d(h_{11} + \tilde{W}_{1/1})}}{e^{\beta_d(h_{11} + \tilde{W}_{1/1})} + e^{\beta_d(h_{12} + \tilde{W}_{2/1})}} = \frac{e^{0.75(3-12.52)}}{e^{0.75(3-12.52)} + e^{0.75(5-14.65)}}$$

$$= 0.52$$

$$P_{2/1} = 0.48$$

$$P_{1/2} = \frac{e^{\beta_d(h_{21} + \tilde{W}_{1/2})}}{e^{\beta_d(h_{21} + \tilde{W}_{1/2})} + e^{\beta_d(h_{22} + \tilde{W}_{2/2})}} = \frac{e^{0.75(3-10.9)}}{e^{0.75(3-10.99)} + e^{0.75(5-14.81)}}$$

$$= 0.80$$

$$P_{2/2} = 0.20$$

f. The estimation of the probabilities of travel $P_{t/i}$ similarly require first the estimation of the expected utilities received from destination choices:

$$\tilde{W}_1 = \frac{1}{\beta_d} \ln \left(e^{\beta_d(h_{11} + \tilde{W}_{1/1})} + e^{\beta_d(h_{12} + \tilde{W}_{2/1})}\right)$$

$$= \frac{1}{0.75} \ln \left(e^{0.75(3 - 12.52)} + e^{1.5(5 - 14.65)}\right) = -8.66$$

$$\tilde{W}_2 = \frac{1}{\beta_d} \ln \left(e^{\beta_d(h_{21} + \tilde{W}_{1/2})} + e^{\beta_d(h_{22} + \tilde{W}_{2/2})}\right)$$

$$= \frac{1}{0.75} \ln \left(e^{0.75(3 - 10.99)} + e^{1.5(5 - 14.81)}\right) = -7.69$$

and consequently,

$$P_{t/1} = \frac{e^{\beta_t(h_1 + \tilde{W}_1)}}{1 + e^{\beta_t(h_1 + \tilde{W}_1)}} = \frac{e^{1.25(7.94 - 8.66)}}{1 + e^{1.25(7.94 - 8.66)}} = 0.29$$

$$P_{t/2} = \frac{e^{\beta_t(h_2 + \tilde{W}_2)}}{1 + e^{\beta_t(h_2 + \tilde{W}_2)}} = \frac{e^{1.25(6.10 - 7.69)}}{1 + e^{1.25(6.10 - 7.69)}} = 0.12$$

g. From the above, we may now estimate the probabilities $P_{jmr/i}$ as

$$P_{jmr/i} = P_{t/i} P_{j/i} P_{m/ij} P_{r/ijm}$$

and

$$P_{111/1} = P_{1/111} P_{1/11} P_{1/1} P_{t/1} = 0.9975 \cdot 0.494 \cdot 0.52 \cdot 0.29$$

$$= 0.0743$$

$$P_{112/1} = P_{2/111} P_{1/11} P_{1/1} P_{t/1} = 0.0025 \cdot 0.494 \cdot 0.52 \cdot 0.29$$

$$= 0.00018$$

$$P_{122/1} = P_{2/112} P_{2/11} P_{1/1} P_{t/1} = 0.0011 \cdot 0.506 \cdot 0.52 \cdot 0.29$$

$$= 0.00008$$

$P_{111/2} = 0.095$; $P_{112/2} = 0.00019$; $P_{122/2} = 0.0000116$;

$P_{121/2} = 0.00104$

$P_{211/1} = 0.01804$; $P_{212/1} = 0.00005$; $P_{221/1} = 0.107$;

$P_{222/1} = 0.0145$;

$P_{211/2} = 0.018$; $P_{212/2} = 0.000072$; $P_{221/2} = 0.0053$;

$P_{222/2} = 0.00072$

h. The demand for travel to the various destinations T_j may now be computed as

$$T = \sum_i N_i \sum_{mr} P_{jmr/i}; \qquad \forall j$$

and

$$T_1 = N_1 [P_{111/1} + P_{112/1} + P_{121/1} + P_{122/1}]$$
$$+ N_2 [P_{111/2} + P_{112/2} + P_{121/2} + P_{122/2}]$$
$$T_2 = N_1 [P_{211/1} + P_{212/1} + P_{221/1} + P_{222/1}]$$
$$+ N_2 [P_{211/2} + P_{212/2} + P_{221/2} + P_{222/2}]$$

i. The demand for travel between the various origin-destination combinations T_{ij} may similarly be computed as

$$T_{ij} = N_i \sum_{mr} P_{jmr/i}; \qquad \forall i, j$$

and

$$T_{11} = N_1 [P_{111/1} + P_{112/1} + P_{121/1} + P_{122/1}];$$
$$T_{12} = N_1 [P_{211/i} + P_{212/1} + P_{221/1} + P_{222/1}]$$
$$T_{21} = N_2 [P_{111/2} + P_{112/2} + P_{121/2} + P_{122/2}];$$
$$T_{22} = N_2 [P_{211/2} + P_{212/2} + P_{221/2} + P_{222/2}]$$

j. The demand for travel between the various origin-destination combinations T_{ij} on given modes may similarly be computed as

$$T_{ijm} = N_i \sum_{r} P_{jmr/i}; \qquad \forall i, j, m$$

and

$$T_{111} = N_1 [P_{111/1} + P_{112/1}]; \qquad T_{112} = N_1 [P_{121/1} + P_{122/1}]$$
$$T_{121} = N_1 [P_{211/1} + P_{212/1}]; \qquad T_{122} = N_1 [P_{221/1} + P_{222/1}]$$
$$T_{211} = N_2 [P_{111/2} + P_{112/2}]; \qquad T_{212} = N_2 [P_{121/2} + P_{122/2}]$$
$$T_{221} = N_2 [P_{211/2} + P_{212/2}]; \qquad T_{222} = N_2 [P_{221/2} + P_{222/2}]$$

1. Finally, the demands for travel from the various origins T_i are equal to

$$T_1 = N_1 P_{t/1}; \qquad T_2 = N_2 P_{t/2}$$

4.9 We have

$$P_{1A1} = \frac{e^{0.1(1.5+4-9)}}{e^{0.1(0)} + e^{0.1(1.5+4-9)} + e^{0.1(1.5+4-12)} + e^{0.1(1.5+5-11)} + e^{0.1(1.5+5-8)}}$$

$$= \frac{0.70}{3.72} = 0.188$$

Similarly, $P_{1A2} = 0.52/3.72 = 0.140$; $P_{1B1} = 0.172$; $P_{1A2} = 0.230$.

The number of travelers to A using the first mode is then 800(0.188) = 150, etc. The results may be represented graphically, in terms of unconditional probabilities of combined choices, and corresponding number of passengers, as in Figure B.6. The results are the same, allowing for rounding errors. The probability that a traveler uses the first mode is then equal to the marginal probability

$$P_{\cdot\cdot 1} = P_{1A1} + P_{1B1} = 0.188 + 0.172 = 0.36$$

Consequently, (0.36) 585 = 288 travelers will use mode 1. Similarly,

$$P_{\cdot\cdot 2} = P_{1A2} + P_{1B2} = 0.37$$

corresponding to 296 passengers on mode 4. Next,

$$P_{1A\cdot} = P_{1A1} + P_{1A2} = 0.328$$

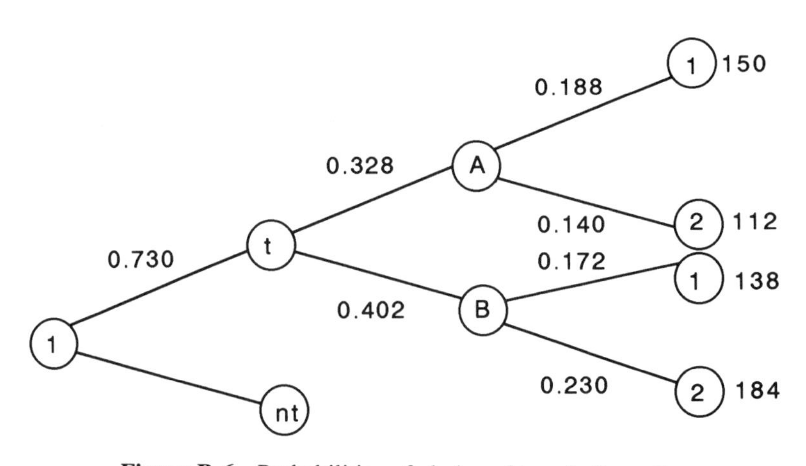

Figure B.6 Probabilities of choice of travel alternatives.

corresponding to 262 passengers going to destination 1. Finally,

$$P_{1B\cdot} = P_{1B1} + P_{1B2} = 0.402$$

corresponding to 322 passengers going to destination 2.

4.10 We have

$$\frac{\partial \tilde{W}_{MR}}{\partial c_{ijmr}} = \frac{\partial}{\partial c_{ijmr}} \frac{1}{\beta_m} \sum_k \sum_l T_{kl} \ln \sum_n e^{\beta_m(h_{kln} + \tilde{W}_{n/kl})}; \qquad \forall\ i, j, m, r \tag{B.14}$$

Consequently,

$$\frac{\partial \tilde{W}_{MR}}{\partial c_{ijmr}} = T_{ij} \frac{1}{\beta_m \sum_m e^{\beta_m(h_{i.jm} + \tilde{W}_{m/ij})}} \frac{\partial}{\partial c_{ijmr}} e^{\beta_m(h_{ijm} + \tilde{W}_{m/ij})}$$

$$= T_{ij} \frac{\beta_m e^{\beta_m(h_{ijm} + \tilde{W}_{m/ij})}}{\beta_m \sum_m e^{\beta_m(h_{ijm} + \tilde{W}_{m/ij})}} \frac{\partial \tilde{W}_{m/ij}}{\partial c_{ijmr}}; \qquad \forall\ i, j, m, r$$

From Formula (4.6), or deriving directly, we also have

$$\frac{\partial \tilde{W}_{m/ij}}{\partial c_{ijmr}} = -P_{r/ijm}; \qquad \forall\ i, j, m, r \tag{B.15}$$

Consequently,

$$\frac{\partial \tilde{W}_{RM}}{\partial c_{ijmr}} = -T_{ij} P_{m/ij} P_{r/ijm} = -T_{ijmr}; \qquad \forall\ i, j, m, r \tag{B.16}$$

and since, as seen above $(\partial \tilde{W}_{RM}/\partial B) = 1$, we indeed obtain Formula (4.19).

5.2 The objective function in terms of the route flows rather than the link flows is

$$\text{Min } F_{(v_a)} = \sum_a \int_0^{\Sigma_i \Sigma_j \Sigma_r (T_{ijr} + S_{ijr})\delta^a_{ijr}} t_a(x)\, dx$$

which, even though the $t_a(x)$'s are strictly increasing functions, is not a strictly convex function, because the T_{ijr}'s intervening in the upper bound of the integral are linearly related through Constraint (5.10b). (See Berge, 1963, for instance.)

5.10

$k\#$	θ	Algorithm step	Link # 1	2	3	4	5	6
		Initialization	$t_1^0 = 5$	$t_2^0 = 15$	$t_3^0 = 6$	$t_4^0 = 8$	$t_5^0 = 7$	$t_6^0 = 8$
		x_a^1	$x_1^1 = 186.3$	$x_2^1 = 101.5$	$x_3 = 112.2$	$x_4 = 51.6$	$x_5 = 112.2$	$x_6 = 115.3$
		Update	$t_1^1 = 103.7$	$t_2^1 = 53.2$	$t_3^1 = 17.0$	$t_4^1 = 9.4$	$t_5^1 = 25.2$	$t_6^1 = 24.4$
1	0.5	Direction	$y^1 = 25.2$	$y^1 = 137$	$y^1 = 237.8$	$y^1 = 2.7$	$y_5^1 = 237.8$	$y_6^1 = 227.5$
		Move	$x_1^2 = 105.7$	$x_2^2 = 119.3$	$x_3^2 = 175$	$x_4^2 = 27.1$	$x_5^2 = 175$	$x_6^2 = 171.4$
		Update	$t_1^2 = 15.2$	$t_2^2 = 87.9$	$t_3^2 = 71.1$	$t_4^2 = 8.1$	$t_5^2 = 114.6$	$t_6^2 = 87.9$
2	0.33	Direction	$y_1^2 = 394.2$	$y_2^2 = 5.8$	$y_3^2 = 0$	$y_4^2 = 144.2$	$y_5^2 = 0$	$y_6^2 = .1$
		Move	$x_1^3 = 200.9$	$x_2^3 = 81.8$	$x_3^3 = 117.3$	$x_4^3 = 65.8$	$x_5^3 = 117.3$	$x_6^3 = 114.9$
		Update	$t_1^3 = 138.5$	$t_2^3 = 31.1$	$t_3^3 = 19.1$	$t_4^3 = 11.6$	$t_5^3 = 28.7$	$t_6^3 = 24$
3	0.25	Direction	$y_1^3 = 2.9$	$y_2^3 = 277.1$	$y_3^3 = 119.9$	$y_4^3 = 0.3$	$y_5^3 = 119.9$	$y_6^3 = 247.3$
		Move	$x_1^4 = 151.4$	$x_2^4 = 130.6$	$x_3^4 = 118.0$	$x_4^4 = 49.4$	$x_5^4 = 118.0$	$x_6^4 = 147.8$
		Update	$t_1^4 = 48.1$	$t_2^4 = 119.7$	$t_3^4 = 19.5$	$t_4^4 = 9.1$	$t_5^4 = 29.2$	$t_6^4 = 52.2$
4	0.2	Direction	$y_1^4 = 291.3$	$y_2^4 = 3.0$	$y_3^4 = 105.7$	$y_4^4 = 58.4$	$y_5^4 = 105.7$	$y_6^4 = 17.1$
		Move	$x_1^5 = 179.4$	$x_2^5 = 105.1$	$x_3^5 = 115.5$	$x_4^5 = 50.9$	$x_5^5 = 115.5$	$x_6^5 = 121.7$
		Update	$t_1^5 = 89.9$	$t_2^5 = 58.9$	$t_3^5 = 18.4$	$t_4^5 = 9.3$	$t_5^5 = 27.4$	$t_6^5 = 28.3$
5	0.16	Direction	$y_1^5 = 64.1$	$y_2^5 = 114.9$	$y_3^5 = 221$	$y_4^5 = 6.6$	$y_5^5 = 221$	$y_6^5 = 192.4$
		Move	$x_1^6 = 160.2$	$x_2^6 = 106.7$	$x_3^6 = 133$	$x_4^6 = 43.3$	$x_5^6 = 133$	$x_6^6 = 133.4$
		Update	$t_1^6 = 59$	$t_2^6 = 61.7$	$t_3^6 = 27.8$	$t_4^6 = 8.7$	$t_5^6 = 42.9$	$t_6^6 = 37.3$
6	0.14	Direction	$y_1^6 = 251.2$	$y_2^6 = 90.9$	$y_3^6 = 57.9$	$y_4^6 = 46.6$	$y_5^6 = 57.9$	$y_6^6 = 45.4$
		Move	$x_1^7 = 173$	$x_2^7 = 104.5$	$x_3^7 = 122.5$	$x_4^7 = 44.0$	$x_5^7 = 122.5$	$x_6^7 = 121.1$
		Update	$t_1^7 = 78.4$	$t_2^7 = 57.9$	$t_3^7 = 21.6$	$t_4^7 = 8.7$	$t_5^7 = 32.9$	$t_6^7 = 27.9$
7	0.125	Direction	$y_1^7 = 114.1$	$y_2^7 = 130.8$	$y_3^7 = 155.1$	$y_4^7 = 14.4$	$y_5^7 = 155.1$	$y_6^7 = 150.3$
		Move	$x_1^8 = 165.6$	$x_2^8 = 107.8$	$x_3^8 = 126.6$	$x_4^8 = 40.3$	$x_5^8 = 126.6$	$x_6^8 = 124.7$
		Update	$t_1 = 66.7$	$t_2 = 63.6$	$t_3 = 23.8$	$t_4 = 8.5$	$t_5 = 36.5$	$t_6 = 30.4$
8	0.11	Direction	$y_1 = 191.4$	$y_2 = 95.8$	$y_3 = 112.8$	$y_4 = 30.7$	$y_5 = 112.8$	$y_6 = 89.3$
		Move	$x_1^9 = 168.5$	$x_2^9 = 106.5$	$x_3^9 = 125.1$	$x_4^9 = 39.2$	$x_5^9 = 125.1$	$x_6^9 = 120.8$
		Update	$t_1 = 71.0$	$t_2 = 61.2$	$t_3 = 23$	$t_4 = 8.5$	$t_5 = 35.1$	$t_6 = 27.7$
9	0.1	Direction	$y_1 = 156.2$	$y_2 = 112.3$	$y_3 = 131.5$	$y_4 = 23.4$	$y_5 = 131.5$	$y_6 = 117.3$
		Move	$x_1^{10} = 167.3$	$x_2^{10} = 107.1$	$x_3^{10} = 125.8$	$x_4^{10} = 37.6$	$x_5^{10} = 125.8$	$x_6^{10} = 126.5$

etc . . .

5.16 The objective function is

$$\min_{(v_1, v_2, v_3)} U' = \int_0^{v_1} (1 + x)\, dx + \int_0^{v_2} (0.5 + 2x)\, dx + \int_0^{v_3} (1 + 0.5x)\, dx$$

Because there are no routes, the constraints may be written directly in terms of link volumes

$$v_1 + v_2 + v_3 = 100$$

$$v_1 \geq 0;\ v_2 \geq 0;\ v_3 \geq 0$$

Replacing v_3 by $(100 - v_1 + v_2)$, the objective function becomes a function of only two variables, and the constraint is eliminated, so that the optimality conditions for unconstrained programs (A.23) may be applied. Setting both partial derivatives of U' with respect to v_1 and v_2 equal to zero, we obtain, after some simplifications, the optimality conditions

$$1.5v_1 + 0.5v_2 = 50$$

$$0.5v_1 + 2.5v_2 = 50.5$$

The solution of this simple linear system is

$$v_1 = 28.5; \; v_2 = 14.5$$

and consequently

$$v_3 = 100 - 28.5 - 14.5 = 57$$

b. We may now check that generalized travel costs are equal on all used routes (here links) from 1 to 2. Replacing in the expression of the link cost functions, we have, as expected:

$$t_1 = 1 + v_1 = 29.5$$

$$t_2 = 0.5 + 2v_2 = 29.5$$

$$t_3 = 1 + 0.5v_3 = 29.5$$

It may be noted that there is in the present case an easier way to find these equilibrium flows. If we assume that *all* three links are used, then the travel times will be equal on all three links, i.e.,

$$1 + v_1 = 0.5 + 2v_2$$

$$0.5 + 2v_2 = 1 + 0.5v_3 = 100 - v_1 - v_2$$

Solving this system of three equations in three unknowns provides the same solution as above. This solution is acceptable, since it verifies the user equilibrium (utility maximization) principle. Note, however, that because link cost functions are linear, Condition (5.5) for strict convexity of the objective function is not met, and that consequently, there may in principle be other solutions in terms of link flows than the one obtained above. In this case, one or two links would be used, with equal costs, and the others would not be used, with higher costs. However, because here all routes consist of a single link, and given the present specification of the link cost functions, it is easy to see that such a situation is not possible, and that consequently, we have obtained the unique network assignment.

5.19

$$\frac{\partial \sum_a \int_0^{\Sigma_i \Sigma_j \Sigma_r (T_{ijr} + S_{ijr})\delta^a_{ijr}} g_a(v)\, dv}{\partial T_{ijr}} = \sum_a \frac{\partial \int_0^{\Sigma_i \Sigma_j \Sigma_r (T_{ijr} + S_{ijr})\delta^a_{ijr}} g_a(v)\, dv}{\partial T_{ijr}}$$

$$= \sum_a g_a \left(\sum_i \sum_j \sum_r (T_{ijr} + S_{ijr})\delta^a_{ijr} \right) \frac{\partial \sum_i \sum_j \sum_r (T_{ijr} + S_{ijr})\delta^a_{ijr}}{\partial T_{ijr}}$$

$$= \sum_a g_a(v_a)\delta^a_{ijr} = g_{ijr}$$

5.21 Replacing in the utility Function (5.46) the T_{ijr}'s by their expression in Formula (5.35b), we get, after substitution of T_0 by $B - \Sigma_i \Sigma_j \Sigma_r T^*_{ijr} c_{ijr}$

$$\tilde{W}_R = B - \sum_a \int_0^{v_a^*} g_a(v)\, dv - \frac{1}{\beta_r} \sum_i \sum_j \sum_r T_{ij} P^*_{r/ij} [\ln T_{ij} + \ln P^*_{r/ij}]$$

$$= B - \sum_a \int_0^{v_a^*} g_a(v)\, dv - \frac{1}{\beta_r} \sum_i \sum_j T_{ij} \ln T_{ij} \sum_r P^*_{r/ij}$$

$$- \frac{1}{\beta_r} \sum_i \sum_j T_{ij} \sum_r P^*_{r/ij} \ln P^*_{r/ij}$$

$$= B + \frac{1}{\beta_r} E_R - \sum_a \int_0^{v_a^*} g_a(v)\, dv$$

$$- \frac{1}{\beta_r} \sum_i \sum_j T_{ij} \sum_r P^*_{r/ij} \left[-\beta_r g^*_{ijr} - \ln \sum_r e^{-\beta_r g^*_{ijr}} \right]$$

$$= B + \frac{1}{\beta_r} E_R - \sum_a \int_0^{v_a^*} g_a(v)\, dv + \frac{1}{\beta_r} \sum_i \sum_j T_{ij} \ln \sum_r e^{-\beta_r g^*_{ijr}}$$

$$+ \sum_i \sum_j \sum_r T^*_{ijr} g^*_{ijr}$$

$$= B + \frac{1}{\beta_r} E_R - \sum_a \int_0^{v_a^*} g_a(v)\, dv + \sum_{ij} T_{ij} \tilde{\Omega}_{m/ij} + \sum_{ijr} T^*_{ijr} g^*_{ijr}$$

where

$$E_R = -\sum_i \sum_j T_{ij} \ln T_{ij} \qquad \text{(B.17)}$$

is the entropy of the given origin-destination flows, and

$$g_{ijr} = \tau t_{ijr} + c_{ijr} \tag{B.18}$$

is as usual the generalized route travel cost.

5.27 If $v_{a_1a_2}$ represents the *unknown* volume going from link a_1 to link a_2, and $g_{12}(v)$ is the intersection delay function of that volume, the R.T.'s U.M. problems, either in the deterministic or probabilistic version, may be refined through the addition in the utility function of the term

$$-\sum_{i} \sum_{a_l \in i^+} \sum_{a_k \in i^-} \int_0^{v_{a_ka_l}} g_{kl}(v + s_{a_ka_l})\, dv$$

in which $s_{a_ka_l}$ is the *given* "general" volume going from link a_k to link a_l, corresponding to the given S_{ij}'s, and of the constraints

$$v_{a_ka_l} = \sum_i \sum_j \sum_r \delta_{ijr}^{a_k} \delta_{ijr}^{a_l} T_{ijr}; \qquad \forall\, i,\ a_k \in i^+,\ a_l \in i^-$$

It may then be shown, from the statement of the KKT conditions, that the user equilibrium principle still holds. However, the problem is significantly enlarged through the presence of the additional variables $v_{a_ka_l}$.

5.28 An additional term

$$\frac{1}{\beta} \sum_i \sum_s T_{is} \ln T_{is}$$

is added to the objective function (5.70). The constraints remain unchanged. At optimality,

$$T_{is} = T_{ij} \frac{e^{-\beta g_{is}^*}}{\sum_{s \in S_j} e^{-\beta g_{is}^*}}; \qquad \forall\, i, j, s$$

where g_{is}^* is the total expected travel time under strategy s from i. Its value is equal to the dual variable of Constraint (5.71).

5.29 The strategies s may be described in terms of the sequences of links available to go from i to j. The strategies and the corresponding values of the δ_s^a's are consequently

$\delta_s^a =$

$a =$	1	2	3	4	5	6
From 1 to 2						
$s = 1$	1					
2		1				1
3			1		1	1
4		1	1		1	1
5	1		1		1	1
6		1	1		1	1
7	1	1	1		1	1
From 1 to 4						
8		1				
9	1			1		
10			1		1	
11	1	1		1		
12		1	1		1	
13	1		1	1	1	
14	1	1	1	1	1	

The results are as follows

Volume v_a^s on link a following strategy s

$s =$	1	2	3	4	5	6	7	8	9	10	11	12	13	14
$a = 1$	92			17	17		9		17		10		9	7
2		18		25		9	13	24			15	13		10
3			12		20	8	10			16		11	11	8
4													43	
5			50				45							
6			115											

Total volume v_a on link a

$a =$	1	2	3	4	5	6
$v_a =$	177	127	96	43	96	115

Average wait from 1 on strategy s (minutes)

$s =$	1	2	3	4	5	6	7	8	9	10	11	12	13	14
$0.5w_{1s} =$	5	3.30	4.20	2	2.27	1.86	1.35	3.36	5	4.26	2	1.86	2.29	1.35

Volume from i on strategy s

$s =$	1	2	3	4	5	6	7	8	9	10	11	12	13	14
$T_{is} =$	92	18	12	42	36	18	32	24	17	16	25	24	20	25

Link travel times

$a =$	1	2	3	4	5	6
$t_a =$	5.2	8.5	5.15	5.3	6.6	9.9

The total travel times (wait plus travel) for the respective strategies are equal to

s	t_s	s	t_s
1	10.13	8	11.8
2	26.7	9	15.4
3	30.8	10	15.9
4	18.1	11	11.3
5	19.1	12	11.8
6	26.7	13	13.4
7	20.9	14	11.4

5.30 For fixed travel times, the STOCH algorithm may be applied *backwards*, i.e., from a given destination j, with g_{ij} in Formula (3.24a) now being equal to the *expected* cost from i to j, estimated from Formula (3.30a) in the MECR algorithm. For variable times, the Frank-Wolfe algorithm may be used, to improve from one iteration (i.e. fixed time solution) to another, as in the deterministic congested case.

6.1 The Hessian is a diagonal block-matrix, where the element $\mathbf{M}_{jj}$ is equal to

$i, j =$	$1, j$	$2, j \cdots$	$m, j \cdots$	n, j
$1, j$	$\dfrac{1}{\beta T_{1j}} + s_j'$	$s_j' \cdots$	$s_j' \cdots$	s_j'
$2, j$	s_j'	$\dfrac{1}{\beta T_{2j}} + s_j'$	$s_j' \cdots$	s_j'
$\cdots$	$\cdots$	$\cdots$	$\cdots$	$\cdots$
i, j	$s_j' \cdots$	$s_j' \cdots$	$s_j' \cdots$	$s_j' \cdots$
$\cdots$	$\cdots$	$\cdots$	$\cdots$	$\cdots$
n, j	s_j'	$s_j' \cdots$	$s_j' \cdots$	$\dfrac{1}{\beta T_{nj}} + s_j'$

A sufficient condition for positive-definiteness of symmetric matrices $[a_{ij}]$ is

$$a_{ii} > \sum_{j \neq i} |a_{ij}| \tag{B.19}$$

All matrices $\mathbf{M}_{jj}$ are symmetric, and the above sufficient condition is then

$$\frac{1}{\beta T_{ij}} + s_j' > \sum_{j \neq i} s_j' = (J - 1)s_j'; \qquad \forall\ i, j$$

or

$$T_{ij} < \frac{1}{(J - 2)\beta s_j'}; \qquad \forall\ i, j$$

The value of β may be arbitrarily set at 1. It is clear that

$$T_{ij} < \underset{i}{\text{Max}}\ (T_i); \qquad \forall\ i, j$$

Thus, for destination cost functions which have a rate of change which is increasing in the demand, and which is not equal to zero at zero demand, it is then sufficient to set the magnitude of the given demands T_i such that

$$\underset{i}{\text{Max}}\ T_i < \left[J\beta\ \underset{j}{\text{Max}}\ \{|s_j'(0)|\} \right]^{-1} \tag{B.20}$$

to guarantee that Condition (B.19) will hold everywhere. Thus, all submatrices M_{jj} will be positive-definite everywhere. Since a diagonal block matrix where all submatrices are positive-definite is itself positive-definite, the Hessian will be positive-definite and consequently, the optimal solution will be unique, in terms of link volumes.

6.4 If the destination costs include a variable term $s_j(T_j)$, as above, the R.T.'s utility function would then be

$$\begin{aligned} \text{Max } U_{(T_{ij}, T_{ijm}, T_0)} = & -\sum_m \sum_{a_m} \int_0^{\Sigma_i \Sigma_j \Sigma_r (T_{ijmr} + S_{ijmr}) \delta^{a_m}_{ijr}} g_a^m(x)\, dx \\ & - \frac{1}{\beta_m'} \sum_m \sum_{ij} T_{ijm} \ln T_{ijm} - \frac{1}{\beta_d'} \sum_{ij} T_{ij} \ln T_{ij} \\ & + \sum_{ijm} T_{ijm} h_{ijm} + \sum_{ij} h_{ij} T_{ij} - \pi \sum_j \int_0^{\Sigma_i T_{ij}} s_j(x)\, dx + T_0 \end{aligned}$$

subject to the same constraints. The KKT conditions may then be obtained from those for the same problem with fixed destination costs, which are developed in the text, with the addition of the corresponding derivatives of the additional integral term, along the lines used in the combined destination and route choice problem.

6.5 Replacing the equilibrium values in Formula (6.4) where $\tilde{W}_{j/i}$ is equal to the minimum travel cost from i to j, we get

$$T_{12} = 150 \frac{e^{-0.05(2 + 1(1.4) + 1(69))}}{e^{-0.05(2 + 1(1.4) + 1(69))} + e^{-0.05(1000000 + 1(24))} + e^{-0.05(4 + 1(9.3) + 1(61))}}$$

$$= 150 \frac{0.0268}{0.0268 + 0 + 0.0244} = 150(0.523) = 79$$

6.6 If the choice of mode precedes the choice of destination in the hierarchical structure representing the individual traveler's decisions, the resulting probabilities of choice would be

$$P_{m/i} = \frac{e^{\beta_m(h_{im} + \tilde{W}_{m/i})}}{\sum_m e^{\beta_m(h_{im} + \tilde{W}_{m/i})}}; \quad \forall\ i, m \tag{B.21}$$

$$P_{j/im} = \frac{e^{\beta_d(h_{ijm} - g_{ijm})}}{\sum_j e^{\beta_d(h_{ijm} - g_{ijm})}}; \quad \forall\ i, j, m \tag{B.22}$$

with

$$\tilde{W}_{m/i} = \frac{1}{\beta_d} \ln \sum_j e^{\beta_d(h_{ijm} - g_{ijm})}; \quad \forall\ i, m \tag{B.23}$$

so that

$$T_{ijm} = T_i P_{m/i} P_{j/im}; \quad \forall\ i, j, m$$

6.7 First, we estimate the derivatives of objective function F in Formula (6.42) with respect to car volumes. We have

$$\frac{\partial}{\partial v_b^c}\left\{\sum_a \int_0^{v_a^c} t_a(y, v_a^t)\, dy\right\} = \frac{\partial}{\partial v_b^c} \int_0^{v_b^c} t_b(y, v_a^t)\, dy$$

$$= t_b(v_b^c, v_a^t); \quad \forall\ b$$

$$\frac{\partial}{\partial v_b^c}\left\{\sum_a \int_0^{v_a^c} t_a(y, 0)\, dy\right\} = \frac{\partial}{\partial v_b^c} \int_0^{v_b^c} t_a(y, 0)\, dy$$

$$= t_b(v_b^c, 0); \quad \forall\ b$$

$$\frac{\partial}{\partial v_b^c}\left\{\sum_a \int_0^{v_a^t} t_a(v_a^c, y)\, dy\right\} = \frac{\partial}{\partial v_b^c}\int_0^{v_b^t} t_b(v_b^t, y)\, dy$$
$$= \int_0^{v_b^t} \frac{\partial}{\partial v_b^c}\{t_b(v_b^t, y)\}\, dy; \qquad \forall\, b$$

However, given the formulation of function t_a,

$$\frac{\partial t_a(v_a^c, v_a^t)}{\partial v_b^c} = \frac{\partial}{\partial v_b^c}\left\{t_a^0\left[1 + 0.15\left(\frac{v_a^c + e_c v_a^t + v_a^0}{K_a}\right)^4\right]\right\}$$
$$= \frac{1}{e_c}\frac{\partial t_a(v_a^c, v_a^t)}{\partial v_b^t}; \qquad \forall\, a, b$$

so that

$$\frac{\partial}{\partial v_b^c}\left\{\sum_a \int_0^{v_a^t} t_a(v_a^c, y)\, dy\right\} = \frac{1}{e_c}\int_0^{v_b^t} \frac{\partial}{\partial v_b^t}\{t_b(v_b^t, y)\}\, dy$$
$$= \frac{1}{e_c}[t_b(v_b^c, v_b^t) - t_b(v_b^c, 0)]; \qquad \forall\, b$$

Finally

$$\frac{\partial}{\partial v_b^c}\left\{\sum_a \int_0^{v_a^t} t_a(0, y)\, dy\right\} = 0; \qquad \forall\, b$$

The other terms in the objective function are not a function of v_b^c, so that their derivatives are zero. Using the various expressions above, we have:

$$\frac{\partial F}{\partial v_b^c} = t_b(v_b^c, v_b^t); \qquad \forall\, b \tag{B.24}$$

Thus, as in the single mode case, the *partial* derivative of the objective function with respect to the car volumes is equal to the car travel cost.

Next, we estimate the derivatives of the objective function with respect to truck volumes.

Similarly

$$\frac{\partial}{\partial v_b^t}\left\{\sum_a \int_0^{v_a^c} t_a(y, v_a^t)\, dy\right\} = \frac{\partial}{\partial v_b^t}\int_0^{v_b^c} t_a(y, v_a^t)\, dy$$
$$= \int_0^{v_b^c} \frac{\partial}{\partial v_b^t}\{t_a(y, v_a^t)\}\, dy; \qquad \forall\, b$$

However, we have

$$\frac{\partial t_a(v_a^c, v_a^t)}{\partial v_b^t} = \frac{\partial}{\partial v_b^t}\left\{ t_a^0 \left[1 + 0.15 \left(\frac{v_a^c + e_c v_a^t + v_a^0}{K_a} \right)^4 \right]^4 \right\}$$

$$= e_c \frac{\partial t_a(v_a^c, v_a^t)}{\partial v_b^c}; \qquad \forall\, a, b$$

so that

$$\frac{\partial}{\partial v_b^t}\left\{ \sum_a \int_0^{v_a^c} t_a(y, v_a^c)\, dy \right\} = e_c \int_0^{v_b^c} \frac{\partial}{\partial v_b^c} \{t_b(y, v_a^c)\}\, dy$$

$$= e_c[t_b(v_b^c, v_b^t) - t_b(0, v_b^c)]; \qquad \forall\, b$$

Also

$$\frac{\partial}{\partial v_b^t}\left\{ \sum_a \int_0^{v_a^c} t_a(y, 0)\, dy \right\} = 0; \qquad \forall\, b$$

$$\frac{\partial}{\partial v_b^t}\left\{ \sum_a \int_0^{v_a^t} t_a(v_a^c, y)\, dy \right\} = t_b(v_b^c, v_b^t); \qquad \forall\, b$$

$$\frac{\partial}{\partial v_b^c}\left\{ \sum_a \int_0^{v_a^t} t_a(0, y)\, dy \right\} = t_b(0, v_b^t); \qquad \forall\, b$$

Combining these various results

$$\frac{\partial F}{\partial v_b^t} = \frac{e_c}{e_c} t_b(v_b^c, v_b^t) = t_b(v_b^c, v_b^t); \qquad \forall\, b \tag{B.25}$$

Since the partial derivatives of the objective function with respect to both modal volumes are the modal link travel times, the problem's solution will, as in the case of a single mode, produce modal route demands which conform to the network equilibrium principle.

6.8 The KKT conditions for Problem (6.61)–(6.62) with respect to the v_a's lead to the same deterministic route demands, Equations (5.7a), since, in terms of these variables, the objective function and the constraints are the same as in Problem (5.10). Consequently, the values of the travel times t_{ijr} have "user equilibrium" values. Similarly, the KKT conditions with respect to the T_{ij}'s lead to Equations (6.67) in Appendix 6.1.

$$T_{ij} = e^{\beta_d(h_{ij} - g_{ij} - \pi s_j)} e^{\lambda_i - 1}; \qquad \forall\, i, j \tag{B.26}$$

in which, similarly, the values of the destination costs s_j have the equilibrium values $s_j(T_j)$. Next, the conditions with respect to the new variables $\tilde{C}_i$ are

$$a_i\gamma(\tilde{W}_i^\gamma + \tilde{W}_i^{\gamma-1} \ln (a_i \tilde{W}_i^\gamma)) - \gamma a_i \tilde{W}_i^{\gamma-1}\lambda_i = 0; \qquad \forall\ i \quad \text{(B.27)}$$

Dividing this equation by the term $\gamma a_i \tilde{W}_i^{\gamma-1}$, we get

$$\tilde{W}_i + \ln (a_i \tilde{W}_i^\gamma) = \lambda_i; \qquad \forall\ i$$

Therefore,

$$e^{\lambda_i} = a_i \tilde{W}_i^\gamma e^{\tilde{W}_i}; \qquad \forall\ i \qquad \text{(B.28)}$$

Replacing this expression in the first equation above, we get

$$\sum_j T_{ij} = T_i = a_i \tilde{W}_i^\gamma \sum_j e^{\tilde{W}_i} e^{\beta_d(h_{ij} - g_{ij} - \pi s_j(T_j))}; \qquad \forall\ i$$

Using the constraint

$$T_i = a_i \tilde{W}_i^\gamma; \qquad \forall\ i$$

we can divide both sides of the above relationship by these equal terms, to obtain:

$$e^{\tilde{W}_i} \sum_j e^{\beta_d(h_{ij} - g_{ij} - \pi s_j(T_j))} = 1; \qquad \forall\ i$$

or

$$e^{\tilde{W}_i} = \frac{1}{\sum_j e^{\beta_d(h_{ij} - g_{ij} - \pi s_j(T_j))}}; \qquad \forall\ i$$

which implies Equation (6.59), as required. Finally, replacing the term $e^{\tilde{W}_i}$ by its expression above in the expression of T_{ij}, we obtain:

$$T_{ij} = a_i \tilde{W}_i^\gamma \frac{e^{\beta_d(h_{ij} - g_{ij} - \pi s_j(T_j))}}{\sum_j e^{\beta_d(h_{ij} - g_{ij} - \pi s_j(T_j))}}; \qquad \forall\ i, j \qquad \text{(B.29)}$$

which is Formula (6.60). It should be noted that the constraints in the LP are no longer all linear, due to the presence of Constraint (6.62b). As a consequence, the feasible region is no longer convex, thus making the problem more difficult to solve. In particular, the partial linearization method may no longer be applied efficiently. Instead, general nonlinear optimizers (e.g., generalized reduced gradient), must be used.

However, for the special value $\gamma = 1$, Constraints (6.62b) are linear, and the partial linearization algorithm may be applied.

6.10 Referring to Figure 6.9, link and route numbering will be as follows

	Link numbers a	
From zone	2	3
To zone 1	1	2
2	—	3

	Route numbers r	
i, j	r	Component a
1, 2	1	1
1, 3	1	2
	2	1, 3

The R.T.'s U.M. problem is

$$\text{Min } U' = \int_0^{v_1} g_1(x)\,dx + \int_0^{v_2} g_2(x)\,dx + \int_0^{v_3} g_3(x)\,dx$$

$$+ \frac{1}{\beta_r}[T_{121} \ln T_{121} + T_{131} \ln T_{131} + T_{132} \ln T_{132}]$$

$$+ \frac{1}{\beta'_d}[T_{12} \ln T_{12} + T_{13} \ln T_{13}] + h_{12}T_{12} + h_{13}T_{13}$$

$$+ \pi\left(\int_0^{T_{12}} s_2(x)\,dx + \int_0^{T_{13}} s_3(x)\,dx\right)$$

such that

$$\sum_j T_{ij} = T_i; \qquad \forall\, i$$

$$\sum_r T_{ijr} = T_{ij}; \qquad \forall\, i, j$$

$$T_{ij} > 0;\ T_{ijr} > 0; \qquad \forall\, i, j, r.$$

The auxiliary route and origin destination volumes at each iteration are given by Formulas (6.13) and (6.14). The initial solution and the solution at the end of the first iteration are thus obtained as

Initialization 1	Links 1	2	3	Destination 2	3
v_a^0	0	0	0	$t_{12}^0 = 1$	$t_{131}^0 = 2;\ t_{132}^0 = 3$
t_a^0	1	2	2	$s_2^0 = 1$	$s_3^0 = 0.5$
T_{12}^0	62	0	0		
$T_{13}^0 = 38$					
T_{131}^0	0	28	0		
T_{132}^0	10	0	10		
v_a^1	72	28	10		

$$T_{12}^0 = \frac{100\,e^{-(1+1)}}{e^{-(1+1)} + e^{-(2+0.5)}} \simeq 62$$

$$T_{13}^0 \simeq 38$$

$$T_{131}^0 = \frac{38\,e^{-2}}{e^{-2} + e^{-3}} = 28$$

$$T_{132}^0 = 10$$

Iteration 1	Links 1	2	3	Destination 2	3
Update t_a^1	2.24	2.38	2	$t_{12}^1 = 2.24$	$t_{131}^1 = 2.38;\ t_{132}^1 = 4.24$
Z_{12}^1	39	0	0	$s_2^1 = 2.24$	$s_3^1 = 1.64$
$Z_{13} = 61$					
Z_{131}^1	0	53	0		
Z_{132}^1	8	0	8		
y_a^1	47	53	8		

$$Z_{12}^1 = \frac{100\,e^{-(2.24+2.24)}}{e^{-(2.24+2.24)} + e^{-(2.38+1.64)}} \simeq 39$$

$$Z_{13}^1 \simeq 61$$

$$Z_{131}^1 = \frac{61\,e^{-2.38}}{e^{-2.38} + e^{-4.24}} \simeq 53$$

$$Z_{132}^1 \simeq 8$$

The step size θ must then be identified through solving

$$\begin{aligned}
\underset{0 \le \theta \le 1}{\text{Min}}\ U'^k = {} & \int_0^{v_1^k + \theta(y_1^k - v_1^k)} g_1(x)\,dx + \int_0^{v_2^k + \theta(y_2^k - v_2^k)} g_2(x)\,dx \\
& + \int_0^{v_3^k + \theta(y_3^k - v_3^k)} g_3(x)\,dx \\
& + \frac{1}{\beta_r}\left[(T_{121} + \theta(Z_{121} - T_{121}))(\ln(T_{121} + \theta(Z_{121} - T_{121}))\right. \\
& + (T_{131} + \theta(Z_{131} - T_{131}))(\ln(T_{131} + \theta(Z_{131} - T_{131})) \\
& \left. + (T_{132} + \theta(Z_{132} - T_{132}))(\ln(T_{132} + \theta(Z_{132} - T_{132}))\right] \\
& + \pi \int_0^{T_{12} + \theta(Z_{12} - T_{12})} s_2(x)\,dx + \pi \int_0^{T_{13} + \theta(Z_{13} - T_{13})} s_3(x)\,dx
\end{aligned}$$

Since $\dfrac{1}{\beta'_d} = \dfrac{1}{\beta_d} - \dfrac{1}{\beta_r} = \dfrac{1}{1} - \dfrac{1}{1} = 0$

and $h_{12} = h_{13} = 0$ and $\pi = 1$

We have

$$\frac{dU'^k(\theta)}{d\theta} = [1 + 0.02(v_1^k + \theta(y_1^k - v_1^k))](y_1^k - v_1^k)$$

$$+ [2 + 0.01(v_2^k + \theta(y_2^k - v_2^k))](y_2^k - v_2^k)$$

$$+ [2 + 0.05(v_3^k + \theta(y_3^k - v_3^k))]\,(y_3^k - v_3^k)$$

$$+ (Z_{121} - T_{121})(\ln\,(T_{121} + \theta(Z_{121} - T_{121})))$$

$$+ (Z_{131} - T_{131})(\ln\,(T_{131} + \theta(Z_{131} - T_{131})))$$

$$+ (Z_{132} - T_{132})\,(\ln\,(T_{132} + \theta(Z_{132} - T_{132})))$$

$$+ (1 + 0.02(T_{12} + \theta(Z_{12} - T_{12})))(Z_{12} - T_{12})$$

$$+ (0.5 + 0.03(T_{13} + \theta(Z_{13} - T_{13})))(Z_{13} - T_{13})$$

$$= (1 + 0.02(72 + \theta(47 - 72)))(47 - 72)$$

$$+ (2 + 0.01(28 + \theta(53 - 28)))(53 - 28)$$

$$+ (2 + 0.05(10 + \theta(8 - 10)))(8 - 10)$$

$$+ (39 - 62)\ln\,(62 + \theta(39 - 62))$$

$$+ (53 - 28)\ln\,(28 + \theta(53 - 28))$$

$$+ (8 - 10)\ln\,(10 + \theta(8 - 10))$$

$$+ (1 + 0.02(62 + \theta(39 - 62)))(39 - 62)$$

$$+ (0.5 + 0.03(38 + \theta(61 - 38)))(61 - 38)$$

$$= 25(1 + 0.02(72 - 25\theta)) - 25(2 + 0.01(28 + 25\theta))$$

$$+ 2(2 + 0.05(10 - 2\theta))$$

$$+ 23\ln\,(62 - 23\theta) - 25\ln\,(28 + 25\theta) + 2\ln\,(10 - 2\theta)$$

$$+ 23(1 + 0.02(62 - 23\theta)) - 23(0.5 + 0.03(38 + 23\theta))$$

Applying the bisection method, we have

$\frac{dU'^1(\theta)}{d\theta}$	a	b	θ^*
	0	1	0.5
26.17	0.5	1	0.75
−15.6	0.5	0.75	0.625
−6.74	0.5	0.625	0.5625
−2.38	0.5	0.5625	0.531

so that $\theta^* \simeq 0.53$. Consequently, the first improved solution is, in terms of origin-destination flows

$$T_{12}^2 = T_{12}^1 + \theta(Z_{12}^1 - T_{12}^1) = 62 + 0.53(39 - 62) \simeq 50$$

$$T_{13}^2 = T_{13}^1 + \theta(Z_{13}^1 - T_{13}^1) = 38 + 0.53(61 - 38) \simeq 50$$

This corresponds to the updated link flows

$$v_1^2 = v_1^1 + \theta(y_1^1 - v_1^1) = 72 + 0.53(47 - 72) = 59$$

$$v_2^2 = v_2^2 + \theta(y_2^2 - v_2^2) = 28 + 0.53(53 - 28) = 41$$

$$v_3^2 = v_3^2 + \theta(y_3^2 - v_3^2) = 10 + 0.53(8 - 10) = 9$$

from which the update link times may be estimated.

$$t_1^2 = 1 + 0.02(59) = 2.18$$

$$t_2^2 = 2 + 0.01(41) = 2.41$$

$$t_3^2 = 2 + 0.05(9) = 2.45$$

6.11 To evaluate the received utility in the combined destination-route case, we first note again that the first two terms in it are the same as in the utility function for the stochastic route choice (5.46). Consequently, the received utility corresponding to them is, according to Formula (5.49a) which was established in Exercise 5.21, is equal to

$$\tilde{W}_R = B - \sum_a \int_0^{\overset{*}{v}_a} g_a(v)\, dv - \frac{1}{\beta_r} \sum_i \sum_j T_{ij}^* \ln T_{ij}^*$$

$$+ \frac{1}{\beta_r} \sum_i \sum_j T_{ij}^* \ln \sum_r e^{-\beta_r \overset{*}{g}_{ijr}} + \sum_i \sum_j \sum_r T_{ijr}^* g_{ijr}^*$$

The received utility corresponding to the other terms is equal to

$$-\frac{1}{\beta'_d} \sum_i \sum_j T_{ij}^* \ln T_{ij}^* + \sum_{ij} h_{ij} T_{ij}^* - \pi \sum_j \int_0^{\Sigma_i \overset{*}{T}_{ij}} s_j(x)\, dx$$

Given the definition of β'_d, we may write

$$-\frac{1}{\beta_r} \sum_i \sum_j T_{ij}^* \ln T_{ij}^* \frac{1}{\beta'_d} \sum_i \sum_j T_{ij}^* \ln T_{ij}^* = -\frac{1}{\beta_d} \sum_i \sum_j T_{ij}^* \ln T_{ij}^*$$

$$= -\frac{1}{\beta_d} \sum_i \sum_j T_i P^*_{j/i} [\ln T_i + \ln P^*_{j/i}]$$

$$= \frac{1}{\beta_d} E_I - \frac{\beta_d}{\beta_d} \sum_i \sum_j T^*_{ij} g^*_{ij} \frac{1}{\beta_d} \sum_i \sum_j T^*_{ij} \ln \sum_j e^{-\beta_d g^*_{ij}}$$

where, for compactness,

$$g_{ij} = h_{ij} - \pi s_j \left(\sum_i T_{ij} \right) + \tilde{W}_{j/i}; \qquad \forall\, i, j \tag{B.30}$$

and

$$E_I = -\sum_i T_i \ln T_i \tag{B.31}$$

Remembering that in this case

$$\tilde{W}_{j/i} = \frac{1}{\beta_r} \ln \sum_r e^{-\beta_r g_{ijr}}; \qquad \forall\, i, j$$

and adding all terms, we have

$$\tilde{W}_{DR} = B + \frac{1}{\beta_d} E_I - \sum_a \int_0^{v_a^*} g_a(v)\, dv + \sum_i \sum_j \sum_r T^*_{ijr} g^*_{ijr}$$

$$- \pi \left[\sum_j \int_0^{\Sigma_i T^*_{ij}} s_j(x)\, dx - \sum_i \sum_j T^*_{ij} s^*_j \right] + \frac{1}{\beta_d} \sum_i T_i \ln \sum_j e^{\beta_d g^*_{ij}}$$

6.12 In the case when route choice is deterministic, the main problem's objective function now does not include the term

$$\frac{1}{\beta_r} \sum_i \sum_j \sum_r T_{ijr} \ln T_{ijr}$$

The partially linearized objective for the subproblem is otherwise the same. The optimality conditions for the subproblem in Algorithm 6.1 may accordingly be shown to be

$$Z^k_{ij} = T_i \frac{e^{\beta_d(h_{ij} - \pi s_j^{k-1} - g_{ij}^{k-1})}}{\sum_j e^{\beta_d(h_{ij} - \pi s_j^{k-1} - g_{ij}^{k-1})}}; \qquad \forall\, i, j \tag{B.32}$$

$$\sum_a g_a(y^k_a) \delta^a_{ijr} - g^{k-1}_{ij} \geq 0; \qquad \forall\, i, j, r \tag{B.33a}$$

$$Z_{ijr}\left[\sum_a g_a(y_a^k)\delta_{ijr}^a - g_{ij}^{k-1}\right] = 0; \qquad \forall\ i, j, r \qquad \text{(B.33b)}$$

Solving the subproblem amounts to distributing the given T_i's from a given origin i to destinations j according to a logit model, as in the probabilistic route choice case. The resulting route flows Z_{ij}, both for the specific travel purpose and for "general" travel, are obtained from a network assignment with fixed travel times, using the MCR algorithm instead of a logit model of route choice. As always, the given "general" route flows S_{ijr}'s must be added to the Z_{ijr}'s. If only the S_{ij}'s are given *and* if the general travelers have the same route utilities as the T_{ij} travelers, the former may then simply be added to the latter and assigned to the MCR together. The other, subsequent operations in the iteration remain the same.

6.18 The objective function is

$$\underset{(v_1, v_1, T_{12}, T_{13})}{\text{Min}} \quad U' = 1\int_0^{v_1} (15 + x)\,dx + 1\int_0^{v_2} (20 + 0.8x)\,dx$$

$$+ \frac{1}{0.01}[T_{12}\ln T_{12} + T_{13}\ln T_{13}]$$

$$+ 1\int_0^{T_{12}} 10x\,dx + 1\int_0^{T_{13}} 15x\,dx \qquad \text{(B.34a)}$$

The constraints are

$$T_{12} + T_{13} = N_1 = 10 \qquad \text{(B.35a)}$$

$$T_{12} = v_1; \qquad T_{13} = v_2 \qquad \text{(B.35b)}$$

and, as usual, the positivity requirements

$$T_{12} > 0; \qquad T_{13} > 0; \qquad v_1 \geq 0; \qquad v_2 \geq 0 \qquad \text{(B.35c)}$$

Since $\beta_r = \infty$, omitting the iteration index, the objective function of the step size problem is:

$$\underset{0 \leq \theta \leq 1}{\text{Min}} \quad U'(\theta) = \int_0^{v_1 + \theta(y_1 - v_1)} g_1(v)\,dv + \int_0^{v_2 + \theta(y_2 - v_2)} g_2(v)\,dv$$

$$+ \frac{1}{\beta_d}\{(T_{12} + \theta(Z_{12} - T_{12})\ln(T_{12} + \theta(Z_{12} - T_{12}))$$

$$+ (T_{13} + \theta(Z_{13} - T_{13})) \ln (T_{13} + \theta(Z_{13} - T_{13}))\}$$

$$+ h_{12}(T_{12} + \theta(Z_{12} - T_{12})) + h_{13}(T_{13} + \theta(Z_{13} - T_{13}))$$

$$+ 10(T_{12} + \theta(Z_{12} - T_{12})) + 15(T_{13} + \theta(Z_{13} - T_{13}))$$

The derivative function to use in the step size determination is

$$\frac{dU'(\theta)}{d\theta} = [15 + v_1 + \theta(y_1 - v_1)](y_1 - v_1)$$

$$+ [20 + 0.8(v_2 + \theta(y_2 - v_2)](y_2 - v_2)$$

$$+ \frac{1}{\beta_d} \{(Z_{12} - T_{12})(1 + \ln (T_{12} + \theta(Z_{12} - T_{12})))$$

$$+ (Z_{13} - T_{13})(1 + \ln (T_{13} + \theta(Z_{13} - T_{13})))$$

$$+ 10(T_{12} + \theta(Z_{12} - T_{12}))(Z_{12} - T_{12})$$

$$+ 15(T_{13} + \theta(Z_{13} - T_{13}))(Z_{13} - T_{13})\}$$

The initial feasible solution and auxiliary solution are

	Link		Destination	
Iteration 0	1	2	2	3
v_a^0	0	0		
t_a^0	15	20	$t_{12}^0 = 15$	$t_{13}^0 = 20$
			$s_2^0 = 0$	$s_3^0 = 0$
T_{12}^0	5.12	0		
T_{13}^0	0	4.88		
v_a^1	5.12	4.88		

$$T_{12}^0 = \frac{10e^{-0.01(15+0)}}{e^{-0.01(15)} + e^{-0.01(20+0)}} = 5.12$$

Iteration 1				
Update t_a^1	20.12	23.90	$t_{12}^1 = 20.12$	$t_{13}^1 = 23.9$
Z_{12}^1	5.64	0	$s_2^1 = 10 \cdot 5.12 = 51.2$;	
Z_{13}^1	0	4.36	$s_3^1 = 15 \cdot 4.88 = 73.2$	
y_a^1	5.64	4.36		

$$Z_{12}^1 = \frac{10e^{-0.01(20.12 + 51.2)}}{e^{-0.01(20.12+51.2)} + e^{-0.01(23.9+73.2)}} = 5.64$$

$$Z_{13}^1 = 4.36$$

Replacing these numerical values in $(dU'/d\theta)$ above:

$$\frac{dU'^{1}(\theta)}{d\theta} = (15 + 5.12 + \theta(5.64 - 5.12))(5.64 - 5.12)$$
$$+ (20 + 0.8(4.88 + \theta(4.36 - 4.88))(4.36 - 4.88)$$
$$+ \frac{1}{0.01}\{(5.64 - 5.12)(1 + \ln(5.12 + \theta(5.64 - 5.12)))$$
$$+ (4.36 - 4.88)(1 + \ln(4.88 + \theta(4.36 - 4.88)))\}$$
$$+ 10(5.12 + \theta(5.64 - 5.12))(5.64 - 5.12)$$
$$+ 15(4.88 + \theta(4.36 - 4.88))(4.36 - 4.88)$$

$\frac{dU'^{1}(\theta)}{d\theta}$	a	b	θ^*
	0	1	0.5
1.81	0.5	1	0.75
−2.67	0.5	0.75	0.625
0.40	0.625	0.75	0.6875
−1.53	0.625	0.6875	0.656

Thus $\theta^* = 0.656$, and the second solution is

$$T_{12}^{2} = T_{12}^{1} + \theta(Z_{12} - T_{12}) = 5.12 + 0.656(5.64 - 5.12)$$
$$\simeq 5.46 \Rightarrow T_{13}^{1} \simeq 4.54$$
$$v_1^2 = 5.12 + (5.64 - 5.12)0.656 \simeq 5.46; \Rightarrow v_2^2 = 4.54$$
$$\Rightarrow t_1^2 = 15 + 5.46 = 20.46; \quad t_2^2 = 20 + 0.8 \cdot 4.54 \simeq 23.63$$

	Link		Destination	
Iteration 2	1	2	2	3
Update t_a^2	20.46	23.63		
Z_{12}^2	5.42	0	$s_2 = 105.46 = 54.6;$	$s_3 = 15 \cdot 454 = 68.1$
Z_{13}^2	0	4.58		
y_a^2	5.42	4.58		

$$Z_{12}^2 = \frac{10e^{-0.01(20.46+54.6)}}{e^{-0.01(20.46+54.6)} + e^{-0.01(23.63+68.1)}} \simeq 5.42$$
$$Z_{13}^2 = 4.58$$

$$\frac{dU'(\theta)}{d\theta} = (15 + 5.46 + \theta(5.42 - 5.46))(5.42 - 5.46)$$

$$+ (20 + 0.8(4.54 + \theta(4.58 - 4.54))(4.58 - 4.54)$$

$$+ \frac{1}{0.01}\{(5.42 - 5.46)(1 + \ln(5.46 + \theta(5.42 - 5.46)))$$

$$+ (4.58 - 4.54)(1 + \ln(4.54 + \theta(4.58 - 4.54)))\}$$

$$+ 10(5.46 + \theta(5.42 - 5.46))(5.42 - 5.46)$$

$$+ 15(4.54 + \theta(4.58 - 4.54))(4.58 - 4.54)$$

$\frac{dU'(\theta)}{d\theta}$	a	b	θ^*
	0	1	0.5
0.02	0.5	1	0.75
−0.01	0.5	0.75	0.625
0.001	0.625	0.75	0.6875
—	0.625	0.6875	0.65625

Thus, $\theta^* = 0.656$

$$T_{12}^3 = 5.46 + 0.656(5.42 - 5.46) = 5.43 \text{ and } T_{13}^3 = 4.57$$

$$t_1^3 = 15 + 5.43 = 20.43; \qquad t_2^3 = 20 + 0.8 \cdot 4.57 = 23.66$$

$$s_2^3 = 10 \cdot 5.43 = 54.3 \qquad s_3^3 = 15 \cdot 4.57 = 68.55$$

Since the change in the values of the T_{1j}'s from the last iteration is less than 1%, the current values may be considered the solution.

An alternative, and simpler way, is possible in the present case, due to the special structure of the network. This would not be possible in the general case, i.e., for any network. We may use constraints (B.35a) to eliminate the link flows, i.e., replace them by the corresponding route flows, as well as replace T_{13} by $(10 - T_{12})$. Objective function (B.34a) accordingly becomes

$$\underset{(T_{12})}{\text{Min}}\ U' = \int_0^{T_{12}} (15 + x)\,dx + \int_0^{10-T_{12}} (20 + 0.8x)\,dx$$

$$+ 100\,[T_{12} \ln T_{12} + (10 - T_{12}) \ln (10 - T_{12})]$$

$$+ 1 \int_0^{T_{12}} 10x\,dx + \int_0^{(10-T_{12})} 15x\,dx \qquad \text{(B.34b)}$$

There are no constraints, since they have been incorporated into the objective function, through the above substitutions. Consequently, this is an unconstrained minimization problem in one variable, which may be solved by setting the first derivative of the objective function at zero. After some manipulations, this leads to the equation

$$T_{12} - 6.08 + 3.73 \ln \left(\frac{T_{12}}{10 - T_{12}} \right) = 0 \tag{B.36}$$

This nonlinear equation may be solved using the fixed point method (Algorithm 5.3), in terms of T_{12}, or Newton's method, Algorithm 5.4. Alternatively, the bisection method may be applied in the range $0 \leq T_{12} \leq 10$. It may also be solved on a spreadsheet, or, as a last resort, graphically, by plotting the variations of the function of T_{12} represented by the left-hand side of the equation in the range from 0 to 10, and visually determining where its value crosses the x axis, i.e, takes on the value zero.

In either case, the solution is

$$T_{12} = 5.43$$

and consequently,

$$T_{13} = 4.57$$

The link volumes are accordingly

$$v_1 = T_{12} = 5.43; \qquad v_2 = T_{13} = 4.47$$

Based on these flows, the link costs are equal to

$$t_1(v_1) = 15 + 5.43 = 20.43$$

$$t_2(v_2) = 20 + 0.8(4.57) = 23.66$$

Similarly, the destination costs are equal to

$$s_1(T_{12}) = 10(5.43) = 54.30$$

$$s_2(T_{13}) = 15(4.57) = 68.55$$

Consequently, the systematic utilities of the two destinations are equal to

$$\tilde{U}_{12} = -\tau t_1 - \pi s_2 = -1(20.43) - 1(54.30) = -74.73$$

$$\tilde{U}_{13} = -\tau t_2 - \pi s_3 = -1(23.66) - 1(68.55) = -92.21$$

Based on these utilities, we would expect that the trip distribution T_{1j} conform to the logit model

$$P_{1j} = \frac{e^{0.01\tilde{U}_{1j}}}{\sum_j e^{0.01\tilde{U}_{1j}}}; j = 1, 2 \tag{B.37}$$

that is,

$$T_{12} = N_1 \frac{e^{0.01\tilde{U}_{12}}}{e^{0.01\tilde{U}_{12}} + e^{0.01\tilde{U}_{13}}} = 10 \frac{e^{-0.01(74.73)}}{e^{-0.01(74.73)} + e^{-0.01(92.21)}} = 5.44$$

This confirms that the model's solution does indeed numerically retrieve the logit demands for the equilibrium utilities. This may be expected from theory, and would of course also be the case with more complex networks.

6.26 In the case of one scheduled mode t and one unscheduled mode c, the total utility received from a single trip on mode m *and* route r/strategy s between the given locations i and j is specified as

$$\tilde{V}_{mr/ij} = \tilde{V}_{m/ij} + \tilde{V}_{r/ijm}; \qquad \forall\ i, j, m$$

where

$$\tilde{V}_{m/ij} = b_{ij} + h_{ijm} + \epsilon_{ijm} = \tilde{U}_{m/ij} + \epsilon_{ijm}; \qquad \forall\ i, j, m$$

and

$$\tilde{V}_{r/ijc} = -\tau t_{ijcr} - c_{ijcr} + \epsilon_{ijcr} = \tilde{U}_{r/ijc} + \epsilon_{ijcr}; \qquad \forall\ i, j, r$$

$$\tilde{V}_{s/ijt} = -\tau t_{its} - \overline{\omega}\omega_{is} - c_{ijt} + \epsilon_{ijts} = \tilde{U}_{s/ijt} + \epsilon_{ijts}; \qquad \forall\ i, s \in S_j$$

where s refers to a given strategy to go from i to j. Accordingly, the probability that an individual traveler between i and j will choose mode m is equal to

$$P_{m/ij} = \frac{e^{\beta_m(\tilde{U}_{m/ij} + \tilde{W}_{m/ij})}}{\sum_m e^{\beta_m(\tilde{U}_{m/ij} + \tilde{W}_{m/ij})}} = \frac{e^{\beta_m(h_{ijm} + \tilde{W}_{m/ij})}}{\sum_m e^{\beta_m(h_{ijm} + \tilde{W}_{m/ij})}}; \qquad \forall\ i, j, m$$

where

$$\tilde{W}_{c/ij} = \frac{1}{\beta_r} \ln \left(\sum_r e^{\beta_r \tilde{U}_{r/ijc}} \right) = \frac{1}{\beta_r} \ln \left(\sum_r e^{-\beta_r(c_{ijcr} + \tau t_{ijcr})} \right); \qquad \forall\ i, j$$

$$\tilde{W}_{t/ij} = \frac{1}{\beta_r} \ln \left(\sum_s e^{\beta_r \tilde{U}_{s/ijt}} \right) = \frac{1}{\beta_r} \ln \left(\sum_s e^{-\beta_r(c_{ijt} + \tau t_{its} + \overline{\omega} w_{is})} \right); \qquad \forall\ i, s \in S_j$$

The joint probability that an individual traveler between i and j will choose car route r is equal to

$$P_{cr/ij} = \frac{e^{\beta_m(h_{ijc} + \tilde{W}_{c/ij})}}{\sum_m e^{\beta_m(h_{ijc} + \tilde{W}_{c/ij})}} \frac{e^{-\beta_r(c_{ijcr} + \tau t_{ijcr})}}{\sum_r e^{-\beta_r(c_{ijcr} + \tau t_{ijcr})}}; \qquad \forall\, i, j, r$$

Similarly, the joint probability that an individual between i and j will choose transit strategy s is equal to

$$P_{ts/ij} = \frac{e^{\beta_m(h_{ijt} + \tilde{W}_{t/ij})}}{\sum_m e^{\beta_m(h_{ijt} + \tilde{W}_{t/ij})}} \frac{e^{-\beta_r(c_{ijt} + \tau t_{ijs} + \overline{\omega} w_{is})}}{\sum_s e^{-\beta_r(c_{ijt} + \tau t_{ijs} + \overline{\omega} w_{is})}}; \qquad \forall\, i, j, s$$

6.27 The R.T.'s U.M. problem in the general case when one made, say t, is scheduled, and one mode c is unscheduled, and route/strategy choice is probabilistic, is

$$\begin{aligned}
\text{Max } U_{MR}(T_0, T_{ijc}, T_{ijts'}, T_{ijcr}, T_{ijt}, w_{is}, v_a^c, v_{as}^t, v_a^t)& \\
= -\tau \sum_m \sum_{a_m} \int_0^{v_{aj}^m} t_a^m(x)\, dx - \overline{\omega} \sum_i \sum_{s \in S_j} w_{is}& \\
- \frac{1}{\beta_m'} \sum_{ijm} T_{ijm} \ln T_{ijm} - \frac{1}{\beta_r} \sum_{i, s \in S_j} T_{its} \ln T_{its}& \\
- \frac{1}{\beta_r} \sum_{ijr} T_{ijcr} \ln T_{ijcr} + \sum_{ijm} h_{ijm} T_{ijm} + T_0&
\end{aligned}$$

where

$$\beta_m' = \frac{\beta_m \beta_r}{\beta_r - \beta_m} \rightarrow \frac{1}{\beta_m'} = \frac{1}{\beta_m} - \frac{1}{\beta_r}$$

subject to

$$\begin{aligned}
v_{as}^t &= f_a w_{is} \delta_s^a; & \forall\, a \in i^-, s \\
v_a^c &= \sum_i \sum_j \sum_r T_{ijr} \delta_{ijr}^a; & \forall\, a \\
\sum_{a \in i^-} v_{as}^t - \sum_{a \in i^+} v_{as}^t &= T_{its}; & \forall\, i, s \in S_j \\
\sum_s v_{as}^t &= v_a^t; & \forall\, a_t \\
\sum_{s \in S_j} T_{its} &= T_{ijt}; & \forall\, i, j
\end{aligned}$$

$$\sum_r T_{ijr} = T_{ijc}; \qquad \forall\, i, j$$

$$\sum_m T_{ijm} = T_{ij} \qquad \forall\, i, j$$

$$v_{aj}^m,\ w_{is},\ T_{ijt},\ T_{ijc},\ T_{ijr},\ T_0 \geq 0, \qquad \forall\, i, j, a, m$$

The solution to this problem will retrieve the aggregate demands corresponding to the probabilities of choice described in the previous exercise, and in which car origin destination travel times have their stochastic route equilibrium values, and transit origin destination travel times have their stochastic strategy equilibrium (expected) values.

6.29 The results are as follows

Equilibrium origin-destination modal demands

$$T_{12c} = 141;\ T_{12t} = 109$$
$$T_{14c} = 81;\ T_{14t} = 69$$

Equilibrium modal link travel times

$$t_1^c = 11.2,\ t_2^c = 15;\ t_3^c = 6.5;\ t_4^c = 8;\ t_5^c = 7.7$$
$$t_1^t = 4.6,\ t_2^t = 8.1;\ t_3^t = 4.3;\ t_4^t = 5.1;\ t_5^t = 6.1$$

Average waiting times: $0.5w_{ij}$

To	$j =$	2	4
From	$i = 1$	5	1.35

Minimum car travel time from i to j

To	$j =$	2	4
From	$i = 1$	11.2	14.2

Expected transit travel time from i to j

To	$j =$	2	4
From	$i = 1$	9.6	10.6

Volume v_{aj}^c on car link a going to destination j

	$j =$	2	4
$a =$	1	141	
	3		81
	5		81

Total car volume v_a^c in link a

$a =$ 1	3	5
141	81	81

Volume v_{as}^t on transit link a following strategy s

$s =$	1	8	10	14
$a =$ 1	109			19
2				28
3				23
4		19	23	
5				

Total transit volume v_a^t on link a

$a =$ 1	2	3	4	5
129	28	23	19	23

Car volume T_{ijc} from i to j

$j =$	2	4
$i =$ 1	141	81

Transit volume T_{is} from i on strategy s

$s =$	1	14
$i =$ 1	109	69

In particular, the equilibrium minimum car travel times t_{ijc} and expected transit travel times t_{is} are respectively equal to the dual values of the second and third set of constraints, respectively. It may be verified that the solution conforms to the probabilities in the preceding exercise, in which the above times are inserted. Taking into account the fact that route/strategy choice is deterministic, and that consequently, the nested logit model becomes a single level multinomial level,

$$T_{12c} = 250 \frac{e^{-0.1(2+1\cdot 11.2)}}{e^{-0.1(2+1\cdot 11.2)} + e^{-0.1(6+1\cdot 9.6)}} = 140$$

$$T_{12c} = 150 \frac{e^{-0.1(3+1\cdot 14.2)}}{e^{-0.1(3+1\cdot 14.2)} + e^{-0.1(8+1\cdot 10.6)}} = 80$$

7.2 With respect to parameter β, the first-order conditions may be formulated as

$$\frac{\partial \Lambda}{\partial \beta} = \frac{\partial}{\partial \beta} \sum_i \sum_j \hat{T}_{ij} \ln P_{j/i} = \sum_i \sum_j \frac{\hat{T}_{ij}}{P_{j/i}} \frac{\partial P_{j/i}}{\partial \beta} \tag{B.38}$$

with

$$\frac{\partial P_{j/i}}{\partial \beta} = \frac{\partial}{\partial \beta}\left[\frac{e^{-\beta(c_{ij} + \tau t_{ij})}}{\sum_j e^{-\beta(c_{ij} + \tau t_{ij})}}\right] = \frac{\partial}{\partial \beta}\left(\frac{e^{-\beta g_{ij}}}{\sum_j e^{-\beta g_{ij}}}\right); \qquad \forall\, i, j \qquad \text{(B.39)}$$

where

$$g_{ij} = c_{ij} + \tau t_{ij} \qquad \forall\, i, j$$

Consequently,

$$\frac{\partial P_{j/i}}{\partial \beta} = \frac{-g_{ij} e^{-\beta g_{ij}} \sum_j e^{-\beta g_{ij}} + e^{-\beta g_{ij}} \sum_j g_{ij} e^{-\beta g_{ij}}}{\left(\sum_j e^{-\beta g_{ij}}\right)^2}$$

$$= -g_{ij} P_{j/i} + P_{j/i} \frac{\sum_j g_{ij}\, e^{-\beta g_{ij}}}{\sum_j e^{-\beta g_{ij}}} = -g_{ij} P_{j/i} + P_{j/i} \sum_j g_{ij} P_{j/i}$$

$$= P_{j/i}\left(-g_{ij} + \sum_j g_{ij} P_{j/i}\right); \qquad \forall\, i, j$$

Therefore, the first-order condition with respect to β may be written

$$\sum_i \sum_j \hat{T}_{ij}\left(-g_{ij} + \sum_j g_{ij} P_{j/i}\right) = 0$$

or, since

$$T_{ij} = \hat{T}_i P_{j/i}; \qquad \forall\, i, j$$

the condition may be written

$$\sum_i \sum_j \hat{T}_{ij} g_{ij} = \sum_i \sum_j g_{ij} \hat{T}_{ij} P_{j/i} = \sum_i \sum_j g_{ij} P_{j/i} \sum_j \hat{T}_{ij}$$
$$= \sum_i \sum_j g_{ij} P_{j/i} \hat{T}_i = \sum_i \sum_j g_{ij} T_{ij}$$

that is, finally

$$\sum_i \sum_j (c_{ij} + \tau t_{ij}) \hat{T}_{ij} = \sum_i \sum_j (c_{ij} + \tau t_{ij}) T_{ij}(\beta, \tau) \qquad \text{(B.40)}$$

Similarly, the condition with respect to τ may be written

$$\sum_i \sum_j t_{ij} \hat{T}_{ij} = \sum_i \sum_j t_{ij} T_{ij}(\beta, \tau) \qquad \text{(B.41)}$$

By combining these equations, the first equation may be replaced by

$$\sum_i \sum_j c_{ij} \hat{T}_{ij} = \sum_i \sum_j c_{ij} T_{ij}(\beta, \tau) \tag{B.42}$$

7.7 The single term in the R.T.'s "partial" utility function (7.32) represents the "entropy" of the distribution of predicted values T_{ij}'s. (See Formula (A.61b) in section 3.6 in Appendix A.) As discussed in that section, the higher the entropy of the distribution of a given total T_i, the more probable it is. Thus, the partial R.T. U.M. problem may be interpreted as insuring that the distribution of predicted values is the most probable.

7.8 Another interpretation of the objective function may also be formulated. In this perspective, the best calibrated model is that which *minimizes* the amount of information contained in the predicted values T_{ij}'s, i.e., which is the least biased, while meeting the other constraints. From information theory (Shannon, 1948), the term

$$\mathrm{I} = \sum_i \sum_j T_{ij} \ln T_{ij} \tag{B.43}$$

measures the amount of information contained in the distribution $\{T_{ij}\}$. This is also the negative of the entropy, as defined in Formula (A.61b). Thus, the partial R.T. U.M. problem may be interpreted as insuring that the distribution of predicted values is the most "neutral," while being compatible with the observed total travel time and cost.

7.9 The calibrated values of $\hat{\beta} = 7.46$ and $\hat{\tau} = 0.77$ imply that the average utility which an individual traveler places on a single trip from i to j is equal to

$$\tilde{U}_{ij} = \alpha_j - 7.46c_{ij} - 0.77t_{ij}; \qquad \forall\, i, j$$

This means that

$$\Delta\tilde{U}_{ij} = -7.46\Delta c_{ij} - 0.7\Delta t_{ij}; \qquad \forall\, i, j$$

For a fixed level of utility $\Delta\tilde{U}_{ij} = 0$ and

$$\frac{\Delta c_{ij}}{\Delta t_{ij}} = -\frac{0.77}{7.46} = 0.103; \qquad \forall\, i, j$$

which implies that a decrease of 1 minute causes an increase of 0.103 dollars. Thus, one hour is worth 0.103(60) = \$6.19.

7.12 It is easy to show, using the KKT conditions, that the solution is of the form

$$T_{ij} = \hat{T}_i \hat{T}_j \lambda_j \mu_i e^{-\beta(c_{ij} + \tau t_{ij})}; \qquad \forall\, i, j \tag{B.44}$$

where the μ_i and λ_j's are the dual variables corresponding to flow consistency constraints on the origin and destination flows, respectively. Using these constraints, we have

$$\sum_i T_{ij} = \hat{T}_j \lambda_j \sum_i \hat{T}_i \mu_i e^{-\beta g_{ij}} = \hat{T}_j; \qquad \forall j \tag{B.45a}$$

$$\sum_j T_{ij} = \hat{T}_i \mu_i \sum_j \hat{T}_j \lambda_j e^{-\beta g_{ij}} = \hat{T}_i; \qquad \forall i \tag{B.45b}$$

These equalities in turn imply

$$\lambda_j = \left[\sum_i \hat{T}_i \mu_i e^{-\beta g_{ij}} \right]^{-1}; \qquad \forall j \tag{B.46a}$$

$$\mu_i = \left[\sum_j \hat{T}_j \lambda_j e^{-\beta g_{ij}} \right]^{-1}; \qquad \forall i \tag{B.46b}$$

together with the two equations for parameters β and τ

$$\sum_i \sum_j \hat{T}_i \hat{T}_j \lambda_j \mu_i e^{-\beta g_{ij}} c_{ij} = \hat{c} \tag{B.47a}$$

$$\sum_i \sum_j \hat{T}_i \hat{T}_j \lambda_j \mu_i e^{-\beta g_{ij}} t_{ij} = \hat{t} \tag{B.47b}$$

7.13 To solve this system of equations in "fixed point" form, start from assumed initial values for β, τ and the λ_j's. Insert them into the second set of equalities for the μ_i's in which every term is now known and determine their numerical values. Using the values for the μ_i's and λ_j's, solve the last two equations for the values of β and τ. Using those values, determine the value of the λ_j's from the first set of equalities. If these values are equal to the previous values, stop. If not, iterate.

This algorithm may be seen as an extension of "Furness" procedure, or "biproportional" matrix adjustment described in Exercise 1.4.

7.18 The M.L. equation, Formula (7.16a), is

$$400(10) + 600(15) = 10 \left(1000 \frac{e^{-10\beta}}{e^{-10\beta} + e^{-15\beta}} \right) + 15 \left(1000 \frac{e^{-15\beta}}{e^{-10\beta} + e^{-15\beta}} \right) \tag{B.48}$$

After simplifications, this leads to

$$e^{-5\beta} = 0.67$$

from which $\hat{\beta} = 0.08$.

The value of the log-likelihood function Λ, according to Formula (7.3b) is equal to

$$\Lambda(\beta) = 400 \ln \left(\frac{e^{-10\beta}}{e^{-10\beta} + e^{-15\beta}} \right) + 600 \ln \left(\frac{e^{-15\beta}}{e^{-10\beta} + e^{-15\beta}} \right)$$

from which

$$\frac{d^2 \Lambda}{d\beta^2} = - \frac{25{,}000\, e^{-25\beta}}{(e^{-15\beta} + e^{-10\beta})^2}$$

and consequently, using Formula (7.44)

$$s_{\hat{\beta}}^2 = - \left(\frac{d^2 \Lambda}{d\beta^2} \right)^{-1}_{\hat{\beta} = 0.08} = 1.66\ 10^{-4}$$

from which $s_{\hat{\beta}} = 0.0129$.

The 95% confidence interval for the true value of parameter β is then

$$0.08 - 1.96(0.0129) \leq \beta \leq 0.08 + 1.96(0.0129)$$

i.e.,

$$0.055 \leq \beta \leq 0.105$$

b. With the above values, the observed value of Λ is equal to

$$\Lambda(0.08) = 400 \ln \left(\frac{e^{-10(0.08)}}{e^{-10(0.08)} + e^{-15(0.08)}} \right)$$

$$+ 600 \ln \left(\frac{e^{-15(0.08)}}{e^{-10(0.08)} + e^{-15(0.08)}} \right) = -674$$

From Formula (7.46)

$$\Lambda(0) = 400 \ln (0.5) + 600 \ln (0.5) = 1000 \ln (1/2) = -693$$

From Formula (7.47)

$$\Lambda(\beta^*) = 400 \ln (400/1000) + 600 \ln (600/1000) = -673$$

The hypothesis that the model may be considered the best and worst model, respectively, may be tested using Formulas (7.49) and (7.50) using these log-likelihood values.

7.19 Since the observed link volumes are not available, alternative formulation (7.41)–(7.42) may be used instead. The problem is then

$$\underset{(T_1, T_2, \beta)}{\text{Min}} \left[\int_0^{T_1} (10 + 0.015x)\, dx + \int_0^{T_2} (15 + 0.01x)\, dx \right] \tag{B.49}$$

such that

$$T_1 + T_2 = 1000 \tag{B.50a}$$

$$T_1 \ln T_1 + T_2 \ln T_2 = \hat{T}_1 \ln \hat{T}_1 + \hat{T}_2 \ln \hat{T}_2 = 5{,}234 \tag{B.50b}$$

The first constraint may be used to eliminate T_2 so that the Lagrangian of the problem is

$$\begin{aligned} L(T_1, \beta) = {} & \int_0^{T_1} (10 + 0.015x)\, dx \int_0^{1000 - T_1} (15 + 0.01x)\, dx \\ & - \beta[T_1 \ln T_1 + (1000 - T_1) \ln (1000 - T_1) - 5{,}234] \end{aligned} \tag{B.51}$$

Taking the partial derivatives of L with respect to T_1 and β provides the calibration equations.

8.1 If τ in Formula (8.1) is replaced with κ, the same derivations can be carried out, provided that definitional relationship (8.24) is changed to

$$v_a^t = \sum_k \sum_j \sum_r \left(\frac{\kappa}{\tau} \phi G_{kjr} + L_{kjr} \right) \rho_{i,j,r}^a; \qquad \forall\, a \tag{B.52}$$

In this form, the KKT conditions for the G_{kjr} would lead to

$$(\omega t_{kjr} - \theta_{kj}) G_{kjr} = 0; \qquad \forall\, k, j, r \tag{B.53a}$$

$$\omega t_{kjr} \geq \theta_{kj}; \qquad \forall\, k, j, r \tag{B.53b}$$

which leads to

$$G_{kj} = \frac{\phi T_j e^{\gamma(g_{kj} - R_k^p - \kappa t_{kj})}}{\sum_k e^{\gamma(g_{kj} - R_k^p - \kappa t_{kj})}}; \qquad \forall\, k, j \tag{B.54}$$

However, as can be seen from the new definition of truck link volumes, this in turn implies that the unit for goods volumes must be changed to the amount carried by the fraction of a truck equivalent to one car, prorated to the ratio of the cost of shipping one unit of goods per unit time to the monetary value of one unit of personal travel time. For instance, if the car equivalency e_t of a truck is equal to 4, and if a truck carries on the average 10,000 pounds of goods, the cost κ of shipping one unit of goods is \$0.5/hr., and the monetary value of one hour of personal travel time τ is \$10, then the unit for the flow of goods is

$$(10{,}000/e_t)(\kappa/\tau) = (10{,}000/4)(0.5/10) = 125 \text{ lb}.$$

Consequently, this means that Z_{kjr} units of goods require the equivalent of

$$G_{kjr}/(\tau/\kappa) = (\kappa/\tau)Z_{kjr}$$

cars to transport. In this example, this would be equal to $G_{kjr}(0.5/10) = 0.05G_{kjr}$ cars.

APPENDIX C

ALPHABETICAL LIST OF SYMBOLS

*	Superscript of optimal, or equilibrium value of a variable
A	Total number of links
a	Index of link
a, b, d	Coefficients in multivariate linear functions
α, β, γ	Parameters in travel demand models
A_j^+	Set of selected (optimal) links, from origin i to destination j
A_m	Total number of links on mode m
a_m	Index of link on mode m
$\boldsymbol{\beta}$	Vector of parameters
$\hat{\beta}$	Estimated value of parameter β
β'	Rescaled parameter in R.T.'s problem
B_{xy}	Aggregate budget of travelers making choices xy
b_{xy}	Individual budget of travelers making choices xy
LP	Lower problem in bilevel optimization
χ	Chi-square statistic
$c(\cdot)$	Criterion function in solution algorithm
c_j	Cost at destination j
c_{nj}	Travel time from traveler n's location to destination j
Cov (X, Y)	Covariance of X and Y
c.p.f.	Cumulative probability distribution function
d	Generic destination choice
δ_a	"Yes/no" decision variable to construct link a
δ_{ijr}^a	Link-route incidence indicator

δ_j	"Yes/No" (0/1) indicator of choice j
δ_a^r	Link-route incidence binary variable for absolute numbering of routes
d_{nj}	Average demand of individual n for alternative j
δ_{nj}	"Yes/No" indicator of choice j by individual n
e	Base of natural logarithms (=2.71828 . . .)
ϵ	Tolerance (precision) for algorithm termination
e_b	Bus-car equivalency
E_s	Entropy of a single value
e_t	Truck-car equivalency
E_u	Entropy of a uniform distribution
E_x	Entropy of distribution of level x choices
ϵ_{xy}	Random component of received utility from having made joint choices xy
$e_{Y/X}$	Elasticity of variable Y with respect to variable X
$E\{X\}$	Expected value (mean) of random variable X
ϕ	Temporal prorating factor for goods movements
$\Phi(x)$	Value at x of the cumulative standard normal distribution function
f_a	Frequency of transit service on link a
f_i	Combined frequency of transit service at node i
G_0	Suppliers' budgetary slack
$g_a(\ )$	Generalized link travel cost (money plus time) function on link a
g_{ij}	Minimum generalized travel cost from location i to location j
G_{lk}^n	Commodity flow from $(n + 1)$th level supplier l to nth level supplier k.
G_x	Freight volumes
H	Hessian of multivariate function
η^2	Entropy ratio
HOV	High-occupancy vehicle
h_x	Constant term in utility specification (attractiveness or bias term)
I	Total number of origin zones
i	Index of origin zone
i^+	Set of links entering node i
i^-	Set of links leaving node i
i.i.d.	Identically and independently distributed
IIA	Independence of irrelevant alternatives
J	Jacobian of several functions
J	Total number of destination zones

j	Index of destination zone
K	Total number of warehouses
k	Generic location of warehouse
K_a	Level of service of link a
k_j	Time conversion factor in destination time function
KKT	Karush-Kuhn-Tucker
L	Likelihood function or Lagrangean
Λ	Log-likelihood function
λ, μ, ν	Dual variables in R.T.'s utility maximization problems
l.h.s.	Left-hand side
L_{ij}	General origin-destination truck volumes
ln	Natural (Neperian) logarithm
M	Total number of modes
m	Index of mode
m	generic mode choice
M.L.	Maximum likelihood
$\text{Max}_{(x)}$	Maximum with respect to x
MCR	Minimum cost route
MECR	Minimum Expected Cost Route
$\text{Min}_{(x)}$	Also Min_x; minimum with respect to x
MSA	Method of Successive Averages
N	Sample size
n	Index of individual traveler in sample
N_i	Number of potential travelers from origin i
$\varnothing$	Means "empty set"
$\in$	Means "belonging to set"
π	Coefficient of destination time in utility function
$P(A)$	Probability of event A
p.d.f.	Probability distribution function
P_a	Probability of choosing link a
Π_i	Product with respect to i
$P_{j/n}$	Probability of choice j by individual n
P_x	(Unconditional) probability of choice x
$P_{x/y}$	(Conditional) probability of choice x given that choice y has already been made
P_{xy}	(Unconditional) joint probability of choices x and y
$P_{xy/z}$	(Conditional) joint probability of choices x and y given that choice z has already been made
$p_m(\,)$	Modal externality function

θ	Step size in linear and partial linearization algorithm
Q_x^l	Goods flows solution for lth auxiliary subproblem
r	Index of route
r	Generic route choice
R.T.	Representative traveler
R_j^1	Expected commodity price at consumer site j
ρ^2	Log-likelihood ratio
R_k^2	Wholesale commodity price at k
ρ_{ijr}^a	Link-route incidence indicator for trucks
r.h.s.	right-hand side
R_{ij}	Number of routes between i and j
r_j	Retail commodity's unit markup at j
R_l^n	Unit commodity price at $(n + 1)$th level supplier l
R_{XY}	Coefficient of correlation between random variables X and Y
$r_{y/x}$	Rate of change of Y with respect to X
S.P.E.	Spatial price equilibrium
S.U.E.	Stochastic user equilibrium
s_j^0	Fixed destination cost at zero volume.
Σ_i	Sum with respect to i
$s_j(\cdot)$	Destination cost function for location j
s_j	Cost at destination j
S_x	Given demands for general-purpose travel
σ_X	Standard deviation of random variable X
$[\sigma_{ij}]$	Variance-covariance matrix
τ	Parameter attached to travel time in utility function
t	t-test statistic
t/i	Choice of traveling from i
T_0	Amount spent on other than travel
t_a^0	Minimum ("free flow") travel time for link a
$t_a(\cdot)$	Travel time function for link a
$t_a^m(\cdot)$	Travel time function for link a of mode m
t_a	Travel time on link a
T_i	Number of travelers from origin i
T_{i0}	Nontravelers in zone i
T_{ij}	Number of travelers from origin i to destination j
t_{ij}	Minimum travel time between i and j
T_{ijm}	Number of travelers using mode m from origin i to destination j

T_{ijmr}	Number of travelers taking route r on mode m from origin i to destination j
T_{ijr}	Volume on route r between i and j
t_{ijr}	Travel time on route r between i and j
T_x^k	Estimate of demand at kth iteration of solution algorithm
t_{nj}	Minimum travel time from traveler n's location to destination j
TS_x	Traveler surplus from choices x and y
SP	Upper problem in bilevel optimization
$\hat{T}_j$	Observed value for demand T_x
U_x	Direct utility of the representative traveler
UE	User equilibrium
U.M.	Utility maximization
$\overline{\omega}$	Coefficient attached to waiting time in utility function
v_a	Volume (flow) on link a
v_a^j	Volume on link a going to destination j
v_a^s	Volume on transit link a using strategy s
v_{am}	Volume on link a of mode m
V_k	Volume at node k
v_a^t	Truck volume on link a in car equivalents
v_a^c	Car volume on link a
v_x^k	Estimate of link flow at kth iteration of solution algorithm
U_{xy}	Direct aggregate utility of choices x and y
$\tilde{V}_{xy}$	Random received utility from having made joint choices x and y
$\forall$	Means "for all values of"
w_{ij}	Expected waiting time from i to j
ω_{is}	Waiting time at node i using strategy s
ω_j	Exponent in destination time function
W_{xy}	Expected value of received utility from making joint choices xy
$\mathbf{X}$	Vector, or matrix
x/y	x conditional on, i.e., given, y
X_{ki}	Characteristic k of origin zone i
$\mathbf{XY}$	Product of two matrices, or vectors, $\mathbf{X}$ and $\mathbf{Y}$
x_{ij}^a	Decision variable for using link a to go from i to j
Y_{kj}	Characteristic k of destination zone j
y_a^k	Solution for link flow in sub (auxiliary) problem at kth iteration
y_{nj}	Observed demand of individual traveler n for travel alternative j.
$\mathbf{y} \cdot \mathbf{x}$	"Dot product" of vectors y and x

Z_x^k	Solution for demand in sub(auxiliary) problem at in kth iteration
Z_x	Auxiliary travel demands in linear subproblem
$\{J\}$	Means the set of all j values
$\lvert x \rvert$	Absolute value of x
$\lVert X \rVert$	Norm, or length of vector X in multidimensional space
$\equiv$	Means "is defined as"

BIBLIOGRAPHY

Aashtiani, H. Z. and T. L. Magnanti. 1981. Equilibria on a congested transportation network. *Siam Journal on Algebraic and Discrete Methods*, 2:213–226.

Abdulaal, M. and L. Leblanc. 1979. Continuous equilibrium network design models. *Transportation Research*, 13B:19–32.

Ahuja, R., J. Orlin and T. Magnanti. 1993. *Network Flows: Theory, Algorithms and Applications*. Englewood Cliffs, NJ: Prentice Hall.

Aiyoshi, E. and K. Shizumu. 1984. A solution method for the static constrained Stackelberg problem via a penalty method. *IEEE Trans. Aut. Control*, AC29-1111–1114.

Amemiya, T. 1978. On a two-step estimation of a multivariate logit model. *Journal of Econometrics*, 8:13–21.

Anandalingham, G. and T. Friesz, eds. 1992. *Hierarchical Optimization*. Basel: J. C. Baltzer.

Anas, A. 1983. Discrete choice theory, information theory, and the multinomial logit and gravity models. *Transportation Research*, 17B:13–23.

Anas, A. 1984. Discrete choice theory and the general equilibrium of employment, housing and travel networks in a Lowry type model of the urban economy. *Environment and Planning*, A-16:1489–1502.

Anas, A. 1988. Statistical properties of mathematical programming models of stochastic network equilibrium. *Journal of Regional Science*, 28:511–530.

Anderson, S., A. DePalma, and J. F. Thisse. 1992. *Discrete Choice Theory of Product Differentiation*. Cambridge, MA: MIT Press.

Armacost, R. and A. Fiacco, 1974. Computational experience in sensitivity analysis for nonlinear programming. *Mathematical Programming*, 6:301–326.

Bard, J. 1982. A grid search algorithm for the bi-level programming problem. *Proceedings of the 14th Annual Meeting of the American Institute for Decision Science*, 256–258.

Bard, J. F. 1983. An algorithm for solving the general bi-level programming problem. *Mathematics of Operations Research*, 8:260–272.

Bard, J. F. and J. E. Falk. 1982. An explicit solution to the multi-level programming problem. *Computers and Operations Research*, 9:77–100.

Bard, J. F. and J. Moore. 1987. A branch and bound algorithm for the bi-level programming problem. Austin, TX: University of Texas, Dept. of Industrial Engineering, working paper.

Batten, D. F. 1982. *Spatial Analysis of Interacting Economies*. Boston: Martinus Nijhoff.

Batten, D. F. and D. E. Boyce. 1985. Spatial interaction and interregional commodity flows. In: P. Nijkamp (ed.), *Handbook of Regional and Urban Economics*, Vol. 1. Amsterdam: Elsevier.

Bazaara, M., H. Sherali and C. Shetty. 1993. *Non-Linear Programming*. 2nd. ed. New York: John Wiley.

Beckmann, M. J., C. B. McGuire and C. B. Winsten. 1956. *Studies in the Economics of Transportation*. New Haven, CT; Yale University Press, Cowles Commission Monograph.

Ben-Akiva, M. 1973. The structure of passenger travel demand models. Cambridge, MA. MIT Dept. of Civil Engineering. Unpublished Ph.D. dissertation.

Ben Akiva, M. and S. Lerman. 1985. *Discrete Choice Analysis: Theory and Application to Travel Demand*. Cambridge: MIT Press.

Ben Ayed, O., D. Boyce, and C. Blair. 1988. A general bi-level linear programming formulation of the network design problem. *Transportation Research*, B-22:311–318.

Ben Ayed, O., C. Blair, D. Boyce and L. LeBlanc. Construction of a real world linear programming model of the highway network design problem. In: Anandalingham, G. and T. Friesz, (eds.), *Hierarchical Optimization*. Basel: J. D. Baltzer, 1992.

Berge, C. 1963. *Topological Spaces*. New York: McMillan.

Bertsekas, D. P. 1982. *Constrained Optimization and Lagrange Multiplier Methods*. New York: Academic Press.

Bialas, W. and M. Karwan. 1984. Two-level programming. *Management Science*, 30(8):1004–1020.

Bishop, Y., S. Feinberg, and P. Holland. 1975. *Discrete multivariate analysis: Theory and practice*. Cambridge, MA: MIT Press.

Bovy, P. H. L. and M. A. Bradley. 1985. Route choice analyzed with stated-preference approaches. *Transportation Research Record*, 1037:11–20.

Boyce, D. E. 1980. A framework for constructing network equilibrium models of urban location. *Transportation Science*, 14(1):77–96.

Boyce, D. E. 1984. Network models in transportation/land use planning. In: M. Florian (ed.), *Transportation Planning Models*. Amsterdam: North-Holland, pp. 221–243.

Boyce, D. E. 1984b. Urban transportation network-equilibrium and design models: Recent achievements and future prospects. *Environment and Planning* A-16:1445–1474.

Boyce, D. E. 1987. Integration of supply and demand models in transportation and location. *Environment and Planning*, A-18:485–489.

Boyce, D., and A. Sen. 1991. Maximum likelihood estimation for travel choice models with endogenous costs. Chicago: Urban Transportation Center, University of Illinois, working paper.

Boyce, D. E., L. LeBlanc, and K. Chon. 1988. Network equilibrium models of urban location and travel choices. *Journal of Regional Science*, 28.

Boyce, D. E., A. Sen, J. Hicks, and Y. Zhang, 1989. Estimation procedures for network equilibrium models of urban location and travel choices. *Paper presented at the North American Meetings of the Regional Science Association*, Santa Barbara, CA. (Available from Urban Transportation Center, University of Illinois at Chicago.)

Boyce, D., D. Boyce, M. Lupa, and Y. Zhang, 1993. Possible schemes for introducing feedback into the four step travel forecasting procedure vs. the equilibrium solution of a combined model. *Paper presented at the 4th National Conference on Transportation Planning Methods*, Daytona Beach, FL. (Available from Urban Transportation Center, University of Illinois at Chicago.)

Boyce, D. E., L. J. LeBlanc, K. S. Chon, Y. J. Lee, and K. T. Lin. 1983. Implementation and computational issues for combined models of location, destination, mode and route choice. *Environment and Planning* A-15:1219–1230.

Braess, D. 1968. Uber ein paradoxon der verkehrsplanung. *Unternehmenforshung*, 12:258–268.

Branston, D. 1976. Link capacity functions: A review. *Transportation Research* 10(4):223–236.

Bureau of Public Roads. 1964. *Traffic Assignment Manual*. Washington, D.C. U.S. Department of Commerce, Urban Planning Division.

Candler, W. and R. Townsley. 1982. A linear two-level programming problem, *Computers and Operations Research*, 9:59–76.

Chao, G. and T. Friesz. 1984. Spatial price equilibrium sensitivity analysis. *Transportation Research*, B-18(6):423–440.

Chen, M. and S. Alfa. 1991. Algorithms for solving Fisk's stochastic traffic assignment model. *Transportation Research*, B-25(6):405–412.

Chen, M. and S. Alfa. 1992. A network design algorithm using a stochastic incremental traffic assignment approach. *Transportation Science* 25(3):215–224.

Chon, K. S. 1982. Testing of combined urban location and travel choice models. Urbana, IL: University of Illinois, Ph.D. thesis, Department of Civil Engineering.

Cohon, J. 1978. *Multiobjective Programming*. Baltimore, MD: John Hopkins University Press.

Constantin, I. 1990. A method for optimizing the frequencies of a transit network. *Paper presented at the ORSA/TIMS Conference*, Philadelphia. (Available from the Centre de Recherche sur les Transports, Université de Montreal.)

Current, J. 1988. The design of hierarchical transportation networks with transportation facilities. *Transportation Science*, 22:270–277.

Dafermos, S. 1968. *Traffic assignment and resource allocation in transportation networks*. Baltimore, MD: Johns Hopkins University. Ph.D. dissertation.

Dafermos, S. C. 1982. Relaxation algorithms for the general asymmetric traffic equilibrium problem. *Transportation Science*, 16(2):231–240.

Daganzo, C. 1979. *Multinomial Probit: The Theory and Its Applications to Demand Forecasting*. New York: Academic Press.

Daganzo, C. F. and Y. Sheffi, 1977. On stochastic models of traffic assignment. *Transportation Science*, 11(3):253–274.

Daganzo, C., and M. Kusnic, 1993. Two properties of the nested logit model. *Transportation Science*, 27(4):395–400.

Daughethy, A., 1988. ed. *Cournot Oligopoly. Characterizations and Applications*. Cambridge: Cambridge University Press.

Davis. G. 1994. Exact solution of the continuous network design problem via stochastic user equilibrium assignment. *Transportation Research*, B-28.

De Cea, J. and J. Fernandez. 1989. Transit assignment to minimal routes: An efficient new algorithm. *Traffic Engineering and Control*, 30(10).

De Cea, J., L. Zubieta, M. Florian, and P. Bunster. 1989. Optimal strategies and optimal routes in public transit assignment models: An empirical comparison. *Traffic Engineering and Control* 30(10).

De Palma, A. and J. F. Thisse, 1993. Competition on Networks. *Transportation Science*, 27(1), (special issue).

Dennis, J. E. Jr., and R. B. Schnabel. 1983. *Numerical Methods for Unconstrained Optimization and Nonlinear Equations*. Englewood Cliffs, NJ: Prentice-Hall.

Deo, N. and C. Pang. 1984. Shortest path algorithms: Taxonomy and annotation. *Networks*. 14:275–323.

Dial, R. B. 1971. A probabilistic multipath traffic assignment algorithm which obviates path enumeration. *Transportation Research*, 5(2):83–111.

Domencich, T., G. Kraft, and P. Valette. 1968. Estimation of urban passenger travel behavior. An economic demand model. *Highway Research Record*, 238:64–78.

Erlander, S. 1977. Accessibility entropy, and the distribution and assignment of traffic. *Transportation Research*, B-11:149–153.

Evans, S. 1976. Derivation and analysis of some models for combining trip distribution and assignment. *Transportation Research*, 9(12):241–246.

Fast, J. D. 1970. *Entropy*. London: Macmillan.

Fiacco, A. 1976. Sensitivity analysis for nonlinear programming using penalty methods. *Mathematical Programming*, 10:287–311.

Fiacco, A. 1983. *Introduction to Sensitivity and Stability Analysis in Non-linear Programming*. New York: Academic Press.

Fisher, M. 1981. The Lagrangian relaxation method for solving integer programming problems. *Management Science*, 27:1–18.

Fisher, M. 1985. An applications-oriented guide to Lagrangian relaxation. *Interfaces*, 15:10–21.

Fisk, C. 1980. Some developments in equilibrium traffic assignment. *Transportation Research*, B-14:243–255.

Fisk, C. 1984. Optimal signal controls on congested networks. *Proceedings of the 9th International Symposium on Transportation and Traffic Theory*, J. Vollmuller and R. Hamerslag. (eds.), Utrecht, VNU Science Press.

Fisk, C. 1986. A conceptual framework for optimal transportation systems planning with integrated supply and demand models. *Transportation Science*, 20(1):37–47.

Fisk, C. and S. Nguyen. 1982. Solution algorithms for network equilibrium models with asymmetric user costs. *Transportation Science*, 16(3):361–381.

Fisk, C. and D. Boyce. 1983. Alternative variational inequality formulations for the network equilibrium-travel choice problem. *Transportation Science*, 17:454–463.

Florian, M. 1977. A traffic equilibrium model of travel by car and public transit modes. *Transportation Science*, 11:166–179.

Florian, M. and H. Spiess. 1982. The convergence of diagonalization algorithms for asymmetric network equilibrium problems. *Transportation Research*, 16B:447–483.

Fortuny-Amat, J. and B. McCarl. 1981. A representation and economic interpretation of a two-level programming problem. *Journal of the Operational Research Society*, 32:783–791.

Fotheringham, A. S. 1988. Consumer site choice and set definition. *Marketing Science*, 7:299–310.

Fotheringham, S. and D. Wong. 1991. The modifiable areal unit problem in multivariate statistical analysis. Buffalo, NY: State University of New York, National Center for Geographic Information and Analysis.

Friedman, J. 1977. *Oligopoly and the Theory of Games*. Amsterdam: North Holland.

Friedman, J. 1983. *Oligopoly Theory*. Cambridge: Cambridge University Press.

Friesz, T. and P. Harker. 1983. Multicriteria spatial price equilibrium network design: Theory and computational results. *Transportation Research*, B-17:411–426.

Friesz, T., R. L. Tobin, T. E. Smith, and P. T. Harker. 1983. A non-linear complementary formulation and solution procedure for the general derived demand network equilibrium problem. *Journal of Regional Science*, 23(3):337–359.

Friesz, T., R. L. Tobin, H. Cho, and N. J. Metha. 1990. Sensitivity analysis-based heuristic algorithms for mathematical programs with variational inequality constraints. *Mathematical Programming*, 48:265–284.

Friesz, T., G. Anandalingam, N. Mehta, K. Nam, S. Shah, and R. Tobin. 1991. The multiobjective network design problem revisited: A simulated annealing approach. Fairfax, VA: George Mason University, Dept. of Systems Engineering.

FTA (Federal Transit Administration). 1977. *User-Oriented Manual for UTPS: An Introduction to Travel Demand Forecasting*. Washington, D.C.

Furness, K. P. 1965. Time function iteration. *Traffic Engineering and Control*, 7(7) 458–460.

Garfinkel, R., and G. Nemhauser. 1972. *Integer Programming*. New York: Wiley.

Gauvin, J. and G. Savard. 1990. The steepest descent method for the non-linear, bilevel programming problem. Technical report available from GERAD, University of Montreal.

Geoffrion, A. M. 1974. Lagrangian relaxation for integer programming, *Mathematical Programming Study*, 2:82–114.

Glover, F. 1989. Tabu search: Part 1. *ORSA Journal on Computing*, 1(3):190–206.

Glover, F. 1990. Tabu Search: Part 2. *ORSA Journal on Computing*, 2(4):4–32.

Glowinski, R., J. Lions, and R. Tremolieres. 1981. *Numerical Analysis of Variational Inequalities*. Amsterdam: North Holland.

Gross, D. and C. M. Harris. 1974. *Fundamentals of Queing Theory*. New York: Wiley.

Gumbel, E. 1958. *Statistics of Extremes*. New York: Columbia University Press.

Harker, P. 1988. Dispersed price equilibrium. *Environment and Planning*, A-20:353–368.

Harker, P. and T. Friesz. 1983. A simultaneous freight network equilibrium problem. *Congressus Numerantium*, 36:365–402.

Harker, P. and J. Pang. 1990. Finite dimensional variational inequality and non-linear complementarity problems. *Environment and Planning*, B48:161–220.

Harris, N. G. 1989. Capacity restraint simulation in a public transport environment. *Traffic Engineering and Control*, 30, C.

Hausman, J. and D. McFadden. 1984. Specification tests for the multinomial logit model, *Econometrica*, 52(5):1219–1240.

Hendrickson, C. 1981. Travel time and volume relationships in scheduled, fixed-route public transportation. *Transportation Research*, A.(15)2:173–182.

Hensher, D. A. 1986. Sequential and full information maximum likelihood estimation of a nested logit model. *The Review of Economics and Statistics*, 68(4):657–667.

Hensher, D. A. and L. W. Johnson. 1981. *Applied Discrete Choice Modeling*. New York: Halstead Press.

Hooke, R. and T. A. Jeeves. 1962. Direct search solution of numerical and statistical problems. *J. Assoc. Computer Machines*, 8(212).

Horowitz, J. L. 1982. Specification tests for probabilistic choice models. *Transportation Research*, 16A(5/6):383–94.

Horowitz, J. 1987. Specification tests for nested logit models. *Environment and Planning*, A-19:395–402.

Horowitz, J. 1991. Modeling the choice set in discrete choice random utility models. *Environment and Planning*, A-23:1237–1246.

Hutcheson, K. 1970. A test for comparing diversities based on the Shannon formula. *Journal of Theoretical Biology*, 29:151–154.

IBM Corporation. 1991. *OSL Library*. White Plains, NY.

ITE (Institute of Traffic Engineers). 1990. *Trip Generation Handbook*. Washington, DC: Institute of Transportation Engineers.

Janson, B. 1991. Dynamic traffic assignment for urban road networks. *Transportation Research*, 25B:143–161.

John, F. 1948. *Extremum Problems with Inequalities as Subsidiary Conditions. Studies and Essays*. New York: Courant Institute.

Josephy, N. H. 1979. Newton's method for generalized equations. Technical summary report No. 1965, Mathematics Research Center, University of Wisconsin at Madison.

Judice, J. and A. Faustino. 1988. The solution of the linear bi-level programming problem using the linear complementarity problem. *Investigacao Operational*, 8:77–95.

Kim, T. J. 1983. A combined land use transportation model when zonal travel demand is endogenously determined. *Transportation Research*, 17B:449–462.

Kim, T. J. 1988. Towards developing a national transportation planning model. *Annals of Regional Science*, 20 SPED: 65–80.

Kim, T. J. 1989. *Integrated Urban Systems Modeling. Theory and Applications*. Boston: Kluwer.

Knudsen, D. C. and A. S. Fotheringham. 1986. Matrix comparison, goodness-of-fit and spatial interaction modeling. *International Regional Science Review*, 10:401–418.

Koppelman, F. and C. Wilmot. 1982. Transferability analysis of disaggregate choice models. *Transportation Research Record*, 895:18–24.

Kraft, G. 1968. *Demand for Intercity Passenger Travel in the Washington-Boston Corridor*. Boston, MA: Systems Analysis and Research Corporation.

Langdon, M. G. 1984. Methods of determining choice probability in utility maximizing multiple alternative models. *Transportation Research*, 18B(3):209–34.

Lasdon, L. and A. Warren. 1978. Generalized reduced gradient software for linearly and non-linearly constrained problems. In: H. Greenberg (ed.), *Design and Implementation of Optimization Software*. Amsterdam: Sijthoff.

LeBlanc, L. J. and D. E. Boyce. 1986. A bilevel programming algorithm for exact solution of the network design problem with user optimal flows. *Transportation Research*, 20B(3):259–265.

Lee, C. K. 1987. Implementation and evaluation of network equilibrium models of urban residential location and travel choices. Urbana, IL. University of Illinois, Department of Civil Engineering, Ph.D. thesis.

Loridan, P. and J. Morgan. 1989a. A theoretical approximation scheme for Stackelberg problems. *Journal of Optimization and Applications*, 61:95–111.

Loridan, P. and J. Morgan. 1989b. New results on approximate solutions in two-level optimization. *Optimization*, 20:819–836.

Lowry, I. 1964. *A Model of Metropolis*. Vol. RM-4035RC. Santa Monica, CA: The Rand Corporation.

Magnanti, T. L. and R. T. Wong. 1984. Network design and transportation planning: Models and algorithms. *Transportation Science*, 18(1):1–55.

Mahmassani, H. and R. Herman. 1984. Dynamic user equilibrium departure time and route choice on idealized networks. *Transportation Science*, 21:89–99.

Manski, C. 1977. The structure of random utility models. *Theory and Decision*, 8:229–254.

Marcotte, P. 1986. Network design problem with congestion effects: A case of bilevel programming. *Mathematical Programming*, 34:142–162.

Marcotte, P. 1983. Network optimization with continuous control parameters. *Transportation Science*, 17:181–197.

Martin, C. and L. Schrage. 1985. Subset coefficient reduction costs for 0–1 integer programming. *Operations Research*, 33:505–526.

McFadden, D. 1973. Conditional logit analysis of qualitative choice behavior. In: P. Zarembka (ed.), *Frontiers in Econometrics*. New York: Academic Press.

McFadden, D. 1981. Econometric Models of probabilistic choice. In: C. Manski and D. McFadden (eds.), *Structural Analysis of Discrete Data with Econometric Applications*. Cambridge, MA: MIT Press.

Miller, T., T. Friesz and R. Tobin. 1992. Heuristic algorithms for delivered price spatially competitive network facility location problems. In: Anandalingham, G. and T. Friesz (eds.), *Hierarchical Optimization*. Basel: J. C. Baltzer.

Moore, E. 1957. The shortest path through a maze. *Proceedings of the International Symposium on the Theory of Switching*. Cambridge, Harvard University Press, pp. 285–292.

Moore, J. and J. Bard. 1987. *An algorithm for the zero-one bi-level programming problem*. Austin, TX: University of Texas, Dept. of Industrial Engineering, working paper.

Murtaugh, B. and M. Saunders. 1987. *MINOS 5 User Guide*. Stanford, CA: Stanford University, Dept. of Operations Research.

Nagurney, A. 1993. *Network Economics: A Variational Inequality Approach*. Boston: Kluwer.

Nash, J. F. 1951. Non-Cooperative Games. *Annals Math*. 45:286–295.

Nemhauser, G. and L. Wolsey. 1988. *Integer and Combinatorial Optimization*. New York: Wiley Interscience.

Nguyen, S. and C. Dupuis. 1984. An efficient method for computing traffic equilibria in networks with asymmetric transportation costs. *Transportation Science*, 18(2):185–202.

Openshaw, S. 1979. A methodology for using models for planning purposes. *Environment and Planning*, A.11:879–896.

Oppenheim, N. 1995a. On the integrability problem in spatial activity systems with externalities. *Regional Science and Urban Economics*.

Oppenheim, N. 1975. A typological approach to individual urban travel behavior prediction. *Environment and Planning*, 8A(2):141–152.

Oppenheim, N. 1976. Disaggregation of an urban population for transportation planning. *International Regional Science Review*, 2(1):88–96.

Oppenheim, N. 1980. *Applied Models in Urban and Regional Analysis*. Englewood Cliffs, NJ: Prentice-Hall.

Oppenheim, N. 1986a. Dynamic forecasting of urban shopping travel. *Transportation Research*, B(15).

Oppenheim, N. 1986b. A Dynamic model of urban retail location and shopping travel. In: Land development simulation and traffic mitigation. *Transportation Research Record*, 1079. Washington, D.C: Transportation Research Board.

Oppenheim, N. 1989. Equilibrium commercial activity and travel distributions: Incorporating endogenous prices and travel costs into the Harris-Wilson model, with some numerical experiments. *Transportation Research* B-25(3):225–242.

Oppenheim, N. 1990. Urban commercial activity and travel: An equilibrium allocation model and some numerical experiments. *Transportation Research Record*, 1229.

Oppenheim, N. 1991a. Discontinuous changes in equilibrium activity and travel patterns. *Papers of the Regional Science Association*, 68.

Oppenheim, N. 1991b. Commercial activity location modeling with endogenous retail prices and shopping travel costs. *Environmental and Planning*, A(10):731–744.

Oppenheim, N. 1993. Equilibrium trip distribution-assignment with variable destination costs. *Transportation Research*. B-27(3):207–217.

Oppenheim, N. 1994a. A combined model of urban personal travel and goods movements. *Transportation Science*, 2(27).

Oppenheim, N. 1995b. *Market Area Analysis and Spatial Demand Modeling: A Discrete Choice Approach*. New York: Chapman and Hall.

Oppenheim, N. 1981. Control of large urban traffic networks. *Proceedings of the 2nd Conference on Issues in Control of Urban Traffic Systems*. New York. Engineering Foundation. 193–199.

Ortega, J. M. and W. C. Rheinboldt. 1970. *Interactivity Solutions of Nonlinear Equations in Several Variables*. New York: Academic Press.

Ortuzar, J. de D., and L. Willumsen. 1990. *Modeling Transport*. New York: Wiley.

Owen, G. 1982. *Game Theory*, 2nd ed. New York: Academic Press.

Papadimitriou, C. and K. Steiglitz. 1982. *Combinatorial Optimization*. Englewood Cliffs, NJ: Prentice-Hall.

Pape, U. 1974. Implementation and efficiency of Moore-algorithms for the shortest route problem. *Mathematical Programming*. 7(2):212–222.

Parker, R. and R. Rardin. 1988. *Discrete Optimization*. New York: Academic Press.

Phiri, P. 1980. Calculation of the equilibrium configuration of shopping facilities sizes. *Environment and Planning*. A-12:983–1000.

Powell, M. A. 1970. Hybrid method for non-linear equations. In: P. Rabinowitz (ed.), *Numerical Methods for Non-linear Algebraic Equations*. London: Gordon and Breach.

Powell, W. and Y. Sheffi. 1982. The convergence of equilibrium algorithms with predetermined step sizes. *Transportation Science*, 61:45–55.

Quandt, R. and W. Baumol. 1966. The demand for abstract transport modes. *Journal of Regional Science*, 6(2):13–26.

Ran, B. and D. Boyce. 1994. *Dynamic Urban Transportation Network Models: Theory and Implications for IVHS*. New York: Springer-Verlag.

Rao, C. 1973. *Linear Statistical Inference and its Applications*, 2nd ed. New York: Wiley.

Rockafellar, R. 1970. *Convex Analysis*. Princeton, NJ: Princeton University Press.

Roy, J. R. and P. F. Lesse. 1986. Modeling product flows under uncertainty. *Environment and Planning*. A-17:1271–1274.

Safwat, K. N. A. and T. L. Magnanti. 1988. A combined trip generation trip distribution model split and traffic assignment model. *Transportation Science* 22(1): 14–30.

Samuelson, P. 1952. Spatial price equilibrium and linear programming. *American Economic Review*, 42:283–303.

Scarf, H. E. 1973. *The computation of economic equilibria*. New Haven: Yale University Press.

Sen, A. 1986. Maximum likelihood estimation of gravity model parameters. *Journal of Regional Science*, 26:461–474.

Sen, A. and S. Soot. 1981. Selected procedures for calibrating the generalized gravity model. *Papers of the Regional Science Association*, 48:165–176.

Sen, A. and Z. Matuszewski. 1987. Properties of maximum likelihood estimates of gravity parameters. *Paper presented at the North American Meetings of the Regional Science Association*, Baltimore.

Shannon, C. E. 1948. A mathematical theory of communication. *Bell System Technical Journal*, 27:379–423, 623–656.

Sheffi, Y. 1985. *Urban Transportation Networks*. Englewood Cliffs, NJ: Prentice Hall.

Sheffi, Y. and W. Powell. 1981. Comparison of stochastic and deterministic traffic assignment over congested networks. *Transportation Research*, B-15B:53–64.

Sherali, H. D., D. Shoyster, and F. H. Murphy. 1983. Stackelberg-Nash-Cournot equilibria: Characterizations and computations. *Operations Research*. 31:253–276.

Small, K. A. and D. Brownstone. 1982. Efficient estimation of nested logit models: An application to trip timing. *Research Memorandum* 296, Economic Research Program, Princeton University.

Smith, M. J. 1979. The existence, uniqueness and stability of traffic equilibria. *Transportation Research*, 13B:293–304.

Smith, T. E. 1985. Remarks on the most-probable-state approach to analyzing probabilistic theories of behavior. *Environment and Planning*, A17:688–695.

Smith, T. E. 1987. A cost-efficiency theory of dispersed network equilibrium. *Environment and Planning*, 19.

Smith, T. E. and Friesz, T. L. 1985. Spatial market equilibrium with flow-dependent supply and demand. *Regional Science and Urban Economics*, 15:181–218.

Sobel, K. L. 1980. Travel demand forecasting by using the nested multinomial logit model. *Transportation Research Record*, 775:48–55.

Spiess, H. 1984. Contributions à la Théorie et aux Outils de Planification des Réseaux de Transports en Commun. (in French). University of Montreal: Department of Computer Science and Operations Research, Ph.D. thesis.

Spiess, H. and M. Florian. 1989. Optimal strategies: A new assignment model for transit networks. *Transportation Research*, B-23(2):83–102.

Stackelberg, H. 1952. *The Theory of the Market*. Oxford: Oxford University Press.

Steenbrink, P. A. 1974. *Optimization of Transport Networks*. Wiley, New York, NY.

Stopher, P. and McDonald, K. 1983. Trip generation by cross classification: An alternative methodology. *Transportation Research Record*, 944:84–91.

Suwansirikul, C., R. Tobin, and T. Friesz. 1987. Equilibrium decomposed optimization: A heuristic for the continuous equilibrium network design problem. *Transportation Science*, 21(4):254–263.

Swait, J. and M. Ben-Akiva. 1987. Incorporating random constraints in discrete choice models of choice set generation. *Transportation Research*, B. 21(2):91–102.

Takayama, T. and G. Judge. 1971. *Spatial and Temporal Price and Allocation Models*. Amsterdam: North Holland.

Tan, H., S. Gershwin, and M. Athans. 1979. *Hybrid optimization in urban traffic networks*. Cambridge, MA: LIDS Technical Report. MIT.

Tobin, R. L. and T. L. Friesz. 1988. Sensitivity analysis for equilibrium network flow. *Transportation Science*, 22(4):242–250.

Tomlin, J. A. and G. S. Tomlin. 1968. Traffic distribution and entropy. *Nature*, 220:974–976.

Transportation Research Board. 1985. *Highway Capacity Manual*. Washington, DC: National Research Council, Special Report 209.

U.S. Bureau of Public Roads. 1964. *Traffic Assignment Manual*. Washington, DC: Dept. of Commerce.

Van Vliet, D. 1981. Selected node-pair analysis in Dial's assignment algorithm. *Transportation Research*, 15B:65–68.

Vanderbilt, B. and S. Louie. 1984. A Monte Carlo simulated annealing approach to optimization over continuous variables. *Journal of Computational Physics*, 56:259–271.

Varian. H. 1992. *Microeconomic Analysis*. 3rd ed. New York: Norton.

Wardrop, J. G. 1952. Some theoretical aspects of road traffic research. *Proceedings, Institution of Civil Engineering, Part II*, 1:325–378.

Warner, S. L. 1962. *Strategic Choice of Mode in Urban Travel. A Study in Binary Choice*. Evanston, IL: Northwestern University Press.

Williams, H. C. W. L. 1977. On the formation of travel demand models and economic evaluation measures of user benefit. *Environment and Planning*, 9A(3):285–344.

Williams, H. C.W. L. and J. Ortuzar. 1982. Behavioral theories of dispersion and misspecification of travel demand models. *Transportation Research* B-16:167–219.

Wilson, A. G. 1967. Entropy maximizing models in the theory of trip distribution, mode split and route split. *Journal of Transportation Economics and Policy*, 3:108–126.

Wilson, A. G. 1970. *Entropy in Urban and Regional Modeling*. London: Pion.

Wilson, A. G., José D. Coelho, Sally M. Macgill, and H. C. W. L. Williams. 1981. *Optimization in Locational and Transport Analysis*. New York: Wiley.

Wonnacott, T. H. and R. J. Wonnacott. 1977. *Introductory Statistics for Business and Economics*. New York: Wiley.

Wooton, H. J. and G. W. Pick. 1967. A model for trips generated by households. *Journal of Transport Economics and Policy*. I(2):137–153.

Zahavi, Y. 1979. *The UMOT Project*. Washington, DC: USDOT, report no. DOT-RSPA-DPB-20-79-3.

Zukhovitsky, S., R. Pollak, and M. Primak. 1973. Concave Multiperson Games: Numerical Methods. *Maketon*, 11–30.

A FINAL NOTE

I sincerely hope that this book has kindled, sustained, or perhaps even expanded your interest in urban travel demand modeling. I believe that the methodology presented here can significantly advance the state of the art in the field. I would welcome any comments or suggestions that you may wish to make, and would be particularly interested in learning about applications of this methodology to specific practical situations.

In addition, I would appreciate your pointing out any typographical or factual errors which you may have found.

In an effort to promote the dissemination and application of this technology, I have developed demonstration software to solve most of the models in the book, in the form of electronic add-on files for execution on two popular computer applications. One is GAMS®, a modeling language, and the other is EXCEL®, a spreadsheet. Both provide interfaces with generic algorithms for linear and non-linear optimization problems, respectively, which may be used for solving small size problems such as those in the illustrative examples.

Further information about this software, as well as other possible forms of technology transfer, is available upon request.

I may be reached at:

The CUNY Institute for Transportation Systems
City College of New York
138th St. and Convent Avenue
New York, N.Y. 10031
Tel: (212) 650 8058
Fax: (212) 650 8374
E-mail: oppenhm@ti-mail.engr.ccny.cuny.edu

I look forward to hearing from you.

AUTHOR INDEX

SUBJECT INDEX